J Manard

errata
p 151
159
201
2?8
+10

p252 Seismicity plot

Civil Engineering and
Engineering Mechanics Series

N. M. Newmark and W. J. Hall, editors

PRENTICE-HALL INTERNATIONAL, INC., *London*
PRENTICE-HALL OF AUSTRALIA, PTY. LTD., *Sydney*
PRENTICE-HALL OF CANADA, LTD., *Toronto*
PRENTICE-HALL OF INDIA PRIVATE LIMITED, *New Delhi*
PRENTICE-HALL OF JAPAN, INC., *Tokyo*

TILTING OF APARTMENT BUILDINGS, NIIGATA (1964)

Fundamentals of Earthquake Engineering

Nathan M. Newmark

Professor and Head
Department of Civil Engineering
University of Illinois
Urbana, Illinois

Emilio Rosenblueth

Professor of Engineering
Universidad Nacional Autónoma
  de México
México, DF

Prentice-Hall, Inc. *Englewood Cliffs, N.J.*

© 1971 by
PRENTICE-HALL, INC.
Englewood Cliffs, N. J.

All rights reserved. No part of this
book may be reproduced in any form
or by any means without permission
in writing from the publisher.

Current printing (last digit):
10 9 8 7 6 5 4 3 2

13-336206-X
Library of Congress Catalog Card Number: 70-150108
Printed in the United States of America

Dedicated to the continuing cooperation between our universities.

CONTENTS

Introduction xv

PART I

Dynamics

1 Simple Linear Systems 3

1.1	Differential Equation	3
1.2	Free Oscillations	4
1.3	Steady-State Vibrations	6
1.4	Transient Disturbances	9
1.5	Numerical Evaluation of Responses	14
1.6	Graphical Evaluation	17
1.7	Analog Evaluation	22
	Problems	24

2 Linear Systems with Several Degrees of Freedom 27

2.1	Equations of Motion	27
2.2	Free Oscillations	32
2.3	Steady-State Vibrations	37
2.4	Transient Disturbances	41
2.5	Numerical Evaluation of Responses to Transient Motions	50
2.6	Effects of Axial Forces	53
	Problems	56

3 Linear Systems with Distributed Mass 61

3.1	Introductory Note	61
3.2	Undamped, Uniform, Shear Beam	62
3.3	Simple Wave Equation	68

3.4	Nonuniform, Undamped, Shear Beams		*73*
3.5	Damped, Uniform, Shear Beams		*73*
3.6	Flexural Beams		*79*
3.7	Frames		*82*
3.8	The Three-Dimensional Wave Equation		*82*
3.9	Body Waves		*83*
3.10	Spherical Body-Waves		*85*
3.11	Cylindrical Body-Waves		*87*
3.12	Wave Reflection and Refraction		*88*
3.13	Surface Waves		*90*
3.14	Group Velocity		*92*
3.15	Soil-Foundation Interaction		*93*
	Problems		*101*

4 *Numerical Computation of Steady-State Responses and Natural Modes* 105

4.1	Introductory Note		*105*
4.2	Discretization		*106*
4.3	Iteration Procedure for Steady-State Vibrations		*117*
4.4	Iteration Procedure for the Fundamental Mode		*124*
4.5	Iteration Procedures for Higher Natural Modes		*130*
	4.5.1	Procedure of Ratio Elimination	*130*
	4.5.2	Procedure of Orthogonal Functions	*130*
4.6	Rayleigh's Method		*135*
4.7	Two Energy Methods Due to Southwell		*138*
4.8	Effects of Changing a Single Mass or a Stiffness Element		*142*
4.9	Step-by-Step Procedures for Steady-State Vibrations		*143*
4.10	Step-by-Step Procedures for Calculation of Natural Modes		*149*
	Problems		*156*

5 *Nonlinear Systems* 161

5.1	Types of Nonlinearity	*161*
5.2	Graphical Evaluation of Responses of Simple Systems	*164*
5.3	Numerical Method	*167*
5.4	One-Dimensional Wave Transmission	*168*
	Problems	*171*

6 *Hydrodynamics* 177

6.1	General Considerations		*177*
6.2	Pressures Against Dams		*179*
	6.2.1	Introductory Note	*179*

	6.2.2	Elementary Solution	*179*
	6.2.3	Inclined Upstream Face	*182*
	6.2.4	Slope Failures in Reservoirs	*184*
	6.2.5	Effects of Water Compressibility	*184*
	6.2.6	Finite Reservoir	*185*
	6.2.7	Effects of Dam Flexibility and Base Rotation	*188*
	6.2.8	Effects of Motion of the Free Surface	*188*
	6.2.9	Natural Modes of Vibration of Water in Reservoirs	*190*
	6.2.10	Vertical Ground Motion	*194*
	6.2.11	Arch Dams	*195*
6.3		Vibration of Liquids in Tanks	*197*
6.4		Vibration of Submerged Structures	*201*
6.5		Tsunamis	*203*
		Problems	*211*

PART II

Earthquake Motions and Structural Responses

7 *Characteristics of Earthquakes* **215**

7.1	Causes of Earthquakes	*215*
7.2	Focus, Magnitude, and Intensity	*217*
7.3	Types of Earth Waves	*220*
7.4	Characteristics of Strong Ground Motions	*225*
7.5	Correlation of Ground Motion Parameters with Magnitude and Focal Distance	*228*
7.6	The Three Translational Components of Ground Motions	*236*
7.7	Rotational Components and Other Space Derivatives of Earthquake Motions	*241*

8 *Seismicity* **247**

8.1	Introductory Note	*247*
8.2	Local Seismicity	*249*
8.3	Regional Seismicity	*260*
8.4	Microregionalization	*265*

9 Probability Distributions of Response Spectral Ordinates 267

9.1	Scope of this Chapter	267
9.2	Stochastic Processes	268
9.3	Idealization of Earthquakes as Segments of White Noise	269
9.4	Idealization of Earthquakes as Stationary Gaussian Processes	276
9.5	Idealization of Earthquakes as Transient Gaussian Processes	283
9.6	Expected Spectral Ordinates for Earthquakes of the Second Type	283
9.7	Earthquake Filtering through Soft Mantles in the Linear Range	286
9.8	Earthquake Filtering through Nonlinear Materials	298
9.9	Computer Simulation of Earthquakes	299
9.10	Distribution of Responses to Earthquakes	302

10 Responses of Linear Multidegree Systems 305

10.1	Introductory Note	305
10.2	Responses to White Noise	308
10.3	Responses to Gaussian Processes and to Earthquakes	313
10.4	Application to Uniform Shear-Beams	313
10.5	Responses to the Simultaneous Action of Several Components of Ground Motion	316
10.6	Flexible Pipes and Tunnels	318

11 Responses of Nonlinear Systems 321

11.1	Introductory Note	321
11.2	Nonlinear Criteria of Failure	321
11.3	Single-Degree Systems with Symmetric, Nonlinear Force-Deformation Relation	323
11.3.1	General Approximate Method of Analysis	323
11.3.2	Approximate Upper Bounds	327
11.3.3	Elastic Systems	328
11.3.4	Elastoplastic Systems	335
11.3.5	Rigid-Plastic Systems	342
11.3.6	Masing-Type Systems	343

11.3.7	Stiffness-Degrading Systems	*348*
11.3.8	A Common Type of Braced Structure	*349*
11.3.9	Effects of Gravity	*350*
11.3.10	Remarks on the Responses of Symmetric, One-Degree Systems	*353*
11.4	Single-Degree Systems with Asymmetric, Force-Deformation Curves	*354*
11.5	Multidegree Structures	*356*
11.6	Rigid-plastic Systems with Distributed Parameters	*359*

12 *Earthquake Effects on Reservoirs* **365**

12.1	Introductory Note	*365*
12.2	Hydrodynamic Pressures on Dams	*365*
12.3	Submerged Structures	*377*
12.4	Tanks	*377*

13 *Behavior of Materials and Structural Components under Earthquake Loading* **381**

13.1	Introductory Note	*381*
13.2	Damping	*382*
13.3	Effects of Rate of Loading	*384*
13.4	High-Cycle Fatigue of Simple Specimens	*393*
13.5	Behavior of Simple Specimens under a Small Number of Loading Cycles	*395*
13.6	Behavior of Structural Components	*399*
	13.6.1 Flexural Members	*399*
	13.6.2 Joints	*408*
	13.6.3 Frames and Continuous Beams	*411*
	13.6.4 Diaphragms	*414*
13.7	Behavior of Complete Structures	*420*
13.8	Human Reaction to Earthquakes	*423*
13.9	Behavior of Soils	*424*
	13.9.1 Dry Cohesionless Soils	*424*
	13.9.2 Partially-Saturated Cohesionless Soils	*429*
	13.9.3 Saturated Cohesionless Soils	*430*
	13.9.4 Saturated Cohesive Soils	*437*
	13.9.5 Partially-Saturated Cohesive Soils	*438*
	13.9.6 Rocks	*438*
	13.9.7 Field Determination of Soil Properties	*438*

PART III

Design

14 Basic Concepts in Earthquake-Resistant Design 443

14.1	Objectives of Earthquake-Resistant Design	443
	14.1.1 Introductory Note	443
	14.1.2 Structural Design	447
	14.1.3 Load Factors and Factors of Safety	447
	14.1.4 Design for a Permissible Probability of Failure	450
	14.1.5 Earthquake Resistant Design	450
14.2	Simple Linear Systems Having Deterministic Parameters	455
14.3	Simple Linear Systems Having Random Parameters	462
14.4	Multidegree Linear Systems	467
14.5	Nonlinear Structures	473

15 Earthquake-Resistant Design of Buildings 477

15.1	Design Spectrum	477
15.2	Base Shear Coefficient	482
15.3	The Calculation of Natural Modes	482
15.4	Shear Distribution	483
15.5	Appendages	488
15.6	Story Torques	490
15.7	Overturning Moments	500
15.8	Drift Limitations	507
15.9	Analysis of Common Structures	514
15.10	Effects of Gravity Forces	516
15.11	Foundation Design	517
15.12	On the Choice of a Structural Solution	518
15.13	Structural Synthesis	530

16 Other Topics in Earthquake-Resistant Design 533

16.1	Inverted Pendulums	533
	16.1.1 General Considerations	533
	16.1.2 Simple Inverted Pendulum	534

16.2	Towers, Stacks, and Stacklike Structures	*537*
16.3	Bridges	*541*
16.4	Retaining Structures	*544*
16.5	Tunnels and Pipes	*546*
16.6	Tanks and Hydraulic Structures	*548*
16.7	Strengthening Damaged Structures	*552*
16.8	Protection against Tsunamis	*561*
16.9	Urban Planning and Relocation	*562*

17 Tests and Observations 565

17.1	Introductory Note	*565*
17.2	Seismoscopes	*566*
17.3	Accelerographs	*569*
17.4	Other Ground-Motion Recording Instruments	*571*
17.5	Measurement of Structural Characteristics	*572*
17.6	Dynamic Testing of Models	*578*

Appendix 1: Calculation of Dynamic Parameters from Natural Modes of Vibration 581

Appendix 2: The Modified Mercalli Intensity Scale 585

Appendix 3: Notation 589

References 593

Index 623

INTRODUCTION

In this text on earthquake engineering we take for granted that the purpose of design in engineering is optimization, and that we deal with random variables. In the past the orthodox viewpoint maintained that the objective of design was to prevent failure; it idealized variables as deterministic. This simple approach is still fruitful when applied to design under only mild uncertainty, and in situations in which the possibility of failure may be contemplated at such a distant future as to be almost irrelevant; but when confronted with the effects of earthquakes, this orthodox viewpoint seems so naïve as to be sterile. In dealing with earthquakes, we must contend with appreciable probabilities that failure will occur in the near future. Otherwise, all the wealth of this world would prove insufficient to fill our needs: the most modest structures would be fortresses. We must also face uncertainty on a large scale, for it is our task to design engineering systems—about whose pertinent properties we know little—to resist future earthquakes and tidal waves—about whose characteristics we know even less.

The next few years will surely diminish our uncertainty, even in what concerns the characteristics of earthquake motions and manifestations. It is unlikely, though, that there will be such a change in the nature of our knowledge to relieve us of the necessity of dealing openly with random variables.

In a way, earthquake engineering is a cartoon of other branches of engineering. Earthquake effects on structures systematically bring out the mistakes made in design and construction, even the minutest mistakes. Add to this the unequivocally dynamic nature of the disturbances, the importance of soil-structure interaction, considerations of human reaction to seismic oscillations and to tsunamis, and the extremely random nature of it all; it is evident that earthquake engineering is to the rest of the engineering disciplines what psychiatry is to other branches of medicine: it is a study of pathological cases to gain insight into the mental structure of normal human beings. This aspect of earthquake engineering makes it challenging and fascinating, and gives it an educational value beyond its immediate objectives. If a civil engineer is to acquire fruitful experience in a brief span of time, expose him to the concepts of earthquake engineering, no matter if he is later not to work in earthquake country.

Owing to these reasons the authors hope that the present book will be useful to civil engineers in general, even if their interest in earthquake resistant design is indirect.

There are some topics either cursorily treated or even omitted in the text. This is done purposely because available information is not sufficient to give specific recommendations. The authors hope that in the coming years research will throw light on these topics. In the meantime, the engineer must use his judgment to make allowances for these unknown factors.

The book has been conceived essentially as a reference work and as a graduate textbook. The reader is assumed to be equipped with a normal undergraduate baggage of mechanics, applied mathematics and structural engineering. With this idea in mind, the book includes the following material: (1) all the dynamics needed to follow and to apply the rest of the text; (2) a description of the characteristics of earthquake motions, mostly in a probabilistic framework, and of behavior of materials when subjected to these disturbances; and (3) recommended design concepts and methods. Appendices cover specialized topics.

A reader familiar with advanced dynamics may skip most of the first part. He might be interested in going through the numerical methods included at the end of Chapters 1, 2, and 5 and in Chapter 4, as well as the coverage of some aspects of hydrodynamics in Chapter 6. If he is interested in direct applications rather than research or the study of special problems, he may skip most of the probabilistic material of Chapters 9 and 10.

Readers interested only in urban construction will lose little by omitting Chapters 6 and 12.

The first part of the book contains several worked-out examples and problems to which answers are provided, for to become proficient in the application of knowledge of dynamics and numerical methods one cannot escape drill. Some of the problems furnish particularly interesting information and insight. The most laborious ones are marked with an asterisk and may well be omitted by the hasty student.

The authors hope that this book will serve as a guide to those who must cope with the effects of earthquakes, will supplement the education of other engineers, and will help open avenues of research.

The authors express their gratitude to all those who have contributed to the completion of this work through countless suggestions for improvement. They are particularly grateful to the following professors of engineering, National Autonomous University of Mexico, for their constructive criticisms and hard work: L. Esteva, A. Flores-Victoria, O. A. Rascón, J. Elorduy, J. Sandoval, and E. Mendoza; to the following professors of engineering, University of Illinois: S. J. Fenves, W. J. Hall, and A. H.-S. Ang; and to Professor C. A. Cornell of the Massachusetts Institute of Technology.

PART 1

Dynamics

1

SIMPLE LINEAR SYSTEMS

1.1 Differential Equation

We shall understand by *simple system* a structure having a single degree of freedom and constant parameters. An example of a simple system is shown schematically in Fig. 1.1.

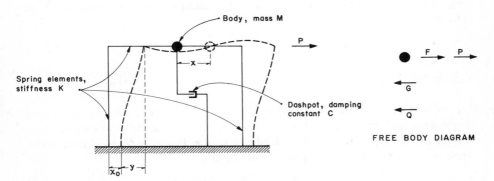

Figure 1.1. Simple system.

Let M denote its mass, Q the force in the spring elements (shear for the example shown), G the force in the damper, x_0 the ground displacement, x the total displacement of the mass, $y = x - x_0$ the displacement of the mass relative to the ground, P an external force acting on the mass, and F the inertia force. We shall take displacements and forces that act on the mass as positive from left to right. In accordance with D'Alembert's principle we may set $-F = G + Q + P$, and, since $F = -M\ddot{x}$, we may write

$$M\ddot{x} - G - Q = P \qquad (1.1)$$

Consider now the case when the damping force is proportional to the rate

4   SIMPLE LINEAR SYSTEMS

of deformation, say $G = -C\dot{y}$, where C is a constant, and the spring force is proportional to the deformation, say $Q = -Ky$, where K is also a constant. Then Eq. 1.1 becomes

$$M\ddot{x} + C\dot{y} + Ky = P \qquad (1.2)$$

Under these conditions the system is governed by a linear differential equation. Hence, we say we are dealing with a *simple linear system*. This type of damping is often referred to as *viscous* or *linear damping*. Quantities C and K are called the damping and spring constants, respectively.

We may write Eq. 1.2 in the more convenient form

$$M\ddot{y} + C\dot{y} + Ky = P - M\ddot{x}_0 \qquad (1.3)$$

Dividing through by M, we obtain

$$\ddot{y} + 2\zeta\omega_1\dot{y} + \omega_1^2 y = \omega_1^2 y_0 - \ddot{x}_0 \qquad (1.4)$$

where $\zeta\omega_1 = C/2M$, $\omega_1^2 = K/M$, and $y_0 = P/K$; that is, y_0 stands for the *static* displacement relative to the base, or the relative displacement which the mass would undergo if the load P were applied at an infinitely slow rate, or if the system consisted exclusively of the spring element, undamped, and massless. Later we shall discuss the significance of ζ.

Depending on the initial conditions and on what we choose for the second member of these equations we can study all the cases which interest us in problems of free and forced vibrations and of transient disturbances of simple linear systems.

1.2 Free Oscillations

It is said that a structure undergoes free vibrations when its base remains motionless and there are no external forces. The second member in Eqs. 1.1–1.4 is then equal to zero, and we have $x = y$. It can be verified that under these conditions the general solution of Eq. 1.4 is

$$y(t) = \Re \{B_1 \exp\left[(-\zeta\omega_1 + i\omega_1')t\right] + B_2 \exp\left[(-\zeta\omega_1 - i\omega_1')t\right]\} \qquad (1.5)$$

(den Hartog, 1956, pp. 37–40) in which t represents time, $\Re$ signifies "the real part of," B_1 and B_2 are arbitrary complex constants, $i^2 = -1$, and

$$(\omega')^2 = \frac{K}{M} - \left(\frac{C}{2M}\right)^2 \qquad (1.6)$$

Equation 1.5 is often written in the more conventional forms

$$y(t) = [\exp(-\zeta\omega_1 t)][a_1 \sin \omega_1' t + a_2 \cos \omega_1' t]$$
$$y(t) = a \exp(-\zeta\omega_1 t)$$

or

$$y(t) = [\exp(-\zeta\omega_1 t)][a_1 \sinh \omega_1'' t + a_2 \cosh \omega_1'' t]$$

[with $(\omega_1'')^2 = -(\omega_1')^2$] depending on whether $(\omega_1')^2$ is positive, zero, or negative, respectively (in which cases ω_1' is real, zero, or imaginary). In these equations a, a_1, and a_2 are arbitrary real constants having units of length.

In the usual case ω_1' is real, and we may rewrite Eq. 1.5 in the form

$$y(t) = a \exp\left[-\zeta\omega_1(t - t_1)\right] \sin\left[\omega_1'(t - t_1)\right] \tag{1.7}$$

where a is an arbitrary real constant having units of length and t_1 is an arbitrary value of t. For a conservative system ($C = 0$), Eqs. 1.5 and 1.7 describe a simple harmonic motion; they describe a damped harmonic motion for dissipative systems ($C > 0$).

The quantity $C_{cr} = 2\sqrt{KM}$ is known as *critical damping*. For dashpot constants C equal to or greater than C_{cr}, the system does not oscillate when it is given a displacement or velocity and allowed to move freely, but creeps back tending to its undeformed state, which it attains after an infinitely long time. For dashpot constants smaller then C_{cr} the system tends to that state through an oscillating motion as described by Eq. 1.7. In practice generally we shall be concerned with cases in which C is less than C_{cr} and usually much less.

It is convenient to define ζ as the ratio C/C_{cr} and to use it as a measure of damping. This quantity is known as *coefficient of damping* or *damping ratio*. In terms of it we can write Eq. 1.6 as

$$\omega_1' = \omega_1\sqrt{1 - \zeta^2} \tag{1.8}$$

and say that ω_1 is the *undamped natural circular frequency* (that is, the natural circular frequency of a system having the same mass and stiffness as the system considered but no dashpot) and ω_1' the *damped* natural circular frequency. The difference between ω_1 and ω_1' is usually small; for example, if ζ is less than 20 percent, ω_1' differs from ω_1 by less than 2 percent.

If the dashpot constant C is greater than the critical value $2\sqrt{KM}$, the damping ratio exceeds unity. In that case ω_1' turns out to be imaginary; that is, the system does not have a real damped natural frequency and does not oscillate in free vibration. In this case the sine function in the solution for free vibrations must be replaced with a hyperbolic function and ω_1' with $\omega_1'' = \sqrt{(-\omega_1')^2}$.

From the natural circular frequency we can compute the *natural frequency* as $\omega_1/2\pi$ or $\omega_1'/2\pi$ as the case may be, and the undamped and damped *natural periods* of vibration as $T_1 = 2\pi/\omega_1$ and $T_1' = 2\pi/\omega_1'$, respectively. The meaning of natural period is illustrated in Fig. 1.2, which represents typical free oscillations of simple systems. Natural frequencies are measured in units such as hertz, ω_1 in radians per second, and T_1 in seconds.

The ratio $y(t)/y(t + T_1')$ is the same as the ratio of amplitudes of successive cycles. From Eq. 1.7 we can derive the expression

$$\frac{y(t)}{y(t + T_1')} = \exp\left(\zeta\omega_1 T_1'\right) \tag{1.9}$$

which has meaning when $\zeta \leq 1$. The natural logarithm of this ratio is called the *logarithmic decrement*. It is equal to $\zeta\omega_1 T_1' = 2\pi\zeta(1 - \zeta^2)^{-1/2}$. Sometimes the

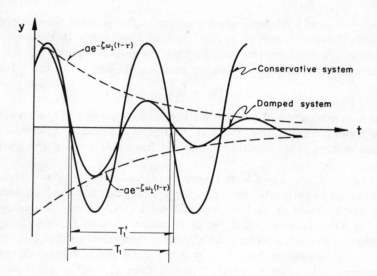

Figure 1.2. Free vibrations of simple systems.

ratio in Eq. 1.9 is used as a measure of damping; thus it is said that an oscillator has a damping of 3:1 when its logarithmic decrement is ln 3.

1.3 Steady-State Vibrations

We shall be interested in steady states of motion resulting from forced vibration in connection with dynamic tests of structures. Consider the case when the ground remains at rest and the external force varies harmonically: $\ddot{x}_0 = 0$, $y_0 = a \sin \omega t$. We may write then the general solution of the corresponding homogeneous equation, which is given by Eq. 1.7, plus the particular solution

$$\frac{y}{a} = B_d \sin(\omega t - \phi) \tag{1.10}$$

where

$$B_d = \left[\left(1 - \frac{\omega^2}{\omega_1^2}\right)^2 + \left(2\zeta\frac{\omega}{\omega_1}\right)^2\right]^{-1/2} \tag{1.11}$$

$$\phi = \tan^{-1}\frac{2\zeta\omega/\omega_1}{1 - \omega^2/\omega_1^2} \tag{1.12}$$

(Blake, 1961); B_d is a dimensionless *response factor*, equal to the ratio of the dynamic to the static displacement response amplitudes, and ϕ is an angular phase shift.

A similar solution applies to the steady state of vibration due to a harmonic ground motion described by $x_0 = a \sin \omega t$ if there is no external force. In this case $\ddot{x}_0 = -a\omega^2 \sin \omega t$, so that

$$-\frac{M}{K}\ddot{x}_0 = a\left(\frac{\omega}{\omega_1}\right)^2 \sin \omega t$$

and it is seen that the second member in Eq. 1.4 is exactly the same as for the case when $y_0 = a \sin \omega t$, save for the factor $(\omega/\omega_1)^2$. Accordingly in this case,

$$B_d = \frac{(\omega/\omega_1)^2}{[(1-\omega^2/\omega_1^2)^2 + (2\zeta\omega/\omega_1)^2]^{1/2}}$$
$$= \left[\left(1 - \frac{\omega_1^2}{\omega^2}\right)^2 + \left(2\zeta\frac{\omega_1}{\omega}\right)^2\right]^{-1/2} \quad (1.13)$$

which is the same expression as for the displacement response factor that corresponds to an external harmonic force, with ω and ω_1 interchanged.

The first part of the solution, the part given by Eq. 1.7, depends on the initial conditions and tends to zero after a sufficiently long time no matter how small the damping ratio. Hence only the part given by Eq. 1.10 is left for the steady state.

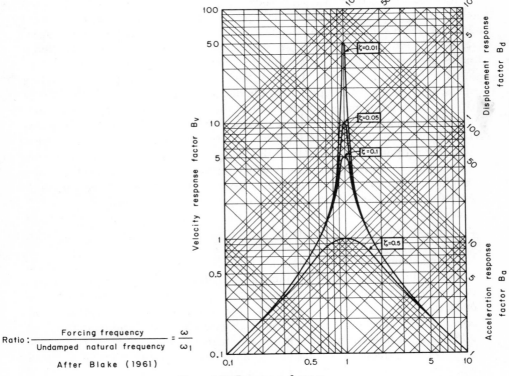

Figure 1.3. Response factors.

The ratios of dynamic to static amplitudes of velocity and of acceleration can be obtained by differentiation of Eq. 1.10 and are respectively equal to $B_v = (\omega/\omega_1)B_d$ and $B_a = (\omega/\omega_1)^2 B_d$. All three response factors are given in Fig. 1.3 for several values of the damping ratio. In this logarithmic representation horizontal lines give values of B_v, 45° lines ascending from left to right give B_d, and descending 45° lines give B_a. All scales are logarithmic.

If ω/ω_1 tends to zero, y approaches y_0; that is, dynamic effects become negligible. When $\omega = \omega_1$ the amplitude of y is $a/2\zeta$. And as ω/ω_1 tends to infinity, the amplitude of y asymptotically approaches $(\omega_1/\omega)^2 a$. We see that the influence of damping is important in the neighborhood of $\omega = \omega_1$.

According to Eq. 1.12, the phase angle ϕ goes from zero to $\pi/2$ as ω/ω_1 changes from zero to one, and from $\pi/2$ to π as this ratio changes from one to infinity. In lightly damped systems, the response is practically in phase with the excitation in the range of excitation frequencies appreciably smaller than the system's natural frequency, and the system's vibrations are practically opposite those of the excitation when ω appreciably exceeds ω_1. The phase shift is $\pi/2$ when $\omega = \omega_1$.

It is said that the forcing function is in *resonance*—in harmonic vibration—with the system when the response factor is a maximum. This obtains when $\omega = \omega_1(1 - 2\zeta^2)^{1/2}$ if the criterion is based on displacement amplitudes, when $\omega = \omega_1$ if it is based on velocity amplitudes, and when $\omega = \omega_1(1 - 2\zeta^2)^{-1/2}$ if it is based on acceleration amplitudes. The resonant value of B_v is $(2\zeta)^{-1}$ and those of B_a and B_d are both $(2\zeta)^{-1}(1 - \zeta^2)^{-1/2}$. For most structures of practical interest, these values are nearly equal to each other, as are the resonant frequencies, since for such structures ζ is very small compared to unity. The differences are less than 2 percent when ζ does not exceed 20 percent.

In forced vibration tests the disturbance is usually induced by a rotating mass. If we neglect the amplitude of the displacement response compared to the arm of the rotating mass, we may assume the external force varying harmonically and proportional to ω^2. Hence if we perform several tests on a simple system keeping the rotating mass and its arm constant, we shall find the displacement response proportional to $\omega^2 B_d$ and therefore to $(\omega/\omega_1)^2 B_d = B_a$. By examining Eq. 1.11 we conclude that the curve which represents the displacement response amplitudes with ω as abscissas, when the system is excited by a force independent of ω, is the same as the curve with $1/\omega$ as abscissas when the system is excited by a force proportional to ω^2, because multiplication of B_d by $(\omega/\omega_1)^2$ is equivalent to substitution of ω/ω_1 by its reciprocal. Such curves are illustrated in Fig. 1.4.

The maximum response amplitude occurs when either ω/ω_1 or ω_1/ω is equal to $(1 - \zeta^2)^{1/2}$ depending on whether the excitation force is frequency independent or proportional to ω^2, respectively (see Eqs. 1.11 and 1.13). The *half-power band width* is the range of frequencies for which the displacement response equals $1/\sqrt{2}$ times the resonant amplitude (see Fig. 1.4). From Eqs. 1.11 and 1.13 for B_d it follows that its maximum equals $(2\zeta)^{-1}(1 - \zeta^2)^{-1/2}$. Hence, the band width in question is equal to the difference in the two roots of ω which make B_d equal to $1/\sqrt{2}$ times this maximum.

For relatively small percentages of damping, as met in most structures of interest, this difference is very nearly equal to $2\zeta\omega_1$. Accordingly we can find the damping ratio from the maximum ordinate in a resonance curve if we know the magnitude of the static displacement response amplitude a or in every case from the band width corresponding to $1/\sqrt{2}$ times the resonant amplitude (Blake, 1961, p.15).

By means of a Fourier analysis, we can compute steady-state responses to forcing functions that are not harmonic functions of time (Sneddon, 1951).

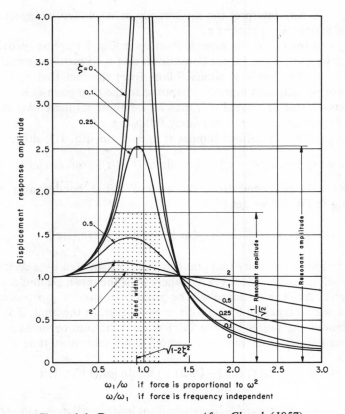

Figure 1.4. Resonance curves. *After Church (1957)*.

Other aspects of forced vibrations and their application to dynamic testing of structures are taken up in Chapter 17 and in Appendix 1.

1.4 Transient Disturbances

Consider the case when there are no external forces and the ground motion consists of a single velocity pulse, that is, the ground acceleration is defined by a Dirac delta function[1]: $\ddot{x}_0 = u\delta(t - \tau)$. Here u is a constant having dimensions

---

[1] This function is defined thus: $\delta(x) = 0$ if $x \neq 0$; $\int_{-x_1}^{0} \delta(x)\,dx = \int_{0}^{x_1} \delta(x)\,dx = \frac{1}{2}$ where x_1 is any positive value. It follows that $\int_{-x_1}^{x_1} \delta(x)dx = 1$ if $x_1 > 0$, and

$$\int_{x_1}^{x_2} f(x)\delta(x - x_3)\,dx = \begin{cases} f(x_3) & \text{if } x_1 < x_3 < x_2 \\ \frac{1}{2}f(x_3) & \text{if } x_3 = x_1 \text{ or } x_2 = x_3 \\ 0 & \text{otherwise} \end{cases}$$

where x_1 and x_2 are any values such that $x_1 < x_2$, and f is any piecewise continuous function, continuous at x_3.

of velocity. We may interpret the disturbance as a sudden change in ground velocity, of magnitude u at time τ.

Since $\ddot{x}_0 = 0$ for $t \geq \tau$, the general solution in Eq. 1.7 applies provided that $\zeta < 1$. In it we must solve for the constants a and t_1 to satisfy the initial conditions $y(\tau) = 0$, $\dot{y}(\tau+) = -u$, which follow from the fact that at $t = \tau+$ the mass has not yet acquired motion in reponse to the change in ground velocity. (Here $\tau+$ is a value of t equal to τ plus an infinitesimal increment in time; in other words, it is an instant immediately following τ.)

The first of these conditions implies $t_1 = \tau$. From Eq. 1.7, then,

$$\dot{y}(t) = a \exp\left[-\zeta\omega_1(t-\tau)\right] \{-\zeta\omega_1 \sin \omega_1'(t-\tau) + \omega_1' \cos[\omega_1'(t-\tau)]\}$$

so that $\dot{y}(\tau+) = a\omega_1'$, and from the second initial condition $a = -u/\omega_1'$. Substituting in Eq. 1.7 we get

$$y(t) = -\frac{u}{\omega_1'} \exp\left[-\zeta\omega_1(t-\tau)\right] \sin\left[\omega_1'(t-\tau)\right] \quad (1.14)$$

Now, any ground velocity diagram can be approximated by a series of steps or pulses each of magnitude $\ddot{x}_0(\tau)\,\Delta\tau$, where $\ddot{x}_0$ is the mean ground acceleration in the interval $\Delta\tau$. The response will be given by a sum of expressions similar to Eq. 1.14 with u replaced by $\ddot{x}_0(\tau)\,\Delta\tau$ in each term. In the limit, if $\dot{x}_0$ is piecewise continuous the pulses become $\ddot{x}_0(\tau)\,d\tau$, and the sum becomes an integral. If the system starts from rest and without deformation at time $t = 0$, then

$$y(t) = -\frac{1}{\omega_1'} \int_0^t \ddot{x}_0(\tau) \exp\left[-\zeta\omega_1(t-\tau)\right] \sin\left[\omega_1'(t-\tau)\right] d\tau$$

$$\dot{y}(t) = -\int_0^t \ddot{x}_0(\tau) \exp\left[-\zeta\omega_1(t-\tau)\right] \cos\left[\omega_1'(t-\tau)\right] d\tau - \zeta\omega_1 y(t) \quad (1.15)$$

$$\ddot{x}(t) = -\omega_1^2 y(t) - 2\zeta\omega_1 \dot{y}(t)$$

For the case of zero damping we have $\zeta = 0$, and the exponential becomes equal to 1.

On occasion we shall find it convenient to deal with another "response," the function $r(t)$, defined by $r^2 = (\omega_1 y)^2 + (\dot{y} + \zeta\omega_1 y)^2$. It will be shown that r represents the radius vector in a phase plane with oblique axes (see Section 1.6, especially Eqs. 1.21). In undamped systems, the total energy referred to the base is $Mr^2/2$, so that r is a measure of the energy per unit mass in the system.

In certain types of studies r is easier to calculate than either y or $\dot{y}$. This is interesting because it follows from the definition of the function that r is never smaller than $\omega_1|y|$ nor smaller than $|\dot{y}|$. Hence, calculation of r gives us an upper bound to the other responses.

We have thus derived expressions that give us the responses to an arbitrary accelerogram. For design we are interested in the maximum numerical value of certain responses, some of which depend only on the natural period and degree of damping of the system. It is therefore of interest to construct curves that represent maximum numerical values of the responses in question as functions of the natural period or natural frequency of simple structures, each curve corresponding to one value of ζ. Such plots are called *response spectra*. Thus the displacement spectrum is defined as $D(T_1)$ or $D(\omega_1) = \max |y(t)|$.

Often, rather than plotting maximum numerical values of the velocity relative to the ground and of the absolute acceleration, one plots *velocity* and *acceleration spectra* which represent, respectively, the quantities $V = \omega_1 D$ and $A = \omega_1^2 D$. The first of these is known as the maximum numerical value of *pseudovelocity* relative to the ground because it results from substituting sin for cos in the integrand of Eq. 1.15 that gives us $\dot{y}$; in the usual range of interest in earthquake-resistant design, V is statistically very close to the actual velocity relative to the ground (Hudson, 1962a), except for very long natural periods. And A, which we may call the maximum numerical value of the absolute *pseudoacceleration*, is very close to the maximum absolute $\ddot{x}$. A represents that part of the absolute acceleration whose product by the mass gives us the maximum spring force; that is, it neglects the part of $\ddot{x}$ whose product by M would give us the force in the damper.

In resorting to displacement, velocity, and acceleration spectra defined as above, what we actually plot is $\max |y(t)|$, but depending on whether we choose to draw D, V, or A spectra, we represent the quantity $\max |y(t)|$ in different manners.

The quantity A is always smaller than or equal to $\max |\ddot{x}|$, and the difference may be important for very rigid systems. Still, use of A in design is sometimes justified not only by considerations of simplicity but also because for some structural materials it gives a better quantity to compare with the structural capacity; the apparent strength increases with the rate of deformation, and this fact partly offsets the omission of the force in the dashpot element.

The use of spectra to represent the maximum responses of a group of simple structures has been most fruitful in the analysis of shock and earthquake effects. We shall make considerable application of spectral representation throughout this text.

There is advantage in representing spectra in logarithmic plots, similar to Fig. 1.3. In such graphs, scales are chosen so that abscissas represent log ω_1 or log T_1, ordinates represent log V, 45° lines in one direction represent log A, and lines perpendicular to these represent log D.

From the foregoing considerations on the response function r, it follows that $r_{\max} = V$ and, further, that $r = |\omega_1 y|$ every time that $|\omega_1 y|$ passes through a maximum. If the variation in successive maxima of $|y|$ is small compared with D, as often the case will be with responses to the motions in which we are interested, $r_{\max}$ will never greatly exceed V. In those cases the r spectrum provides a close upper bound to the velocity spectrum.

Considerable work has been done in applied dynamics and theory of communications on the basis of *Fourier spectra* rather than the response spectra described above. The Fourier spectrum of a ground motion is defined as

$$F(\omega) = \int_0^s \ddot{x}_0(t) \exp[-i\omega t] \, dt = \int_0^s \ddot{x}_0(t) \cos \omega t \, dt - i \int_0^s \ddot{x}_0 \sin \omega t \, dt \quad (1.16)$$

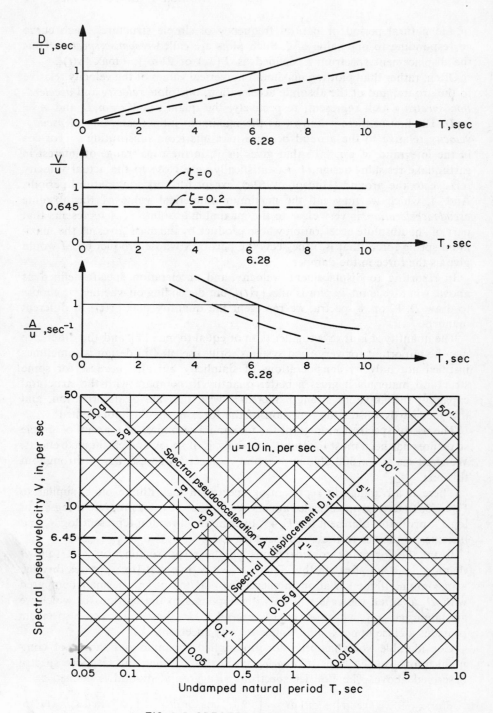

FIG. 1.5 SPECTRA FOR VELOCITY PULSE

where s is the duration of ground motion (Hudson, 1962a). It follows that the Fourier amplitude spectrum is

$$|F(\omega)| = \left\{ \left[\int_0^s \ddot{x}_0(t) \cos \omega t \, dt \right]^2 + \left[\int_0^s \ddot{x}_0(t) \sin \omega t \, dt \right]^2 \right\}^{1/2} \quad (1.17)$$

so that $|F(\omega)|$ is equal to $r(s)$ when $\zeta = 0$. Accordingly, $|F(\omega)| = r_{max}$; further, $|F(\omega)|$ is nearly always smaller than V. (In these comparisons we take $\omega = \omega_1$.)

The square of $|F(\omega)|/s$ is known as the *power spectral density* for which we find numerous applications in the theory of random motions.

Fourier spectra contain all the information concerning the ground motion because their Fourier transform restores that motion. Further, the power spectral density finds direct application in prediction of fatigue damage. But in structural design against earthquakes, response spectra are probably more significant and are usually preferred in describing ground motions, as these spectra give maximum values of responses directly.

In certain applications it is useful to resort to *residual spectra* (Rubin, 1961). These are defined as the maximum numerical values of responses after the ground motion has ended. Residual velocity spectra for zero damping coincide with Fourier spectra because in that case $|F(\omega)| = r(s)$ and $r(s) = \max |\dot{y}(t)|$ for $t \geq s$.

Example 1.1. Calculate the response spectra that correspond to a velocity pulse of magnitude u.

Solution. Substitution of $\ddot{x}_0 = u\delta(t - t_1)$ in Eq. 1.15 gives

$$y = -\frac{u}{\omega_1'} \exp\left[-\zeta \omega_1 (t - t_1)\right] \sin\left[\omega_1'(t - t_1)\right]$$

$$\dot{y} = u \exp\left[-\omega_1(t - t_1)\right] \{\eta \sin\left[\omega_1'(t - t_1)\right] - \cos\left[\omega_1'(t - t_1)\right]\}$$

where $\eta = \zeta(1 - \zeta^2)^{-1/2}$. If we equate $\dot{y}$ to zero, solve for t, substitute in the expression for y, and multiply by ω_1, we find $V = u \exp(-\eta \cos^{-1} \zeta)$. This result is plotted in Fig. 1.5 for $\zeta = 0$ and 0.2.

Example 1.2. Calculate the undamped response spectrum that corresponds to the accelerogram $\ddot{x}_0 = a$ when $0 < t \leq t_0$ and $\ddot{x}_0 = 0$ elsewhere (Fig. 1.6).

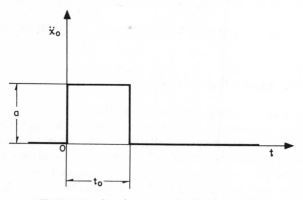

Figure 1.6. Accelerogram for Example 1.2.

14 SIMPLE LINEAR SYSTEMS Chap. 1

Solution. Substitution in the first part of Eq. 1.15 with $\zeta = 0$ gives

$$y = -\frac{a}{\omega_1} \int_0^{t'} \sin\left[\omega_1(t-\tau)\right] d\tau$$

where $t' = \min(t, t_0)$. From here,

$$V = \begin{cases} \dfrac{2a}{\omega_1} & \text{if } T \leq 2t_0 \\ \left|\dfrac{2a}{\omega_1} \sin \dfrac{\omega_1 t_0}{2}\right| & \text{if } T \geq 2t_0 \end{cases}$$

The result is shown in Fig. 1.7.

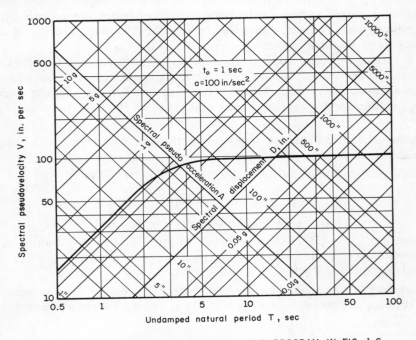

FIG. 1.7 UNDAMPED SPECTRA FOR ACCELEROGRAM IN FIG. 1.6

1.5 Numerical Evaluation of Responses

The ground motions which concern us are irregular to such an extent that analytical evaluation of structural responses must be ruled out. Of the various other approaches available, first we shall describe the use of numerical methods.

The form of Eq. 1.15 lends itself little to numerical evaluation because the upper limit of the integrals t appears in the integrands. If we used the expressions as they stand we would have to discard the values of y and $\dot{y}$ computed for previous instants and start anew every time we introduced an increment in t. Because usually we shall be interested in maxima of $|y|$, this procedure would require an impractically large number of computations.

We can surmount the difficulty in various ways. Perhaps the most obvious of these consists in expanding the function $\exp[-\omega_1(t-\tau)]\sin\omega_1(t-\tau)$ in the expression for y. Thus,

$$y(t) = -\frac{\exp(-\zeta\omega_1 t)}{\omega_1}\bigg[\sin(\omega_1 t)\int_0^t \ddot{x}_0(\tau)\exp(\zeta\omega_1\tau)\cos(\omega_1'\tau)\,d\tau$$
$$+\cos(\omega_1 t)\int_0^t \ddot{x}_0(\tau)\exp(\zeta\omega_1\tau)\sin(\omega_1'\tau)\,d\tau\bigg]$$

We can effect a similar transformation on the expressions for $\dot{y}$. In these forms cumulative summation permits evaluation of the integrals at successive instants.

This step is the basis for one of the first practical numerical methods found in the literature for computing spectral ordinates. The method was proposed by M.P. White (1942).

The approach described has two main drawbacks. First, it is limited to linear systems. Second, unless damping is very small and the duration of motion not too large, the exponentials under the integral signs increase to very large values causing serious loss of accuracy.

In view of these objections it is usually preferable to use a numerical method that deals directly with the responses rather than one which attempts explicit evaluation of Duhamel's integrals. One such method is described below (Newmark, 1959). It is adaptable to nonlinear systems with any number of degrees of freedom.

For a system governed by Eqs. 1.2–1.4 the method is applied as follows.

1. Let the values of y, $\dot{y}$, and $\ddot{y}$ be known at time $t = t_i$; identify them by the subscript i. Let $t_{i+1} = t_i + \Delta t$. Assume the value of $\ddot{y}_{i+1}$.
2. Compute $\dot{y}_{i+1} \doteq \dot{y}_i + (\ddot{y}_i + \ddot{y}_{i+1})\Delta t/2$.
3. Compute $y_{i+1} \doteq y_i + \dot{y}_i\Delta t + (\tfrac{1}{2} - \beta)\ddot{y}_i(\Delta t)^2 + \beta\ddot{y}_{i+1}(\Delta t)^2$.
4. Compute a new approximation to $\ddot{y}_{i+1}$ from Eq. 1.4:

$$\ddot{y} = -2\zeta\omega_1\dot{y} - \omega_1^2(y - y_0) - \ddot{x}_0$$

where y_0 is the static displacement relative to the base.
5. Repeat steps 2–4 beginning with the newly computed $\ddot{y}_{i+1}$ or with an extrapolated value, until a satisfactory degree of convergence is attained.

Step 2 is consistent with a straight-line approximation to $\ddot{y}$ in the interval considered. In step 3 taking $\beta = \tfrac{1}{4}$ would be consistent with a straight-line variation of $\ddot{y}$ in that same interval while $\beta = \tfrac{1}{6}$ corresponds to a parabolic variation. If we choose β greater than $\tfrac{1}{8}$ we are assured that the method is stable only if it converges. In general a choice of β between $\tfrac{1}{4}$ and $\tfrac{1}{6}$ is satisfactory from all points of view, including that of accuracy; the lower limit assures stability even in the generalized version of the present method applicable to systems with several degrees of freedom (see Section 2.7).

With $\Delta t = 0.1 T_1$ and $\beta = \frac{1}{6}$ the rate of convergence is such that after the third cycle of successive approximations to $\ddot{y}_{i+1}$ the error (due to using a finite number of cycles) does not exceed 1 part in 1000 (if the first approximation is taken assuming $M = 0$), and errors in the natural period computed by this method are of the order of 2 to 3 percent. Greater accuracy and faster convergence are attained by shortening the ratio $\Delta t/T_1$.

The main assets of the method described are its adaptability to nonlinear and multidegree-of-freedom systems and its efficiency, especially when used in digital computers. As with other convenient numerical methods most of the time spent in processing recorded data goes into putting them in a form that can be fed the computer, unless automatic equipment is available for the purpose (Hudson, 1962a).

In order to save a few cycles when one desires high accuracy, there is always the possibility of using an extrapolation procedure, apparently due to Aitken (1937, quoted by Crandall, 1956). When applying the present method of analysis there is some advantage in resorting to extrapolation if computations are carried out on a desk calculator. This is debatable in the case of digital computers because usually it will be found simpler to shorten the time interval or run a larger number of cycles than to introduce a subroutine.

Superficial study of the expressions used in the numerical method suggests the convenience of combining them so as to eliminate the need for successive approximations. Indeed $\ddot{y}_{i+1}$ is uniquely determined once $\ddot{x}_{0,i+1}$ and the pertinent quantities at $t = t_1$ are known. Although there are computational advantages in this variation, it implies a loss of generality because different sets of expressions would be required when dealing with nonlinear systems.

Other practical methods for computation of structural responses, generally suited for the solution of initial-value problems, are described and discussed in works such as those of Salvadori and Baron (1952). Among them, Fox's method for damped systems and Noumerov's method for conservative systems are especially powerful. Both are also applicable to nonlinear systems having one degree of freedom.

Example 1.3. Compute the responses of the system defined by $K = 36$ lb/in., $M = 4$ lb sec²/in., $\zeta = 0.2$ to the ground motion defined by $\ddot{x}_0 = -(30 \text{ in./sec}^3)t$ for $0 \leq t \leq 0.4$ sec, and $\ddot{x}_0 = 0$ outside this range.

Solution. From the structural characteristics it follows that $\omega_1 = \sqrt{36/4} = 3$ rad/sec, $\zeta\omega_1 = 0.2 \times 3 = 0.6$ sec^{-1}, $T_1 = 6.2832/3 = 2.0944$ sec. We shall adopt $\Delta t = 0.2$ sec $\cong 0.1 T_1$ and $\beta = 0.2$. Hence $\dot{y}_{i+1} \doteq \dot{y}_i + 0.1(\ddot{y}_i + \ddot{y}_{i+1})$, $y_{i+1} \doteq y_i + 0.2\dot{y}_i + 0.012\ddot{y}_i + 0.008\ddot{y}_{i+1}$. Also, $\ddot{y}_{i+1} = -1.2\dot{y}_{i+1} - 9y_{i+1} - \ddot{x}_{0,i+1}$.

Computations are summarized in Table 1.1 up to $t = 0.6$ sec and may be continued in the same manner for larger values of t. Accelerations in successive cycles corresponding to the same value of t were found from the values of $\dot{y}$ and y obtained at the end of the preceding cycle, except $\ddot{y} = 7.534$ and -5.550 in./sec² for $t = 0.4-$ and 0.6 sec, respectively, which were found by the delta squared method of extrapolation. Notice that the discontinuity in $\ddot{x}_0$ at $t = 0.4$

TABLE 1.1. SOLUTION OF EXAMPLE 1.3

t (sec)	$\ddot{x}_0$ (in./sec²)	$\ddot{y}$ (in./sec²)	$\dot{y}$ (in./sec)	y (in.)
0	0	0	0	0
0.2	−6	5.000	0.5000	0.04000
		5.040	0.5040	0.04032
		5.033	0.5033	0.04026
		5.034	0.5034	0.04027
0.4−	−12	8.000	1.8078	0.26536
		7.442	1.7510	0.26079
		7.534	1.7602	0.26163
		7.533	1.7601	0.26162
0.4+	0	−4.467	1.7601	0.26162
0.6	0	−6.000	0.7134	0.51204
		−5.464	0.7670	0.51633
		−5.550	0.7584	0.51564

sec introduces an equal discontinuity of opposite sign in $\ddot{y}$ at the same instant.

In this example we would have obtained more accurate results had we used $\beta \simeq \frac{1}{6}$ rather than $\frac{1}{5}$, but this contention should not be generalized.

1.6 Graphical Evaluation

Graphical methods of analysis are usually attractive because they give a clear picture of physical processes.

The motions in which we are interested are nearly always quite complicated; hence graphical methods are neither practical nor reasonably accurate in analysis of structural responses to earthquakes. But this does not detract from their value in aiding to visualize structural behavior under earthquake excitation.

Of the various graphical methods available we shall describe one in particular because it is adaptable to nonlinear systems. In its more general version this graphical construction is known as the *gyrogram* (Rojansky, 1948; Jacobsen and Ayre, 1958, Chap. 1). The method is carried on the *phase plane* which is defined by the Cartesian coordinates $y, \dot{y}/\omega_1$.

Consider first a conservative (undamped) simple linear system. According to Eq. 1.4, when there are no external forces, the system is governed by the differential equation

$$\ddot{y} + \omega_1^2 y = -\ddot{x}_0 \qquad (1.18)$$

Now, any ground velocity diagram may be approximated as a series of steps: $\dot{x}_0 = \sum_{i=0}^{j} u_i$ if $t_j < t < t_{j+1}$, and $\dot{x}_0 = \sum_{i=0}^{j-1} u_i + u_j/2$ if $t = t_j$ (Fig. 1.8a). The

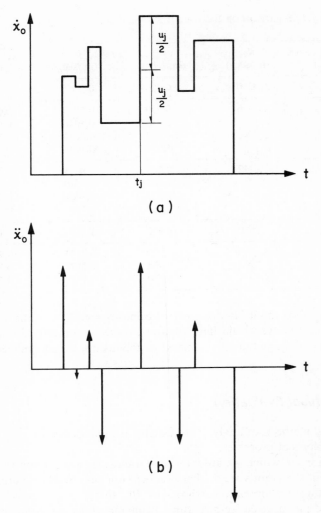

Figure 1.8. Approximation to arbitrary ground motion.

corresponding accelerogram is a series of pulses, represented schematically in Fig. 1.8b: $\ddot{x}_0 = \sum_i u_i \delta(t - t_i)$. When t does not coincide with any of the t_i's, the second member in Eq. 1.18 is nil, so that according to Eq. 1.7 the response of the system is given by

$$y = a \sin[\omega_1(t - \tau)]$$

and (1.19)

$$\frac{\dot{y}}{\omega_1} = a \cos[\omega_1(t - \tau)]$$

where a and τ are constants depending on the values specified for y and $\dot{y}$ at the beginning of the interval. Equations 1.19 constitute the parametric form of a circular arc in the phase plane (Fig. 1.9). The arc's center is at the origin, and its radius is $[y^2 + (\dot{y}/\omega_1)^2]^{1/2}$. (This is the same as r/ω_1 when r is defined as in Section 1.4 for $\zeta = 0$.) The angle described by the radius vector during any interval Δt is $\omega_1 \Delta t$ in the clockwise direction provided Δt lies entirely between two successive values of t_i.

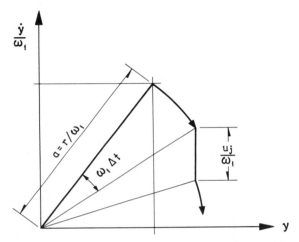

Figure 1.9. Phase-plane representation of Eq. 1.19.

When t passes from $t_i -$ to $t_i +$, y does not change by more than an infinitesimal amount while $\dot{y}$ changes by $-u_i$. Therefore each increment u_i in ground velocity is reflected as a straight vertical segment of direction opposite that of the increment in ground velocity, and of length u_i/ω_1, its initial point coinciding with the terminal point of the arc for the interval $i - 1, i$ and the segment's terminal point marking the beginning of the arc for the interval $i, i + 1$.

The response spectral ordinate for a given ω_1 is equal to the maximum abscissa (in absolute value) drawn in the graphical construction. The corresponding Fourier amplitude spectrum ordinate equals ω_1 times the radius vector to the terminal point in the graph.

Usually it will be more accurate to approximate the ground velocity diagram by means of inclined straight lines and the accelerogram by the corresponding stepped functions: $\ddot{x}_0 = a_i$ if $t_i < t \leq t_{i+1}$. If we make the second member of Eq. 1.18 constant, the general solution will be equal to y as given by the first part of Eq. 1.19 plus a particular solution of Eq. 1.18 such as $y = -\ddot{x}_0/\omega_1^2$:

$$y = a \sin [\omega_1(t - \tau)] - \frac{\ddot{x}_0}{\omega_1^2}$$

$$\frac{\dot{y}}{\omega_1} = a \cos [\omega_1(t - \tau)]$$

(1.20)

Again we have the parametric equations for a circular arc with the radius as in Eq. 1.19 and the same angle $\omega_1 \Delta t$ described by the radius, but the arc's

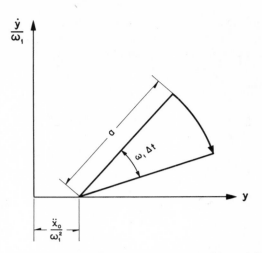

Figure 1.10. Phase-plane representation of Eqs. 1.20.

center lies now on the y-axis at a distance $\ddot{x}_0/\omega_1^2$ from the origin and $\ddot{x}_0$ is the ordinate to the stepped accelerogram for the interval under consideration. Figure 1.10 illustrates this type of graphical construction.

In a linearly damped system the solution with $\ddot{x}_0 = 0$ (corresponding to one interval where the accelerogram is replaced by pulses), instead of being the parametric equations of a circular arc in the phase plane, is an oblique logarithmic spiral with focus at the origin (see Eq. 1.7):

$$y = a \exp\left[-\zeta\omega_1(t-\tau)\right] \sin\left[\omega_1'(t-\tau)\right]$$
$$\frac{\dot{y}}{\omega_1} = (1-\zeta^2)^{1/2} a \exp\left[-\zeta\omega_1(t-\tau)\right] \cos\left[\omega_1'(t-\tau)\right] - \zeta y \quad (1.21)$$

(Fig. 1.11). The angle described by the radius vector during a time interval Δt is now $\omega_1' \Delta t$. The shape of the spiral depends only on the parameter ζ because a change in τ is equivalent to a rotation of the spiral. Hence there is need to construct only one curve for a given damping coefficient, and once this is done use of tracing paper makes the graphical construction as simple as for an undamped system.

Again if we approximate the accelerogram by means of a stepped function we shall have merely to add $-\ddot{x}_0/\omega_1^2$ to the expression for y, leaving the one for $\dot{y}/\omega_1$ unchanged. This is equivalent to shifting the spiral focus in each interval to a point on the y-axis a distance $\ddot{x}_0/\omega_1^2$ from the origin.

Another approach in the graphical analysis of damped systems, which can be easily generalized to nonlinear systems, consists in taking the term $2\zeta\omega_1\dot{y}$ in Eq. 1.4 as constant during a short time interval and subtracting it from both sides of the equation. This is equivalent to solving Eq. 1.18 with the second member equal to $-\ddot{x}_0 - 2\zeta\omega_1\dot{y}$, constant within each interval. The value of $\dot{y}$ can be taken equal to the relative velocity at the beginning of each interval, or to an estimated average for the interval whose representation is to be drawn,

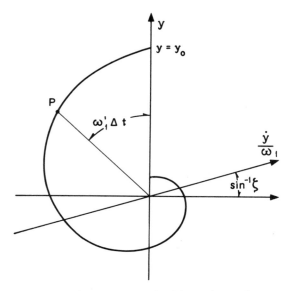

Figure 1.11. Phase plane method for a damped system. *After Jacobsen and Ayre (1958).*

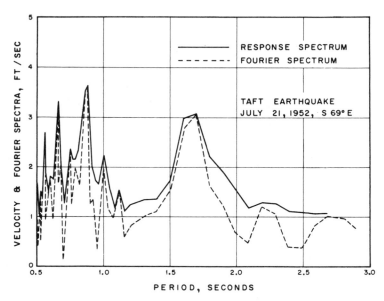

Figure 1.12. Typical shapes of response and Fourier spectra. *After Hudson (1962a).*

or to such an average as obtained by successive approximations. In any case the path described by a point in the phase plane is once more a series of circular arcs each with center on the y-axis at a distance $\ddot{x}_0/\omega_1^2 + 2\zeta\dot{y}/\omega_1$ from the origin.

Example 1.4. Is the slope dv/dT continuous at maxima and minima of D?

Solution. A gradual variation in the natural period brings about a gradual change in the shape of a ground motion's gyrogram. Study of the possibilities of these changes discloses that dv/dT must be continuous at points where $V(T)$ is maximum but may be (and usually is) discontinuous at points of minima since the branch of the gyrogram for which $\max|y|$ occurs may change abruptly when T is varied continuously. The outer branch may then be tending toward the origin while the inner one may be moving away from the origin. In this case F, and hence V, first descend and then increase, thus giving place to a minimum where the slope is discontinuous. On the other hand the Fourier amplitude spectrum necessarily has a continuous derivative with respect to T for all values of the natural period except $T = 0$. Typical shapes of response and Fourier amplitude spectra are compared in Fig. 1.12.

1.7 Analog Evaluation

The simplest analog to a mechanical system is a scaled model of the system. Dynamic testing of models is treated in Chapter 17. The next most obvious analog is another mechanical system. The torsion pendulum was used in the calculation of the first published set of earthquake spectra (Biot, 1941).

Mechanical analogs are objectionable in that their degree of damping cannot be controlled adequately. As a consequence the spectra that for some time formed the basis of practically all rational studies of earthquakes and that were assumed to correspond to undamped systems actually contained varying degrees of damping. This situation became apparent when electrical analogs were used for computation of spectra (Housner and McCann, 1949). Use of electrical

TABLE 1.2. EQUIVALENCES BETWEEN MECHANICAL AND ELECTRICAL SYSTEMS

Mechanical	Electrical	
	Force-voltage	Force-current
Node (point)	Closed loop	Node
Force	Voltage	Current
Mass	Inductance	Capacitance
Dashpot constant	Resistance	Conductivity
Flexibility*	Capacitance	Inductance

*Reciprocal of stiffness.

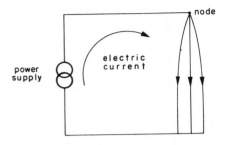

FORCE-CURRENT ANALOG

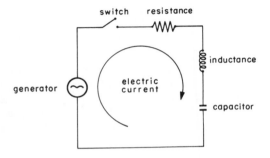

FORCE-VOLTAGE ANALOG

Figure 1.13. Electrical analogs to mechanical system in Fig. 1.1.

analogs in computation of responses to ground motions is quite trustworthy (Hudson, 1962a).

Electrical systems analogous to the simple structure of Fig. 1.1 are shown in Fig. 1.13. Table 1.2 gives equivalences between the mechanical and the two electrical systems. The analogy between the three systems stems from the fact that they are all governed by the same differential equation.

Further information can be obtained from the extensive literature on the subject (Soroka, 1961). Simple electrical analogs can be easily built or acquired commercially.

Electric analogs yield results which are less accurate than those of numerical methods; yet they can be built to give a satisfactory degree of accuracy, and they supply quick answers. Preference for one or the other method depends on the equipment available, personal inclination, and the degree of accuracy desired.

PROBLEMS

1.1. Derive the following expressions for natural periods of simple systems.
   a. $T_1 = 2\pi\sqrt{J/K_t}$ for an undamped torsion pendulum, where $J =$ mass polar moment of inertia and $K_t =$ torsional rigidity of supporting wire.
   b. $T_1 = 2\pi\sqrt{L/g}$ for a mathematical pendulum, where $L =$ distance from support to center of mass and $g =$ acceleration of gravity. (*Hint:* express J and the restoring moment in terms of the pendulum mass.)
   c. $T_1 = 2\pi\sqrt{L/g}$ for an undamped physical pendulum under small oscillations, where $L =$ distance from support to center of percussion.

1.2. Show that the following expression gives the ground motion, $x_0(t)$, as a function of the record obtained by a seismograph (simple linearly damped system):

$$x_0(t) = x_0(0) + \dot{x}_0 t + b_t\{y(0) + [\dot{y}(0) + 2\zeta\omega_1 y(0)]t - y(t)$$
$$- (2\zeta\omega_1 + \omega_1^2 t)\int_0^t y(\tau)\,d\tau + \omega_1^2 \int_0^t y(\tau)\tau\,d\tau\}$$

where b is the reciprocal of the seismograph magnification factor and $y(t)$ is the ordinate of the record at time t.

1.3. The system in Fig. 1.14 consists of a rigid body supported by a rigid, massless bar joined to its support by a linear spring without damping. Let W be the weight of the body and W_1 the buckling load of the bar-spring system. Neglecting rotational inertia and assuming small deflections, show that the natural period of the system is $T_1 = T_0(1 - W/W_1)^{-1/2}$ where T_0 is the natural period computed neglecting the action of gravity.

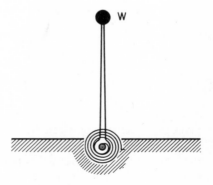

Figure 1.14. System described in Problem 1.3.

1.4. Calculate analytically the Fourier and response spectra corresponding to $\zeta = 0$ and to $\zeta = 0.2$ for the ground motion defined by $\ddot{x}_0 = a \sin 2\pi t/t_0$ when $0 < t \leq t_0$ and $\ddot{x}_0 = 0$ when $t \geq t_0$. (Use trial and error to determine the instant of maximum response for $\zeta = 0.2$.)

Ans. See Fig. 1.15.

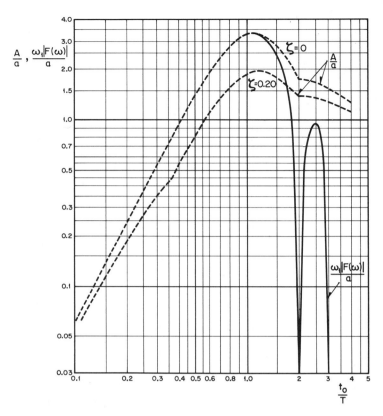

Figure 1.15. Problem 1.4. *After Veletsos and Newmark (1964) for A/a.*

1.5. A mass is allowed to touch a massless, undamped, linearly elastic system (say a very light simply supported beam). It is then suddenly released. Using a graphical construction on the phase plane, show that the maximum deflection of the system under the mass equals twice the static deflection.

1.6. Repeat Examples 1.2 and 1.3 using the numerical and graphical procedures described in this chapter. For Example 1.2 take $\zeta = 0.1$. (Take time intervals equal to about 1/10 the natural period in question.) Compare results.

2

LINEAR SYSTEMS WITH SEVERAL DEGREES OF FREEDOM

2.1 Equations of Motion

We shall be concerned in this chapter with systems having constant parameters, with lumped rigid masses joined to each other and to the ground by linear, massless springs and dashpots. Each mass may have up to six degrees of freedom, three of which correspond to translations and three to rotations. Most problems in earthquake-resistant design are simplified without undue error by assuming that each mass has but one degree of freedom; even in problems involving torsion in buildings, the mass taken at each floor is rarely assigned more than two degrees of freedom in translation and one in rotation about a vertical axis.

In order to describe configurations of the system—that is, its displacements and rotations—we shall need as many linearly independent quantities as there are degrees of freedom. These quantities are usually called *generalized displacements*. We shall denote them by x_r. Out of similarity with established usage for Cartesian coordinates, we may regard a generalized displacement as the product of a scalar and a vector, and designate the latter a *generalized coordinate*. A change in the magnitude of a generalized displacement implies, then, a change in the scalar. We may choose as coordinates the three displacements of the centroid and the three rotations about the principal axes of inertia for each mass, or sets of linear combinations of these quantities.

The displacement scalars may be arranged in a column following any conventional order. When this is done for a system having N degrees of freedom, the displacement scalars constitute an N-dimensional column vector, which we shall denote by **x** and which we shall call a *generalized configuration* of the system. Every state of the system can be represented by one such vector provided the base is assumed fixed. Cases in which the base undergoes displacements will be taken up later, beginning with Section 2.4.

Every set of internal and external forces and moments acting on the structure may be represented by a single column vector, which we shall call *generalized force*. Each term in the vector stands for the component of the generalized force on the corresponding coordinate, arranged in the same order as the terms in the generalized configuration.

Suppose that the rth generalized displacement x_r is forced to take a finite value simultaneously keeping all the other displacements equal to zero. Then spring forces and moments (generalized spring forces) will appear in all spring elements whose lengths change in the process. Associated with this change and acting along the sth coordinate there will be a generalized spring force which we shall denote by Q_{rs}. If the value given to x_r is unity, Q_{rs} is called the sth *stiffness influence coefficient* corresponding to the rth degree of freedom, and we shall denote it by K_{rs}. The matrix of K_{rs} values, ordered in the same manner as the x's is usually known as the *stiffness matrix*. We shall denote it by $\mathbf{K}$.

The reciprocal $\mathbf{K}^{-1}$ of the stiffness matrix is commonly known as the *flexibility matrix* of the system. This matrix can be regarded as an ordered table of the displacements induced by unit forces. According to the principle of reciprocal relations, both $\mathbf{K}$ and $\mathbf{K}^{-1}$ are symmetric.

In a similar way, if we give a finite value $\dot{x}_r$ (a velocity) to the rate of change of x_r, keeping the derivatives of all the other displacements with respect to time equal to zero, the dashpots will develop a generalized force whose terms we shall denote by C_{rs}. If $\dot{x}_r$ is unity, we may call C_{rs} a *damping influence coefficient*. The matrix $\mathbf{C}$ of these influence coefficients, ordered in the same manner as the x's and the K's is known as the *damping matrix*.

By giving the acceleration $\ddot{x}_r$ a unit value while keeping all other second derivatives of the generalized displacements with respect to time equal to zero, we produce a set of inertia forces, or a generalized inertia force whose sth component we shall denote by M_{rs}. Their matrix, $\mathbf{M}$, again in the same order, is called the *inertia matrix*.

As an example consider the two-story building in Fig. 2.1. We shall assume that each story has a single degree of freedom, corresponding to the displacements x_1, x_2. Every configuration of the building is associated with a vector

$$\mathbf{x} = \begin{vmatrix} x_1 \\ x_2 \end{vmatrix}$$

In this case we have associated each degree of freedom to the mass M_1 or M_2 of one story. The diagonal matrix of these masses, tabulated in the same order as the x's, constitutes the *inertia matrix:*

$$\mathbf{M} = \begin{vmatrix} M_1 & 0 \\ 0 & M_2 \end{vmatrix}$$

The inertia matrix is diagonal only when we choose as coordinates quantities proportional to the displacements of the centroid of each mass and the rotations of the mass about its principal inertia axes. It is then called *mass matrix*. When the x's include displacements and rotations along the three axes, we associate with the mass of the ith body, M_i, all three displacement components and with

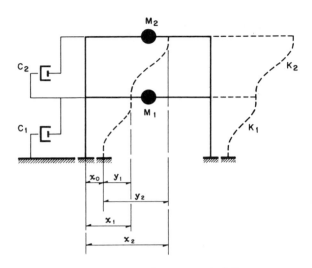

Figure 2.1 Two-story shear building.

each of its principal polar moments of inertia we associate the corresponding rotations about those axes.

Had we chosen as generalized coordinates in the foregoing example the elongations of the springs, say with generalized displacements $x_{01} = x_1$ and $x_{12} = x_2 - x_1$, the inertia matrix would have taken the form

$$\mathbf{M} = \begin{vmatrix} M_1 + M_2 & M_2 \\ M_2 & M_2 \end{vmatrix}$$

In every case the inertia matrix can be written almost from inspection by giving each generalized displacement x_i a unit acceleration while keeping all others equal to zero. The inertia force that appears then along the ith coordinate, with opposite sign, gives us the jth term in the ith row of the inertia matrix. If $\mathbf{F}$ denotes the column vector of generalized inertia forces (that is, forces and moments), we shall always have $\mathbf{F} = -\mathbf{M}\ddot{\mathbf{x}}$.

For the structure shown in Fig. 2.1, a unit displacement of floor 1, while floor 2 is kept in its original position, introduces the shears K_1 and K_2 in the first and second stories, respectively, hence $K_{11} = K_1 + K_2$ and $K_{12} = -K_2$. A unit displacement of the second floor induces only a shear equal to K_2 in the second story, so that $K_{21} = -K_2$ and $K_{22} = K_2$. Accordingly,

$$\mathbf{K} = \begin{vmatrix} K_1 + K_2 & -K_2 \\ -K_2 & K_2 \end{vmatrix}$$

The flexibility matrix for systems similar to the one in Fig. 2.1 can be written by inspection. Indeed, a unit horizontal force at floor 1 produces displacements K_1^{-1} at both floors, while a unit horizontal force applied at the second floor produces the displacements K_1^{-1} and $K_1^{-1} + K_2^{-1}$ at floors 1 and 2, respectively.

Therefore,
$$\mathbf{K}^{-1} = \begin{vmatrix} K_1^{-1} & K_1^{-1} \\ K_1^{-1} & K_1^{-1} + K_2^{-1} \end{vmatrix}$$

This is the inverse of the stiffness matrix given above. The same simplicity operates in the computation of the flexibility matrix in other systems that lend themselves to analysis by the method of forces. Computation of the stiffness matrix is simpler in systems that lend themselves to analysis by the method of displacements.

Also for the example in Fig. 2.1,
$$\mathbf{C} = \begin{vmatrix} C_1 + C_2 & C_2 \\ -C_2 & C_2 \end{vmatrix}$$

Inertia, stiffness, flexibility, and damping matrices can be written immediately for structures of a common type that is often called *simply coupled*. In these systems mass M_r is connected by a spring and a dashpot element to only M_{r-1} and M_{r+1}, except M_1 which is connected thus to the ground and to M_2, and the last mass, say M_N, which is only connected in this manner to M_{N-1} (Fig. 2.2), and the system is statically determinate. Other masses may also be

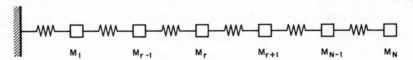

Figure 2.2. Statically-determinate, one-dimensional system.

connected to a support, in which case the system is said to be *closely coupled* or *close coupled*, and it is statically indeterminate. Each mass has a single degree of freedom, and all degrees correspond to displacement or rotation in a single direction. Simply coupled systems may be regarded as special cases of close-coupled ones. Examples of simply coupled systems include discrete "shear-beam" buildings, that is, structures in which the deformation of each story is assumed to depend only on the horizontal shear at that story. (Such is the case of the building shown in Fig. 2.1.) Choosing as generalized displacements the absolute displacements, the inertia matrix for closely coupled systems is the diagonal matrix

$$\mathbf{M} = \begin{vmatrix} M_1 & & & \\ & M_2 & & \\ & & \ldots & \\ & & & M_N \end{vmatrix}$$

If the system is statically determinate, the stiffness matrix is

$$\mathbf{K} = \begin{vmatrix} K_1 + K_2 & -K_2 & 0 & 0 & \cdots & 0 \\ -K_2 & K_2 + K_3 & K_3 & 0 & \cdots & 0 \\ 0 & -K_3 & K_3 + K_4 & -K_4 & \cdots & 0 \\ 0 & 0 & -K_4 & K_4 + K_5 & \cdots & 0 \\ 0 & 0 & 0 & -K_5 & \cdots & 0 \\ \cdots \\ 0 & 0 & 0 & 0 & \cdots & K_N \end{vmatrix}$$

And the flexibility matrix is

$$\mathbf{K}^{-1} = \begin{vmatrix} K_1^{-1} & K_1^{-1} & K_1^{-1} & K_1^{-1} & \cdots & K_1^{-1} \\ K_1^{-1} & K_1^{-1} + K_2^{-1} & K_1^{-1} + K_2^{-1} & K_1^{-1} + K_2^{-1} & \cdots & K_1^{-1} + K_2^{-1} \\ K_1^{-1} & K_1^{-1} + K_2^{-1} & \sum_{r=1}^{3} K_r^{-1} & \sum_{r=1}^{3} K_r^{-1} & \cdots & \sum_{r=1}^{3} K_r^{-1} \\ \cdots \\ K_1^{-1} & K_1^{-1} + K_2^{-1} & \sum_{r=1}^{3} K_r^{-1} & & \cdots & \sum_{r=1}^{N} K_r^{-1} \end{vmatrix}$$

As in simple structures we may apply D'Alembert's principle to each body. We apply it with respect to forces and moments in each of the reference directions. Thus we obtain as many equations of the form

$$M_r \ddot{x}_r + \sum_s C_{rs} \dot{x}_s + \sum_s K_{rs} x_s = P_r$$

as there are degrees of freedom, where P_r is the component of external forces in the direction of x_r. The system of these equations can be written in matrix form

$$\mathbf{M}\ddot{\mathbf{x}} + \mathbf{C}\dot{\mathbf{x}} + \mathbf{K}\mathbf{x} = \mathbf{P}$$

where $\mathbf{P}$ is the column vector of external forces.

Methods are available for solving these systems of equations (Crandall and McCalley, 1961), but they are cumbersome. For most cases in earthquake engineering a simplified method suffices in which the problem is first solved neglecting the presence of dashpots, and damping is later taken into account in an approximate manner. Lack of precise data on damping does not usually justify a more refined treatment. Therefore we shall be content to drop the second term in the left-hand member of the foregoing expression and first attempt to solve the matrix equation

$$\mathbf{M}\ddot{\mathbf{x}} + \mathbf{K}\mathbf{x} = \mathbf{P} \tag{2.1}$$

We derived this expression for the case when the ground does not move. In order to generalize our result to systems that rest on a base having several degrees of freedom, we introduce the column vector $\mathbf{x}_0$, which is the set of static displacements due to base motion. In other words, $\mathbf{x}_0(t)$ consists of the set of values $x_{0i}(t)$ which would be equal to the displacement in the ith coordinate of the system at time t as a response to the imposed set of base motions if these occurred at an infinitely slow rate. As for a simple system, let $\mathbf{y} = \mathbf{x} - \mathbf{x}_0$; that is, $\mathbf{y}$ stands for the set of structural displacements referred to the static configuration imposed by the base motions. In a closely coupled system, $\mathbf{y}$ consists of the terms $x_i - x_0$ where x_0 now represents the base displacement, since the induced static displacement would be equal at all points to the displacement of the base. Then Eq. 2.1 becomes $\mathbf{M\ddot{x}} + \mathbf{Ky} = \mathbf{P}$ or

$$\mathbf{M\ddot{y}} + \mathbf{Ky} = \mathbf{P} - \mathbf{M\ddot{x}}_0 \tag{2.2}$$

We shall use either of these forms depending on the type of problem and structure with which we are dealing. And we shall use the flexibility matrix in solving certain problems.

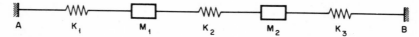

Figure 2.3. Statically-indeterminate, one-dimensional system.

Consider, as an example of a structure whose base may deform, the system in Fig. 2.3, in which the motions of supports A and B, x_a and x_b, respectively, are not necessarily equal. The inertia matrix and the column vector $\mathbf{x}$ are the same as for the example shown in Fig. 2.1. Straightforward analysis gives us

$$\mathbf{x}_0 = \begin{vmatrix} x_{01} \\ x_{02} \end{vmatrix}$$

where

$$x_{01} = \frac{x_a(K_2^{-1} + K_3^{-1}) + x_b K_1^{-1}}{K_1^{-1} + K_2^{-1} + K_3^{-1}}$$

$$x_{02} = \frac{x_a K_3^{-1} + x_b(K_1^{-1} + K_2^{-1})}{K_1^{-1} + K_2^{-1} + K_3^{-1}}$$

If supports A and B were constrained to undergo equal displacements at all times, we would have

$$\mathbf{x}_0 = \begin{vmatrix} x_a \\ x_a \end{vmatrix} = \begin{vmatrix} x_b \\ x_b \end{vmatrix}$$

$$\mathbf{y} = \begin{vmatrix} x_1 - x_a \\ x_2 - x_a \end{vmatrix}$$

and would deal with the problem as though the system rested on a rigid support.

2.2 Free Oscillations

Here we are interested in free vibrations of the system, for which $\mathbf{P} = \mathbf{M\ddot{x}}_0 = 0$ and $\mathbf{y} = \mathbf{x}$. From Eq. 2.1,

$$\mathbf{M\ddot{x}} + \mathbf{Kx} = 0 \tag{2.3}$$

A structure is said to vibrate in one of its *natural modes* when its (time-dependent) free displacements can be put in the form

$$\mathbf{x}(t) = \mathbf{z}_n \theta_n(t)$$

in which subscript n denotes the order of the mode, $\mathbf{z}_n$ is the mode shape which does not vary wiith time, and θ_n is a scalar function. In other words, the system oscillates in one of its natural modes if the base remains motionless, all the system's masses describe a synchronous motion, and the shape of the configuration does not depend on time although its magnitude varies with t, as prescribed by the nth function θ.

Substitution of the last expression in Eq. 2.3 leads to

$$\mathbf{M}\mathbf{z}_n \ddot{\theta}_n + \mathbf{K}\mathbf{z}_n \theta_n = 0$$

which is separable into

$$\ddot{\theta}_n + \omega_n^2 \theta_n = 0 \tag{2.4}$$

and

$$(\mathbf{K} - \omega_n^2 \mathbf{M})\mathbf{z}_n = 0 \tag{2.5}$$

where ω_n is a constant for the system and is independent of time.

We can write the general solution of Eq. 2.4 as

$$\theta_n = \sin \omega_n(t - t_n)$$

where we have dropped an arbitrary multiplier, because θ_n will be multiplied by the set of amplitudes $\mathbf{z}_n$, which contain just such arbitrary factors. The frequency, circular frequency, and period of θ_n are defined in terms analogous to those used for a simple conservative system. They constitute the nth natural frequency, and so on, of the multidegree-of-freedom system. Since the amplitude of θ_n is equal to one, that of the rth coordinate in the nth mode is equal to the value of the rth term in the vector $\mathbf{z}_n$, say z_{rn}.

The homogeneous matrix Eq. 2.5 admits nontrivial solutions only when

$$\det (\mathbf{K} - \omega_n^2 \mathbf{M}) = 0 \tag{2.6}$$

This is an Nth-degree equation in ω_n^2 (if the system has N degrees of freedom), called the *characteristic* equation of the system. Its solution provides N positive roots, and the square roots of these yield the corresponding natural frequencies. We take these arbitrarily as positive because a change in the sign of ω_n amounts to introducing the constant factor -1 in the expression for θ_n, and this is immaterial.

We have said that all N roots are positive. In some systems it is possible to have one or more roots equal to zero. However, this cannot be the case in the structures with which we are concerned, since at least one mass must be connected to the ground by a spring element, and a natural frequency equal to zero would imply a stressless rigid-body translation, which the connection to the ground prevents.

Substitution of ω_n^2 in Eq. 2.5 gives a homogeneous Nth-order matrix equation in $\mathbf{z}_n$. We may therefore arbitrarily choose the value of any of the terms in this vector and find the shape of the nth natural mode by solving the matrix equation; accordingly the scale of the mode is fixed arbitrarily. We do this for every value of ω_n^2 and thus find the shapes of all the natural modes. Methods for numerical computation of natural modes are described in Section 2.3. Experimental procedures are dealt with in Chapter 17.

Conventionally we order the results in correspondence to increasing ω's; the *fundamental*, or first natural mode, corresponds to the lowest natural circular frequency ω_1.

Once the natural modes have been computed it is possible to substitute them into expressions of the form

$$\mathbf{x}_n(t) = \mathbf{z}_n \sin \omega_n (t - t_n)$$

The *characteristic roots* ω_n^2 are also known as *critical values* or *eigenvalues*. The set of mode amplitudes, $\mathbf{z}_n$, constitutes the nth *characteristic function* or *eigenvector*.

It can be shown that natural modes form a *complete orthogonal* set with either $\mathbf{M}$ or $\mathbf{K}$ as weighting matrix. Orthogonality implies that

$$\mathbf{z}_m^T \mathbf{M} \mathbf{z}_n = 0 \quad \text{when} \quad \omega_m \neq \omega_n \quad (2.7)$$

where the superscript T signifies that the matrix or vector in question is transposed. Equivalently, from Eq. 2.5,

$$\mathbf{z}_m^T \mathbf{K} \mathbf{z}_n = 0 \quad \text{when} \quad \omega_m \neq \omega_n \quad (2.8)$$

Equation 2.8 can be written in the form $\mathbf{F}_m^T \mathbf{z}_n = 0$ when $\omega_m \neq \omega_n$. We see that orthogonality can be interpreted as the condition that the inertia forces associated with the mth natural mode do not perform work when displaced through the configuration of the nth natural mode if $\omega_m \neq \omega_n$. Hence the term orthogonality, since the set of inertia forces $\mathbf{F}_m$ may be regarded as an nth-dimensional vector perpendicular to $\mathbf{z}_n$.

When two or more natural modes have the same natural frequency, the system may undergo free vibrations in that natural frequency with any configuration that results from a linear combination of the natural modes in question. Accordingly we may choose any such configurations to define those natural modes. In particular, if we adopt the convention that such combinations be orthogonal with $\mathbf{M}$ or $\mathbf{K}$ as weighting factor, we can replace the condition $\omega_m \neq \omega_n$ in Eqs. 2.7 and 2.8 with simply $m \neq n$.

The fact that the natural modes constitute a *complete* set means that any configuration $\mathbf{x}$ that satisfies the boundary conditions can be expressed as a linear combination of natural modes:

$$\mathbf{x} = \sum_n a_n \mathbf{z}_n \quad (2.9)$$

In this formula the dimensionless factors a_n are known as *participation coefficients*, especially when the modes have been normalized previously according to some convention. (A commonly used normalizing convention consists in choosing the scale of $\mathbf{z}_n$ such that $\mathbf{z}_n^T \mathbf{M} \mathbf{z}_n = 1$ in some arbitrary units.)

Since
$$z_n \sin \omega_n(t - t_n)$$
are solutions of Eq. 2.3, linearity of the differential equations of motion ensures that any linear combination of such terms

$$\mathbf{x}(t) = \sum_n a_n \mathbf{z}_n \sin \omega_n(t - t_n) \tag{2.10}$$

also satisfies Eq. 2.3; that is, it constitutes a solution to the problem of free vibrations. Now, in view of Eq. 2.9, $\mathbf{x}(t)$ as given by Eq. 2.10 is the general solution of the problem of free vibrations.

In Eq. 2.10 there is a time shift t_n for each natural mode. These quantities as well as the coefficients a_n must be chosen in every individual problem so as to satisfy the initial conditions.

Damped systems do not in general have natural modes of vibration in the real domain ("classical" normal modes) although we can extend to them the concept of natural modes by working with complex mode shapes. In any such complex "natural mode" the various masses in the system move out of phase, so that in the domain of real numbers the mode shape changes from one instant to the next, and it is not a classical natural mode.

The necessary and sufficient condition for a damped linear system to have classical natural modes is that there exist a transformation which diagonalizes the three matrices $\mathbf{C}$, $\mathbf{K}$, and $\mathbf{M}$ (Caughey, 1960). (Notice that a transformation can always be found which diagonalizes any two of these matrices.) The natural modes of the damped system coincide then with those of a structure without damping, having the same inertia and stiffness matrices as the original system.

This condition holds, in particular, if $\mathbf{C}$ is a linear combination of $\mathbf{K}$ and $\mathbf{M}$, that is, if $\mathbf{C} = a\mathbf{M} + b\mathbf{K}$, where a and b are constants. (This case was studied by Rayleigh, 1945, pp. 130–131.) We may say then that sufficient conditions for a damped linear system to have classical modes are that (1) there be a damping element in parallel with every spring element, and dashpot constants be proportional to the corresponding spring stiffnesses, or (2) there be a damping element connecting every mass to the base, and dashpot constants be proportional to the corresponding masses, or (3) there be damping elements resulting from any linear combination of 1 and 2. In these cases the damping ratio of the nth natural mode is given by

$$\zeta_n = \frac{a}{2\omega_n} + \frac{b\omega_n}{2}$$

(Morduchow, 1961).

In most of the materials and structures that we meet in practice, the damping ratios are nearly independent of the natural frequency, so much so that a system in which ζ_n is independent of n is said to have *structural damping*. The condition cannot be met by combining linearly the two simple cases of 1 and 2, except for systems having two degrees of freedom (see the foregoing expression for ζ_n). Therefore, when we wish to specify mechanical constants for a system whose response to a ground motion is to be computed directly (that is, not using modal analysis), we can make only the first two natural modes have equal damping ratios if we specify dashpot constants that are linear combinations of 1 and 2. More complicated distributions of dashpot constants are required to give three or more equal damping ratios. Methods are available for computing dashpot constants, spring stiffnesses, and masses in a system whose natural modes, frequencies, and damping ratios are specified (see Appendix 1). These methods may be used when we wish to postulate a dynamic system that will give accurate results. It is also possible to stipulate slightly nonlinear systems in which the percentage of equivalent linear damping is the same for all natural frequencies and is independent of the amplitude of vibration (Rosenblueth and Herrera, 1964).

In most applications it suffices to go through a modal analysis, in which the system is treated as undamped. Damping ratios are specified on the basis of empirical information and later incorporated to reduce computed modal responses; these are then combined according to an approximate relationship.

Example 2.1. Assume the following values for the structure of Fig. 2.1. Weights of bodies concentrated at floors 1 and 2: 3924 and 1962 metric tons; stiffnesses: $K_1 = 60$ ton/cm, $K_2 = 40$ ton/cm. Compute the natural modes and periods and verify the orthogonality condition.

Solution. $M_1 = 4$ ton sec^2/cm, $M_2 = 2$ ton sec^2/cm. Hence,

$$\mathbf{M} = \begin{vmatrix} 4 & 0 \\ 0 & 2 \end{vmatrix} \text{ton sec}^2/\text{cm} \qquad \mathbf{K} = \begin{vmatrix} 100 & -40 \\ -40 & 40 \end{vmatrix} \text{ton/cm}$$

According to Eq. 2.6 we must equate to zero the determinant of

$$\begin{vmatrix} 100 - 4\omega_n^2 & -40 \\ -40 & 40 - 2\omega_n^2 \end{vmatrix}$$

Hence,

$$8\omega_n^4 - 360\omega_n^2 + 4000 - 1600 = 0$$

or

$$\omega_n^4 - 45\omega_n^2 + 300 = 0$$

whence

$$\omega_n^2 = 22.5 \mp \sqrt{506.25 - 300}$$
$$= 22.5 \mp 14.3615 = 8.1385, 36.8615 \text{ (rad/sec)}^2$$
$$\omega_1 = 2.86 \text{ rad/sec} \qquad \omega_2 = 6.08 \text{ rad/sec}$$
$$T_1 = \frac{2\pi}{2.86} = 2.20 \text{ sec} \qquad T_2 = \frac{2\pi}{6.08} = 1.03 \text{ sec}$$

Following Eq. 2.5, for the first mode we get

$$\begin{vmatrix} 100 - 4 \times 8.1385 & -40 \\ -40 & 40 - 2 \times 8.1385 \end{vmatrix} \begin{vmatrix} z_1 \\ z_2 \end{vmatrix} = 0$$

or
$$67.446z_1 - 40z_2 = 0$$
and
$$-40z_1 + 23.723z_2 = 0$$

We may choose $z_2 = 1$ cm, from which $z_1 = 0.5931$ cm in both equations. For the second mode,

$$\begin{vmatrix} 100 - 4 \times 36.8615 & -40 \\ -40 & 40 - 2 \times 36.8615 \end{vmatrix} \begin{vmatrix} z_1 \\ z_2 \end{vmatrix} = 0$$

or
$$-47.446z_1 - 40z_2 = 0$$
and
$$-40z_1 - 33.723z_2 = 0$$

With $z_2 = 1$ cm we find $z_1 = -0.8431$ cm.
Orthogonality of these modes is verified:
$$4 \times 0.5931(-0.8431) + 2 \times 1 \times 1 = 0$$

2.3 Steady-State Vibrations

Consider the case when the ground remains motionless while a system of external forces acts on the structure varying in time as simple harmonic functions, all forces with the same circular frequency, say ω, and all in phase. The forces may be decomposed along the N coordinates proportional to the generalized displacements x_r that define the structural configurations. The vector of external forces in Eq. 2.1 can then be written in the form $\mathbf{P} = \mathbf{b} \sin \omega t$, in which $\mathbf{b}$ is the vector of force amplitudes, listed in the same order as the terms in the vector $\mathbf{x}$.

To solve Eq. 2.1 with this second term we try a solution of the form

$$\mathbf{x}(t) = \sum_n a_n \mathbf{z}_n \sin \omega t \tag{2.11}$$

Now according to Eq. 2.5, $\mathbf{K}\mathbf{z}_n = \omega_n^2 \mathbf{M}\mathbf{z}_n$. If we substitute this relationship and Eq. 2.10 in Eq. 2.1 and divide both members by $\sin \omega t$ we get

$$\mathbf{M} \sum_n (\omega_n^2 - \omega^2) a_n \mathbf{z}_n = \mathbf{b} \tag{2.12}$$

since
$$\ddot{\mathbf{x}}(t) = -\omega^2 \mathbf{x}(t)$$

Premultiplying both members of Eq. 2.11 by $\mathbf{z}_m^T$ we obtain

$$\sum_n (\omega_n^2 - \omega^2) a_n \mathbf{z}_m^T \mathbf{M} \mathbf{z}_n = \mathbf{z}_m^T \mathbf{b}$$

But owing to Eq. 2.7, only that term remains for which $m = n$. Thus,
$$(\omega_m^2 - \omega^2)a_m \mathbf{z}_m^T \mathbf{M} \mathbf{z}_m = \mathbf{z}_m^T \mathbf{b}$$
and
$$a_m = \frac{1}{\omega_m^2 - \omega^2} \frac{\mathbf{z}_m^T \mathbf{b}}{\mathbf{z}_m^T \mathbf{M} \mathbf{z}_m}$$

We have proved that a solution of the form of Eq. 2.11 exists and that
$$\mathbf{x}(t) = \sum_n \frac{1}{\omega_n^2 - \omega^2} \frac{\mathbf{z}_n^T \mathbf{b} \mathbf{z}_n}{\mathbf{z}_n^T \mathbf{M} \mathbf{z}_n} \sin \omega t \tag{2.13}$$

We may write the general solution as the sum of $\mathbf{x}$ as given by Eq. 2.13 and terms that depend on the initial conditions. The latter terms damp out with time provided the system has any positive amount of damping.

Comparing with the corresponding expression for a simple conservative system (Eq. 1.7 with $\zeta = 0$) we notice that each natural mode behaves much as a simple system in every respect. In keeping with the remarks made previously on approximate treatment of damping in structures with several degrees of freedom, we may account for damping in Eq. 2.13 by writing it in the form
$$\mathbf{x}(t) = \sum_n (B_d)_n \frac{1}{\omega_n^2} \frac{\mathbf{z}_n^T \mathbf{b} \mathbf{z}_n}{\mathbf{z}_n^T \mathbf{M} \mathbf{z}_n} \sin(\omega t - \phi_n) \tag{2.14}$$

in which
$$(B_d)_n = \left[\left(1 - \frac{\omega^2}{\omega_n^2}\right)^2 + \left(2\zeta_n \frac{\omega}{\omega_n}\right)^2 \right]^{-1/2}$$
$$\phi_n = \tan^{-1} \left(\frac{2\zeta_n \omega / \omega_n}{1 - \omega^2/\omega_n^2} \right)$$

and $\zeta_n =$ damping ratio in the nth natural mode.

The fraction that appears in Eq. 2.14 can be computed in a simple manner for special cases. Consider for example closely coupled systems. For these it is easily shown that Eq. 2.14 becomes
$$x_r = \sum_n (B_d)_n \frac{1}{\omega_n^2} \frac{\sum_i b_i z_{in}}{\sum_i M_i z_{in}^2} z_{rn} \sin(\omega t - \phi_n) \tag{2.15}$$

where z_{in} is the amplitude of the nth natural mode at the ith body. This expression applies even when there is stiffness coupling of a higher order between the masses.

A similar solution applies to the steady state of vibration due to a harmonic ground motion of the base when there are no external forces. Let
$$\mathbf{x}_0 = -\mathbf{a} \sin \omega t$$
Proceeding with $\mathbf{y}$ in Eq. 2.2 as we did with $\mathbf{x}$ in Eq. 2.1 and setting $\mathbf{P} = 0$ we find
$$\mathbf{y}(t) = \sum_n (B_d)_n \frac{\omega^2}{\omega_n^2} \frac{\mathbf{z}_n^T \mathbf{M} \mathbf{a} \mathbf{z}_n}{\mathbf{z}_n^T \mathbf{M} \mathbf{z}_n} \sin(\omega t - \phi_n) \tag{2.16}$$
to replace Eq. 2.14.

Numerical methods suitable for computation of steady-state responses are treated in Chapter 4.

Example 2.2. Calculate the steady-state response of the system described in Example 2.1 when acted upon at the second floor, in the direction of x, by a force

$$P(t) = 2 \sin 3t$$

expressed in metric tons and seconds; damping ratio in both natural modes is 10 percent.

Solution. In order to apply Eq. 2.14 we compute the following quantities.

$$\sum_i b_i x_{i1} = \sum_i b_i x_{i2} = 2 \times 1 = 2 \text{ ton cm}$$

$$\sum_i M_i x_{i1}^2 = 4 \times 0.5931^2 + 2 \times 1^2 = 3.403 \text{ ton cm sec}^2$$

$$\sum_i M_i x_{i2}^2 = 4(-0.8431)^2 + 2 \times 1^2 = 4.843 \text{ ton cm sec}^2$$

$$(B_d)_1 = \left[\left(1 - \frac{9}{8.1385}\right)^2 + 0.2^2 \times \frac{9}{8.1385}\right]^{-1/2} = 4.25$$

$$(B_d)_2 = \left[\left(1 - \frac{9}{36.8615}\right)^2 + 0.2^2 \times \frac{9}{36.8615}\right]^{-1/2} = 1.21$$

$$\phi_1 = \tan^{-1} \frac{0.2 \times 3/2.86}{1 - 9/8.1385} = 2.039 \text{ rad}$$

$$\phi_2 = \tan^{-1} \frac{0.2 \times 3/6.06}{1 - 9/36.8615} = 0.130 \text{ rad}$$

$$x_1 = \frac{4.25 \times 2 \times 0.5931}{8.1385 \times 3.403} \sin(3t - 2.039)$$

$$+ \frac{1.21 \times 2(-0.8431)}{36.8615 \times 4.843} \sin(3t - 0.130)$$

$$= 0.185 \sin(3t - 2.100)$$

$$x_2 = \frac{4.25 \times 2 \times 1}{8.1385 \times 3.403} \sin(3t - 2.039)$$

$$+ \frac{1.21 \times 2 \times 1}{36.8615 \times 4.843} \sin(3t - 0.130)$$

$$= 0.304 \sin(3t - 2.025)$$

Example 2.3. A rectangular slab (Fig. 2.4) that carries a uniformly distributed load of 0.6 ton/m² including its own weight and measures 6.3 × 40.3 m is supported transversely on six equal frames, spaced 8 m on centers. The stiffness of each frame is 5 ton/cm and its mass is negligible. Longitudinally there are two edge frames, 6 m center to center, with a stiffness of 20 ton/cm each and negligible mass. At the first interior frame a horizontal, harmonically oscillating force is applied, transversely to the slab, with an amplitude of 200 ton and a frequency of 2 Hz. Compute the maximum shear produced in the most highly stressed frame; assume that the slab is infinitely rigid in its own plane and that damping is negligible.

40 LINEAR SYSTEMS WITH SEVERAL DEGREES OF FREEDOM   Chap. 2

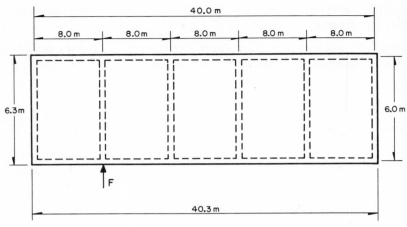

$F = 200 \sin 12.56\, t \text{ (ton)}$

Figure 2.4. Example 2.3.

Solution. We choose as generalized displacements a transverse and a longitudinal translation of the centroid of the slab and a rotation about a vertical axis through that centroid. Denoting these displacements by x_1, x_2, and x_3, respectively, we find

$$\mathbf{x} = \begin{vmatrix} x_1 \\ x_2 \\ x_3 \end{vmatrix} \qquad \mathbf{M} = 15.5 \begin{vmatrix} 1 & 0 & 0 \\ 0 & 1 & 0 \\ 0 & 0 & 138.7 \end{vmatrix} \qquad \mathbf{K} = 3000 \begin{vmatrix} 1 & 0 & 0 \\ 0 & 1.33 & 0 \\ 0 & 0 & 198.7 \end{vmatrix}$$

where the units are meters and tons.

The natural frequencies are found from the characteristic equation

$$\det \begin{vmatrix} 3000 - 15.5\omega^2 & 0 & 0 \\ 0 & 4000 - 15.5\omega^2 & 0 \\ 0 & 0 & 198.7 \times 3000 - (138.7 \times 15.5)\omega^2 \end{vmatrix} = 0$$

whence,

$$\omega_1^2 = 193 \qquad \omega_2^2 = 258, \qquad \omega_3^2 = 277 \text{ (rad/sec)}^2$$

$$\mathbf{z}_1 = \begin{vmatrix} 1 \\ 0 \\ 0 \end{vmatrix} \qquad \mathbf{z}_2 = \begin{vmatrix} 0 \\ 1 \\ 0 \end{vmatrix} \qquad \mathbf{z}_3 = \begin{vmatrix} 0 \\ 0 \\ 1 \end{vmatrix}$$

Thus, the modes are not coupled.

To enter Eq. 2.14 we need

$$\mathbf{b} = \begin{vmatrix} 200 \\ 0 \\ 2400 \end{vmatrix}$$

$$\mathbf{z}_1^T\mathbf{b}\mathbf{z}_1 = \begin{vmatrix} 200 \\ 0 \\ 0 \end{vmatrix} \quad \mathbf{z}_2^T\mathbf{b}\mathbf{z}_2 = 0 \quad \mathbf{z}_3^T\mathbf{b}\mathbf{z}_3 = \begin{vmatrix} 0 \\ 0 \\ 2400 \end{vmatrix}$$

$$\mathbf{z}_1^T\mathbf{M}\mathbf{z}_1 = 15.5, \quad \mathbf{z}_3^T\mathbf{M}\mathbf{z}_3 = 15.5 \times 138.7 = 2150$$

Since $\omega^2 = 4(2\pi)^2 = 158$ (rad/sec)^2, substitution in Eq. 2.13 gives us

$$\mathbf{x}_{max} = \frac{1}{193 - 158} \cdot \frac{1}{15.5} \begin{vmatrix} 200 \\ 0 \\ 0 \end{vmatrix} + \frac{1}{277 - 158} \cdot \frac{1}{2150} \begin{vmatrix} 0 \\ 0 \\ 2400 \end{vmatrix}$$

$$= \begin{vmatrix} 0.369 \\ 0 \\ 0.00938 \end{vmatrix}$$

The most highly stressed transverse frame is the exterior one closest to the point of load application, since the translation and rotation components in $\mathbf{x}_{max}$ have the same signs as in the vector $\mathbf{b}$. Hence the amplitude of the maximum shear is

$$Q_{max} = (0.369 + 0.00938 \times 20)500$$
$$= 278.5 \text{ ton}$$

2.4 Transient Disturbances

Here we shall consider responses to arbitrary ground motions—that is, cases governed by Eq. 2.2 in which $\mathbf{P} = 0$, and $\ddot{\mathbf{x}}_0$ is an arbitrary vector function of time. In order to compute the displacements of the system we shall devote our attention first to a certain set of base displacements that would induce the static accelerations

$$\ddot{\mathbf{x}}_0 = \mathbf{u}\,\delta(t - \tau)$$

where $\mathbf{u}$ is a vector of constant values having the units of $\dot{\mathbf{x}}_0$. The corresponding displacements relative to $\mathbf{x}_0$ have the form given by Eq. 2.8 for all $t \geq \tau$ since $\ddot{\mathbf{x}}_0 = 0$ when $t > \tau$

$$\mathbf{y}(t) = \sum_n a_n \mathbf{z}_n \sin \omega_n(t - t_n) \tag{2.17}$$

Differentiating with respect to time we obtain

$$\dot{\mathbf{y}}(t) = \sum_n a_n \omega_n \mathbf{z}_n \cos \omega_n(t - t_n)$$

These expressions satisfy the conditions

$$\mathbf{y}(\tau) = 0, \quad \dot{\mathbf{y}}(\tau+) = -\mathbf{u}$$

(which are derived from $\mathbf{x}_0(\tau) = 0$, $\dot{\mathbf{x}}_0(\tau-) = 0$, and $\dot{\mathbf{x}}_0(\tau+) = \mathbf{u}$). If we take $t_n = \tau+$, we obtain

$$\sum_n a_n \omega_n \mathbf{z}_n = -\mathbf{u}$$

Premultiplying the last equation by $z_m^T M$ we get

$$-z_m^T M u = \sum_n a_n \omega_n z_m^T M z_m$$

Using Eq. 2.7 and changing m to n we arrive at

$$a_n = -\frac{z_n^T M u}{\omega_n z_n^T M z_n}$$

Substituting a_n in Eq. 2.17 gives us

$$y(t) = -\sum_n \frac{1}{\omega_n} \frac{z_n^T M u}{z_n^T M z_n} z_n \sin \omega_n (t - \tau)$$

Notice that the fraction in this expression is a scalar. Hence we may change the order of multiplication and write

$$y(t) = -\sum_n \frac{1}{\omega_n} \frac{z_n z_n^T M}{z_n^T M z_n} u \sin \omega_n (t - \tau)$$

In order to consider an arbitrary $\ddot{x}_0$ we take $\ddot{x}_0(\tau)\, d\tau$ for u and integrate with respect to time; this is permissible in view of the linearity of the problem. This gives us, finally,

$$y(t) = -\sum_n \frac{1}{\omega_n} \frac{z_n z_n^T M}{z_n^T M z_n} \int_0^t \ddot{x}_0(\tau) \sin[\omega_n(t-\tau)]\, d\tau \qquad (2.18)$$

The response is thus the sum of terms similar to the responses of simple undamped systems, one for each natural mode of vibration, and having the same natural frequency as the corresponding mode. These terms are multiplied by the dimensionless square matrices of coefficients

$$H_n = \frac{z_n z_n^T M}{z_n^T M z_n}$$

At the same time we may write

$$x_0(t) = \sum_k i_k x_k(t)$$

where i_k is a column vector whose ith term i_{ki} is equal to the static displacement x_i induced by a unit base displacement in the kth degree of freedom of the base, and $x_k(t)$ is the displacement of the base in its kth degree of freedom at time t. Thus Eq. 2.18 may be written as

$$y(t) = \sum_k \sum_n H_n i_k y_k(t, T_n) \qquad (2.19)$$

where $y_k(t, T_n)$ represents the displacement relative to the ground, at time t, of a simple system with natural period T_n, resting on a base that undergoes the motion $x_k(t)$. Often, $y_k(t, T_n)$ is known as an *excitation coefficient* for displacement.

Equivalently we may write $H_n i_k$ as $a_{nk} z_n$ with

$$a_{nk} = \frac{z_n^T M i_k}{z_n^T M z_n}$$

Here a_{nk} may be called the kth *participation coefficient* of the nth natural mode.

Sec. 2.4 TRANSIENT DISTURBANCES 43

With this notation Eq. 2.18 becomes

$$\mathbf{y}(t) = \sum_k \sum_n a_{nk} \mathbf{z}_n y_k(t, T_n) \tag{2.20}$$

The advantage of the form of Eq. 2.19 lies in that the vector $\mathbf{H}_n \mathbf{i}_k$ need be computed only once for a given structure and then used for any arbitrary ground motion. Equation 2.20 on the other hand makes it simpler to write formulas for combinations of responses in the various natural modes.

Equation 2.19 admits another interesting interpretation. The term i_{ik}, representing the static structural displacement x_i due to a unit base displacement x_k, may be regarded as the influence coefficient for the action of the structure on the base along the kth degree of freedom of the base (i.e., minus the reaction of the ground against the structure, with the same point of application and line of action as the coordinate x_k of base motions). But the product $-\omega_n^2 \mathbf{z}_n^T \mathbf{M}$ is the row vector of the amplitudes of the inertia forces acting on the system when it oscillates in nth natural mode because $-\omega_n^2 \mathbf{z}_n^T$ is the row vector of acceleration amplitudes in that mode. Now the product of the ith term in this vector by the corresponding term in $\mathbf{i}_k$ gives us, by Betti's theorem, the amplitude of the reaction at the base along the kth degree of freedom of the base. We shall denote that amplitude, with minus sign, by the symbol Q_{kn} and the row vector of these amplitudes by $\mathbf{q}_n^T$. Then we may rewrite Eq. 2.18 in the form

$$\mathbf{y}(t) = -\sum_n \frac{1}{\omega_n^3} \frac{\mathbf{z}_n \mathbf{q}_n^T}{\mathbf{z}_n^T \mathbf{M} \mathbf{z}_n} \int_0^t \ddot{\mathbf{x}}_k(\tau) \sin[\omega_n(t-\tau)]\, d\tau \tag{2.21}$$

where $\mathbf{x}_k$ is the column vector of base displacements.

With the same limitations as in the cases considered previously we may introduce the effects of damping in Eqs. 2.18–2.21 by writing ω_n' for ω_n and inserting the factor $\exp(-\zeta_n \omega_n t + \zeta_n \omega_n \tau)$ in the integrands, or by stating that $y(t, T_n)$ refers to a simple system having a natural period T_n and damping factor ζ_n.

In the special case of closely coupled systems resting on a rigid base these expressions become

$$y_r(t) = -\sum_n \frac{1}{\omega_n} \frac{\sum_i M_i z_{in}}{\sum_i M_i z_{in}^2} z_{rn} \int_0^t \ddot{x}_0(\tau) \sin[\omega_n(t-\tau)]\, d\tau$$

$$= -\sum_n \frac{1}{\omega_n^3} \frac{Q_n}{\sum_i M_i z_{in}^2} z_{rn} \int_0^t \ddot{x}_0(\tau) \sin[\omega_n(t-\tau)]\, d\tau$$

$$H_{rn} = \frac{\sum_i M_i z_{in}}{\sum_i M_i z_{in}^2} z_{rn}$$

$$i_{ik} = 1$$

$$a_n = \frac{\sum_i M_i z_{in}}{\sum_i M_i z_{in}^2}$$

$$y_r(t) = \sum_n a_n z_{rn} y_k(t, T_n)$$

44 LINEAR SYSTEMS WITH SEVERAL DEGREES OF FREEDOM Chap. 2

If the closely coupled system rests on two or more supports that may perform unequal motions, Eq. 2.18 becomes

$$y_r(t) = -\sum_k \sum_n \frac{\sum_i M_i z_{in} \int_0^t \ddot{x}_k(\tau) \sin[\omega_n(t-\tau)]\, d\tau}{\omega_n \sum_i M_i z_{in}^2} z_{rn}$$

$$= -\sum_k \sum_n \frac{Q_{kn} z_{rn}}{\omega_n^3 \sum_i M_i z_{in}^2} \int_0^t \ddot{x}_k(\tau) \sin[\omega_n(t-\tau)]\, d\tau$$

In every linear conservative system the quantities that have the greatest interest in design are linear functions of the displacements relative to the ground, y, and hence linear functions of the absolute accelerations $\ddot{x}$. These relationships are approximately correct for linear dissipative systems. Hence it is a simple task to compute responses pertinent in design once expressions have been found for y and $\ddot{x}$. According to Eq. 2.20, for a conservative system we may write

$$\ddot{\mathbf{y}}(t) = \sum_k \sum_n \frac{-a_{nk}}{\omega_n^2}(-\omega_n^2 \mathbf{z}_n) y_k(t, T_n) \tag{2.22}$$

where the term $-a_{nk}/\omega_n^2$ may be regarded as participation coefficients for acceleration and $-\omega_n^2 \mathbf{z}_n$ as the vector of accelerations in the nth natural mode; $y_k(t, T_n)$ is still the nth excitation coefficient for displacement. Alternatively we may retain a_{nk} as participation coefficient and call $y_k(t, T_n)/\omega_n^2$ the corresponding excitation coefficient for acceleration.

Now let $\mathbf{l} = \mathbf{\Lambda y}$ be the column vector of design responses (such as story shears, overturning moments, bending moments, stresses, relative displacements between consecutive floors in a building, or in general, generalized forces or generalized deformations) that are of interest, where $\mathbf{\Lambda}$ denotes the (usually rectangular) matrix of the corresponding influence coefficients. For the nth mode we have the set of amplitudes $\mathbf{l}_n = \mathbf{\Lambda z}_n$. Therefore Eq. 2.20 gives us

$$\mathbf{l}(t) = \mathbf{\Lambda} \sum_k \sum_n a_{nk} \mathbf{z}_n y_k(t, T_n) + \mathbf{\Lambda x}_0 \tag{2.23}$$

as response to an arbitrary set of ground motions. A similar expression may be written in terms of the simple-system accelerations $\ddot{y}_k$. The second term in the right-hand member of Eq. 2.23 is zero and may be dropped if the base displacements are such that they would not produce strains in the structure when applied statically.

In elementary structural design, criteria of failure are often based on the maximum absolute values of responses such as $\mathbf{l}$. Let $\mathbf{L}$ denote the vector of maximum absolute values of the responses in question. If $\mathbf{l}(t)$ has been computed from Eq. 2.23 at a number of sufficiently close instants, the maximum absolute values can be located with sufficient accuracy. But if the ground motion is defined only by its displacement spectrum, the foregoing derivations allow us no more than to calculate an upper bound to the desired maxima. Thus

$$\mathbf{L} \leq \sum_k \sum_n |a_{nk} \mathbf{\Lambda z}_n| D_k(T_n) + |\mathbf{\Lambda x}_0|_{\max} \tag{2.24}$$

where $D_k(T_n)$ denotes the ordinate of the displacement spectrum for the kth component of base motion at an abscissa equal to T_n, and the sign $\leq$ is to be interpreted in the sense that no term in the vector of the left-hand side of the expression exceeds the corresponding term in the right-hand side.

Similar expressions may be written for **L** in terms of velocity and acceleration spectral ordinates

$$\mathbf{L} \leq \sum_k \sum_n \left|\frac{a_{nk}}{\omega_n} \mathbf{\Lambda z}_n\right| V_k(T_n) + |\mathbf{\Lambda x}_0|_{\max} \qquad (2.25)$$

$$\mathbf{L} \leq \sum_k \sum_n \left|\frac{a_{nk}}{\omega_n^2} \mathbf{\Lambda z}_n\right| A_k(T_n) + |\mathbf{\Lambda x}_0|_{\max} \qquad (2.26)$$

Here the subscript and argument of V and A have the same meaning as for the displacement spectra. With the usual approximations, viscous damping may be included in the spectral ordinates. Whether these upper bounds are adequate for design, or improved estimates must be searched for, depends on many factors; often Eqs. 2.24–2.26 are all that is required.

Inequalities similar to Eqs. 2.24–2.26 can be written in terms of residual spectral values and referred to the response after the motion has ended. For undamped structures this is equivalent to writing Eq. 2.25 in terms of Fourier spectral ordinates. The expected values of L^2 at an arbitrary unspecified instant $t = s$ are obtained by adding the squares rather than the absolute values of the Fourier spectral ordinates.

Example 2.4. Write an explicit expression for the relative displacement between floors in the upper story of the structure of Example 2.1 as response to an arbitrary ground motion. Assume that damping is 10 percent of critical in both natural modes.

Solution. The matrix $\mathbf{\Lambda}$ associated with relative displacements in the two stories is

$$\begin{vmatrix} 1 & 0 \\ -1 & 1 \end{vmatrix}$$

We are interested in the second term of the vector **l** of relative displacements; in the nth natural mode the amplitude of this relative displacement is

$$l_{2n} = z_{2n} - z_{1n}$$

In order to apply Eq. 2.23 we must compute the coefficients a_{nk}—say a_1 and a_2—since the ground has a single degree of freedom. Using the corresponding expression for one-dimensional systems and the two natural modes computed in Example 2.1, we obtain

$$a_1 = \frac{4 \times 0.5931 + 2 \times 1}{4 \times 0.5931^2 + 2 \times 1^2} = 1.280$$

$$a_2 = \frac{4(-0.8431) + 2 \times 1}{4(-0.8431)^2 + 2 \times 1^2} = -0.282$$

whence $l_2(t) = 1.280(1 - 0.5931)[y(t, 2.20)] - 0.282(1 + 0.8431)[y(t, 1.04)]$
$= 0.521[y(t, 2.20) - y(t, 1.04)]$

The displacements relative to the ground, $y(t, 2.20)$ and $y(t, 1.04)$, are to be computed as responses to the ground motion in question in simple systems having 10 percent of critical damping and natural periods of 2.20 and 1.04 sec, respectively. The corresponding damped circular frequencies are $(1 - 0.01)^{1/2}$ times ω_1 and ω_2, or 2.85 and 6.03 rad/sec, respectively. The values of $\zeta_n\omega_n$ are

$$0.1 \times 2.86 = 0.286$$

and

$$0.1 \times 6.06 = 0.606 \text{ sec}^{-1}$$

respectively. Therefore,

$$y(t, 2.20) = -\frac{1}{2.85}\int_0^t \ddot{x}_0(\tau) \exp\left[-0.286(t-\tau)\right] \sin\left[2.85(t-\tau)\right] d\tau$$

$$y(t, 1.04) = -\frac{1}{6.03}\int_0^t \ddot{x}_0(\tau) \exp\left[-0.606(t-\tau)\right] \sin\left[6.03(t-\tau)\right] d\tau$$

An upper bound to $L_2 = \max |l_2(t)|$ can be found as described in the text: $L_2 \leq 0.52 | D(2.20) + D(1.04)|$, where the D's represent the ordinates of the spectrum of the ground motion for 0.1 of critical damping and periods equal to 2.20 and 1.04 sec, respectively.

Example 2.5. Find the absolute displacements at time $t = 1.5$ sec of masses 1 and 2 of the system shown in Fig. 2.3 as responses to the ground motions

$$\ddot{x}_a = -\ddot{x}_b = 1.2 \text{ in./sec}^2$$

when $0 < t \leq 1$ sec and $\ddot{x}_a = \ddot{x}_b = 0$ outside this range. Take

$$K_1 = 2K_2 = 3K_3 = 6 \text{ kip/in.}$$
$$M_1 = 2M_2 = 4 \text{ kip sec}^2/\text{in.}$$

and in both natural modes, $\zeta = 0$.

Solution. Our first step consists in computing the static displacements. From the expresions given in Section 2.1, we obtain

$$x_{01} = \frac{(2+3)x_a + x_b}{1+2+3} = \frac{1}{6}(5x_a + x_b)$$

$$x_{02} = \frac{3x_a + (1+2)x_b}{1+2+3} = \frac{1}{6}(3x_a + 3x_b)$$

$$\mathbf{x}_0 = \frac{1}{6}\begin{vmatrix} 5x_a + x_b \\ 3x_a + 3x_b \end{vmatrix}$$

The specified ground motions give us

$$\ddot{\mathbf{x}}_0 = \begin{vmatrix} 0.8 \\ 0 \end{vmatrix}$$

when $0 < t \leq 1$ sec, and

$$\ddot{\mathbf{x}}_0 = 0 \qquad \mathbf{x}_0 = \begin{vmatrix} 0.8t - 0.4 \\ 0 \end{vmatrix}$$

when $t \geq 1$ sec.

Next we compute the natural modes of vibration. To this end we write

$$\mathbf{M} = \begin{vmatrix} 4 & 0 \\ 0 & 2 \end{vmatrix} \qquad \mathbf{K} = \begin{vmatrix} 9 & -3 \\ -3 & 5 \end{vmatrix}$$

According to Eq. 2.16 we must solve the expression

$$\det \begin{vmatrix} 9 - 4\omega_n^2 & -3 \\ -3 & 5 - 2\omega_n^2 \end{vmatrix} = 0$$

or

$$8\omega_n^4 - 38\omega_n^2 + 45 - 9 = 0$$

which gives us

$$\omega_n^2 = 2.3750 \mp 1.0680 = 1.3070, 3.4430 \text{ (rad/sec)}^2$$

Therefore, $\omega_1 = 1.1431$ and $\omega_2 = 1.8555$ rad/sec. Substituting in Eq. 2.5 and solving we obtain

$$\mathbf{z}_1 = \begin{vmatrix} 0.7953 \\ 1.0000 \end{vmatrix} \qquad \mathbf{z}_2 = \begin{vmatrix} -0.6287 \\ 1.0000 \end{vmatrix}$$

If we use Eq. 2.18 to compute $\mathbf{y}$, we must first evaluate

$$\mathbf{H}_1 = \frac{\begin{vmatrix} 0.7953 \\ 1.0000 \end{vmatrix} |0.7953 \quad 1.0000| \begin{vmatrix} 4 & 0 \\ 0 & 2 \end{vmatrix}}{|0.7953 \quad 1.0000| \begin{vmatrix} 4 & 0 \\ 0 & 2 \end{vmatrix} \begin{vmatrix} 0.7953 \\ 1.0000 \end{vmatrix}}$$

$$= \frac{\begin{vmatrix} 2.5300 & 1.5906 \\ 3.1812 & 2.0000 \end{vmatrix}}{4 \times 0.7953^2 + 2 \times 1.0000^2} = \begin{vmatrix} 0.558 & 0.351 \\ 0.704 & 0.442 \end{vmatrix}$$

In like manner,

$$\mathbf{H}_2 = \begin{vmatrix} 0.442 & -0.351 \\ -0.704 & 0.559 \end{vmatrix}$$

Therefore,

$$\mathbf{y}(1.5) = -\frac{1}{1.1431} \begin{vmatrix} 0.558 & 0.351 \\ 0.704 & 0.442 \end{vmatrix} \int_0^1 \begin{vmatrix} 0.8 \\ 0 \end{vmatrix} \sin[1.1431(1.5 - \tau)] \, d\tau$$

$$- \frac{1}{1.8555} \begin{vmatrix} 0.442 & -0.351 \\ -0.704 & 0.558 \end{vmatrix} \int_0^1 \begin{vmatrix} 0.8 \\ 0 \end{vmatrix} \sin[1.8555(1.5 - \tau)] \, d\tau$$

$$= -\frac{1}{1.3070} \begin{vmatrix} 0.4464 \\ 0.5632 \end{vmatrix} (\cos 0.5716 - \cos 1.7147)$$

$$- \frac{1}{3.4430} \begin{vmatrix} 0.3546 \\ -0.5632 \end{vmatrix} (\cos 0.9278 - \cos 2.7823) = - \begin{vmatrix} 0.494 \\ 0.171 \end{vmatrix}$$

48 LINEAR SYSTEMS WITH SEVERAL DEGREES OF FREEDOM Chap. 2

Finally,
$$\mathbf{x}(1.5) = \mathbf{x}_0(1.5) + \mathbf{y}(1.5)$$
$$= \begin{vmatrix} 0.800 \\ 0 \end{vmatrix} - \begin{vmatrix} 0.494 \\ 0.171 \end{vmatrix} = \begin{vmatrix} 0.306 \\ -0.171 \end{vmatrix} \text{in.}$$

Instead of using Eq. 2.18 we could have resorted to Eqs. 2.19, 2.20, or 2.21. Either could have been more convenient depending on the number of ground motions to consider and nature of the responses to compute.

Example 2.6. The ground that supports the structure in Example 2.3 (Fig. 2.4) undergoes a motion that is transverse to the structure and such that the base of one exterior frame suddenly acquires a velocity of 20 cm/sec while the rest of the column bases remain motionless. Compute the shear in the most highly stressed frame 0.1 sec after the ground motion has begun.

Solution. In Example 2.3 we chose as generalized coordinates the transverse and longitudinal displacements of the centroid of the slab and the rotation of the slab about a vertical axis through the centroid. We are interested in frame shears. Hence we must construct the matrix of influence coefficients as a three-column matrix for which the term in the kth row of the ith column represents the shear in the kth frame due to a unit value of the ith coordinate. This is the case if we decide to consider separately the stresses induced statically by the base motions; otherwise we would have to add one column for each degree of freedom of the base. We select the separate treatment for dynamic and static effects, choose to write the transverse frames first, and reserve the last two rows for the long frames. Then,

$$\Lambda = \begin{vmatrix} 5 & 0 & 100 \\ 5 & 0 & 60 \\ 5 & 0 & 20 \\ 5 & 0 & -20 \\ 5 & 0 & -60 \\ 5 & 0 & -100 \\ 0 & 20 & 60 \\ 0 & 20 & -60 \end{vmatrix} 10^2$$

The first column in the matrix was derived by giving x_1 a unit value while keeping x_2 and x_3 equal to zero. The displacement in each frame, relative to the ground, was found next. This displacement was multiplied by the corresponding frame stiffness. The two remaining columns were determined similarly.

It is convenient here to use Eq. 2.21. To this end we compute $\mathbf{q}_n^T$. In this particular example the natural modes contain the three degrees of freedom uncoupled. Therefore, $\mathbf{q}_n^T$ is simply the transposed nth column of Λ:

$$\mathbf{q}_1^T = |5, 5, 5, 5, 5, 5, 0, 0| 10^2$$
$$\mathbf{q}_2^T = |0, 0, 0, 0, 0, 0, 20, 20| 10^2$$
$$\mathbf{q}_3^T = |100, 60, 20 - 20 - 60, -100, 60, -60| 10^2$$

Next we must compute the integral in Eq. 2.21. Now, from statics we immediately find

$$\ddot{\mathbf{x}}_k(t) = \begin{vmatrix} 0.20 \\ 0 \\ 0 \\ 0 \\ 0 \\ 0 \\ 0 \\ 0 \end{vmatrix} \delta(t)$$

and from the solution of Example 2.3, $\omega^2_{1,2,3} = 193, 258,$ and 277 (rad/sec)^2. Therefore,

$$\int_0^t \ddot{\mathbf{x}}_k(\tau) \sin[\omega_n(t-\tau)]\, d\tau = \begin{vmatrix} 0.20 \\ 0 \\ 0 \\ 0 \\ 0 \\ 0 \\ 0 \\ 0 \end{vmatrix} \sin(0.1\omega_{1,2,3})$$

The products of $\mathbf{q}_n^T$ and the corresponding integrals become, respectively, $100 \sin 1.3893 = 98.4$, 0, and $2000 \sin 1.6646 = 1.991$.

From the solution of Example 2.3, $\mathbf{z}_1^T \mathbf{M} \mathbf{z}_1 = 15.5$ and $\mathbf{z}_3^T \mathbf{M} \mathbf{z}_3 = 2150$. We now enter Eq. 2.21:

$$\mathbf{y}(0.1) = \frac{1}{193^{3/2}} \begin{vmatrix} 1 \\ 0 \\ 0 \end{vmatrix} \frac{98.4}{15.5} + \frac{1}{277^{3/2}} \begin{vmatrix} 0 \\ 0 \\ 1 \end{vmatrix} \frac{1991}{2150} = \begin{vmatrix} 0.00237 \\ 0 \\ 0.0002552 \end{vmatrix}$$

Let $\mathbf{l}_d$ denote the column vector of dynamic shears in the frames. Then,

$$\mathbf{l}_d = \mathbf{\Lambda}\mathbf{y} = \begin{vmatrix} 3.737 \\ 2.716 \\ 1.965 \\ 0.675 \\ -0.346 \\ -1.367 \\ 1.531 \\ -1.531 \end{vmatrix}$$

50 LINEAR SYSTEMS WITH SEVERAL DEGREES OF FREEDOM Chap. 2

From statics we find the static shears

$$l_s = \begin{vmatrix} -4.98 \\ 3.68 \\ 2.34 \\ 1.00 \\ -0.34 \\ -1.68 \\ 2.11 \\ -2.11 \end{vmatrix}$$

The total shears are $l_d + l_s$. The maximum shear in absolute value is also the maximum term in this vector and is therefore equal to $|2716 + 3680| = 6.386$ ton.

2.5 Numerical Evaluation of Responses to Transient Motions

The numerical procedure described in Section 1.5 can be extended to systems having several degrees of freedom. One group of methods consists in computing the natural modes and evaluating the response in each mode as for a simple system. This can be accomplished by evaluating the Duhamel integral (by White's procedure, direct evaluation of responses, graphical construction in the phase plane, or use of a single-degree analog computer) and combining the responses in the various modes at each instant of interest.

A second group of methods, which is ordinarily more convenient, consists in dealing directly with the responses of the system without breaking them up into those which correspond to the natural modes. The analysis can be performed in fairly elaborate electric analog computers (Murphy, Bycroft, and Harrison, 1956). It can also be carried out numerically.

Numerically, in a simple system we had to compute the forces acting on the mass in terms of given displacements and velocities of the mass relative to the ground. In multidegree systems we must be able to compute the forces acting on all the masses in terms of all the displacements and velocities. This is done easily by use of the stiffness and damping matrices. Accordingly, once these matrices are available the methods become essentially identical with those for simple systems (Newmark, 1962).

In this generalized form, the criteria of convergence and stability cited for single-degree-of-freedom systems in terms of their natural period must be expressed as dependent on the shortest natural period of the multidegree system.

> Example 2.7. An undamped system with two degrees of freedom has the following stiffness and inertia matrices.

Sec. 2.5 EVALUATION OF RESPONSES TO TRANSIENT MOTIONS 51

$$\mathbf{K} = \begin{vmatrix} 10 & 1 \\ 1 & 5 \end{vmatrix} \text{ton/cm} \qquad \mathbf{M} = \begin{vmatrix} 2 & 0 \\ 0 & 1 \end{vmatrix} \text{ton sec}^2/\text{cm}$$

The system is connected to the base in such a way that static displacement

TABLE 2.1, EXAMPLE 2.7

t sec	Q_1 ton	$\ddot{x}$ cm/sec²	$\dot{x}$ cm/sec	x_1 cm	$x_1 - x_0$ cm	Q_2 ton	$\ddot{x}$ cm/sec²	$\dot{x}$ cm/sec	x_2 cm	$x_2 - x_0$ cm	x_0 cm
0	0	0	0	0	0	0	0	0	0	0	0
0.2	2.540	1.350	0.135	0.0090	-0.2310	1.380	1.500	0.150	0.0100	-0.2300	0.24
0.2	2.546	1.270	0.127	0.0085	-0.2315	1.386	1.380	0.138	0.0092	-0.2308	0.24
0.2	2.546	1.273	0.127	0.0085	-0.2315	1.386	1.386	0.138	0.0092	-0.2308	0.24
0.4	4.548	+2.300	0.484	+0.0662	-0.4138	2.468	2.100	0.486	+0.0693	-0.4107	0.48
0.4	4.548	2.274	0.481	0.0660	-0.4140	2.455	2.468	0.523	0.0718	-0.4082	0.48
0.4	4.548	2.274	0.481	0.0660	-0.4140	2.455	2.455	0.522	0.0717	-0.4083	0.48
0.4	4.548	2.274	0.481	0.0660	-0.4140	2.455	2.455	0.522	0.0717	-0.4083	0.48
0.6	5.585	2.700	0.978	0.2105	-0.5095	2.960	3.200	1.088	0.2301	-0.4899	0.72
0.6	5.581	2.793	0.987	0.2111	-0.5089	2.967	2.960	1.064	0.2285	-0.4915	0.72
0.6	5.580	2.790	0.987	0.2111	-0.5089	2.966	2.967	1.065	0.2286	-0.4914	0.72
0.6	5.580	2.790	0.987	0.2111	-0.5089	2.966	2.966	1.065	0.2286	-0.4914	0.72
0.8	5.409	2.900	1.556	0.4650	-0.4950	2.790	2.980	1.660	0.5010	-0.4590	0.96
0.8	5.423	2.704	1.536	0.4637	-0.4963	2.798	2.790	1.641	0.4997	-0.4603	0.96
0.8	5.422	2.711	1.537	0.4638	-0.4962	2.797	2.798	1.642	0.4998	-0.4602	0.96
0.8	5.422	2.711	1.537	0.4638	-0.4962	2.797	2.797	1.642	0.4998	-0.4602	0.96
1.0	4.104	2.150	2.023	0.8216	-0.3784	1.977	2.200	2.142	0.8802	-0.3198	1.20
1.0	4.111	2.052	2.013	0.8210	-0.3790	1.985	1.977	2.120	0.8787	-0.3213	1.20
1.0	4.111	2.055	2.014	0.8210	-0.3790	1.985	1.985	2.121	0.8787	-0.3213	1.20
1.0	4.111	2.055	2.014	0.8210	-0.3790	1.985	1.985	2.121	0.8787	-0.3213	1.20
1.2	1.931	0.950	2.315	1.2575	-0.1825	0.712	0.700	2.390	1.3341	-0.1059	1.44
1.2	1.930	0.965	2.316	1.2576	-0.1824	0.712	0.712	2.391	1.3341	-0.1059	1.44
1.2	1.930	0.965	2.316	1.2576	-0.1824	0.712	0.712	2.391	1.3341	-0.1059	1.44
1.4	-0.653	-0.320	2.381	1.7316	0.0516	-0.735	-0.800	2.382	1.8165	0.1365	1.68
1.4	-0.652	-0.326	2.380	1.7315	0.0515	-0.735	-0.735	2.388	1.8169	0.1369	1.68
1.4	-0.652	-0.326	2.380	1.7315	0.0515	-0.735	-0.735	2.388	1.8169	0.1369	1.68
1.6	-3.083	-1.500	2.197	2.1932	0.2732	-2.026	-2.100	2.104	2.2707	0.3507	1.92
1.6	-3.080	-1.541	2.193	2.1929	0.2729	-2.029	-2.026	2.111	2.2712	0.3512	1.92
1.6	-3.080	-1.540	2.193	2.1929	0.2729	-2.029	-2.029	2.111	2.2712	0.3512	1.92
1.8	-4.830	-2.500	1.789	2.5943	0.4343	-2.869	-2.900	1.618	2.6471	0.4871	2.16
1.8	-4.836	-2.415	1.797	2.5949	0.4349	-2.871	-2.869	1.621	2.6473	0.4873	2.16
1.8	-4.836	-2.418	1.797	2.5949	0.4349	-2.871	-2.871	1.621	2.6473	0.4873	2.16
2.0	-5.547	-2.800	1.275	2.9034	0.5034	-3.069	-3.000	1.034	2.9132	0.5132	2.40
2.0	-5.549	-2.773	1.278	2.9036	0.5036	-3.068	-3.069	1.027	2.9127	0.5127	2.40
2.0	-5.549	-2.774	1.278	2.9036	0.5036	-3.068	-3.068	1.027	2.9127	0.5127	2.40

TABLE 2.1, EXAMPLE 2.7 (Continued)

t sec	Q_1 ton	$\ddot{x}$ cm/sec²	$\dot{x}$ cm/sec	x_1 cm	x_1-x_0 cm	Q_2 ton	$\ddot{x}$ cm/sec²	$\dot{x}$ cm/sec	x_2 cm	x_2-x_0 cm	x_0 cm
2.2	-10.156	-5.200	0.481	3.0875	0.9275	-5.332	-5.460	0.174	3.0408	0.8808	2.16
2.2	-10.165	-5.078	0.493	3.0883	0.9283	-5.337	-5.332	0.187	3.0417	0.8817	2.16
2.2	-10.165	-5.083	0.493	3.0883	0.9283	-5.337	-5.337	0.186	3.0417	0.8817	2.16
2.4	-12.578	-6.900	-0.705	3.0731	1.1531	-6.386	-6.200	-0.968	2.9665	1.0465	1.92
2.4	-12.617	-6.289	-0.644	3.0772	1.1572	-6.383	-6.386	-0.987	2.9652	1.0452	1.92
2.4	-12.615	-6.309	-0.646	3.0770	1.1570	-6.383	-6.383	-0.986	2.9652	1.0452	1.92
2.4	-12.615	-6.308	-0.646	3.0770	1.1570	-6.383	-6.383	-0.986	2.9652	1.0452	1.92
2.6	-12.388	-6.200	-1.897	2.8225	1.1425	-5.958	-6.000	-2.224	2.6429	0.9629	1.68
2.6	-12.388	-6.194	-1.896	2.8225	1.1425	-5.959	-5.958	-2.220	2.6432	0.9632	1.68
2.6	-12.388	-6.194	-1.896	2.8225	1.1425	-5.959	-5.959	-2.220	2.6432	0.9632	1.68
2.8	-9.573	-4.300	-2.945	2.3320	0.8920	-4.155	-4.100	-3.206	+2.0925	0.6525	1.44
2.8	-9.540	-4.787	-2.994	2.3288	0.8888	-4.150	-4.155	-3.212	2.0921	0.6521	1.44
2.8	-9.541	-4.770	-2.992	2.3289	0.8889	-4.150	-4.150	-3.211	2.0921	0.6521	1.44
2.8	-9.541	-4.770	-2.992	2.3289	0.8889	-4.150	-4.150	-3.211	2.0921	0.6521	1.44
3.0	-4.687	-2.500	-3.719	1.6502	0.4502	-1.376	-1.400	-3.766	1.3853	0.1853	1.20
3.0	-4.698	-2.343	-3.703	1.6513	0.4513	-1.378	-1.376	-3.764	1.3854	0.1854	1.20
3.0	-4.698	-2.349	-3.704	1.6513	0.4513	-1.378	-1.378	-3.764	1.3854	0.1854	1.20
3.2	1.090	0.800	-3.859	0.8845	-0.0755	1.748	1.700	-3.732	0.6255	-0.3345	0.96
3.2	1.106	0.545	-3.884	0.8828	-0.0772	1.748	1.748	-3.727	0.6259	-0.3341	0.96
3.2	1.105	0.553	-3.883	0.8829	-0.0771	1.748	1.748	-3.727	0.6259	-0.3341	0.96
3.2	1.105	0.553	-3.883	0.8829	-0.0771	1.748	1.748	-3.727	0.6259	-0.3341	0.96
3.4	6.608	3.600	-3.468	0.1377	-0.5823	4.506	4.700	-3.082	-0.0649	-0.7849	0.72
3.4	6.629	3.304	-3.438	0.1357	-0.5843	4.515	4.506	-3.101	-0.0662	-0.7862	0.72
3.4	6.628	3.314	-3.439	0.1358	-0.5842	4.515	4.515	-3.100	-0.0661	-0.7861	0.72
3.4	6.628	3.314	-3.439	0.1358	-0.5842	4.515	4.515	-3.100	-0.0661	-0.7861	0.72
3.6	10.578	5.400	-2.568	-0.4718	-0.9518	6.251	6.900	-1.958	-0.5799	-1.0599	0.48
3.6	10.589	5.289	-2.579	-0.4725	-0.9525	6.277	6.251	-2.023	-0.5842	-1.0642	0.48
3.6	10.589	5.299	-2.577	-0.4725	-0.9525	6.277	6.277	-2.020	-0.5841	-1.0641	0.48
3.6	10.589	5.299	-2.577	-0.4725	-0.9525	6.277	6.277	-2.020	-0.5841	-1.0641	0.48
3.8	12.259	6.200	-1.427	-0.8760	-1.1160	6.612	6.800	-0.712	-0.8591	-1.0991	0.24
3.8	12.264	6.130	-1.434	-0.8764	-1.1164	6.618	6.612	-0.731	-0.8603	-1.1003	0.24
3.8	12.264	6.132	-1.434	-0.8764	-1.1164	6.618	6.618	-0.730	-0.8603	-1.1003	0.24
4.0	11.323	5.600	-0.260	-1.0441	-1.0441	5.454	5.400	0.472	-0.8821	-0.8821	0
4.0	11.319	5.661	-0.255	-1.0437	-1.0437	5.453	5.454	0.477	-0.8817	-0.8817	0
4.0	11.319	5.660	-0.255	-1.0437	-1.0437	5.453	5.453	0.477	-0.8817	-0.8817	0
4.2	10.705	5.350	0.846	-0.9836	-0.9836	5.330	5.300	1.549	-0.8691	-0.8691	0
4.2	10.705	5.352	0.846	-0.9836	-0.9836	5.329	5.330	1.552	-0.8689	-0.8689	0

of the base does not induce strains in the system. Compute the responses to the ground motion $x_0 = 1.2t$ (x_0 in cm, t in sec), if $0 \leq t \leq 2$ and $x_0 = 4.8 - 1.2t$, if $2 \leq t \leq 4$, with $x_0 = 0$ outside this range.

Solution. The problem is solved in Table 2.1, where we have taken $\Delta t = 0.2$ sec and $\beta = \frac{1}{6}$. Computational details are similar to those expounded in connection with Example 1.3.

2.6 Effects of Axial Forces

Gravity loads, among several other factors, may combine with changes in the geometry of structures to originate nonlinear behavior in systems whose constituent materials behave linearly. These effects are often substantial. It is desirable to set forth methods that will allow their explicit consideration.

Refer first to a structure that is subjected to a statically applied set of forces that produce effects insensitive to geometric changes, plus another set of forces, also applied statically, whose effects (generalized forces) are proportional to the deformations. A simple example of this sort is found in a beam subjected to a set of lateral forces plus a set of axial loads; the latter may act at intermediate points along the beam axis, and their points of application may follow the beam in its deformation. Let the second set of forces be proportional to a scalar, say λ. In the general case we may write the equation of equilibrium as

$$\mathbf{Kx} = \mathbf{P} + \lambda \mathbf{Nx} \qquad (2.27)$$

where $\mathbf{K}$ is the stiffness matrix, $\mathbf{x}$ the vector of deformations or displacements, $\mathbf{P}$ the vector of transverse forces, and $\mathbf{N}$ a square matrix that depends on the pattern of the second set of loads.

When the effects of a single set of axial loads are going to be analyzed, λ may be taken equal to 1 and the matrix $\mathbf{N}$ found as follows. Assume $\mathbf{K} = 0$ and take any component x_i of the displacement vector equal to -1 and all other components equal to 0. The generalized forces required to maintain equilibrium constitute the ith column of $\mathbf{N}$. Repeat for every i.

If all axial loads are multiplied by a load factor, without changing the magnitude of applied forces whose effects are insensitive to geometric changes, that factor is λ.

When the first set of loads does not exist, Eq. 2.27 becomes

$$(\mathbf{K} - \lambda \mathbf{N})\mathbf{x} = 0 \qquad (2.28)$$

This expression governs the phenomenon of buckling of the system. Its similarity with Eq. 2.5, which governs the free vibrations of the system, allows several statements to be made immediately. Thus, Eq. 2.28 admits nontrivial solutions only for a set of characteristic values of λ that produce the buckling loads. The corresponding solutions of the equation are the system's natural buckling modes. These form a complete orthogonal set, with weighting factor $\mathbf{N}$. Methods for calculation of natural buckling modes and the corresponding loads are essentially the same as those used for calculation of natural modes of vibration and the corresponding frequencies. However, negative roots of λ may result; when this situation arises it means that the critical loads have a sign opposite the one assumed.

Similarity of Eq. 2.27 with the equation that governs steady-state, forced harmonic motion of an undamped system (Eq. 2.1 with $\mathbf{b} \sin \omega t$ substituted for $\mathbf{P}$) illustrates that essentially the same methods of analysis apply to calculation of responses to this type of dynamic responses and to calculation of the deformations of structures subjected to combined transverse and axial forces.

The inclusion of slenderness effects in the motion of a system acted upon by forces varying with time, may be achieved in a process similar to the one that led to Eqs. 2.27 and 2.28. These effects may be determined from the general expression

$$\mathbf{M}\ddot{\mathbf{x}} + \mathbf{C}\dot{\mathbf{x}} + (\mathbf{K} - \lambda\mathbf{N})\mathbf{x} = \mathbf{P} \qquad (2.29)$$

which replaces the corresponding one in Section 2.1. Here every matrix and every vector in the equation, as well as λ and the vector of ground displacements $\mathbf{x}_0$, may be functions of time. The numerical method presented in Section 2.4 for the analysis of linear systems under transient disturbances applies to systems governed by Eq. 2.29 with the simple device of replacing the stiffness matrix with the one in parenthesis in this expression.

Steady-state vibrations, too, may be treated through this device provided that $\lambda\mathbf{N}$ is not a function of time. (When λ varies with time, say as a constant plus a term proportional to $\sin \omega t$, the structural response is generally no longer sinusoidal. Various approximate schemes may be set up to analyze the problem, making use of energy methods for example.)

Free vibrations of undamped systems under static axial loads are of special interest. For this case the equation of motion, derived directly from Eq. 2.29, is

$$\mathbf{M}\ddot{\mathbf{x}} + (\mathbf{K} - \lambda\mathbf{N})\mathbf{x} = 0$$

Proceeding as in the derivation of Eq. 2.5, we conclude that the natural modes of vibration of the system, $\mathbf{z}_n$, satisfy

$$(\mathbf{K} - \lambda\mathbf{N} - \omega_n'^2\mathbf{M})\mathbf{z}_n' = 0 \qquad (2.30)$$

where the prime is used to distinguish the corresponding natural modes and circular frequencies from those of a system in which $\lambda = 0$. Again replacement of the stiffness matrix with $\mathbf{K} - \lambda\mathbf{N}$ allows application of the methods covered in Section 2.2.

A specially simple situation arises when the nth natural mode of vibration of a system for which $\lambda = 0$ has the same shape as the nth natural buckling mode; m and n may or may not be equal to each other. Under these conditions it is convenient to premultiply Eqs. 2.5 and 2.30 by $\mathbf{z}_n^T$ after writing $\mathbf{z}_n$ for $\mathbf{x}$ and λ_m for λ in Eq. 2.28. We obtain

$$\omega_n^2 = \frac{\mathbf{z}_n^T \mathbf{K} \mathbf{z}_n}{\mathbf{z}_n^T \mathbf{M} \mathbf{z}_n}$$

$$\lambda_m = \frac{\mathbf{z}_n^T \mathbf{K} \mathbf{z}_n}{\mathbf{z}_n^T \mathbf{N} \mathbf{z}_n}$$

and

$$\mathbf{z}_n^T \mathbf{K} \mathbf{z}_n - \lambda \mathbf{z}_n^T \mathbf{N} \mathbf{z}_n - \omega_n'^2 \mathbf{z}_n^T \mathbf{M} \mathbf{z}_n = 0$$

Sec. 2.6   EFFECT OF AXIAL FORCES   55

Dividing the last expression by $z_n^T K z_n$ we see that

$$\omega_n'^2 = \omega_n^2 \left(1 - \frac{\lambda}{\lambda_m}\right) \quad (2.31)$$

In many problems it is only worth taking slenderness effects in the calculation of the fundamental frequency of vibration because λ_2 is many times greater than λ_1 and because the fundamental mode of vibration may account for the most significant contributions to dynamic responses. This is usually true in seismic analysis when slenderness effects are essentially associated with gravity loads. In these cases it may be sufficient to correct ω_1 for such effects in accordance with Eq. 2.31, taking $m = 1$, even if the shape of the fundamental mode of vibration does not coincide exactly with that of the fundamental buckling mode.

If, superimposed on the axial loads, there is a set of static forces whose effects are insensitive to geometric changes, the total deformation may be obtained by solving Eq. 2.27, where **P** denotes these forces, and adding the solution of the dynamic problem (Eqs. 2.29–2.31) in which the static forces in question have been ignored.

Example 2.8. Discuss the effects of gravity loads on the vibrations of the system shown in Fig. 2.5a.

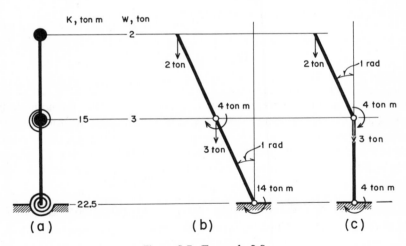

Figure 2.5. Example 2.8.

Solution. We choose the rotation at joints 1 and 2 as generalized coordinates. It follows immediately that

$$\mathbf{K} = \begin{vmatrix} 22.5 & 0 \\ 0 & 15 \end{vmatrix} \text{ton m} \quad \mathbf{M} = \begin{vmatrix} 44 & 8 \\ 8 & 8 \end{vmatrix} \frac{1}{9.81} \text{ton m sec}^2$$

By deforming the system as in Fig. 2.5b and taking the joints as hinges we find the need for static moments of 14 and 4 ton m at joints 1 and 2, respectively.

56   LINEAR SYSTEMS WITH SEVERAL DEGREES OF FREEDOM   Chap. 2

These values give the first column in matrix **N**. The moments in Fig. 2.5c give the second column:

$$\mathbf{N} = \begin{vmatrix} 14 & 4 \\ 4 & 4 \end{vmatrix} \text{ ton m}$$

For the condition stipulated, $\lambda = 1$. Hence,

$$\mathbf{K} - \lambda\mathbf{N} = \begin{vmatrix} 8.5 & -4 \\ -4 & 11 \end{vmatrix} \text{ ton m}$$

Notice that this same matrix could have been obtained directly by retaining the joint stiffness and vertical loads while giving the system the deformations

$$\mathbf{x} = \begin{vmatrix} 1 \\ 0 \end{vmatrix} \text{m} \quad \text{and} \quad \mathbf{x} = \begin{vmatrix} 0 \\ 1 \end{vmatrix} \text{m}$$

The generalized forces required to preserve equilibrium would have resulted in the first and second columns, respectively, of $\mathbf{K} - \lambda\mathbf{N}$.

Proceeding in the usual manner we find that, if we neglect the effects of gravity forces,

$$\mathbf{z}_1 = \begin{vmatrix} 1 \\ 0.34 \end{vmatrix} \quad \mathbf{z}_2 = \begin{vmatrix} 1 \\ -4.35 \end{vmatrix}$$

and $\omega_{1,2}^2 = 4.73, 23.91$ (rad/sec)^2.

On the other hand, solution of Eq. 2.29 results in

$$\mathbf{z}'_1 = \begin{vmatrix} 1 \\ 0.51 \end{vmatrix} \quad \mathbf{z}'_2 = \begin{vmatrix} 1 \\ -3.97 \end{vmatrix}$$

and $\omega'^2_{1,2} = 1.32, 19.68$ (rad/sec)^2.

Had the buckling problem been solved first, we would have found $\lambda_{1,2} = 1.37, 6.13$. Assuming (incorrectly) that $\mathbf{z}_{1,2}$ had, respectively, the same shapes as the first and second buckling modes, Eq. 2.30 would apply with $m = n$, to give

$$\omega'_1 = 4.73\sqrt{\frac{1-1}{1.37}} = 1.28 \text{ rad/sec}$$

$$\omega'_2 = 23.91\sqrt{\frac{1-1}{6.13}} = 21.88 \text{ rad/sec}$$

The fact that λ is considerably smaller than λ_1 makes this approximate treatment give reasonably good results in the present case.

PROBLEMS

2.1.* Compute the natural periods and modes of vibration and corresponding damping ratios for the systems in Fig. 2.6.

Ans. (a) $\mathbf{z}_1^T = (1.0, 1.657, 2.177)$, $T = 0.41$ sec, $\zeta = 3.8$ percent.
$\mathbf{z}_2^T = (1.0, 0.917, -1.947)$, $T = 0.16$ sec, $\zeta = 9.5$ percent.

* Solution of problems marked with an asterisk is lengthy.

Chap. 2 PROBLEMS 57

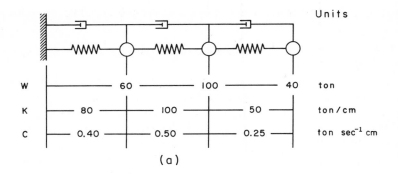

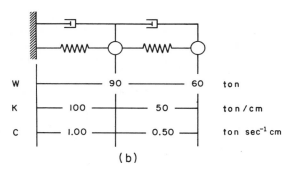

Figure 2.6. Problem 2.1.

$$z_3^T = (1.0, -0.474, 0.171), \quad T = 0.10 \text{ sec}, \quad \zeta = 15.2 \text{ percent}.$$
(b) $z_1^T = (1.0, 2.186), \quad T = 0.30 \text{ sec}, \quad \zeta = 10.5 \text{ percent}.$
$\quad\;\; z_2^T = (1.0, -0.686), \quad T = 0.14 \text{ sec}, \quad \zeta = 22.4 \text{ percent}.$

2.2. The structure shown in Fig. 2.7 consists of an infinitely rigid plate that weighs 12 kip and has uniform thickness. It is supported by a steel H-column fixed at the base. The cross-sectional moment of inertia of the column, in the direction of motion, is $I = 500$ in.^{4}, and the area of its web is $A_w = 10$ in.^{2} Compute the natural frequencies of the system in the plane of the drawing including effects of rotary inertia and

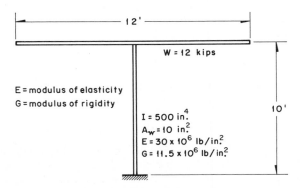

Figure 2.7. Problem 2.2.

shear deformations and neglecting the mass of the column and effects of gravity forces.

Ans. $f_1 = 4.03, f_2 = 16.71$ Hz.

2.3. What are the natural frequencies in the preceding problem if effects of gravity forces are included?

Ans. $f_1 = 4.01, f_2 = 16.71$ Hz

2.4. The corners of a massless, infinitely rigid, square plate, whose side measures 6 m, rest on four equal massless columns. Each column has a moment of inertia of 100,000 cm⁴ in all directions, a modulus of elasticity $E = 200$ ton/cm², a height of 3 m, and a fixed base. At the center of one edge of the plate there is a concentrated mass of 0.1 ton sec²/cm (Fig. 2.8). Find the natural frequencies of the system.

Ans. $f_1 = 2.30, f_2 = 3.00, f_3 = 5.55$ Hz.

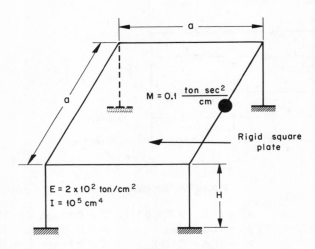

Figure 2.8. Problem 2.4.

2.5. Find the undamped natural frequencies for the system in Fig. 2.9.

Ans. $f_1 = 1.82, f_2 = 4.94$ Hz.

2.6. Compute the steady-state responses of the systems in Problem 2.1 to the force $P = 2 \sin \omega t$ applied at the last mass, where $\omega = 1.2 \omega_1$. ω_1 is the fundamental circular frequency of the system in question, P is in metric tons, and t is in seconds. Neglect damping.

Ans. For the system in Fig. 2.6a, $\mathbf{x}^T = (-0.1068, -0.1701, -0.1796) \sin \omega t$.
For the system in Fig. 2.6b, $\mathbf{x}^T = (-0.0666, -0.1218) \sin \omega t$.

2.7. In the system of Problem 2.4 a force (10 metric ton) sin $2\omega_1 t$ acts on M parallel to the edge on which the mass is located. Compute the steady-state amplitudes of oscillation of the mass.

Ans. $x = 0.099$ cm, $\theta = -5.03 \times 10^{-4}$ rad.

2.8. Compute the response of the system in Fig. 2.6a to the ground motion $x_0 = 3t$ (x_0 in centimeters and t in seconds) for $0 \leq t \leq 1$ sec; $x_0 = 0$ for $t \leq 0$; $x_0 = 3$ cm for $t \geq 1$ sec. Assume that all points of the support describe the same motion. Take $\Delta t = 0.1$ sec. Include effects of damping in the computations.

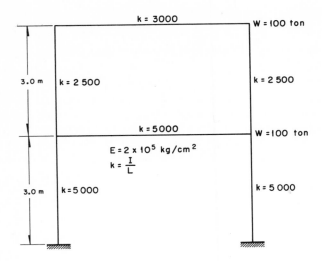

Figure 2.9. Problem 2.5.

2.9. Describe the motion of M in the system of Problem 2.4 as response to the ground motion $\dot{x}_0 = 0$ for $t \leq 0$ and $\dot{x}_0 = u$ for $t > 0$.

2.10. The ground motion is $\ddot{x}_0 = a \sin \omega t$ when $0 \leq t \leq t_0$ and $\ddot{x}_0 = 0$ when $t \geq t_0$. Calculate the spectra corresponding to $\zeta = 0$ and 0.2, respectively. Next take $a = 6$ cm/sec² and $t_0 = 1.0$ sec and compute an upper limit for the maximum total shear in the first story columns of the structure in Problem 2.5 assuming a damping ratio of 0.2 in both natural modes.

Ans. $Q_{max} = 2.25$ ton

2.11.* The system in Fig. 2.10 is a two-span continuous beam. Support A undergoes the motion $x_a = 0.5 \sin 20t$ for one full sine wave and is at rest for all other values of t (x_0 and t are in centimeters and seconds, respectively). Support B describes the same motion as A, but it does so 0.1 sec later, and support C does the same 0.2 sec after support A. Find the maximum bending moment in the beam, at support B, as a response to the specified ground motion. Neglect shearing strains, damping, and the mass of the beam.

Ans. 6.73 ton m

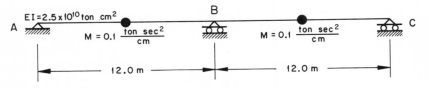

Figure 2.10. Problem 2.11.

2.12. Repeat the calculations in Example 2.7 taking the displacements of points 0 and 1 as generalized displacements.

3

LINEAR SYSTEMS WITH DISTRIBUTED MASS

3.1 Introductory Note

For the systems that we studied in Chapter 2 we assumed that the masses were *lumped* (i.e., concentrated) at discrete points or at rigid bodies and joined to each other and to the ground by massless springs and dashpots. Here we shall deal with systems whose masses are distributed. Again our treatment will be restricted to systems whose parameters do not change with time.

We may regard the lumped-parameter approach as an approximation to distributed-parameter systems. In principle, if the distributed mass is treated as concentrated at a sufficiently large number of points and the connecting elements are stipulated correctly, behavior of a discrete structure can be made to approach that of the corresponding continuous system as closely as desired, as will be discussed in Chapter 4. But in many cases replacing the continuous system with one that has lumped masses complicates the solution of dynamic problems. Here we shall present those cases of special interest where there is advantage in preserving the nature of distributed-parameter systems.

Usually the more accurate idealizations of actual systems consist in treating them as having distributed parameters. In most cases, then, the lumped-parameter systems of Chapter 2 constitute simplifications of the idealized systems.

Lumping of mass is normally achieved by ascribing a tributary volume to each of the discrete points where the lumping is desired. Often, a preferable means in one-dimensional problems consists of using equivalent concentrations of either mass or inertia forces at the discrete points (Newmark, 1943). If distributed inertia forces are replaced in this manner with concentrations, the principle of reciprocal relations ceases to apply rigorously (except in very simple cases), and natural modes are no longer found to be orthogonal. Therefore, the advantages of using equivalent concentrations of inertia forces are sometimes lost if we have to compute natural modes of vibration. These advantages can be exploited fully when we deal directly with numerical computations of responses.

Use of equivalent concentrations of mass is usually less accurate than use of equivalent concentrations of inertia forces, but the former is not objectionable from the point of view of natural-mode orthogonality. These lumping techniques are also of interest when we establish the parameters of electric analogs. The matter of lumping is taken up in detail in Section 4.2.

Instead of lumping the mass at discrete points, it is sometimes advantageous to idealize the distributed-parameter system as consisting of rigid bars, plates, or three-dimensional bodies, joined to each other by massless elements in which the entire flexibility has been concentrated. This is the case when we wish to take rotational inertia into account.

Energy methods described in Chapter 4 are especially suited for the analysis of systems having distributed mass and flexibility and whose parameters have not been lumped.

Distributed-parameter systems may be viewed as having an infinite number of degrees of freedom and hence as limiting cases of those treated in Chapter 2. The ordinary differential equations that we established in terms of matrices become now partial differential equations, in which the independent variables are time and the space coordinates. The relations derived in Chapter 2 are valid for systems having distributed parameters provided we replace the matrices with the corresponding functions of space coordinates and the summation signs with the corresponding integrals.

3.2 Undamped, Uniform, Shear Beam

The simplest distributed-parameter structure is a nondimensional, close-coupled, linear, undamped system with uniform mass and stiffness per unit length. The motion is governed by the partial differential equation

$$m \frac{\partial^2 x}{\partial t^2} - k \frac{\partial^2 x}{\partial X^2} = p \tag{3.1}$$

where m is the mass density (mass per unit length or per unit volume), x the displacement at a point of abscissa X and at time t, k the stiffness, and p the distributed load (load per unit length or unit volume) applied in the direction of x.

One such system is the shear beam, a structure whose change in slope at every section is proportional to the shear acting at the section. In this case, m and p are measured per unit length, k is the shear stiffness, X is measured along the beam axis, and x is perpendicular to X. The behavior of tall buildings can often be approximated by this differential equation.

Equation 3.1 may be derived easily as follows. With reference to Fig. 3.1, let dX denote the length of an infinitesimal segment of the shear beam. Assume that the slope $\partial x/\partial X$ is proportional to the average shearing stress on the cross section:

Sec. 3.2 UNDAMPED, UNIFORM, SHEAR BEAM 63

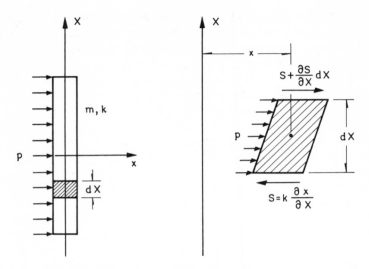

Figure 3.1. One-dimensional, close-coupled, linear, undamped system with uniform mass and stiffness.

$$\frac{\partial x}{\partial X} = \frac{S}{k}$$

Here S denotes the transverse shear at the section considered. The difference between S at the top and bottom of the infinitesimal segment is $(\partial S/\partial X)\, dX$. According to D'Alembert's principle, this must be in equilibrium with the sum of the external force acting on that segment, $p\, dX$, and the corresponding inertia force, $-(\partial^2 x/\partial t^2) m\, dX$:

$$\frac{\partial S}{\partial X} dX + p\, dX - m \frac{\partial^2 x}{\partial t^2} dX = 0$$

but

$$S = k \frac{\partial x}{\partial X}$$

Substituting and dividing through by dX we obtain Eq. 3.1.

Other examples of systems whose motion Eq. 3.1 governs are:

1. An isotropic (or orthotropic homogeneous, or horizontally layered) semiinfinite medium undergoing horizontal displacements. Here m is the mass per unit volume, p acts horizontally and is uniformly distributed in horizontal planes, and k is the modulus of rigidity.

2. The same medium undergoing vertical displacements. In an isotropic medium k will stand for $E(1-v)/(1-v-2v^2)$, where $E=$ modulus of elasticity, $v=$ Poisson's ratio, and x will represent displacements along the X axis. The effective or bulk modulus quoted in this paragraph is

derived easily. Let σ_i and ϵ_i denote principal stress and strain, respectively, in the direction i, and let $i = 1$ correspond to the vertical direction. From symmetry, $\sigma_2 = \sigma_3$ and $\epsilon_2 = \epsilon_3$. Since there is to be no horizontal displacement, $\epsilon_2 = 0$. But

$$\epsilon_2 = \frac{\sigma_1 + \sigma_2}{E}v - \frac{\sigma_2}{E}$$

Hence,

$$\frac{\sigma_1}{E}\left[v - \frac{\sigma_2}{\sigma_1}(1-v)\right] = 0$$

or

$$\frac{\sigma_2}{\sigma_1} = \frac{v}{1-v}$$

Now,

$$\epsilon_1 = \frac{\sigma_1}{E} - \frac{2\sigma_2}{E}v$$

$$= \frac{\sigma_1}{E}\left(1 - \frac{2v^2}{1-v}\right)$$

$$= \left(\frac{\sigma_1}{E}\right) \cdot \left(\frac{1-v-2v^2}{1-v}\right)$$

and

$$\frac{\sigma_1}{1} = \frac{E(1-v)}{1-v-2v^2}$$

3. A slender bar undergoing longitudinal displacements. In this case, instead of k we should write E, the modulus of elasticity, and again m is the mass per unit volume. (The one-dimensional solution holds if we neglect radial displacements, which is a reasonable simplification when the cross-sectional dimensions of the bar are small compared with the length of the waves under consideration, but may introduce serious errors for very short wavelengths.)

4. A slender bar undergoing torsional oscillations. We replace m with the cross-sectional polar moment of inertia, k with the torsional stiffness, x with rotation of the bar, and p with the torque per unit length. (Again this treatment is satisfactory only if the cross-sectional dimensions are small compared with the wavelength.)

5. A taut string under uniform and constant horizontal projection of the tension to which it is subjected. Again m is the mass and p the transverse load, both per unit length, but k must be replaced with the horizontal component of the string tension.

The general solution of Eq. 3.1 can be expressed either as a series development in terms of the natural modes of vibration of the shear beam or in the form of the wave equation. We can derive the first type of solution by the method of separation of variables, as follows.

The homogeneous equation, obtained by setting $p = 0$ in Eq. 3.1, is

$$\frac{\partial^2 x}{\partial t^2} - v^2 \frac{\partial^2 x}{\partial X^2} = 0 \qquad (3.2)$$

where $v^2 = k/m$. If we assume that $x(X, t)$ can be written as the product of a function of X alone and a function of t alone,

$$x = z_n \theta_n(t) \qquad (3.3)$$

we can write Eq. 3.2 as

$$z_n \ddot{\theta}_n - v^2 z_n'' \theta_n = 0$$

where the primes denote differentiation with respect to X. Hence,

$$\frac{\ddot{\theta}_n}{\theta_n} = v^2 \frac{z_n''}{z_n} = -\omega_n^2$$

where ω_n is arbitrary. The last expression can be broken into

$$\frac{\ddot{\theta}_n}{\theta_n} = -\omega_n^2 \qquad (3.4)$$

and

$$\frac{z_n''}{z_n} = -\frac{\omega_n^2}{v^2} \qquad (3.5)$$

Equation 3.4 states that ω_n does not depend on X, and Eq. 3.5 states that it does not depend on t; therefore ω_n is a constant.

The general solution of Eq. 3.4 is, then,

$$\theta_n = \sin \omega_n(t - t_n) \qquad (3.6)$$

(save for an arbitrary factor that will be included in z_n). Here t_n is arbitrary. The general solution of Eq. 3.5 is

$$z_n = A_n \sin \frac{\omega_n}{v}(X - a_n) \qquad (3.7)$$

where A_n and a_n are arbitrary. Substituting Eqs. 3.6 and 3.7 in 3.3 we obtain

$$x = A_n \sin \left[\frac{\omega_n}{v}(X - a_n)\right] \sin [\omega_n(t - t_n)] \qquad (3.8)$$

This form describes the nth natural mode of vibration of the system because it is a solution that satisfies the definition of natural modes given in Chapter 2. The corresponding natural circular frequency is ω_n; together with the constant a_n it is determined in each case so as to satisfy the boundary conditions. The arbitrary constant A_n defines the amplitude of vibration, and t_n is an arbitrary time shift.

By combining linearly as many solutions of the form of Eq. 3.8 as required, we obtain the general solution of Eq. 3.2. The amplitudes and time shifts can be found in such a way as to satisfy any initial configuration of displacements and velocities that complies with the boundary conditions, because any such configuration can be expressed as a linear combination of the natural modes. The general solution of Eq. 3.1 is obtained by adding a particular solution of Eq. 3.1 to this general solution of Eq. 3.2.

As examples consider first the case of a taut string of length L, whose end supports are fixed against displacement. If we take $X = 0$ at one end, the boundary conditions are

$$x(0, t) = x(L, t) = 0$$

Since $\omega_n = 0$ would give us a trivial solution, we are interested in the case when $a_n = j\pi v/\omega_n$ in Eq. 3.8, where j is either zero or an integer, so as to satisfy the condition at $X = 0$. The cases $j \neq 0$ are not interesting because they give the same solution as $j = 0$. We shall therefore take $a_n = 0$. In order to satisfy the condition at $X = L$ we must have $\omega_n = n\pi v/L$, where $n = 1, 2, 3, \ldots$. Therefore, the natural periods of vibration are

$$T_1 = \frac{2L}{v} \qquad T_n = \frac{1}{n}T_1$$

The mode shapes are

$$z_n = A_n \sin \frac{n\pi v X}{L}$$

Now consider a shear beam of length L, fixed at the base and free at the opposite end. With $X = 0$ at the base,

$$x(0, t) = 0 \qquad x'(L, t) = 0 \tag{3.9}$$

where the prime denotes differentiation with respect to X. (The second boundary condition in Eq. 3.9 comes from the requirement that $S = 0$ at $X = L$.) From Eq. 3.8, then, $a_n = 0$ and

$$\omega_n = \frac{(2n-1)\pi v}{2L} \qquad n = 1, 2, 3, \ldots \tag{3.10}$$

Therefore,

$$T_1 = \frac{4L}{v} \qquad T_2 = \frac{1}{3}T_1 \qquad T_3 = \frac{1}{5}T_1, \ldots \tag{3.11}$$

The distribution of S along the beam in each natural mode is obtained by differentiating the displacement with respect to X and multiplying by k. The corresponding amplitudes of the transverse shear result from taking $\sin \omega_n(t - t_n) = 1$:

$$S_n = A_n k \frac{\omega_n}{v} \cos \frac{\omega_n}{v} X$$

or

$$S_n = \frac{(2n-1)\pi A_n k}{2L} \cos \frac{\omega_n}{v} X \tag{3.12}$$

The shapes and shear distributions in the first three modes are shown in Fig. 3.2.

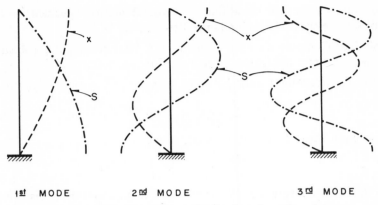

1ˢᵗ MODE   2ⁿᵈ MODE   3ʳᵈ MODE

Figure 3.2. Shapes and shear distributions in a shear beam.

Shear beams have been proposed as idealizations of tier buildings (Westergaard, 1933) and of soil layers (Jacobsen, 1964). It is of interest to compute the participation coefficients when the ground describes an arbitrary motion. Replacing the matrix products with integrals in the appropriate expression of Section 2.4, we have

$$x(t) = -\sum_n \frac{a_n}{\omega_n} \sin\left(\frac{\omega_n}{v}X\right) \int_0^t \ddot{x}_0(\tau) \sin[\omega_n(t-\tau)]\, d\tau \qquad (3.13)$$

where

$$a_n = \frac{\int_0^L m \sin[(\omega_n/v)X]\, dX}{\int_0^L m \sin^2[(\omega_n/v)X]\, dX} = \frac{4}{(2n-1)\pi} \qquad (3.14)$$

and $\ddot{x}_0$ is the base motion. We can always obtain the shear distribution as a function of time by adding the modal contributions at every instant at which we are interested. Thus, if we wish, we can draw the shear envelope. Often, however, it will be satisfactory to arrive at an upper limit to this envelope. We can compute such an upper limit by using the method of Chapter 2 for the transverse shears caused by a ground motion that is specified by its spectrum. We need merely add the amplitudes of S, in all the natural modes, at each elevation.

Example 3.1. Compute an upper limit to the shear envelope in a shear beam whose base undergoes a sudden change of velocity.

Solution. The undamped velocity spectrum for a velocity pulse of magnitude u is $V = u$ (see Example 1.1 and Fig. 1.6). According to Eqs. 3.13 and 3.14 and noticing that $S = k\, \partial x/\partial X$ we find

$$S \leq \frac{4uk}{\pi v} \sum_{n=1}^{\infty} \frac{1}{2n-1} \left|\cos\frac{(2n-1)\pi X}{2L}\right|$$

This series diverges. However, by writing the expression for each term as a function of time we can show that the ground motion specified would not produce infinite shears in the undamped structure.

Example 3.2. Compute an upper limit to the shear envelope in a shear beam whose base is at rest and is then given a constant acceleration a.

Solution. The spectrum is now $V = a/\omega$. Hence,

$$S \leq \frac{8aLm}{\pi^2} \sum_{n=1}^{\infty} \frac{1}{(2n-1)^2} \left| \cos \frac{(2n-1)\pi X}{2L} \right|$$

The maximum shear occurs at the base, where

$$S = \frac{8aLm}{\pi^2} \sum_{n=1}^{\infty} \frac{1}{(2n-1)^2} = aLm$$

Hence, the series converges at all elevations. The envelope shear distribution is shown in Fig. 3.3.

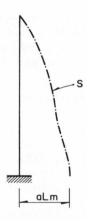

Figure 3.3. Example 3.2.

3.3 Simple Wave Equation

The second form of the general solution to the differential equation of a shear beam not subjected to external loads is

$$x = F_1\left(\frac{X}{v} - t\right) + F_2\left(\frac{X}{v} + t\right) \tag{3.15}$$

where F_1 and F_2 are any functions admitting second derivatives. Validity of this relation is verified easily by substituting it in Eq. 3.2. Again we obtain the general solution of Eq. 3.1 by adding to Eq. 3.15 a particular solution of 3.1. Equation 3.15 is known as the *simple wave equation* or merely the *wave equation*.

In Eq. 3.15, F_1 represents a shear wave traveling in the direction of X, and F_2 represents a shear wave traveling in the opposite direction. The waves do not change shape as they travel along the shear beam, and their velocity v is independent of the wave shape. These traits are peculiar to systems governed by Eq. 3.2.

In all systems governed by Eqs. 3.1 and 3.2, such as those mentioned in Section 3.2, v represents the velocity of propagation of the corresponding waves. Thus, oscillations along the X axis in a homogeneous medium, or in one that is stratified in layers perpendicular to X, propagate in the direction of this axis with a velocity

$$v = v_p = \sqrt{\frac{E(1-\nu)}{\rho(1-\nu-2\nu^2)}}$$

while the velocity of transverse waves is

$$v = v_s = \sqrt{\frac{\mu}{\rho}}$$

where $\mu =$ modulus of rigidity and $\rho =$ mass per unit volume.

At a boundary where motion is prevented a wave suffers reflection, changing sign and direction without change in magnitude. Thus the displacements x_1 (X, t) given by Eq. 3.15 become $-x_1(2a - X, t)$ when reflected at a boundary located at $X = a$. At the boundary the reflected wave cancels the incoming motion exactly, giving $x(a, t) = 0$ for all t.

At a free end the reflection takes place without change in sign. $x_1(X, t)$ becomes $x_1(2a - X, t)$ if there is a free end at $X = a$. Therefore the derivative of the incoming wave with respect to X is canceled exactly by the derivative of the reflected wave at $X = a$, and there is no resultant shear at the free end. The displacements of the free end will be twice x_1.

Consider a stretched string that is plucked into a triangular wave close to a fixed end, as shown in Fig. 3.4a. Upon release, two waves will form, traveling in opposite directions (Fig. 3.4b). Reflection at the fixed end will cancel the displacements of the incoming wave, as depicted in Fig. 3.4c. Sometime later the reflected wave will have traveled back, as in Fig. 3.4d. Wave travel under similar initial conditions in the neighborhood of the free end of a shear beam is illustrated in Fig. 3.5a–d.

At the interface between two sections of a shear beam having different mass and stiffness per unit length, continuity and equilibrium conditions must be fulfilled (Westergaard, 1933). If x_1 denotes the incoming wave, x_2 the refracted wave, x_3 the reflected wave, and m_0 and k_0 refer to the first section, and m and k refer to the second section, continuity requires that

$$x_1 + x_3 = x_2$$

at the interface. Equilibrium requires that, also at the interface,

$$k_0\left(\frac{\partial x_1}{\partial X} + \frac{\partial x_3}{\partial X}\right) = k\frac{\partial x_2}{\partial X}$$

Let the interface be located at $X = 0$. Then these conditions are met by

$$x_2\left(\frac{X}{v}, t\right) = (1 + \alpha) x_1\left(\frac{X}{v_0}, t\right) \tag{3.16}$$

70 LINEAR SYSTEMS WITH DISTRIBUTED MASS

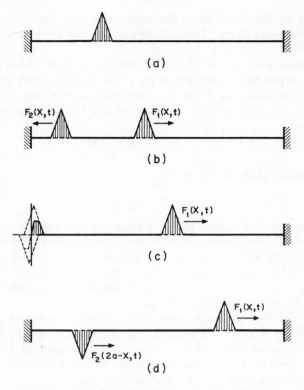

Figure 3.4. Triangular wave in a stretched string.

$$x_3\left(\frac{X}{v_0}, t\right) = \alpha x_1\left(\frac{-X}{v_0}, t\right) \tag{3.17}$$

where the reflection constant α is given by

$$\alpha = \frac{m_0 v_0 - mv}{m_0 v_0 + mv} \tag{3.18}$$

The fixed-end and free-end conditions studied above constitute special cases of this solution corresponding respectively to $\alpha = 1$ and -1.

The problem can be visualized better by having recourse to the method of *characteristics*. In it we take t as the abscissas and X as the ordinates, and we draw lines (called characteristics) that represent solutions to the expression $x =$ constant. As long as the mass and stiffness per unit length do not change, these are straight lines. Their slope is the velocity of propagation of shear waves. The lines mark the travel of arbitrarily chosen characteristic points in the wave shape. At an interface a characteristic splits into the one that corresponds to the refracted wave and into one for the reflection. The reflection has a slope equal to that of the original characteristic but with opposite sign.

Consider for example the case of a horizontal layer of depth H on bedrock. Assume that a velocity pulse travels upward as a plane, horizontal shear wave.

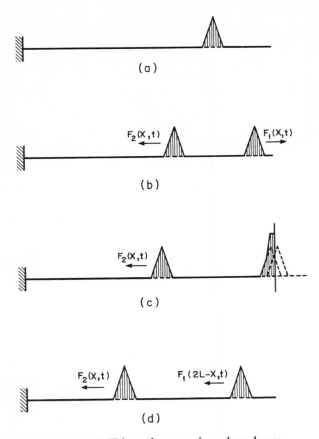

Figure 3.5. Triangular wave in a shear beam.

The characteristics that mark arrival of the pulse are shown in Fig. 3.6 for the case $\rho = \rho_0$, $\mu = \mu_0/9$, from which $\alpha = 0.5$, where subscript 0 refers to the rock. The same figure shows schematically the accelerations and velocities that would be observed at the ground surface, compared with those one would find at the bedrock surface if the layer were not present. Amplitudes in the latter case would be taken to be those of the wave itself, due to reflection at the free surface.

The nth "image" of the pulse at the ground surface occurs at a time $(2n + 1)H/v$ after the pulse arrives at the interface. The change in ground velocity for the nth image is $(1 + \alpha)(-\alpha)^n u$ where u is the change in velocity at the rock surface without the overlying formation.

In the case of an arbitrary, horizontal shear wave that would produce the displacement $x_0(t)$ at the rock surface if this surface were free, we obtain for the actual free ground surface

$$x(t) = (1 + \alpha) \sum_{n=0}^{\infty} (-\alpha)^n x_0 \left[t - (2n + 1)\frac{H}{v} \right] \qquad (3.19)$$

72   LINEAR SYSTEMS WITH DISTRIBUTED MASS   Chap. 3

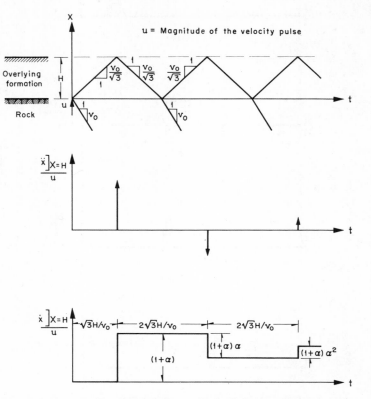

Figure 3.6. Example of the method of characteristics.

This expression has been confirmed experimentally in the case of a shallow alluvial formation resting on rock (Takahashi and Hirano, 1941).

A graphical construction using the method of characteristics has been proposed (Westergaard, 1933) for finding horizontal shears in buildings idealized as shear beams with distinct segments along their height. The method is useful for some buildings provided that the ground motion in question is sufficiently simple.

The foregoing problem can be solved in terms of natural modes of vibration instead of dealing directly with shear waves, but the solution is more elaborate. Conversely, problems that are solved more conveniently by using a modal analysis can also be dealt with by introducing wave analysis. Each natural mode of vibration of a uniform shear beam may be viewed as the superposition of a sinusoidal wave and its successive reflections. Thus, the fundamental mode of a cantilever shear beam is a sine wave whose length is four times the beam length, and the fundamental period (first part of Eq. 3.11) is equal to four times what it takes for the wave to travel along the beam in one direction.

3.4 Nonuniform, Undamped Shear Beams

In the derivation of Eq. 3.1 we assumed that both k and m were constant throughout the shear beam. When this is not the case we find

$$\frac{\partial x}{\partial X} = \frac{\partial k}{\partial X}\frac{\partial x}{\partial X} + k\frac{\partial^2 x}{\partial X^2}$$

and, by the same reasoning as we applied in deducing Eq. 3.1,

$$m\frac{\partial^2 x}{\partial t^2} - k'\frac{\partial x}{\partial X} - k\frac{\partial^2 x}{\partial X^2} = p \tag{3.20}$$

where $k' = \partial k/\partial X$. We see that Eq. 3.1 applies even if m is a function of X, but that a variable stiffness per unit length requires the addition of the term $k'\,\partial x/\partial X$. This would be the case, for example, for the gravity dam shown in Fig. 3.7.

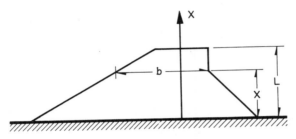

Figure 3.7. Gravity dam.

One case of practical interest is that of a horizontal soil layer whose unit mass is uniform and whose stiffness varies linearly with depth. Proceeding as above we find that its natural modes of vibration can be expressed in closed form in terms of Bessel functions, and we can compute its natural frequencies from the roots of an expression involving Bessel functions (Ambraseys, 1959).

Ambraseys (1959) has shown that there is little error in the computed natural frequencies when we replace the variable stiffness with a constant stiffness equal to its mean value; the error is less than 10 percent even for a ratio of maximum to minimum μ as high as 25. Hence in most applications a nonuniform soil mantle can be analyzed reasonably well as a uniform shear beam.

The same idealization, of uniform unit mass and linearly varying stiffness, may serve as an acceptable approximation to many tier buildings.

3.5 Damped, Uniform Shear Beams

Two types of problems deserve our attention in connection with the dynamic behavior of linear viscoelastic solids: steady-state vibrations and wave transmission. In both, external forces are zero except at the boundaries. Hence, in a shear beam D'Alembert's principle gives the following equation of equilibrium, to be satisfied in both types of problems,

$$\frac{\partial s}{\partial X} = \rho \frac{\partial^2 x}{\partial t^2} \qquad (3.21)$$

where s denotes shearing stress, X is measured along the beam axis, ρ is the density, x denotes transverse displacement, and t is time. With appropriate interpretations, this and the following equations apply also to plane stress and plane strain problems, and to other problems such as the torsional or longitudinal vibrations of slender bars, the plane wave transmission in stratified soil, and the vibrations of taut strings.

Consider first the case of steady-state vibrations, under which every beam element is subjected to a strain that can be put in the form

$$\frac{\partial x}{\partial X} = a \sin \omega t$$

where a is not a function of t. Linear behavior ensures that the corresponding stress will also be a sinusoidal function of time, with some phase shift relative to the strain. Hence we may also write the stress in the form

$$s = b \sin \omega t + c \cos \omega t$$

or

$$s = \mu \left(1 + \frac{Q}{\omega} \frac{\partial}{\partial t}\right) \frac{\partial x}{\partial X} \qquad (3.22)$$

where $\mu = b/a$ may be called the "static modulus of rigidity," $Q = c/b$, and ω is always taken as positive.

Combining with Eq. 3.21 we arrive at the differential equation

$$\mu \left(1 + \frac{Q}{\omega} \frac{\partial}{\partial t}\right) \frac{\partial^2 x}{\partial X^2} = \rho \frac{\partial^2 x}{\partial t^2} \qquad (3.23)$$

Comparison with the equation that governs the motion of a system with a single degree of freedom subjected to ground vibration (Section 1.2) shows that every particle of the viscoelastic body behaves as a simple system with a percentage of critical damping equal to $Q/2$.[1] (This conclusion holds even for three-dimensional vibrations of isotropic viscoelastic solids.)

Substituting in Eq. 3.23 we verify that the following expression constitutes a solution

$$x(X, t) = d \exp(-\beta X) \sin \left[\omega \left(t - t_1 - \frac{X}{v}\right)\right] \qquad (3.24)$$

where d is a constant, β is called the *attenuation* and is usually a function of ω, t_1 is an arbitrary time shift, and v is the *phase velocity* of waves with circular

---

[1] The conclusion is rigorously true under steady-state vibrations. Under free vibrations it is also true when Q is proportional to ω as well as for certain other special conditions, but it gives a satisfactory approximation in all cases when $Q/2\mu \ll 1$.

frequency ω. Performing this substitution and solving separately for the sine and cosine terms we obtain expressions for v and β in terms of μ/ρ and Q. The solution represents a steady train of waves traveling in the positive X direction. By changing the sign of X we obtain a train of waves traveling in the opposite sense. The linear combination or Fourier analysis of solutions of the type of Eq. 3.24 can be made to satisfy any boundary condition $x(0, t)$ in a semiinfinite beam. By combining with solutions for waves traveling in the negative X direction we can also satisfy any initial conditions $x(X, 0)$, $\dot{x}(X, 0)$ and deal with finite beams.

One case having special interest concerns a semiinfinite beam subject to the initial and boundary conditions[2]

$$\ddot{x}(X, 0) = 0 \text{ for } X > 0 \qquad \ddot{x}(0, t) = u\,\delta(t)$$

where u is a constant having units of velocity and δ is the Dirac delta function. The corresponding Fourier spectrum is constant and equal to u. Accordingly,

$$\ddot{x}(X, t) = \frac{u}{\pi} \int_0^\infty \exp(-\beta X) \cos\left[\omega\left(t - \frac{X}{v}\right)\right] d\omega \qquad (3.25)$$

where again β and v are functions of ω. Through a change of variables it can be shown that

$$\ddot{x}(\alpha X, \alpha t) = \left(\frac{1}{\alpha}\right) \ddot{x}(X, t) \qquad (3.26)$$

where α is an arbitrary positive factor. Hence it is necessary only to compute the pulse shape at one station or at one instant to be in possession of the complete solution for the semiinfinite beam.

Over a considerable range of ω's a very satisfactory idealization of the behavior of several materials, including most metals, rocks, soils, and plastics, is the "constant-Q" hypothesis. From the substitution of Eq. 3.24 in 3.23 we find that in this case β is proportional to ω and v is independent of ω. The percentage of damping derived from free or forced vibrations is $Q/2$, also independent of ω.

For a constant-Q material Eq. 3.25 gives

$$\ddot{x}(X, t) = \frac{u}{\pi} \frac{(\beta/\omega)X}{(\beta/\omega)^2 X^2 + (t - X/v)^2} \qquad (3.27)$$

(which satisfies Eq. 3.26), where β/ω and v are material constants. The solution represents a pulse symmetric about $t = X/v$. If we substitute $t = 0$, we find $\ddot{x}$ different from zero at all finite values of X; this implies an infinitely rapid transmission of the pulse front.

Another assumption (Sezawa, 1927b) that has found favor and limited experimental confirmation (Iida, 1937) treats the material as a Sezawa or Voigt body. In this idealization, Q is proportional to ω. If we write

$$Q = \frac{\mu'}{\mu} \omega$$

---

[2] See Hunter (1960) for an ample treatment of the subject of wave transmission in viscoelastic media.

where μ' is a material constant having units of kg cm^{-2} sec, Eq. 3.23 becomes

$$\left(\mu + \mu'\frac{\partial}{\partial t}\right)\frac{\partial^2 x}{\partial X^2} = \rho\frac{\partial^2 x}{\partial t^2} \tag{3.28}$$

The same expression can be derived from inspection of the rheologic model in Fig. 3.8, where every infinitesimal element is assumed to behave as a Kelvin body—a rigid body joined to the closest elements by linear springs and dashpots in parallel. This approach shows that Eq. 3.28 is valid for the material in question under an arbitrary variation of x and is not limited to steady-state vibrations.

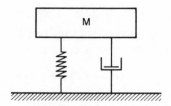

Figure 3.8. Kelvin body.

Dividing both members of Eq. 3.28 by ρ we arrive at

$$\frac{\partial^2 x}{\partial t^2} - \left(v_0^2 + \frac{\mu'}{\mu}\frac{\partial}{\partial t}\right)\frac{\partial^2 x}{\partial X^2} = 0 \tag{3.29}$$

where $v_0^2 = \mu/\rho$ as in the undamped case. The phase velocity in this medium tends to v_0 as ω tends to zero.

Equation 3.29 is separable. If we write

$$x = z(X)\theta(t)$$

and solve the resulting differential equation, we obtain

$$z_n = A_n \sin\frac{\omega_n}{v_0}(X - a_n)$$

as in Eq. 3.7, and

$$\theta_n = \exp\left(-\frac{\omega_n^2 \mu'}{2\mu}t\right)\sin\omega_n' t \tag{3.30}$$

where

$$\omega_n' = \omega_n\sqrt{1 - \left(\frac{\omega_n \mu'}{2\mu}\right)^2} \tag{3.31}$$

Thus, the shape of the natural modes is unaffected by damping.

This is in agreement with the conclusion derived in Chapter 2, to the effect that if linear damping is due to dashpots in parallel with the elastic spring elements, and dashpot constants are proportional to the spring stiffnesses, the shape of the natural modes does not depend on the amount of damping. The percentage of damping, according to Eq. 3.31, is

$$\zeta_n = \frac{\omega_n \mu'}{2\mu} \quad (3.32)$$

Also in agreement with the conclusion of Chapter 2 and with the comments made in connection with Eq. 3.23, the degree of damping is proportional to the frequency of vibration. When $\omega_n > 2\mu/\mu'$, the amount of damping exceeds the critical value, and the system does not oscillate in free vibration. This situation occurs for sufficiently high frequency in all systems having Sezawa damping.

Transmission of an instantaneous pulse in a Sezawa medium has been solved approximately (Kanai, 1950; Rosenblueth, 1951) by using Eq. 3.25 under the assumption that $v = v_0$, independently of ω. With the initial and boundary conditions specified again as an instantaneous velocity pulse at the end of a semiinfinite beam, this approximate solution is

$$\ddot{x} = \frac{u}{4t\sqrt{\pi(\mu'/2\rho)t}} \left\{ (v_0 t + X) \exp\left[-\frac{(v_0 t - X)^2}{2(\mu'/\rho)t}\right] - (v_0 t - X) \exp\left[-\frac{(v_0 t + X)^2}{2(\mu'/\rho)t}\right] \right\} \quad (3.33)$$

Numerical solutions are also possible, whether using this Fourier transform or a Laplace transform, and can lead to results as accurate as desired.

A more attractive approach for the study of wave transmission through rock and soil formations, making full use of digital computers, consists in discretizing the strata into a closely coupled group of masses joined to each other by spring and dashpot elements (Ang and Rainer, 1964; Ang, 1966; Costantino, 1967; Idriss and Seed, 1968; J. K. Minami and Sakurai, 1969; Whitman, 1969). This lends itself to the study of the phenomenon in soils having any variation of parameters with depth, and even nonlinear characteristics (see Chapter 5). Through a proper arrangement of dashpots, the system can be made to approach any dependence of Q on ω (see Appendix 1).

In most papers using lumped-parameter models to study multiple wave reflection in soil layers overlying bedrock, it is assumed that the motion of bedrock is the same as though the soil were not present. This is consistent with the assumption that bedrock is infinitely rigid. It omits consideration of "radiation damping," that is, of the energy lost through partial refraction of the waves in the soil back into bedrock. This objection can be overcome by replacing the bedrock halfspace with a massless dashpot having a constant equal to $\sqrt{\mu\rho}$ per unit area of interface (Rosenblueth and Elorduy, 1969b). The artifice is consistent with the assumption that bedrock behaves linearly, is homogeneous, and has no internal damping. Sezawa-type internal damping of bedrock may be incorporated by assigning the dashpot an appropriate mass per unit area of interface.

If the numerical model is terminated within bedrock the boundary condition is such that it permits passage of waves without reflection. Appropriate choice of terminal-lumped parameters to satisfy this condition has been presented by

Newmark (1968) based on an electrical analog that also fulfills this condition (Smith, 1958).

Essentially the same remarks apply to the use of finite elements in this type of analysis, for which useful computer programs have been developed (Khanna, 1969). We note in passing that the use of quadrilateral elements (Wilson, 1969) in these problems has advantages over the more common use of triangular elements. Quadrilateral elements provide stresses at their periphery while triangular elements do this at their centroids, so that the fact that stress continuity must be satisfied at interfaces and zero stress satisfied at the ground requires a much finer mesh with triangular elements than with quadrilateral elements.

Both lumped-parameter and finite-element models permit analyzing soil-rock systems with arbitrary geometry of interfaces. However, it is as yet practical to analyze only two-dimensional problems.

The dashpot representation of bedrock is only approximate in two-dimensional problems, and there is need to introduce two dashpots at each node or finite element, one for P and one for S waves. By introducing this cutoff sufficiently far from geometric irregularities, however, we may reduce the corresponding errors as much as we wish.

If μ'/μ is constant or, in general if, for a given frequency, Q is the same along a beam even if μ and ρ vary from one segment to another, reflection and refraction at interfaces take place without change in phase, and the principles stated in connection with the same phenomena in elastic media apply without modification. For example, the ground acceleration due to a unit velocity pulse, at the surface of a layer having $H = 100$ ft, $v_0 = 500$ ft/sec, and $\mu' = \mu/(2000$ sec^{-1}), which is the average deduced from laboratory tests on certain clays (Rosenblueth, 1951) and which rests on a rock formation such that at the interface the reflection constant α is $\frac{1}{2}$, has been computed in this manner and is shown by the dashed line in Fig. 3.9. The full line was obtained under the assumption that both the rock and the soil obey the constant-Q hypothesis, again taking $\sqrt{\mu/\rho} = 500$ ft/sec and $\alpha = \frac{1}{2}$, with $Q/2 = 0.04$. Although the figure does not show a pronounced difference in the shapes of the first pulses, Sezawa's assumption does give, in proportion, a more drastic attenuation of the high-frequency components, and this affects the earthquake spectra that may be calculated for the surface of soft ground (see Chapter 9).

The bulk of experimental evidence favors the adoption of the constant-Q hypothesis for most materials over the range of significant ω's (Knoppof, 1959). Only certain shales and a few other materials, such as paraffin, follow Sezawa's assumption to a satisfactory degree. However, the wave velocity for very low frequencies can be expected to be considerably lower than the velocity throughout most of the range of ω's. The assumption that v remains constant as ω goes to zero is responsible for the infinite velocity of transmission of instantaneous pulses in both the Sezawa and the constant-Q models. Hillier has found a phase velocity for polythene at $\omega = 0$ equal to less than one half those which correspond to the range of 1 to 20 kHz (quoted by Hunter, 1960), while

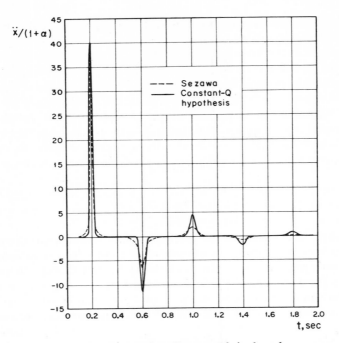

Figure 3.9. Acceleration diagram of single pulse in viscoelastic ground.

over this range v varies only about 10 percent. From the ratios of static to dynamic moduli for a certain volcanic clay (Rosenblueth, 1951), we conclude that at several hundred cycles per second, v is of the order of 1.5 to 2.0 times the velocity associated with $\omega = 0$, and at that relatively high frequency v is not very sensitive to ω. Now, the solution for the transmission of an instantaneous pulse is not too sensitive to the manner in which the attenuation approaches zero as the frequency tends to zero. Hence, according to H. Kolsky, good experimental confirmation is obtained under the assumption that β is still proportional to ω but that v varies as a constant plus a small term proportional to the logarithm of ω (Hunter, 1960). Better results are obtained by integrating Eq. 3.25 numerically after substitution of experimentally determined values of $\beta(\omega)$ and $v(\omega)$.

3.6 Flexural Beams

For sufficiently small frequencies of vibration, many structures (e.g., chimney stacks) and many structural members can be idealized adequately as beams whose deformations depend only on bending moments, neglecting the influence of shear, damping, and rotary inertia. Under these assumptions we can set up

80 LINEAR SYSTEMS WITH DISTRIBUTED MASS Chap. 3

the differential equation of motion using D'Alembert's principle. For the case of small displacements we obtain

$$m\frac{\partial^2 x}{\partial t^2} + \frac{\partial^2}{\partial X^2}\left(EI\frac{\partial^2 x}{\partial X^2}\right) = p \tag{3.34}$$

where

m = mass per unit length
x = displacement perpendicular to beam axis
t = time
E = modulus of elasticity
I = cross-sectional moment of inertia
X = coordinate along beam axis
p = external load per unit length

Equation 3.34 is known as the Bernoulli–Euler formulation of the problem. It has been solved for a number of cases of free and forced vibrations and transient disturbances (Jacobsen and Ayre, 1958). Some of these occur frequently in practice, and examples will be given here.

Consider the free vibrations of a uniform beam, that is, one in which m and EI are independent of X and for which $p = 0$. Equation 3.24 becomes

$$m\frac{\partial^2 x}{\partial t^2} + EI\frac{\partial^4 x}{\partial X^4} = 0$$

with constant coefficients. Proceeding as for the shear beam we find the shape of the nth natural mode given by

$$z_n = A_n \sinh[\lambda_n(X - a_n)] + B_n \sin[\lambda_n(X - b_n)] \tag{3.35}$$

where A_n, B_n, a_n, and b_n are constants having units of length. The last two constants and the ratio A_n/B_n depend on the boundary conditions. Further, the parameter $\lambda_n^4 = \omega_n^2 m/EI$ and the natural circular frequencies ω_n also depend on the end conditions. Equation 3.35 can also be written as

$$z_n = A'_n \sinh \lambda_n X + A''_n \cosh \lambda_n X + B'_n \sin \lambda_n X + B''_n \cos \lambda_n X \tag{3.36}$$

where A'_n, A''_n, B'_n, and B''_n are constants, any three of which depend on the boundary conditions, and the fourth one is arbitrary.

For example, in a cantilever beam, we must meet the conditions $x = 0$, $\partial x/\partial X = 0$ at the fixed end (say, at $X = 0$), and $\partial^2 x/\partial X^2 = 0$, $\partial^3 x/\partial X^3 = 0$ at the free end (say, at $X = L$). From Eq. 3.36 we find that these conditions are met if λ_n is a root of

$$\cos \lambda L \cosh \lambda L = -1 \tag{3.37}$$

and the coefficients of the hyperbolic and trigonometric functions must satisfy relations

$$A'_n = -B'_n \qquad A''_n = -B''_n$$
$$\frac{A''_n}{A'_n} = \frac{\cos \lambda_n L + \cosh \lambda_n L}{\sin \lambda_n L - \sinh \lambda_n L} \tag{3.38}$$

The first six roots of Eq. 3.37 are $\lambda_n L = 1.875, 4.694, 7.855, 10.996, 14.137$, and 17.279. For high n, $\lambda_n L \doteq (n - \tfrac{1}{2})\pi$.

To improve the accuracy of the differential equation we may introduce terms to account for shearing distortions and rotary inertia using ordinary strength of materials. This leads to what is known as a *Timoshenko beam*. Its natural periods are always longer than those of the purely flexural beam treated above. The importance of the contributions of the additional terms increases with the order of the natural mode of vibration and decreases with the beam slenderness. While the Bernoulli–Euler idealization gives infinite velocities for infinitesimal flexural wavelengths in beams, Timoshenko's theory gives finite velocities for all wavelengths.

The corresponding differential equation and exact solutions are available for certain cases, such as the uniform cantilever (Jacobsen and Ayre, 1958; Sutherland and Goodman, 1951). Results for cantilever beams are shown in Fig. 3.10. If only the fundamental period is of interest, the energy methods described in Chapter 4 are well suited for these systems. Usually the limit obtained by the method of Southwell and Dunkerley is quite accurate. (See Section 4.7.)

In most practical applications we face variable cross section and variable mass per unit length. Hence, for most applications the use of numerical methods on the discretized beam will be the appropriate procedure to obtain natural modes and responses.

Viscous damping can be introduced in the computations, and some problems

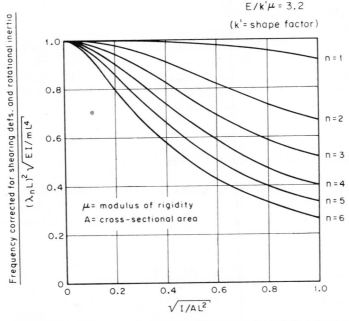

Figure 3.10. Exact solution for the uniform cantilever Timoshenko beam. *After Sutherland and Goodman (1951).*

in damped flexural-beam vibrations have been solved (Wieckowski, 1958), but again only the simplest cases are amenable to analytic treatment. Usually numerical methods will be more in order. Yet, when it is a matter only of computing natural modes and periods of vibration, good results will normally ensue from the assumptions that the shape of the natural modes is unaffected by damping and that the natural periods can be obtained by correcting the ones computed without damping, under the hypothesis that the degree of damping is independent of the frequency of vibration. The same assumptions allow us to compute the steady-state responses to harmonic oscillation and the response in transient states using modal analysis.

3.7 Frames

Problems involving vibration of framed structures without sidesway (i.e., with joints fixed against translation) are common in connection with machinery-excited oscillations but not in earthquake engineering. Usually the natural periods associated with such motions are too short to be excited appreciably by earthquakes. Natural modes of these systems and of continuous beams can be computed by establishing the solution to the differential equation of prismatic-beam vibrations with arbitrary linear end conditions and computing the conditions necessary to satisfy continuity requirements. Efficient methods for accomplishing this have been published (Veletsos and Newmark, 1957) and include the tabulated information required.

In most earthquake-engineering problems, sufficient accuracy is attained by lumping masses in question at floor elevations and neglecting their rotary inertia due to rotation about horizontal axes. (See Chapter 4 for further discussion of this matter.) The flexural members are treated as massless, and we may then use the method due to Goldberg, Bogdanoff, and Moh (1959) or any other appropriate procedure among those described in Chapter 4.

3.8 The Three-Dimensional Wave Equation[3]

Consider a linearly elastic, isotropic, homogeneous body with mass density ρ. Consider also a Cartesian system of coordinates X_1, X_2, and X_3. And let x_1, x_2, and x_3 denote the corresponding displacements, p_{ii} the normal stresses, and p_{ij} the shearing stresses. Under motion, at points not subjected to external forces, the body satisfies the equation

$$\rho \frac{\partial^2 x_i}{\partial t^2} = \frac{\partial p_{ii}}{\partial X_i} + \frac{\partial p_{ij}}{\partial X_j} + \frac{\partial p_{ki}}{\partial X_k} \tag{3.39}$$

which is derived immediately from D'Alembert's principle.

---

[3] This and the following six sections are based predominantly on Bullen (1953). See also Ewing, Jardetsky, and Press (1957).

Now let

$$\theta = \frac{\partial x_i}{\partial X_i} + \frac{\partial x_j}{\partial X_j} + \frac{\partial x_k}{\partial X_k} \qquad (3.40)$$

This is the *cubical dilatation*, that is, the volume change per unit volume of material, and is independent of the choice of coordinates.

We may describe linear elastic behavior of an isotropic medium by the expression

$$p_{ij} = \lambda \theta \delta_{ij} + 2\mu \frac{\partial x_i}{\partial X_j} \qquad (3.41)$$

where λ and μ (the latter often represented by G) are Lamé's constants, characteristic of the medium in question, and δ_{ij} is Kronecker's delta ($=-1$ if $i=j$; $=0$ if $i \neq j$).

Combining Eqs. 3.40 and 3.41 it is possible to arrive at the equation of motion in terms of displacements:

$$\rho \frac{\partial^2 x_i}{\partial t^2} = (\lambda + \mu) \frac{\partial \theta}{\partial X_i} + \mu \nabla^2 x_i \qquad (3.42)$$

where ∇^2 is Laplace's operator $\partial^2/\partial X_1^2 + \partial^2/X_2^2 + \partial^2/\partial X_3^2$. We may regard this expression as a generalized Eq. 3.1 with $p = 0$.

Equation 3.42 was derived under the assumptions of linear elastic behavior, small displacements, small velocities, and absence of external forces. Transformations of the equation to lift these restrictions are well known. Solutions of Eq. 3.42 for special cases give rise to the classical waves of seismology, among which are the types of waves we discuss subsequently.

Specific ideal substances result from assigning Lamé's constants special values. Thus, $\mu = \infty$ gives rise to a *rigid body*; if $0 < \mu < \infty$ the substance is a *perfect solid*; if $\mu = 0$ it is a perfect *fluid*.

Lamé's constants are often replaced with other parameters, related to them as follows:

and
$$\lambda = \frac{\nu E}{(1+\nu)(1-2\nu)} \qquad \mu = \frac{E}{2(1+\nu)}$$
$$E = \frac{\mu(3\lambda + 2\mu)}{\lambda + \mu} \qquad \nu = \frac{\lambda}{2(\lambda + \mu)} \qquad (3.43)$$

where E is Young's modulus, ν is Poisson's ratio, and $\mu = G$ is known as the modulus of rigidity.

3.9 Body Waves

Let us differentiate Eq. 3.41 with respect to X_i, make $i = 1, 2$, and 3, and add the results. In accord with Eq. 3.40 we obtain

$$\rho \frac{\partial^2 \theta}{\partial t^2} = (\lambda + 2\mu)\nabla^2 \theta \qquad (3.44)$$

Now we apply the operator **curl** to Eq. 3.42:

$$\rho \frac{\partial^2}{\partial t^2} \mathbf{curl}\, x_i = (\lambda + \mu)\, \mathbf{curl}\, \frac{\partial \theta}{\partial X_i} + \mu \nabla^2\, \mathbf{curl}\, x_i$$

But the components of **curl** $\partial \theta / \partial X_i$ are of the form

$$\frac{\partial}{\partial X_i} \frac{\partial \theta}{\partial X_j} - \frac{\partial}{\partial X_j} \frac{\partial \theta}{\partial X_i}$$

and are therefore nil. Hence,

$$\rho \frac{\partial^2}{\partial t^2} \mathbf{curl}\, x_i = \mu \nabla^2\, \mathbf{curl}\, x_i \qquad (3.45)$$

Comparing Eqs. 3.2 and 3.44 we conclude that a *dilatational* (or *irrotational*) disturbance θ is transmitted through the substance with velocity

$$v_p = \sqrt{\frac{\lambda + 2\mu}{\rho}} = \sqrt{\frac{(1-\nu)E}{(1-\nu-2\nu^2)\rho}} \qquad (3.46)$$

where the second form is derived from the first two parts of Eq. 3.43. Similarly, Eq. 3.45 states that a *rotational* (or *equivoluminal*) disturbance is transmitted with velocity

$$v_s = \sqrt{\frac{\mu}{\rho}} \qquad (3.47)$$

We notice that $v_p > v_s$. The two types of waves are called, respectively, the *primary* or *P* waves and the *secondary* or *S* waves.

Secondary waves may be plane polarized. In seismology, when an *S* wave is polarized so that all particle motion takes place in a horizontal direction, it is called an *SH* wave. When all the motion takes place in vertical planes that contain the direction of wave travel, the wave is called *SV*.

At points sufficiently distant from the source of a disturbance the waves may be regarded as plane. This is relevant to many problems in seismology, for the distance to the station is often great compared with the dimensions of the source. In this case the displacements associated with *P* and *S* waves are *longitudinal* and *transverse*, respectively, to the direction of propagation.

For plane waves we may choose X_i to coincide with the direction of propagation. Then $x_j = x_k = 0$, and Eq. 3.42 degenerates into Eq. 3.2 with the appropriate changes in notation. The same happens with Eq. 3.45 if we make $x_i = x_k = 0$.

We can construe all body waves in the interior of a homogeneous isotropic substance to be formed of a *P* and an *S* group of waves that traverse the substance independent of each other. The same is not always true of waves in the neighborhood of free surfaces or interfaces between substances of different characteristics, where Rayleigh, Love, and other types of waves describe the motion; the simplest types are considered in Section 3.10.

When the first *P* and *S* waves to arrive follow the same path, which is almost always approximately true, the time that elapses between their arrivals at a station equals $X/v_s - X/v_p$, where X is the focal distance of that station. Hence this time is used commonly by seismologists to locate the foci of earthquakes and other disturbances.

Much of the mathematics involved in the treatment of body waves simplifies when we adopt Poisson's relation $\lambda = \mu$. The relation implies that $\nu = \frac{1}{4}$ (see Eq. 3.43) and $v_p = v_s\sqrt{3}$ (see Eqs. 3.46 and 3.47). This assumption holds with good accuracy for most of the rocks in the earth's crust. But we should not apply it without discrimination. For example, certain clays with an average water content above 300 percent have given ratios of v_p/v_s of the order of 15 (Figueroa, 1964), which implies $\nu \simeq 0.498$.

Note that v_p and v_s are independent of the frequency of the waves. Hence once the P and S waves become nearly plane, they are no longer distorted or dispersed as they travel through an elastic, homogeneous medium. This is true under the assumption that the medium is not dissipative. In viscoelastic media the attenuation and the wave velocity, or at least one of these quantities, are functions of the frequency and hence of the wavelength. Hence even plane waves change in shape as they travel. There are strong indications that the constant-Q hypothesis (see Section 3.6), which gives rise to an attenuation proportional to the frequency of vibration, adequately reflects the behavior of most of the rock formations through which seismic waves travel, for both P and S waves (Gutenberg, 1958). Roughly, the attenuation $\beta(\omega)$ is found to be $300/T$ for both types of waves, where the factor 300 is in units of sec/km and T is the period of the waves in question.

The constant-Q hypothesis leads to wave velocities that are independent of ω. For rocks, as for plastics, a more realistic idealization preserves the proportionality between attenuation and frequency, but it allows for a mild increase in wave velocity with frequency. This dependence of velocity on ω gives rise to a group velocity that differs from the velocity of the waves proper, as will be discussed in Section 3.14.

3.10 Spherical Body Waves

Consider the case of spherically symmetrical waves—for example, those which emanate from a spherical cavity in an infinite medium subjected to an interior, time-dependent, uniformly distributed pressure. Symmetry ensures that only P waves will be produced. We have seen that the dilatation satisfies Eq. 3.44. In the present problem, introducing spherical coordinates and making use of spherical symmetry and of the first part of Eq. 3.46, the differential equation can be put in the form

$$\frac{\partial^2(R\theta)}{\partial t^2} = v_p^2 \frac{\partial^2(R\theta)}{\partial R^2} \tag{3.48}$$

where R denotes distance from the spherical center. This is of the same form as the one-dimensional wave equation (Eq. 3.2). Hence we can write its most gen-

eral solution as
$$\theta = R^{-1}[f(R - v_p t) + g(R + v_p t)] \tag{3.49}$$
where f and g are arbitrary functions twice differentiable and represent, respectively, an outgoing and an incoming wave.

The dilatation is equal to the sums of the normal strains in the radial and in both the circumferential directions. The former is equal to $\partial x/\partial R$, where x denotes radial displacement, and the circumferential strains are both equal to x/R, as can be established from simple considerations on geometry. Consequently,
$$\theta = \frac{\partial x}{\partial R} + 2\frac{x}{R} \tag{3.50}$$

Now for simplicity, write $f = F''$ and $g = G''$. Equations 3.49 and 3.50 are solved by
$$x = \frac{1}{R}(F' + G') - \frac{1}{R^2}(F + G) \tag{3.51}$$

From here we can obtain the radial particle velocity and acceleration by differentiating with respect to time:
$$\dot{x} = \frac{v_p}{R}(F'' - G'') - \frac{v_p}{R^2}(F' - G') \tag{3.52}$$
$$\ddot{x} = \frac{v_p^2}{R}(F''' + G''') - \frac{v_p^2}{R^2}(F'' + G'') \tag{3.53}$$

We see that at small distances from the center the particle displacements, velocities, and accelerations vary as R^{-2} while at long distances they decay as R^{-1}.

Jeffreys (quoted by Bullen, 1953, pp. 75-76) has solved the problem of a spherical cavity subjected to an internal, uniformly distributed pressure that varies as a step function of time. Let a denote the radius of the cavity. We suppose that there is zero displacement everywhere for $t < 0$ and that the cavity receives an internal pressure $pH(t)$, where p is a constant and H is Heaviside's unit step function.[4] At any point a distance $R > a$ from the center of the cavity the displacement remains nil until the instant $t_1 = (R - a)/v_p$. For $t - t_1$ Jeffreys has shown that, if Poisson's relation holds ($\nu = 0.25$),

$$x = \frac{pa^3}{4\mu R^2}$$
$$\times \left\{1 + \left[\left(\frac{R}{a} - \frac{1}{2}\right)\sqrt{2}\sin\left(\frac{2\sqrt{2}\,v_p t_2}{3a}\right) - \cos\left(\frac{2\sqrt{2}\,v_p t_2}{3a}\right)\right]\exp\left(-\frac{2v_p t_2}{3a}\right)\right\}$$

where $t_2 = t - t_1$. The solution is of the form of Eq. 3.51. When R/a is large and t_2 is not too great, this expression gives
$$x = \frac{\sqrt{2}\,pa^2}{4\mu R}\sin\left(\frac{2\sqrt{2}\,v_p t_2}{3a}\right)\exp\left(-\frac{2v_p t_2}{3a}\right) \tag{3.54}$$

---

[4] This function is defined as $H(t) = 0$ for $t < 0$, $H(t) = 1$ for $t > 0$, and $H(0) = \frac{1}{2}$. It may be regarded as the integral of Dirac's delta.

This is a damped sine wave, which is damped so rapidly that it appears nearly as a single swing.

Study of Eq. 3.42 or of Eq. 3.45 and comparison with Eq. 3.44 show that the displacements in a purely rotational disturbance are transmitted in accordance with a law similar to that for the dilatation in an irrotational wave. Hence, the corresponding displacements, velocities, and accelerations decay as R^{-1}.

Jeffreys (quoted by Bullen, 1953, pp. 76–77) has also obtained the solution for the wave that is emitted by the spherical cavity considered above, when its surface is subjected to a shearing stress that varies as $pH(t)$ times the cosine of the angle between the radius to each point on the spherical surface and an arbitrary diametral plane. At large distances from the center and again provided that the time elapsed since the first arrival of the S wave is not too great, the solution is similar to Eq. 3.54: a rapidly damped sine wave that approximates a single swing.

The period of the P wave according to Eq. 3.54 is $3.4a/v_p$. That of the S wave just considered is $3.6a/v_s$. The solutions are displayed graphically in Fig. 3.11. Their consideration is important for the understanding of the characteristics of earthquakes.

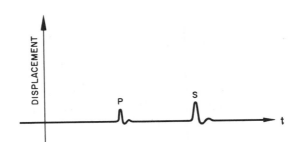

Figure 3.11. Single swings at large distances from the source. *After Bullen (1953)*.

3.11 Cylindrical Body Waves

Under conditions of cylindrical symmetry and with the use of cylindrical coordinates, Eq. 3.44 becomes

$$\frac{\partial^2 \theta}{\partial t^2} = v_p \left(\frac{\partial^2 \theta}{\partial R^2} + \frac{1}{R} \frac{\partial \theta}{\partial R} \right)$$

This equation does not admit solutions of the type $f(R \pm v_p t)$. It is solved, on the other hand (Bullen, 1953, pp. 68–69), by

$$\theta = [A_1 J_0(\kappa R) + A_2 Y_0(\kappa R)] \sin v_p(t - t_1) \tag{3.55}$$

where J_0 and Y_0 are Bessel functions of the first and second kinds, respectively, and A_1, A_2, κ, and t_1 are arbitrary constants. By properly selecting these constants and linearly combining solutions of the type of Eq. 3.55, any initial and boundary conditions may be satisfied.

At short distances from the source, θ varies approximately as $\ln R$. At large distances from the center the following asymptotic approximations are useful

$$J_0(\kappa R) = \sqrt{\frac{2}{\pi \kappa R}} \cos\left(\kappa R - \frac{\pi}{4}\right) + 0[(\kappa R)^{-3/2}]$$

$$Y_0(\kappa R) = \sqrt{\frac{2}{\pi \kappa R}} \sin\left(\kappa R - \frac{\pi}{4}\right) + 0[(\kappa R)^{-3/2}]$$

It follows that at large distances from the original disturbance the dilatation decays essentially as $R^{-1/2}$. The same applies directly to the displacements in rotational waves. The latter may include displacements parallel to the axis of the cylinder, as well as in the circumferential direction while satisfying the condition of symmetry.

Notice that the foregoing considerations have not taken into account wave attenuation due to internal damping. Consideration of cylindrical waves is important in connection with earthquake ground-surface motions at large focal distances.

3.12 Wave Reflection and Refraction

Laws which govern the reflection and refraction of body waves at interfaces can be deduced from considerations of continuity and equilibrium. The resulting expressions of the wave amplitudes are lengthy except for the simplest cases.

Relations between the reflected and refracted wave directions and the direction of the incident wave are particularly simple and of special interest in the case of elastic waves that strike a plane or nearly plane interface. Indeed, according to Snell's law (which may be derived in a number of ways), the sine of the angle that the direction of propagation of any wave (whether incident, reflected, or refracted) makes with the normal to the interface between two elastic media is proportional to that wave's velocity of propagation. For example, with reference to Fig. 3.12, if ϕ_{p1} denotes the angle that an incident P wave forms with the normal to the interface, ϕ_{p2} and ϕ_{p3} the corresponding angles for the refracted and reflected P waves, and ϕ_{s2} and ϕ_{s3} for the refracted and reflected SV waves, respectively (with motion occurring parallel to the plane of the drawing), these angles are related by the expression

$$\phi_{p1} = \phi_{p3}$$

$$\frac{\sin \phi_{p1}}{v_{p1}} = \frac{\sin \phi_{p2}}{v_{p2}} = \frac{\sin \phi_{s2}}{v_{s2}} = \frac{\sin \phi_{s3}}{v_{s3}}$$

where $v_{p1}, v_{p2}, \ldots$ denote the corresponding wave velocities.

For a more specific example, assume that $\phi_{p1} = 30°$, that $v_{p1}/v_{p2} = 2$, and that Poisson's relation applies to both media. In this case $\phi_{p2} = 14.5°$, $\phi_{s2} =$

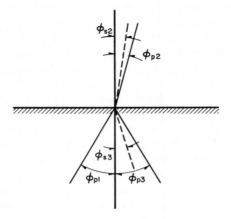

Figure 3.12. Refracted and reflected P and S waves.

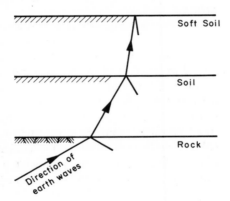

Figure 3.13. Wave refraction in horizontally-stratified media.

8.3°, and $\phi_{s3} = 16.8°$. This is the case depicted in Fig. 3.12. Reflections and refracted transmissions through several layers of different properties are shown in Fig. 3.13.

When the incident wave is of the *SH* type (with motion occurring perpendicularly to the plane of the drawing), the reflected and refracted waves are also *SH*. For certain combinations of moduli of rigidity and densities in the two media there exists one (and only one) value of the angle of incidence for which there is no reflected wave. If the moduli of rigidity and the densities do not satisfy that specific relation, there is always a reflected wave. However, if the shear-wave velocity in the second medium exceeds that in the medium from which the wave proceeds, there is a range of angles of incidence for which no refracted wave emerges. Instead, arrival of the *SH* wave gives rise, in the second medium, to an interface movement whose amplitude decreases exponentially from the boundary and which travels with the velocity of the incident wave projected on the interface. In this case there is total reflection of the incident wave, usually with a change of phase.

An incident P wave may give rise to both P and SV, but not to SH, reflected and refracted waves. For normal and grazing incidences, no SV waves are produced, but waves of this type are reflected in all other cases. When Poisson's relation holds, there are two values of the angle of incidence for which there is no reflected P wave, and for all intermediate values of the incidence angle the reflected P wave is very small. If Poisson's relation does not hold exactly, there is always a reflected P wave, but over a wide range of angles the energy associated with it is less than half of that in the reflected SV wave.

For an incident SV wave, reflections and refractions can be SV or P but not SH. The remarks made about the absence or small magnitude of the reflected P wave relative to the SV wave when the incident disturbance was of the P type apply in reciprocal manner to the case when it is of the SV type. And, as in the case of an incident SH wave, there are conditions of total reflection with a usual change of phase and an exponentially decaying disturbance in the second medium near the interface.

Usually an incident P or S wave induces both P and S reflected and refracted waves, save for large angles between the incident wave and the normal to the interface. However, a wave that travels normal to the interface gives rise only to waves of the same kind as itself. And in all cases the most important component of the refracted motion is of the same kind as the incident wave. These considerations, coupled with the facts that wave velocity is usually an increasing function of depth and that thick, soft deposits are usually bounded by nearly horizontal surfaces [(the free surface and the top of bedrock) for they become eroded otherwise], make it plain that most of the consequences of multiple reflection phenomena can be studied adequately under the assumption that we deal with vertically traveling S waves, which was adopted in Section 3.3 (see Fig. 3.12).

The remarks that have been presented in connection with wave reflection and refraction, which coincide with the principles of geometrical optics (with the proper generalizations because in seismic waves we deal with disturbances of the P type in addition with the S waves, which are the only ones that occur in optics), do not apply when the interface has a curvature that is large compared with the radius of the incident wave front. As in optics, there is wave diffraction and, as in optics, the elementary ray theory does not apply. For the same reason the ray treatment is not valid near the focus of an earthquake.

3.13 *Surface Waves*

Consider a homogeneous, isotropic, elastic halfspace limited by the plane $X_1 X_2$, above which we assume there is vacuum. We seek a steady-state solution of Eq. 3.42 that is largely confined to the surface, and such that at every instant the displacements of all particles lying on an axis parallel to OX_2 are equal. That is, we wish to find a solution of the form $x_i = f_i(X_3) \sin \omega(X_1/v_r - t)$ where $f_{1,2,3}$ are rapidly decreasing functions of X_3 (taken positive into the half-space), ω is the circular frequency of the waves in question, and v_r is their velocity.

Rayleigh succeeded in finding such a solution. In it v_r satisfies the condition

$$\left(2 - \frac{v_r^2}{v_s^2}\right)^2 = 4\left(1 - \frac{v_r^2}{v_p^2}\right)^{1/2}\left(1 - \frac{v_r^2}{v_s^2}\right)^{1/2}$$

and $x_2 = 0$. If Poisson's relation holds, the above relationship gives $v_r = 0.92\, v_s$ and

$$x_1 = a\,[\exp(0.85\kappa X_3) - 0.58 \exp(0.39\kappa X_3)] \sin(X_1 - v_r t)$$
$$x_3 = a[0.85 \exp(0.85\kappa X_3) - 1.47 \exp(0.39\kappa X_3)] \cos(X_1 - v_r t)$$

where $\kappa = \omega/v_r$, and a is a constant. (The relation between v_r and v_s for values of ν other than 0.25 can be found in Duke, 1969.)

Putting X_3 equal to zero we see that during passage of the disturbance, a surface particle describes the ellipse

$$x_1 = 0.42a \sin \beta \qquad x_3 = -0.62a \cos \beta \qquad (3.56)$$

where β is a parameter that decreases as time increases. The ellipse is therefore described in a retrograde fashion (Fig. 3.14). This type of motion is known as *Rayleigh waves*.

The conclusion $x_2 = 0$ contradicts observations. Earthquake surface waves normally exhibit a horizontal shear component traveling horizontally. Consequently there must exist another type of surface wave, not covered by Rayleigh's solution. Love showed that surface waves of the type sought exist when the ground consists of an upper stratum underlayed by a semiinfinite medium in which the velocity of shear waves is greater than it is in the stratum. The

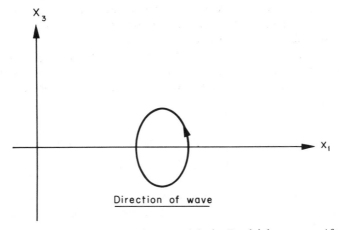

Figure 3.14. Path of a surface particle in Rayleigh waves. *After Bullen (1953).*

velocity of Love waves lies between these two shear velocities and is a function of the frequency. Using primes to denote the stratum, the velocity of Love waves v_l can be found from

$$\mu\left(1 - \frac{v_l^2}{v_s^2}\right)^{1/2} - \mu'\left(\frac{v_l^2}{v_s'^2} - 1\right)^{1/2} \tan \kappa H \left(\frac{v_l^2}{v_s'^2} - 1\right)^{1/2} = 0 \qquad (3.57)$$

where $\kappa = \omega/v_l$. We see that as $\kappa \to 0$, or when we deal with long waves, $v_l \to v_s$, and as $\kappa \to \infty$, for short waves, $v_l \to v_s'$.

Solutions are available for Rayleigh, Love, and other types of waves under a variety of stratification conditions. In many such solutions the velocity of wave propagation is a function of the wave frequency. When this happens, unless we are dealing with sinusoidal, steady-state conditions, we find that the shape of a disturbance changes as it travels along the medium in question. Sharp disturbances become trains of waves, each train containing oscillations of essentially equal frequency. Further, the velocity of a group of waves under these conditions differs from the velocity of an individual wave. This type of dispersion does not necessarily combine in additive manner with the dispersion due to internal damping[5] and accounts partly for the increase in duration of earthquake motions with focal distance.

3.14 Group Velocity

We have seen that in viscoelastic materials, wave velocities are functions of the frequency of the waves. Even in a perfectly elastic solid, Love waves, among others, travel with a velocity that depends on the frequency and hence, ordinarily, on wavelength. The phenomenon, known as dispersion, gives rise to reinforcement and interference of waves having nearly the same velocities. This causes the appearance of clusters of waves of essentially equal wavelengths. The location of these clusters in space moves with a velocity, called *group velocity*, that differs from the velocities of the waves.

Some idea of the effect of dispersion in this context may be gleaned from the study of the combination of two one-dimensional waves of the same amplitude but slightly different frequencies and velocities. Let us consider, then, the combined wave

$$x = a \sin \kappa(X - vt) + \sin (\kappa + \Delta\kappa)[X - (v + \Delta v)t]$$

where $\kappa v = \omega$, the circular frequency. This we can write in the form

$$x = 2a \sin \left(\frac{2\kappa + \Delta\kappa}{2} X - \frac{2\omega + \Delta\omega}{2} t\right) \cos \left(\frac{\Delta\kappa}{2} X - \frac{\Delta\omega}{2} t\right)$$

The sine function in this expression represents a wave with a frequency and a length equal to the averages of the original waves. The cosine function is a

---

[5] This remark is proved for waves traveling along a linearly damped cylindrical rod (Hunter, 1960).

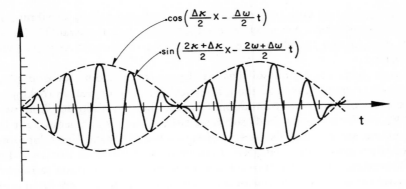

Figure 3.15. Combined wave.

very long wave that envelops the motions corresponding to the first factor, as shown by the dashed line in Fig. 3.15. This envelope moves in the direction of the waves with a velocity equal to $(\Delta\omega/2)/(\Delta\kappa/2)$ or $\Delta\omega/\Delta\kappa$. In the limit, when there is a continuous spectrum of wave frequencies, we may write for the group velocity

$$v_g = \frac{d\omega}{d\kappa} = v + \kappa \frac{dv}{d\kappa}$$

or, introducing the symbol $\Lambda = 2\pi/\kappa$ for wavelength,

$$v_g = v - \Lambda \frac{dv}{d\Lambda}$$

Only when v does not depend on the wavelength does the group velocity coincide with the wave velocity, and no clusters develop.

It can be shown that, when dispersed waves undergo reflection and refraction at an interface, the angles that the corresponding paths form with the interface are functions of the individual wave velocities as for nondispersed waves, while the velocities of transmission of energy follow the law of the group velocities.[6]

3.15 Soil-Foundation Interaction

The same contact stresses between soil and foundation that may be held responsible for earthquake effects on structures also cause deformations in the soil, especially in the vicinity of every structural foundation. The phenomenon constitutes one form of dynamic soil-structure interaction. It is also known in the literature as "energy feedback to the ground," "foundation yielding," and "foundation compliance." It has received considerable attention with a

---

[6] A more thorough explanation of the matter of group velocity, based on a Fourier integral representation of dispersed waves, is found in Bullen (1953), pp. 58–66, 93–95, and 107–108.

view to application both to seismic problems and to the study of vibrations of machine foundations. Yet no entirely satisfactory solution is available for cases other than circular foundations, even under the assumption of perfectly elastic soil behavior.

A rigid body resting on soil has six degrees of freedom: for example, an up-and-down motion, torsion about a vertical axis, two degrees in rocking, and two degrees of horizontal translation. Suppose that the responses in all modes were known for a massless body subjected either to an instantaneous pulse or to a harmonic, steady-state disturbance along each component. Then appropriate use of either convolution integrals, Laplace (Sandi, 1960), or Fourier (Monge and Rosenberg, 1964) transforms would permit calculation of the responses of any structure of linear behavior resting on a rigid foundation supported in turn by a soil of linear behavior.

Most of the solutions available concern a rigid plate, either circular or rectangular, resting on an isotropic, homogeneous, linearly elastic halfspace, under steady-state vibration and have been obtained assuming that the distribution of contact stresses is the same as under static loading, independently of the frequency of vibration. Actually the distribution of contact stresses depends on the frequency. Lysmer (1965) has succeeded in solving the problem of a rigid plate under steady-state vertical oscillation taking into account the proper distribution of contact stresses. To this end he has taken the solution for a flexible plate that applies a vibratory uniform pressure on the ground (Sung, 1953). By subtracting the effects of a smaller concentric plate he has obtained the responses to a ring that applies uniformly distributed vibratory pressures; by replacing the rigid plate with a set of 20 concentric rings and equating their vertical displacements at every instant he has obtained a numerical solution.

Using a somewhat similar approach, Elorduy (1967) has developed a method applicable to the vibrations of a rigid plate of arbitrary shape resting on an elastic halfspace. He makes use of the known solution for the free-field effects of a vertical (Pekeris, 1955) or a horizontal (Chao, 1960) concentrated pulse applied at a point of the free surface of the elastic halfspace. He then solves two sets of simultaneous equations to satisfy the boundary condition at the base of the plate. Elorduy's application to rectangular plates is beset with the simplifying assumption that the phase lag between force and displacement is the same at all points of contact between the plate and the halfspace. Nevertheless, his solution for the oscillations of a square plate agrees well with the solution due to Kobori (1962), which was obtained by a different procedure.

Elorduy's approach, after removing the simplifying assumption and incorporating an explicit consideration of coupling between vertical and horizontal displacements, can give results as accurate as desired for plates of arbitrary shape. However, as in Lysmer's treatment, the method gives rise to sets of very ill-conditioned equations in some range of the variables. This difficulty was obviated by Robertson (1966) through a transformation of the integral equation from which these sets of equations are derived. He was thus able to arrive at the exact solution for the vertical oscillations of a rigid circular plate on an elastic halfspace. His method can be adapted to the analysis of the rocking, torsional, and translational oscillations of rigid circular plates and to the vibrations of infinitely long rigid band plates. However, it is not applicable in any form to finite square or rectangular plates.

Tajimi (1969) has been able to solve the problem of rocking and translational oscillations of a rigid, circular, cylindrical pier embedded in an elastic stratum when both the stratum and the pier rest on an elastic halfspace.

A comparison of the exact solution for a rigid circular plate on an elastic halfspace with the solution based on the same distribution as under static conditions shows that the latter is satisfactory up to and somewhat beyond the resonant frequency, but not much beyond. For very high frequencies the solution obtained by assuming a static pressure distribution even predicts an equivalent negative damping, which makes it unacceptable. In the study of the vibration of machine foundations, such high frequencies are often of interest; in problems of earthquake-resistant design this is not necessarily the case. Since many problems have been solved only under the simplifying assumption in question, we shall retain it in the presentation of some solutions.

Our lack of concern with very high frequencies stems from the following consideration. It is well known that soil-foundation interaction may affect the fundamental mode and period of vibration appreciably but that its effects are small on the second mode and period and negligible on the higher harmonics. As an illustration consider a flexural two-mass system. Let the flexibilities be concentrated at the base and at the first mass, the masses be equal to each other, the flexibilities also be equal to each other, and the masses be equally spaced. If we introduce a spring at the foundation to simulate rocking, with the same flexibility as the spring elements at the joints, the fundamental period will increase 36 percent while the second natural period increases 8 percent. Indeed, it follows from the orthogonality of natural modes that if the fundamental mode of vibration of a building is a straight line, there can be no base overturning moment in any of the higher modes (Bielak, 1969) and hence these are not affected by the possibility of interaction with rocking motion of the base. Since the fundamental mode is almost always approximately straight, interaction can rarely have an important effect on the higher modes and periods. [This conclusion is apparently contradicted in papers by Parmelee (1967 and 1969), but the corresponding solutions fail to take into account vibration in other natural modes when analyzing the response in any given mode.]

Now, the fundamental period of the soil-structure system is not smaller than that of an infinitely rigid structure resting on the same soil and having the same masses and geometry as the structure in question. Because the second natural period in buildings is of the order of one half to one third of the fundamental (except when soil-foundation interaction is such as to make the fundamental mode much more significant than the harmonics), we are not interested in an accurate evaluation of the phenomenon of foundation compliance much beyond a frequency equal to about twice the first resonant frequency associated with a rigid block resting on soil, and usually not much beyond the first resonant frequency.

In principle, once the solutions were available for instantaneous pulses or steady-state disturbances, integral transforms would solve every problem of interest. The approach would be impractical, however, and would preclude analyzing nonlinear structures. A more attractive even if only approximate treatment replaces the soil with a virtual mass fixed to the foundation, a massless spring, and a massless dashpot in parallel with the spring. The three parameters must be defined for every degree of freedom and may be so placed as to include correctly coupling between the various degrees. In this manner we have no difficulty in applying the standard methods of analysis for multidegree systems to a new system, whose degrees of freedom include those of the structure proper plus six of the foundation, and we may even deal with nonlinear structural behavior.

A rigorous treatment of this sort would require having two of the parameters in every degree of freedom vary with the frequency of vibration because we would have to adjust for two quantities at each frequency: the amplitude of response and its phase shift with respect to a harmonic excitation. If, as proposed, we take the parameters as independent of frequency, we must fulfill certain conditions. In a simple system, as we saw in Chapter 1, the response at low frequency is essentially sensitive to the spring constant. Hence, if our model is to cover a range of low frequencies, the spring stiffnesses must coincide with the values derived from static loading. (In a real soil this is to be interpreted as a rapid, quasistatic loading in which consolidation and creep are not given the opportunity to occur to an appreciable extent.) In the ranges of the resonant frequencies the dynamic magnifications of responses are sensitive only to the percentages of damping; these ranges will fix the dashpot constants. For high frequencies, only the masses are significant. Lysmer points out that, as the frequency of excitation tends to infinity, the wavelengths of the disturbances emanating from the foundation tend to zero; hence the virtual masses must also tend to zero, and if we wish our solution to hold for all possible frequencies, we must take the virtual mass in every natural mode as zero.

Reasoning along these lines and adjusting to the exact solution we mentioned earlier for the vertical oscillations of a circular plate, so as to minimize the error in the amplitude of the responses to a harmonic force applied at the center of the plate, Lysmer proposes the following parameters for this degree of freedom

$$K = \frac{4}{1 - \nu} \mu r \tag{3.58}$$

$$C = \frac{0.85 K r}{v_s} \tag{3.59}$$

where K is the spring constant, ν and μ are Poisson's ratio and modulus of rigidity, r is the radius of the plate, C the dashpot constant, and v_s the velocity of shear waves in the soil ($\sqrt{\mu/\rho}$). The spring constant in Eq. 3.58 is that for static loading. The dashpot constant in Eq. 3.59 is chosen such that in the entire range of possible Poisson ratios, $0 \leq \nu \leq 0.5$ and forcing frequencies $0 \leq \omega \leq \infty$, the computed amplitude of the response does not differ from the exact solution by more than about 30 percent; in the range of greatest interest, it differs by less than 20 percent. The phase change between the force and the response is automatically approximated also in a rough manner.

The model described is the simplest that replaces the soil with a small number of elements having parameters independent of the frequency and yet gives the correct order of magnitude of the responses. But the condition that the model be acceptable for very high frequencies causes a loss of accuracy in the lower frequency range, and this loss is unnecessary in the analysis of responses to earthquakes. By introducing a virtual mass of soil we have one additional parameter that permits a better adjustment over a limited range of frequencies. When we do this, the computed responses will be smaller than in the absence of the virtual mass if we retain the dashpot constant as given by Eq. 3.59. Hence we must compensate by adopting a smaller dashpot constant. The following constants (Nieto, Rosenblueth, and Rascón, 1965) give response amplitudes that check with the "exact" solution [which assumes the same contact stress distribution as under static loading (Sezawa, 1927a; Reissner, 1936; Arnold, Bycroft, and Warburton, 1955; Richart, 1962)] within a few percent at least up to forcing frequencies equal to twice that of resonance: K as in Eq. 3.58, the virtual mass equal to that of a cylindrical body of soil having the same base as the plate and a height h equal to 0.27 times the square root of the base area A (Fig. 3.16), and a dashpot constant

$$C = \frac{0.64 Kr}{v_s}$$

The latter can be put in the more convenient form

$$C = \frac{1.35 Kh}{v_s} \tag{3.60}$$

A comparison with the "exact" solution is shown in Fig. 3.17.

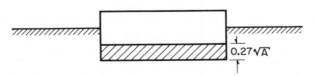

Figure 3.16. Virtual mass in vertical oscillations of circular plate.

Using a similar type of adjustment together with available information on spring constants and solutions for circular and rectangular rigid plates, Table 3.1 has been constructed (Nieto, Rosenblueth, and Rascón, 1965; Barkan, 1962). It is a partial list of stiffnesses, virtual masses, and dashpot constants for various degrees of freedom of plates of these shapes.

The positions of the springs and dashpots are important to reflect the proper coupling between various degrees of freedom. Owing to symmetry, in circular and rectangular plates with uniformly distributed mass, there is coupling only between the rocking and transverse-displacement degrees. In plates of other

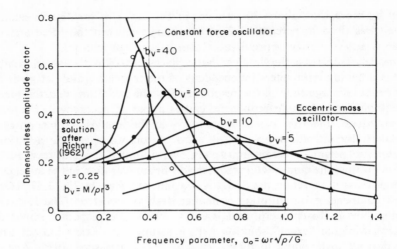

Figure 3.17. Comparison of responses of circular plates to vertical excitation.

TABLE 3.1. STIFFNESSES, VIRTUAL MASSES, AND DASHPOT CONSTANTS

Degree of freedom	Height of soil prism	Dashpot constant	Stiffness	
			Circular base	Rectangular base*
Vertical	$0.27\sqrt{A}$	$5.42\sqrt{K\rho h^3}$	$4\mu r/(1-\nu)$	$E\sqrt{A}\,c_s/(1-\nu^2)$
Horizontal	$0.05\sqrt{A}$	$41.1\sqrt{K\rho h^3}$	$5.8\pi\mu r(1-\nu^2)/(2-\nu)^2$	$E\sqrt{A}\,k_T/(1-\nu^2)$
Rocking†	$0.35\sqrt{A}$	$0.97\sqrt{K\rho h^5}$	$2.7\mu r^3 (\nu = 0)$	$EIk_\phi/\sqrt{A}(1-\nu^2)$
Torsion	$0.25\sqrt{A}$	$3.76\sqrt{K\rho h^5}$	$16\mu r^3/3$	$1.5EJk_T/\sqrt{A}(1-\nu^2)$

Aspect ratio	c_s	k_T					k_ϕ‡
		$\nu = 0.1$	0.2	0.3	0.4	0.5	
1	1.06	1.00	0.938	0.868	0.792	0.704	1.984
1.5	1.07	1.01	0.942	0.864	0.770	0.692	2.254
2.0	1.09	1.02	0.945	0.870	0.784	0.686	2.510
3.0	1.13	1.05	0.975	0.906	0.806	0.700	2.955
5.0	1.22	1.15	1.050	0.950	0.850	0.732	3.700
10.0	1.41	1.25	1.160	1.040	0.940	0.940	4.981

*Coefficients c_s, k_T, and k_ϕ tabulated in subsequent columns.
†Take moments of inertia with respect to axis at soil-foundation interface.
‡Rocking parallel to long side.

shapes or with other mass distributions, there may be coupling with other degrees of freedom or among all six of them. The same situation sometimes stems from asymmetric distribution of stiffnesses in the superstructure.

SOIL-FOUNDATION INTERACTION

Comparisons (Nieto, Rosenblueth, and Rascón, 1965) are shown in Figs. 3.18–3.20 between the response amplitudes obtained from the models described in Table 3.1 and the "exact" solutions for steady-state harmonic excitation (Sung, 1953; Richart, 1962). We notice that the agreement for horizontal vibrations is comparable to that for vertical oscillations in Fig. 3.17. Agreement is adequate for torsional and rocking motion throughout most of the range of excitation frequencies covered in the figures, except in the neighborhood of the

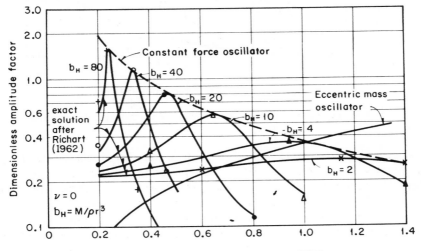

Figure 3.18. Comparison of responses of circular plates to horizontal excitation.

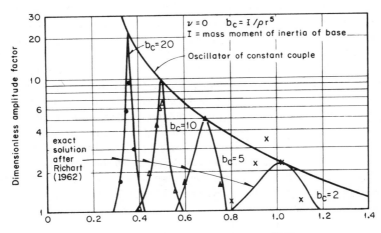

Figure 3.19. Comparison of responses of circular plates to rocking excitation.

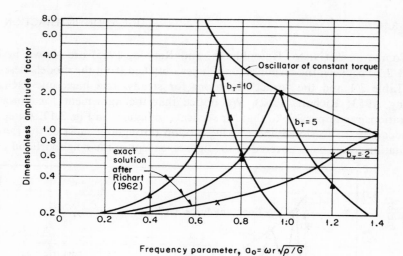

Figure 3.20. Comparison of responses of circular plates to torsional excitation.

resonant values when these are very small or very large. The discrepancy is important in these short intervals and should not be disregarded in the analysis of machine foundations or in the calculation of responses to earthquakes having well-defined, prevailing frequencies when these frequencies lie close to the rocking or torsional natural frequencies of the machine foundations. For most purposes in earthquake-resistant design, however, these discrepancies may well be overlooked because they affect only the contributions of short intervals in the entire range of significant frequencies of the motion.

Matters would improve if we varied one or two parameters in the models as a function of frequency. No doubt this should be done in the cases of narrow-band excitation that we quoted in the foregoing paragraph. Apparently, we could always proceed in this manner when using modal analysis. By trial and error or iteration we could find the values of parameters giving the best adjustment in the neighborhood of the natural frequencies of the soil-structure system and recompute these frequencies in terms of those parameters. But modal analysis does not apply strictly when we include soil-structure interaction because the combined system lacks classical natural modes. Hence, if we resort to modal analysis at all, great refinements are unwarranted. And if we wish to attain great accuracy there will be little advantage in adopting the simplified models proposed in this article, and we shall do well to return to the "exact" solutions. These allow us to compute the *transfer functions* of the system (its responses to instantaneous pulses), from which we can find the effects of various types of earthquakes on systems of linear behavior, as will be done in Chapters 9 and 10.

Ordinarily, analysis of pronouncedly nonlinear systems with soil-structure interaction will be formulated validly in terms of the models that Table 3.1 proposes, since nonlinearity will ensure that a vast range of frequencies will enter into play.

For other shapes of foundation the constants K for vertical oscillations are obtained readily by making reasonable assumptions about the contact pressure distribution, using charts (Newmark, 1947) to find the settlement of various points as though the foundation were flexible and to compute the foundation's average contact pressure and average settlement. Ordinarily the ratio of the two will give a satisfactory approximation to K. For example, under a circular plate subjected to a central vertical load the obviously wrong assumption that the contact pressure is uniform gives an error of only 5 percent (Timoshenko and Goodier, 1951). The spring constants that correspond to rocking oscillations can be obtained in similar fashion, while those for torsional and horizontal motions require integration of Cerrutti's equation for displacements at the ground surface. Once K has been obtained, the data in Table 3.1 can be used as a guide to estimate the dashpot constant and the virtual mass of soil. Studies are needed to allow reasonable estimates to be made of these parameters for deep, compensated foundations and for foundations on piles.

Numerical solutions have been obtained using high-speed computers for specific two-dimensional cases using lumped-parameter models and finite elements (Parmelee, 1969; Wilson, 1969). Some solutions correspond to surface foundations on a halfspace; others correspond to a foundation on a soil layer that in turn rests on a bedrock halfspace (Whitman, 1969), to partially compensated foundations (J. K. Minami and Sakurai, 1969), to a circular pier in a layered halfspace (Tajimi, 1969), and to foundations on point bearing piles (Penzien, Scheffey, and Parmelee, 1964; Kobori, Minai, and Inoue, 1969). Essentially the same remarks apply as the ones made on the problem of multiple wave reflection (Section 3.5) concerning "radiation damping" and the correct specification of boundary conditions where the soil or rock is assumed to terminate.

PROBLEMS[7]

3.1*. Compute the fundamental period of a cylindrical chimney stack of steel with circular cross section 6 ft in diameter, whose height is 90 ft, and whose thickness is $\frac{1}{2}$ in. (Fig. 3.21). Neglect shear deformations, rotary inertia, damping, gravity effects, and soil-foundation interaction.

Ans. 0.406 sec.

3.2. The unit weight and modulus of elasticity of a soil formation are 2.0 ton/m³ and 2×10^5 ton/m². Compute the velocities of dilatational, rotational, and Rayleigh waves in this material. Assume that Poisson's relation applies.

Ans. $v_p = 1085$ m/sec, $v_s = 626$ m/sec, $v_r = 576$ m/sec.

3.3*. A 30-m layer of the material specified in Problem 3.2 rests on what may be idealized as a semiinfinite rock formation having a unit weight of 2.8 ton/m³, a modulus of elasticity of 3×10^6 ton/m², and a Poisson's ratio of 0.25. Compute the

---

[7] Solution of problems marked with an asterisk is lengthy.

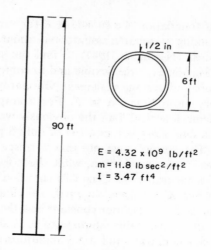

Figure 3.21. Problem 3.1.

velocity of Love waves in the upper mantle. Also, the first three natural periods of horizontal motion in the upper mantle assuming (a) no damping, (b) $\zeta = 0.06$ independent of frequency, and (c) Sezawa damping with $\zeta = 0.06$ for the fundamental mode. Assume that the shear modulus does not depend on the frequency of vibrations.

Ans. $v = 1380$ m/sec.

Period	a (sec)	b (sec)	c (sec)
T_1	0.1916	0.1919	0.1919
T_2	0.0638	0.0639	0.0649
T_3	0.0383	0.0383	0.0415

3.4. For the conditions in (b) and (c) of Problem 3.3 plot the ground-surface accelerations that correspond to the first arrival of a shear wave that reached the interface as the motion $\ddot{x} = a\delta(t)$.

Ans. See Fig. 3.22.

3.5*. Using the data in Table 3.1, find the maximum amplitudes of vibration of a rigid cube resting on soil, subjected to the following harmonic excitations with frequency of 45 rad/sec: (1) a vertical force applied at the center of the cube, with amplitude of 20 ton; (2) a horizontal force, parallel to faces of the cube and applied at the center of its base with amplitude of 20 ton; (3) a couple contained in a vertical plane about the center of the cube; (4) a torque about a vertical axis, with amplitude of 20 ton m. Take the side of the cube to measure 2 m, its unit weight to be 2 ton/m³, that of the soil 1.8 ton/m³, and the modulus of rigidity of the soil and Poisson's ratio 500 ton/m² and 0.5, respectively. Neglect coupling between the four degrees of freedom. (Let $z_1, \ldots, z_4$ denote, respectively, the amplitudes of the following disturbances, vertical displacement of the center, lateral base displacement, angle of tilt in rocking motion, and rotation about a vertical axis.)

Ans. $z_1 = 7.37$ mm, $z_2 = 14.69$ mm, $z_3 = 0.0143$ rad, $z_4 = 0.029$ rad, all other z's are zero.

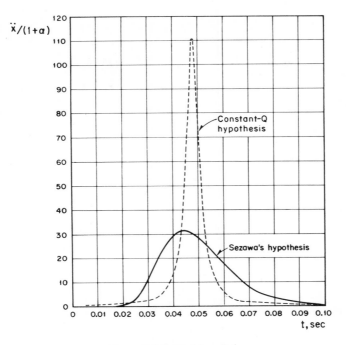

Figure 3.22. Problem 3.4.

4

NUMERICAL COMPUTATION OF STEADY-STATE RESPONSES AND NATURAL MODES

4.1 Introductory Note

Of systems that have distributed mass, only the simplest allow analytical computation of steady-state responses and natural modes of vibration. We devote the next section to numerical discretization of such systems for we have expounded general procedures for the analysis of discrete systems. Moreover, such procedures are especially appropriate for digital computers. There are other means for solving vibration problems in systems with distributed parameters. We may regard such means as adaptations of methods suited for the solution of partial differential equations in general. Mostly for the sake of brevity we shall treat only those methods that use discretization even though other approaches on occasion prove preferable. We shall make some exceptions by presenting certain energy methods and the method of matrices of transmission because of their advantages in a wide class of problems. The latter method lends itself well for use in digital computers.

Replacing the system with one that has lumped parameters does not ensure that we can expediently compute its steady-state or free oscillations by the methods we described in Chapter 2, unless we choose the new system so that it has a small number of degrees of freedom or unless it is so simple that discretizing the original system is unnecessary. The labor involved in solving any system with a moderately large number of degrees of freedom by straightforward methods is so great that the possibility of arriving at a solution by these means has no more than academic interest. Use of numerical methods is mandatory, except in near-trivial cases, whether the discrete system is specified as such or stems as the result of discretization.

A diversity of numerical procedures have been proposed for the calculation of critical vectors and eigenvalues of discrete systems. We may regard most procedures as variations or systematizations of two basic methods: Vianello's iteration method and Holzer's table. In both, the calculation of natural modes of vibration constitutes a special case of the calculation of steady-state responses for the condition in which excitation forces and displacements are nil. Even Rayleigh's method may be viewed as a particular case of the first cycle in the iteration method.

Both groups of methods have been generalized to give direct treatments of systems with distributed parameters. And both groups have been presented in the literature in diverse guises: in terms of elementary algebra, in matrix notation, as solutions of partial differential equations, and as methods for solving integral and integrodifferential equations.

Under the circumstances it has seemed wise to attempt a unified treatment of these numerical methods in the present chapter. In essence we shall be concerned with only two methods. One we shall call the *iteration method;* it constitutes a formal version of Vianello's procedure. The other we shall call *the method of transmission matrices,* and we may view it as a generalization of Holzer's table.

One other diversification would seem warranted. Calculation of higher modes of vibration by the iteration method requires a sweeping process. This may consist of imposing restrictions on the system itself so that it cannot vibrate in the lower modes; this is accomplished by deflating the inertia, stiffness, and damping matrices. Or it may consist in eliminating the lower modes from the assumed shape of the one to be computed. Only the second approach is presented here. The reader is referred to the work of Crandall and McCalley (1961, esp. pp. 28.32–28.38) for description of the first procedure, as well as for general-purpose methods especially suited for machine computation of systems having a very large number of degrees of freedom.

4.2 Discretization

There are various ways in which we may effect discretization to solve a system with distributed parameters. The most obvious approach consists of lumping these parameters. If we lump only distributed masses, we are left with a system that has a finite number of degrees of freedom that can be dealt with by the methods of Chapter 2 and whose natural modes of vibration and steady-state responses can be computed by some of the numerical methods of this chapter. Such systems are often too complex for computational purposes, and there may be advantages in lumping the flexibilities and the distributed damping elements as well, by concentrating them at discrete springs and dashpots. Replacement of the original system with one having lumped parameters reduces the number of degrees of freedom and introduces errors in the natural modes and periods: the errors are generally an increasing function of the order of the mode.

Lumping has an advantage over other methods in that it furnishes a picture of the object to be analyzed. On the other hand, it is often less accurate than

the other methods that require about the same amount of numerical work.
First consider a system governed by the one-dimensional wave equation

$$m \frac{\partial^2 x}{\partial t^2} - k \frac{\partial^2 x}{\partial X^2} = p \qquad (4.1)$$

(see Section 3.2), for example, a conservative, uniform, shear beam, where m is the mass per unit length, x is the displacement perpendicular to the beam axis, t is the time, k is the stiffness of the unit of length, X is a coordinate measured along the beam axis, and p is the external force per unit length.

Suppose we lump the mass at discrete points spaced uniformly a distance ΔX, joined by massless springs. The mass concentrated at each point will replace the distributed mass between centerpoints of the springs; hence it will equal $M = m \Delta X$ at all interior bodies and $M/2$ at a free end.

We take the spring stiffnesses identical to those of the original system because, if we disregard the masses and distributed forces, there is no difference between the original system and its discrete model. Hence the stiffness of each spring will be $K = k/\Delta X$.

We replace the external forces, distributed between spring centerpoints, with a concentrated force P equal to ΔX times the mean value of p in that interval: $P = \bar{p} \Delta X$. At a free end, $P = \bar{p} \Delta X/2$.

Accordingly, in idealizing the continuous system as discrete, we replace Eq. 4.1 with the system of equations

$$M\ddot{x}_i + K(x_i - x_{i+1}) + K(x_i - x_{i-1}) = P_i \qquad (4.2)$$

except at a free end, say at $i = N$, where we substitute $M/2$ for M. Equation 4.2 is derived from D'Alembert's principle applied to the ith point of lumped mass.

We may put the sytem of equations in the form

$$M\ddot{x}_i + K(-x_{i-1} + 2x_i - x_{i+1}) = P_i \qquad (4.3)$$

(again except at free ends) that makes it clear that lumping of masses is equivalent to replacing Eq. 4.1 with a first-order, finite-difference approximation. Higher-order approximations, use of Fox's procedure, or other improvements on the finite-difference approximation produce methods of solution more powerful than the one implied in the system of Eq. 4.3. At the same time they are accompanied by a loss of the physical picture provided by lumping.

We may use the difference-differential Eq. 4.3 to compute natural modes by setting $P_i = 0$. The boundary condition at a fixed end, say at $i = 0$, is $x_0 = 0$, and at a free end, say at $i = N$, it is such that $K(x_N - x_{N+1})$ does not enter Eq. 4.2. Alternatively we may regard the series of N bodies as one half of a system of $2N - 1$ bodies with $2N$ springs and fixed at both ends, provided we confine our attention to symmetric modes. If we do this, we shall have all equations identical with Eq. 4.3, we do away with the exception at $i = N$, and we introduce the restriction $x_{N+1} = x_{N-1}$. We may adapt Hildebrand's solution for a string that has $2N - 1$ uniformly spaced beads and is fixed at both ends (Hildebrand, 1952, pp. 254–55) to a shear beam with $N - 1$ equal interior, equally spaced bodies, each with mass M equal to the mass of the system divided by N, plus an end body with mass $M/2$ (the remaining mass $M/2$ may be regarded as rigidly fixed to the base), to give

$$\omega_n = 2\sqrt{\frac{K}{M}} \sin \frac{(2n-1)\pi}{4N}$$

and

$$z_{in} = C_n \sin \frac{(2n-1)\pi i}{2N}$$

where ω_n is the nth natural circular frequency, z_{in} is the amplitude of oscillation of the ith mass in the nth natural mode, and C_n is an arbitrary coefficient.

We may write L, length of the shear beam, for $N\,\Delta X$, and X for $i\,\Delta X$. Then, as N tends to infinity, this solution approaches the exact values

$$\omega_n = \frac{(2n-1)\pi}{2L}\sqrt{\frac{k}{m}}$$

and

$$z_n(X) = C_n \sin \frac{(2n-1)\pi x}{2L}$$

Table 4.1 compares the natural circular frequencies of a continuous system with those for the corresponding discrete systems having various numbers of masses. The percentage errors in ω_n increase with n and decrease with N. The same applies to percentage errors in C_n. Information in the table may serve as a guide when we select the spacing of lumped masses which are to replace continuous systems similar to shear beams.

TABLE 4.1. COMPARISON OF NATURAL FREQUENCIES OF SHEAR BEAMS

	ω_n in terms of $N^{-1}\sqrt{K/M} = (1/L)\sqrt{k/m}$						Error (%)				
n	$N=1$	$N=2$	$N=3$	$N=4$	$N=5$	$N=\infty$	$N=1$	$N=2$	$N=3$	$N=4$	$N=5$
1	1.414	1.530	1.552	1.560	1.564	1.571	10.0	2.6	1.2	0.7	0.5
2	—	3.697	4.243	4.444	4.541	4.712	—	21.5	10.0	0.7	3.6
3	—	—	5.795	6.650	7.071	7.854	—	—	26.2	15.3	10.0
4	—	—	—	7.846	8.910	10.996	—	—	—	28.7	19.0
5	—	—	—	—	9.875	14.137	—	—	—	—	30.6

Duncan (1952) has shown that the error in the natural frequencies, involved in this elementary type of mass lumping, in systems that have a single degree of freedom at each point (such as a shear beam or a Bernoulli–Euler flexural beam) is always of the order of the square of the spacing between the points of mass lumping, provided the end conditions are properly taken into account; otherwise the error is of the order of this spacing. We verify in Table 4.1 that errors vary indeed as N^{-2} approximately. This property may be used to advantage in extrapolating from the solutions for two different spacings.

Sec. 4.2   DISCRETIZATION   **109**

In flexural beams vibrating in a principal plane, lumping the mass at discrete points along the beam axis yields a system without rotary inertia. It is possible to lump the masses at discrete rods perpendicular to the beam axis (Fig. 4.1a). If the spacing between rods is ΔX and the cross-sectional moment of inertia of the beam is I, the polar moment of inertia of each rod should be $I\rho\,\Delta X$, where ρ is the mass per unit volume in the beam. As the mass in each rod should be equal to $A\rho\,\Delta X$, where A = cross-sectional area of beam, we may regard a rod as formed by a massless, rigid stem perpendicular to the beam axis, with two concentrated bodies, each having a mass $A\rho\,\Delta X/2$ and lying from the axis at distances equal to the radius of gyration of the cross section (Fig. 4.1b). If the massless beam segments that remain between rods are assigned the proper shear flexibility, the resulting system is a discretized Timoshenko beam.

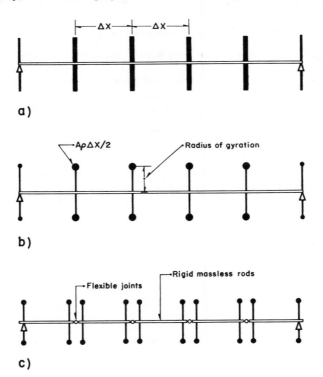

Figure 4.1. Discretized flexural beam.

Flexibilities may also be concentrated at discrete points, to give systems as the one shown in Fig. 4.1c. This step reduces the accuracy of analysis further but may well be justified in many cases, as its equations of motion can be set up in the form of simple difference-differential equations.

In building frames it is common to assume the masses concentrated at floor levels and to neglect vertical accelerations and rotary inertia about horizontal axes. Each floor is assigned a mass equal to that of its own weight, plus the live

110 NUMERICAL COMPUTATION OF STEADY-STATE RESPONSES   Chap. 4

mass it carries, and half the mass of the columns, walls, partitions, and windows in the stories immediately above and below that floor. This practice introduces insignificant errors in normal multistory frames, except in those for which floor rotation due to overturning is important. In such cases it is advisable to take the masses as lumped at beam-column intersections. The errors are objectionable in low buildings having flexible beams. In a single-story building, for example, vertical displacements at quarter points of beam spans, due to lateral forces,

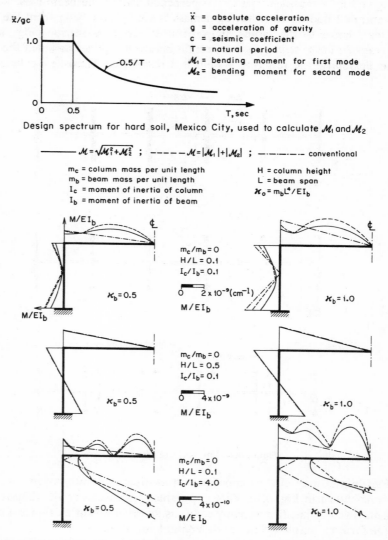

Figure 4.2. Comparison of dynamic response according to conventional and exact methods. Massless columns. *After Esteva (1965).*

may even exceed the lateral displacements; moreover, the deflection curves of the columns may differ markedly from the symmetrical shape we imply when we allocate one half of the story mass to the roof.

Sezawa and Kanai (1932) and Esteva (1965) have investigated the problem especially as it applies to single-story frames. Some of Esteva's results are given in Figs. 4.2 and 4.3. These results are useful mostly as guides to decide when we may safely disregard vertical accelerations due to lateral forces. In very regular frames the results are directly applicable.

In tall buildings it is frequently important to take into account the vertical displacement caused by foundation rocking and by axial deformations of the columns, both due to overturning moments. A simple, approximate way of doing

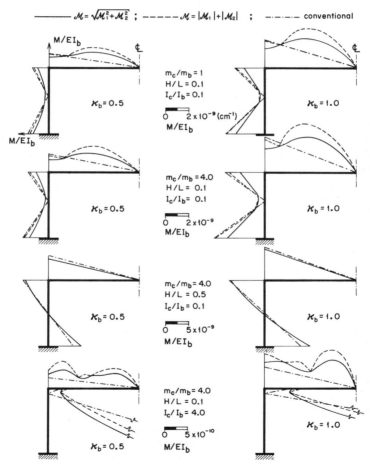

Figure 4.3. Comparison of dynamic response according to conventional and exact methods. Columns with important mass distributed. *After Esteva (1965)*.

this consists in assuming that all the masses are concentrated at the girder-column intersection.

On the other hand it is often desirable to work with a number of degrees of freedom smaller even than the one associated with lumping the mass at every floor level. This is particularly true when treating coupled translation and torsion modes. We may achieve the reduction by concentrating the mass at only some of the floor levels. Let i_k denote the floor levels at which we wish to take the mass as concentrated, while $i = 0, 1, 2, \ldots$ stands for any floor level. Then the mass at elevation i_k will be the kth concentration of mass statically equivalent to the actual distribution in the sense of statical equivalence used later in this section (roughly, the total mass between midpoints of elevations i_{k-1} and i_{k+1}). If we idealize the building as a shear beam, the stiffness of the spring element between elevations i_k and i_{k+1} will be

$$K_{k,k+1} = \frac{1}{\sum_{i=i_k}^{i_{k+1}-1} \frac{1}{K_{i,i+1}}}$$

where $K_{i,i+1}$ is the stiffness of the spring element between floors i and $i + 1$. The relationship is derived immediately from the condition that relative displacements between elevations i_k and i_{k+1} must be the same in both systems under any set of lateral loads applied statically at elevations $i_{k,\ k+1}$. A comparable relation holds for torsional stiffnesses.

Next consider the simplest of two-dimensional systems—the uniform, conservative, isotropic membrane. Substitution of concentrated masses for the surface-distributed mass and of a system of orthogonal, elastic strings for the membrane replaces the Laplace operator ∇^2 in the differential equation

$$\rho \ddot{x} + S\nabla^2 x = w$$

with the first-order, five-point finite difference operator

$$\nabla^2 \doteq \frac{1}{h^2} \left\{ \begin{array}{c} \text{\textcircled{1}} \\ \text{\textcircled{1}} \quad \text{\textcircled{-4}} \quad \text{\textcircled{1}} \\ \text{\textcircled{1}} \end{array} \right\}$$

In the above expressions h is the mesh size, S is the membrane tension per unit width, ρ is the mass per unit area, and w is the external force per unit area. In this case we find, again, that an improved finite-difference treatment may yield considerably more accurate results than the first-order approximation and hence than the equivalent lumped-parameter approximation.

Discrete lumped-mass systems have been proposed, and solved through use of computers, for a variety of dynamic problems, including even those in three-dimensional, axisymmetric continua (Ang and Rainer, 1964; Ang, 1966). Finite-element methods for two- and three-dimensional problems have also been developed to a high degree of sophistication (Clough, 1960b; Clough, 1965; Argyris, 1965; Costantino, 1967; Wilson, 1969). These schemes are most useful even though the number and type of simultaneous equations resulting from discretization of three-dimensional media is usually so great that only the large digital computers can handle them effectively, unless the problem can be treated as one or two dimensional owing to symmetry.

In one-dimensional systems such as beams, arches, and frames, greater accuracy can be achieved by discretizing the forces and otherwise treating the system as with distributed parameters than by replacing the specified system with one having lumped parameters. In the more accurate procedure, we compute distributed inertia forces at the points along the axis of the member at which the external distributed loads are specified. Next we approximate the sum of external and inertia forces by suitable, simple functions, such as broken, straight lines or parabolas that pass through the points at which the distributed forces are known (Fig. 4.4a), and we replace the distributed forces with statically equivalent concentrated forces that are numerically equal to the reaction of simply supported beams but have signs opposite those of these reactions (Fig. 4.4b).

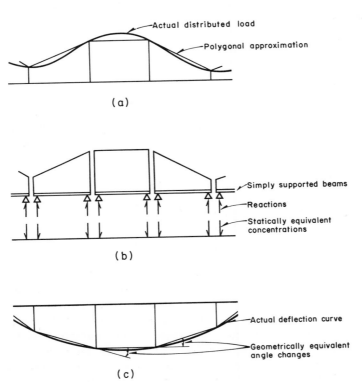

Figure 4.4. Equivalent concentrations in numerical procedure for computing bending moments and deflections.

We compute bending moments and curvatures, due to the concentrated loads, in the distributed-parameter member at the same points at which we had computed the concentrations. Then we treat the distributed curvatures as distributed loads, using if we wish the concept of the conjugate beam, and thus find concentrated angle changes *geometrically equivalent* to the curvatures (Fig. 4.4c). Finally we obtain slopes and deflections by adding these angle changes, provided we satisfy the boundary conditions.

The method is described in detail elsewhere (Newmark, 1943; Godden, 1965). It will suffice to summarize the formulas that provide equivalent concentrated loads for polygonal and parabolic approximations and illustrate their application to the calculation of moments and deflections in one specific example.

Let the distributed load p be specified at the uniformly spaced points ... a, b, c, ... ; let h denote the spacing between these points along the axis of abscissas; let P_{ab}, P_{ba} ... denote the equivalent concentrations at point a of span ab, point b of span ab, ... Then, for a polygonal approximation to p,

$$P_{ab} = \frac{h}{6}(2p_a + p_b) \tag{4.4}$$

For a parabolic approximation, that is, if we replace $p(x)$ by a second-degree parabola through points (a, p_a), (b, p_b), and (c, p_c),

$$P_{ab} = \frac{h}{24}(7p_a + 6p_b - p_c) \tag{4.5}$$

and

$$P_{ba} = \frac{h}{24}(3p_a + 10p_b - p_c) \tag{4.6}$$

By changing the subscripts in these expressions we obtain the formulas for the equivalent concentrations at points b and c. Now let $P_b = P_{ba} + P_{bc}$. For the polygonal approximation we get

$$P_b = \frac{h}{6}(p_a + 4p_b + p_c) \tag{4.7}$$

and for the parabolic approximation

$$P_b = \frac{h}{12}(p_a + 10p_b + p_c) \tag{4.8}$$

If the distributed load p were exactly a series of straight lines or a second-degree parabola, application of the proper equivalent concentration formulas would give exact values of the average shear in the segments between points at which p is specified and exact values of the bending moments at these points. The errors in bending moments, introduced by the approximation to p, are of the order of h^2 in the polygonal approximation, of the order of h^3 in Eqs. 4.5–4.7, and of the order of h^4 in Eq. 4.8. The latter are usually negligible in comparison with errors due to other sources and can be ignored in almost all cases in which the node points are spaced sufficiently close to each other so that the computed bending moments may be regarded as described adequately.

When seeking an extremely accurate answer, we may resort to an extrapolation procedure that requires obtaining the solution for three or four different spacings and expressing the errors as power series of h. The procedure is greatly expedited through the use of certain coefficients (Salvadori and Baron, 1952, pp. 75–81).

The polygonal approximation leads to results identical with those of first-order finite differences and with those obtained by concentrating the loads and masses at discrete points. The advantage of the polygonal approximation is that the concentrations can be computed in a systematic and relatively accurate manner at the points we choose, while other criteria for lumping loads and masses that place the lumped quantities at centroids of the segments either lead to lumping at awkward points or sacrifice accuracy by lumping at locations that do not coincide strictly with the centroids.

Use of the parabolic approximation for computation of the fundamental mode of vibration leads to results that are so much more accurate, as compared with the polygonal approximation, that it is almost always advantageous. Yet, its use in problems in which more than one natural mode must be computed has a drawback because the distributed-parameter system is replaced implicitly with a different discrete system in each natural mode. Hence, the orthogonality relation is lost. To obviate this difficulty one should resort to the straight-line approximation (Eq. 4.4 and Eq. 4.7) or apply Eq. 4.5 at even-numbered node points and Eq. 4.8 at odd-numbered nodes. The scheme is unnecessary in certain special cases, as when it is desired to compute only the first two modes of a symmetric system.

Example 4.1. Compute the bending moments and deflections for the simply supported beam in Fig. 4.5a. Use the parabolic approximation. Neglect shearing deformations.

Solution. The solution is given in Fig. 4.5b in a conveniently systematized tabulation. Lines 1 and 2 contain the data. Line 3 was computed from Eq. 4.8 as the statement of the problem did not require computation of shears; if these had been requested, we would have used Eq. 4.6 to obtain concentrations at both sides of each node point in order that their partial sums give us the shearing forces at these points.

Line 4 contains the average shears in the segments between node points (hence the overbar over V_1). Subscript 1 is intended to indicate that these shears are tentative because the reactions have not been computed, and $\bar{V}_1$ at the first segment was merely estimated.

The cumulative sum of $\bar{V}_1 h$ gives us the tentative bending moments in line 5. We see that the boundary condition at the right-hand support ($M = 0$) is not satisfied ($M = $ bending moment). We conclude that the $\bar{V}_1$'s are in error because they correspond to an erroneous left-hand reaction. Changing this reaction is equivalent to introducing a constant corrective term in $\bar{V}_1$ and hence a straight-line correction in M_1. The correction is displayed in line 6, and the corrected moments ($M_1 + M_c$) in line 7.

Line 8 gives us the curvatures $\alpha = M/EI$ at node points. The geometrically

116 NUMERICAL COMPUTATION OF STEADY-STATE RESPONSES Chap. 4

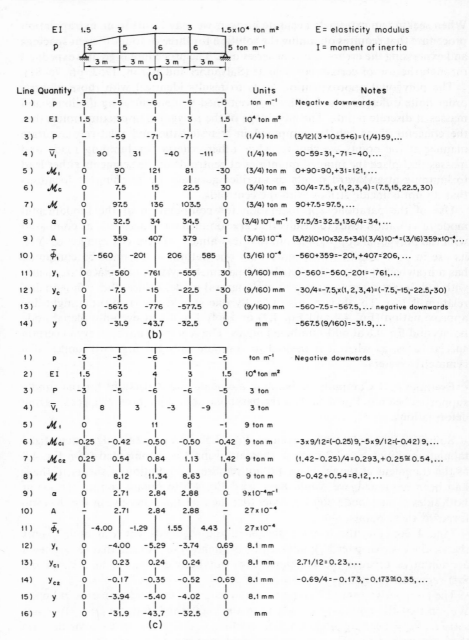

Figure 4.5. Example 4.1.

equivalent concentrations A obtained by replacing p and P in Eq. 4.8 with α and A, respectively, are given in line 9.

Proceeding as with the shears we obtain the tentative average slopes $\bar{\phi}_1$ in segments between node points (line 10). As with the bending moments, the corresponding tentative deflections are $y_1 = \sum \phi_1 h$ (line 11). These receive a straight-line correction (line 12). And lines 13 and 14 give us the final deflections.

An alternate solution is shown in Fig. 4.5c. In this solution we take the equivalent force concentrations as $P = ph$, compute the bending moments as in the standard solution, and add a correction to them, equal at each point to $ph^2/12$. It can be easily shown that the two variants of the parabolic approximation are equivalent, provided h is constant, there are no discontinuities in p, none in its derivative, and the end moments zero are (Godden, 1965, pp. 41–43). A similar simplification is used in computing the deflections from the curvatures.

4.3 Iteration Procedure for Steady-State Vibrations

Consider a discrete, linear, conservative system on a fixed base, subjected to the harmonic external force disturbance $\mathbf{P} = \mathbf{b} \sin \omega t$. According to Sections 2.1 and 2.2 the system complies with the matrix differential equation

$$\mathbf{M}\ddot{\mathbf{y}} + \mathbf{K}\mathbf{y} = \mathbf{P} \qquad (4.9)$$

Let $y = \mathbf{a} \sin \omega t$. Substitute $\mathbf{P}$ and $\mathbf{y}$ in Eq. 4.9, and divide throughout by $\sin \omega t$:

$$(-\omega^2 \mathbf{M} + \mathbf{K})\mathbf{a} = \mathbf{b} \qquad (4.10)$$

Add $\omega^2 \mathbf{M} \mathbf{a}$ to both members and premultiply them by $\mathbf{K}^{-1}$:

$$\mathbf{a} = \mathbf{K}^{-1}\mathbf{b} + \mathbf{K}^{-1}\omega^2 \mathbf{M}\mathbf{a} \qquad (4.11)$$

The first member represents the total displacement amplitudes; the first term in the right-hand member is the vector of static response amplitudes; and $-\omega^2 \mathbf{M}\mathbf{a}$ is the set of inertia force amplitudes, so that we may call the second term in the right-hand member the vector of *dynamic displacement* amplitudes.

The iteration method to be described is essentially the one that, in various versions, has come to be known as Vianello's, Stodola's, Volterra's or Newmark's. In this method we compute the static amplitudes $\mathbf{K}^{-1}\mathbf{b}$, estimate the vector $\mathbf{a}$, compute the inertia forces $-\omega^2 \mathbf{M}\mathbf{a}$ and the corresponding dynamic displacements, and substitute in Eq. 4.11 to obtain a new approximation to $\mathbf{a}$. Beginning with the latter, we repeat the process until we achieve a sufficiently high degree of convergence. The method converges to the correct answer provided that $\omega < \omega_1$.

To prove the last contention, notice that the ith approximation to $\mathbf{a}$, call it $\mathbf{a}^i$, may be developed in terms of the natural modes of vibration, let us say

$$\mathbf{a}^i = \sum_n \mathbf{a}_n^i \mathbf{z}_n$$

and the same applies to the static displacements, so that we can write

$$\mathbf{K}^{-1}\mathbf{b} = \sum_n c_n \mathbf{z}_n$$

where the c_n's are constants.

Since $\mathbf{z}_n$ satisfies Eq. 4.10 with $\mathbf{b} = 0$, it also satisfies the relation

$$\frac{1}{\omega_n^2}\mathbf{z}_n = \mathbf{K}^{-1}\mathbf{M}\mathbf{z}_n$$

Substituting the last three expressions in Eq. 4.11 and writing $\mathbf{a}^i$ in the right-hand member and $\mathbf{a}^{i+1}$ in the left-hand member, we obtain

$$\sum_n a_n^{i+1} \mathbf{z}_n = \sum_n \left(c_n + a_n^i \frac{\omega^2}{\omega_n^2} \right) \mathbf{z}_n$$

whence,

$$a_n^{i+1} = c_n + a_n^i \frac{\omega^2}{\omega_n^2} \tag{4.12}$$

so that, if a_n^0 denotes the nth-mode coefficient which corresponds to the assumed displacements $\mathbf{a}^0$ we have

$$a_n^{i+1} = c_n \sum_{j=0}^{i} \left(\frac{\omega^2}{\omega_n^2}\right)^j + a_n^0 \left(\frac{\omega^2}{\omega_n^2}\right)^i \tag{4.13}$$

For this expression to remain finite as i tends to infinity, a sufficient condition is that ω be smaller than all the ω_n's; that is, $\omega < \omega_1$. If a_n is arbitrary, the condition is also a necessary one unless $c_1 = 0$.

If $\omega < \omega_1$, the sum in Eq. 4.13 tends to $(1 - \omega^2/\omega_n^2)^{-1}$, and the term $a_n^0(\omega^2/\omega_n^2)^i$ tends to zero independently of a_n^0 as i tends to infinity. Hence a_n^{i+1} tends to $c_n(1 - \omega^2/\omega_n^2)^{-1}$. But if we let

$$\mathbf{a} = \sum_n a_n \mathbf{z}_n$$

and if we choose $\mathbf{a}^0$ correctly, so that $\mathbf{a}^0 = \mathbf{a}$, we have $a_n^{i+1} = a_n^i = a_n$, and Eq. 4.13 gives us

$$a_n = \frac{c_n}{1 - \omega^2/\omega_n^2} \tag{4.14}$$

Consequently, if $\omega < \omega_1$, a_n^i converges to a_n, and $\mathbf{a}^i$ converges to $\mathbf{a}$. We have shown that for the iteration procedure to converge it suffices that the frequency of the excitation forces be smaller than the fundamental frequency, and it converges then to the correct answer. This boundary may be raised if the static displacements due to the excitation forces contain no component of the fundamental mode of vibration.

By taking $n = 1$ in Eq. 4.14 we have an expression for the coefficient of the first mode. If we take

$$\mathbf{a}^0 = \frac{\mathbf{K}^{-1}\mathbf{b}}{1 - \omega^2/\omega_1^2}, \quad \mathbf{a}^0 - \mathbf{K}^{-1}\mathbf{b} = \frac{\mathbf{K}^{-1}\mathbf{b}}{\omega_1^2/\omega^2 - 1} \tag{4.15}$$

we start with an assumed set of displacements for which the coefficient of the first mode exactly satisfies Eq. 4.14. In this case Eq. 4.13 gives us

Sec. 4.3   ITERATION PROCEDURE FOR STEADY-STATE VIBRATIONS   119

$$a_1^{i+1} = c_1 \sum_{j=0}^{i} \left(\frac{\omega^2}{\omega_1^2}\right)^j + a_1^0 \left(\frac{\omega^2}{\omega_1^2}\right)^i$$

$$= \frac{c_1[1 - (\omega^2/\omega_1^2)^i]}{1 - \omega^2/\omega_1^2} + \frac{c_1(\omega^2/\omega_1^2)^i}{1 - \omega^2/\omega_1^2}$$

$$= \frac{c_1}{1 - \omega^2/\omega_1^2} = a_1^0$$

Consequently the series does not diverge even if $\omega > \omega_1$. Through this artifice, then, we extend the range of convergence up to $\omega < \omega_2$. When $\omega < \omega_1$, use of Eq. 4.15 yields a more rapidly converging process than would result, in general, from an arbitrary choice of $\mathbf{a}^0$. In the latter case Eq. 4.15 may be applied to advantage though ω_1 be only known in an approximate manner.

We may also extend the range of convergence or accelerate the rate, as the case may be, through use of the delta-squared extrapolation method (Crandall, 1956). Its repeated use is especially indicated when $\omega_1 < \omega < \omega_2$, to counteract numerical and round off errors which necessarily introduce small components of the first mode.

The foregoing treatment is applicable without essential modification when the excitation includes base motion. The static and dynamic components of displacement should merely incorporate the corresponding contribution. This remark applies even when the supports on which the system rests do not move as a rigid body.

Damped systems in which dashpot constants are stipulated may be treated in the same way except that separate accounts of the sine and cosine terms must be carried (or of the real and imaginary parts of the solution if the exponential version of trigonometric functions is preferred, as in Eq. 1.5). If the percentage of damping is specified in each natural mode, the response may be computed as for a conservative system and modified as for a damped system with a single degree of freedom.

Example 4.2. The bar in Fig. 4.6a is subjected to an axial alternating force that, in units of metric tons, centimeters, and seconds, is given by $P = 10 \sin 2t$. Compute its steady-state response.

Solution. First we shall replace the bar with two concentrated masses joined to each other and to the support by massless springs, as shown in lines 1 and 2, Fig. 4.6b. The same figure depicts the solution for this discrete system. Line 3 gives the static spring forces. (Here and in the succeeding lines the units are such that the values shown for each quantity are the corresponding amplitudes: if forces and deformations were desired as functions of time, we would multiply the units shown by $\sin 2t$.) Line 4, the spring elongations, was obtained by dividing line 3 by line 1. Line 5, the static displacements, resulted from adding the spring elongations from left to right. Line 6, the tentative dynamic displacements, was obtained by applying to x_s the dynamic factor of Eq. 4.15 with $\omega_1 = 1.711$ rad/sec, which comes from Table 4.1. The total displacements relative to the base, $x^0 = x_s + x_d^0$, appear in line 7. The product $\omega^2 M x^0$ gives us the inertia forces F in line 8. Their sum supplies the dynamic forces Q_d (line 9).

120 NUMERICAL COMPUTATION OF STEADY-STATE RESPONSES Chap. 4

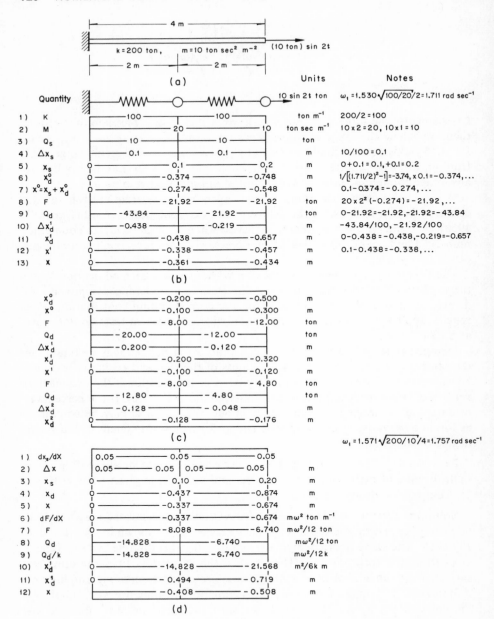

Figure 4.6. Example 4.2.

Dividing by K we obtain the dynamic spring elongations, Δx_d^1 in line 10. The revised total displacements $x^1 = x_s + x_d^1$, appear in line 12. Comparison with line 7 gives a basis for deciding whether computations have been carried out to a satisfactory stage of convergence or additional cycles are in order. In the latter case the delta-squared extrapolation scheme is usually advantageous. Line 13 gives the total displacements resulting from several additional cycles.

Because ω exceeds ω_1, had we started with an arbitrary set of x_d, the procedure would have diverged. The situation is illustrated in Fig. 4.5c.

In Fig. 4.6d the system is treated with distributed mass and flexibility, and only the inertia forces and deformations are replaced with equivalent concentrations, in the same manner as a flexural beam was treated in Example 4.1. Thus, from the elongation per unit length, dx_s/dX in line 1, we compute the equivalent concentrations in line 2 by using Eq. 4.4 (although the fact that dx_s/dX is uniform makes these results identical with the ones which Eq. 4.7 would yield). By adding the concentrations from left to right we arrive at the static displacements x_s in line 3 which do not differ from the values in Fig. 4.6b. The fundamental circular frequency is now 1.771 rad/sec computed using the procedure described in Section 4.3, and we use this value to estimate x_d in accordance with Eq. 4.15 (line 4). The distributed inertia forces, dF/dX in line 6, are concentrated through use of Eqs. 4.5 and 4.6 in line 7 and do differ from the values in Fig. 4.6b, as dF/dX is not a linear function of X. The rest of the computations in Fig. 4.6d do not require further explanation; the last line contains the results arrived at after several cycles. The use of additional cycles would not have been advisable, as the procedure would begin to diverge due to the introduction of roundoff errors.

TABLE 4.2 COMPARISON OF SOLUTIONS IN FIGS. 4.6(b) AND (d) WITH EXACT SOLUTION FOR EX. 4.2.

Type of solution	$x_{X=L/2}$	$x_{X=L}$
Ex. 4.2(b)	−0.361	−0.434
Ex. 4.2(c)	−0.408	−0.508
Exact	−0.404	−0.506

In Table 4.2 we compare the solutions of Figs. 4.6b and 4.6d with the exact values, which were found through direct solution of the differential equation of motion (Section 3.2). We see that mass discretization has introduced appreciable errors in exchange for only a small saving in labor. The ratio of the errors introduced by the two approaches increases rapidly with a decrease in the distance between points of mass lumping or node points.

Example 4.3. The square slab in Fig. 4.7a weighs 40 metric tons; this weight is uniformly distributed. The slab rests on massless structural elements along its periphery. The stiffnesses of these elements are $K_{AB} = 20$ ton/cm and $K_{BC} = K_{CD} = K_{DA} = 10$ ton/cm. The ground undergoes harmonic oscillations parallel to AB and CD of amplitude 1.2 cm and period 0.5 sec. Calculate the response of the system.

122   NUMERICAL COMPUTATION OF STEADY-STATE RESPONSES   Chap. 4

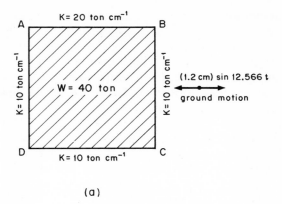

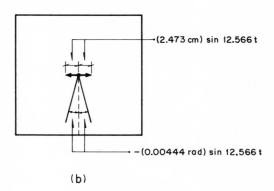

Figure 4.7. Example 4.3.

Solution. From consideration of symmetry we know that only two degrees of freedom will be excited. We choose the first of these degrees to be translation of the center of gravity parallel to the direction of motion, and the second degree to be the rotation of the slab about a vertical axis through that center. Then the inertia matrix is

$$\mathbf{M} = \begin{vmatrix} 4.08 & 0 \\ 0 & 68 \end{vmatrix}$$

in units of metric tons, meters, and seconds. In the same units,

$$\mathbf{K} = \begin{vmatrix} 3{,}000 & 5{,}000 \\ 5{,}000 & 25{,}000 \end{vmatrix}$$

From here,

$$\mathbf{K}^{-1} = \begin{vmatrix} 50 & -10 \\ -10 & 6 \end{vmatrix} 10^{-5}$$

Sec. 4.3   ITERATION PROCEDURE FOR STEADY-STATE VIBRATIONS

We shall arbitrarily assume

$$\mathbf{a}^0 = \begin{vmatrix} 0.020 \\ 0.002 \end{vmatrix}$$

Applying the equivalent of Eq. 4.11 for ground motions $\mathbf{x} = \mathbf{x}_0 + \mathbf{K}^{-1}\omega^2\mathbf{M}\mathbf{x}$ or, if $\mathbf{x} = \mathbf{a}\sin\omega t$ and $\mathbf{x}_0 = \mathbf{a}_0\sin\omega t$, then $\mathbf{a} = \mathbf{a}_0 + \mathbf{K}^{-1}\omega^2\mathbf{M}\mathbf{x}$. Therefore,

$$\mathbf{a}^1 = \begin{vmatrix} 0.012 \\ 0 \end{vmatrix} + \begin{vmatrix} 50 & -10 \\ -10 & 6 \end{vmatrix} 10^{-5} \times 12.566^2 \begin{vmatrix} 4.08 & 0 \\ 0 & 68 \end{vmatrix} \begin{vmatrix} 0.020 \\ 0.002 \end{vmatrix}$$

$$= \begin{vmatrix} 0.01629 \\ 0 \end{vmatrix}$$

Now we enter the improved $\mathbf{a}^1$ into the same expression and obtain a further improved vector. After additional cycles we arrive at

$$\mathbf{a} = \begin{vmatrix} 0.02473 \\ -0.00444 \end{vmatrix}$$

The steady-state response consists, therefore, of a displacement (2.473 cm) sin 12.566t parallel to the ground motion, and a rotation (−0.00444 rad) sin 12.566t about a vertical axis through the centroid (Fig. 4.7b).

Example 4.4. Solve the foregoing example if there are dashpots in the vertical planes as the spring elements. The dashpot constants are $C_{AB} = 0.02$ ton sec/cm and $C_{BC} = C_{CD} = C_{DA} = 0.01$ ton sec/cm.

Solution. We must solve for the response as a time function

$$\mathbf{x} = \mathbf{x}_0 + \mathbf{K}^{-1}\mathbf{C}(\dot{\mathbf{x}} - \dot{\mathbf{x}}_0) + \omega^2\mathbf{K}^{-1}\mathbf{M}\mathbf{x}$$

By writing

$$\mathbf{x} = \mathbf{a}\sin\omega t + \mathbf{a}'\cos\omega t$$

and separately equating sine and cosine terms we find

$$\mathbf{a} = \mathbf{a}_0 - \omega\mathbf{K}^{-1}\mathbf{C}\mathbf{a}' + \omega^2\mathbf{K}^{-1}\mathbf{M}\mathbf{a}$$

and

$$\mathbf{a}' = -\omega\mathbf{K}^{-1}\mathbf{C}(\mathbf{a} - \mathbf{a}_0) + \omega^2\mathbf{K}^{-1}\mathbf{M}\mathbf{a}'$$

Now,

$$\mathbf{C} = \begin{vmatrix} 3 & 5 \\ 5 & 25 \end{vmatrix}$$

in units of metric tons, meters, and seconds. Hence

$$\omega\mathbf{K}^{-1}\mathbf{C} = \begin{vmatrix} 0.0126 & 0 \\ 0 & 0.0126 \end{vmatrix}$$

Substituting in the expressions for a and a' and choosing arbitrarily

$$\mathbf{a}^0 = \begin{vmatrix} 0.012 \\ 0 \end{vmatrix} \qquad \mathbf{a}'^0 = \begin{vmatrix} -0.003 \\ -0.003 \end{vmatrix}$$

we get

$$\mathbf{a}^1 = \begin{vmatrix} 0.012 \\ 0 \end{vmatrix} + \begin{vmatrix} 0.0126 & 0 \\ 0 & 0.0126 \end{vmatrix} \begin{vmatrix} -0.003 \\ -0.003 \end{vmatrix} + \begin{vmatrix} -0.322 & -1.074 \\ -0.064 & 0.644 \end{vmatrix} \begin{vmatrix} 0.012 \\ 0 \end{vmatrix}$$

$$= \begin{vmatrix} 0.01583 \\ -0.00081 \end{vmatrix}$$

$$\mathbf{a}'^1 = \begin{vmatrix} 0.0126 & 0 \\ 0 & 0.0126 \end{vmatrix} \begin{vmatrix} 0.012 - 0.012 \\ 0 - 0 \end{vmatrix} + \begin{vmatrix} 0.322 & -1.074 \\ -0.064 & 0.644 \end{vmatrix} \begin{vmatrix} -0.003 \\ -0.003 \end{vmatrix}$$

$$= \begin{vmatrix} 0.00226 \\ -0.00174 \end{vmatrix}$$

Consequently, at the end of the first cycle we find the steady-state response consisting of a displacement $(1.583 \text{ cm}) \sin 12.566t - (0.226 \text{ cm}) \cos 12.556t = (1.60 \text{ cm}) \sin (12.566t - 0.142)$, plus a rotation about the vertical axis through the centroid $(-0.00081 \text{ rad}) \sin 12.566t - (0.00174 \text{ rad}) \cos 12.566t = (0.00192 \text{ rad}) \sin (12.566t - 2.007)$.

4.4 Iteration Procedure for the Fundamental Mode[1]

The method to be described is based on the Stodola–Vianello iteration procedure but is much improved through the use of the Schwartz quotient (for a definition of the quotient see Crandall, 1956) and the possibility of finding close upper and lower limits to the natural frequencies.

We may regard the first few steps of the procedure as an adaptation of the one described in Section 4.3 for steady-state responses of conservative systems. The rest of the procedure will be justified later.

To compute the fundamental mode take the following steps.

1. Assume the configuration of this mode, say $\mathbf{z}^0$.

2. Compute $\mathbf{F}^0/\omega^2 = \mathbf{M}\mathbf{z}^0$, where $\mathbf{F}^0 =$ inertia forces corresponding to the assumed configuration. (These forces must be computed in terms of ω^2 because the natural circular frequencies are unknown at this stage.)

3. Compute the configuration $\mathbf{z}^1$ (again in terms of ω^2) which these inertia forces would induce in the structure if they acted statically: $\mathbf{z}^1/\omega^2 = \mathbf{K}^{-1}\mathbf{F}^0/\omega^2$.

4. Compute the set of ratios

$$\lambda_r^1 = \frac{z_r^0}{z_r^1/\omega^2}$$

[1] The procedure is essentially the one proposed by Newmark (1943).

where r denotes the rth degree of freedom. If all the λ's are equal, we have found a natural mode, for in that case $\lambda_r^1 = \omega^2$,

$$\mathbf{z}^0 = \mathbf{z}^1 = \mathbf{K}^{-1}\mathbf{F}^0 = \omega^2 \mathbf{K}^{-1}\mathbf{M}\mathbf{z}^0$$

and so ω must be a natural circular frequency and $\mathbf{z}^0$ satisfies Eq. 2.5. Under a wide range of conditions, if not all the λ's are equal, we have found an upper and a lower bound for ω_1, for as we shall show, under those conditions

$$\min \lambda_r^1 \leq \omega_1^2 \leq \max \lambda_r^1 \qquad (4.16)$$

5. We can improve the upper bound, since as we shall also show,

$$\omega_1^2 = \frac{\mathbf{z}^{0T}\mathbf{M}(\mathbf{z}^1/\omega^2)}{(\mathbf{z}^1/\omega^2)^T\mathbf{M}(\mathbf{z}^1/\omega^2)} \qquad (4.17)$$

We shall show further that the best possible estimate of ω_1^2, on a least-squares criterion, is precisely the second member in Eq. 4.17. This quantity is known as the *Schwartz quotient*.

6. If not all λ's are equal, repeat the process beginning with a configuration proportional to $\mathbf{z}^1/\omega^2$. The successive cycles give increasingly accurate estimates of $\mathbf{z}_1$, $\lim_{i\to\infty} \mathbf{z}^i = \mathbf{z}_1$, and hence $\lim_{i\to\infty} \lambda_r^i = \omega_1^2$.

7. If desired, accelerate the convergence through use of the delta-squared extrapolation scheme.

If by some accident $\mathbf{z}^0$ contained no component of the fundamental mode, the procedure would theoretically converge to the second natural mode. This is a remote situation (except in symmetric systems, in which we can detect it easily). But even if the assumed configuration contained no component of $\mathbf{z}_1$, such a component would normally creep in in successive cycles due at least to roundoff errors.

To prove the inequalities Eqs. 4.16 and 4.17 notice that Eq. 4.12 becomes

$$a_n^{i+1} = a_n^i \frac{\omega^2}{\omega_n^2}$$

when c_n is zero. Therefore the first cycle of the iteration procedure we have just described transforms the assumed configuration

$$\mathbf{z}^0 = \sum_n a_n \mathbf{z}_n \qquad (4.18)$$

into

$$\mathbf{z}^1 = \sum_n a_n \frac{\omega^2}{\omega_n^2} \mathbf{z}_n \qquad (4.19)$$

Now premultiply both members of both of these expressions by $\mathbf{z}_1^T \mathbf{M}$. It follows from the orthogonality of natural modes that

$$\mathbf{z}_1^T \mathbf{M} \mathbf{z}^0 = a_1 \mathbf{z}_1^T \mathbf{M} \mathbf{z}_1 \qquad (4.20)$$

and

$$\mathbf{z}_1^T \mathbf{M} \mathbf{z}^1 = a_1 \frac{\omega^2}{\omega_1^2} \mathbf{z}_1^T \mathbf{M} \mathbf{z}_1 \qquad (4.21)$$

From the definition of λ_r^1,

$$\frac{\min \lambda_r^1}{\omega^2} z_r^1 \leq z_r^0 \leq \frac{\max \lambda_r^1}{\omega^2} z_r^1$$

Hence, if z_r^0 has the same sign as z_r^1 for all r,

$$\frac{\min \lambda_r^1}{\omega^2} \mathbf{z}_1^T \mathbf{M} \mathbf{z}^1 \leq \mathbf{z}_1^T \mathbf{M} \mathbf{z}^0 \leq \frac{\max \lambda_r^1}{\omega^2} \mathbf{z}_1^T \mathbf{M} \mathbf{z}^1$$

as $\mathbf{M}$ is positive definite. Using Eqs. 4.20 and 4.21, Eq. 4.16 follows immediately. A sufficient condition for its validity is that the assumed configuration must have the same signs everywhere as the fundamental mode of vibration.

Resorting again to Eqs. 4.18 and 4.19 we may write the second member of Eq. 4.17, divided by ω_1^2, as

$$\frac{\sum_n (a_n/\omega_1^2) \mathbf{z}_n^T \mathbf{M} \sum_m (a_m/\omega_m^2) \mathbf{z}_m}{\sum_n (a_n/\omega_n^2) \mathbf{z}_n^T \mathbf{M} \sum_m (a_m/\omega_m^2) \mathbf{z}_m}$$

In view of the orthogonality of natural modes, this is the same as

$$\frac{\sum_n (a_n/\omega_1 \omega_n)^2 \mathbf{z}_n^T \mathbf{M} \mathbf{z}_n}{\sum_n (a_n/\omega_n^2)^2 \mathbf{z}_n^T \mathbf{M} \mathbf{z}_n}$$

Every term in the numerator and in the denominator is positive, and every term in the numerator is at least as great as the corresponding one in the denominator, since $\omega_1 \leq \omega_n$. Consequently this expression is not smaller than 1 and Eq. 4.17 is proved.

The contention that the second member in Eq. 4.17 gives the best estimate of ω_1^2 on a least-squares criterion rests on the definition of the error vector as $\mathbf{z}^1 - \mathbf{z}^0$. Its squared amplitude, with the masses as weighting factors, is

$$(\mathbf{z}^1 - \mathbf{z}^0)^T \mathbf{M} (\mathbf{z}^1 - \mathbf{z}^0)$$

and may be regarded as the sum of squared errors. We may now write $\mathbf{z}^1$ as $\omega^2(\mathbf{z}^1/\omega^2)$ where the vector in parentheses does not vary with ω. Equating to zero the derivative of the foregoing expression for the sum of squared errors with respect to ω^2, we obtain for ω^2 precisely the second member of Eq. 4.17. This expression, then, is such that it minimizes the sum of the squared differences between the assumed and the computed configuration. In this sense it is the best estimate of ω_1^2. Often a few cycles of iteration may be saved by using various extrapolation techniques to estimate the configuration of the fundamental mode.

The procedure described can be applied without essential changes to the calculation of natural buckling modes. (Special techniques, however, are necessary if the axial load exceeds the fundamental buckling load in numerical value, since buckling loads may be negative in some structures.) Hence it is also especially suited for vibration problems involving the action of static axial forces.

Adaptation of this method to problems that involve linear damping is accomplished by keeping separate accounts of sine and cosine terms.

Sec. 4.4   ITERATION PROCEDURE FOR THE FUNDAMENTAL MODE   **127**

Example 4.5. Using the present iteration procedure find the fundamental mode and period of vibration of the system analyzed in Example 4.3 (Fig. 4.7a).

Solution. In Example 4.3 we found the inertia and flexibility matrices, in units of metric tons, meters, and seconds:

$$\mathbf{M} = \begin{vmatrix} 4.08 & 0 \\ 0 & 68 \end{vmatrix} \quad \mathbf{K}^{-1} = \begin{vmatrix} 50 & -10 \\ -10 & 6 \end{vmatrix} 10^{-5}$$

Let us take

$$\mathbf{z}^0 = \begin{vmatrix} 2 \\ -1 \end{vmatrix}$$

Then,

$$\frac{\mathbf{F}^0}{\omega^2} = \begin{vmatrix} 4.08 & 0 \\ 0 & 68 \end{vmatrix} \begin{vmatrix} 2 \\ -1 \end{vmatrix} = \begin{vmatrix} 8.16 \\ -68.00 \end{vmatrix}$$

$$\frac{\mathbf{z}^1}{\omega^2} = \begin{vmatrix} 50 & -10 \\ -10 & 6 \end{vmatrix} \begin{vmatrix} 8.16 \\ -68.00 \end{vmatrix} 10^{-5} = \begin{vmatrix} 1.0880 \\ -0.4896 \end{vmatrix} 10^{-2}$$

so

$$\boldsymbol{\lambda}^1 = \begin{vmatrix} 183.8 \\ 204.2 \end{vmatrix}$$

Thus we have found an approximate shape for the fundamental mode and can already state that

$$183.8 \text{ (rad/sec)}^2 < \omega_1^2 < 204.2 \text{ (rad/sec)}^2$$

or

$$0.440 \text{ sec} < T_1 < 0.464 \text{ sec}$$

Indeed, from Eq. 4.17,

$$\omega_1^2 < 199.6 \text{ (rad/sec)}^2$$

and

$$T_1 > 0.445 \text{ sec}$$

and we might be content to take this value for the fundamental period.

We begin a second cycle with

$$\mathbf{z}^1 = \begin{vmatrix} 2.2 \\ -1.0 \end{vmatrix}$$

which are rounded-off numbers proportional to those of $\mathbf{z}^1/\omega^2$ found at the end of the first cycle. This gives

$$\frac{\mathbf{z}^2}{\omega^2} = \begin{vmatrix} 1.1288 \\ -0.4978 \end{vmatrix}$$

$$\boldsymbol{\lambda}^2 = \begin{vmatrix} 194.9 \\ 200.9 \end{vmatrix}$$

whence

$$0.443 \text{ sec} < T_1 < 0.450 \text{ sec}$$

128 NUMERICAL COMPUTATION OF STEADY-STATE RESPONSES Chap. 4

and, from Eq. 4.17, $T_1 < 0.445$ sec. This coincides with the exact answer, $\omega_1 = 14.123$ rad/sec, to three significant figures. The exact solution gives

$$\mathbf{z}_1 = \begin{vmatrix} 2.287 \\ -1.000 \end{vmatrix}$$

from which the result found in the second cycle differs by 4 percent (after adjusting one value of **z**). Free vibrations consist, therefore, of a displacement (2.287 cm) sin 14.123t and a rotation (-0.01 rad) sin 14.123t about a vertical axis through the centroid.

Example 4.6. Compute the fundamental frequency of vibration of the tapered beam in Fig. 4.8a. The mass per unit length and the width of the beam vary linearly between the supports.

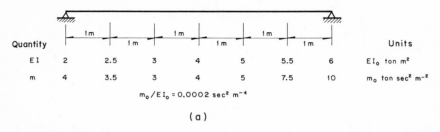

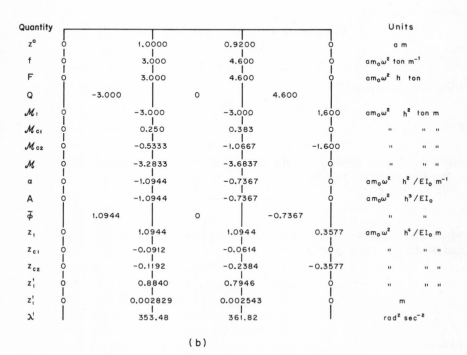

Figure 4.8 (a and b). Example 4.6, first mode.

Solution. For this purpose we choose to take three segments of equal length in one half of the beam and compute forces, moments, and deflections at the end points of these segments. The first cycle is worked out in Fig. 4.8b. We begin by computing the mass per unit length and the product EI at the interior mesh points and then enter the first cycle proper. The assumed deflections z^0 are expressed in arbitrary units of length a. From these we compute, in terms of the unknown ω^2, the inertia forces per unit length $f/\omega^2 = mz^0$. Next we use the alternate solution exemplified in Fig. 4.5c to arrive at the bending moments and again at the deflections z^1 that mark the end of the first cycle. The line of λ^1 serves as a basis for deciding whether additional cycles are in order.

After several cycles we find $\omega_1^2 = 358.3$ (rad/sec)^2. The solution with six equal segments gives $\omega_1^2 = 341.3$ (rad/sec)^2. Since we have only used Eq. 4.8 to find equivalent concentrations, we have errors of the order of h^4 in ω_1^2. If two values of ω_1^2 are computed with m and n equal segments, respectively, the assumption that the errors are proportional to h^4 leads immediately to the extrapolation formula

$$\omega_1^2 = \frac{m^4(\omega_1^2)_m - n^4(\omega_1^2)_n}{m^4 - n^4}$$

where $(\omega_1^2)_{m,n}$ are the corresponding computed values of ω_1^2. If $m = 2n$,

$$\omega_1^2 = \frac{16(\omega_1^2)_{2n} - (\omega_1^2)_n}{15}$$

Using this extrapolation criterion we find $\omega_1^2 = 340.1$ (rad/sec)^2, or $\omega_1 = 18.44$ rad/sec, which differs from the exact solution by 0.01 percent. Ordinarily such high accuracy is unwarranted; the solution with three equal segments would probably have been adequate.

In this example we introduced linear corrections to satisfy the boundary conditions of zero end moments and no deflection. The method has distinct advantages over the more orthodox one of computing reactions and end slopes because it allows a more systematic treatment, an automatic verification of the order of magnitude of the computed values (because the deflections which are estimated are based on judgment), and the use of smaller numbers in the computations, because the correction configurations can be made quite small.

Appropriate linear corrections should be introduced for other end conditions, such as those associated with instability or indeterminateness under static loads.

In the calculation of higher modes it often happens that one or more node points fall near points of zero deflection. (In flexural systems and in shear beams, points of zero deflection coincide with inflection points.) In such cases, convergence may be hampered seriously because a small change in the deflection curve from one cycle to the next may introduce large percentage errors in the deflections of those points, or even change their sign. When this happens it is advisable to disregard the node points in question in the computation of the λ's. In the foregoing example, an odd number of segments was chosen purposely to obviate this situation.

4.5 Iteration Procedures for Higher Natural Modes

Two procedures will be described in the present section for the calculation of higher natural modes of vibration. They are discussed in detail by Godden (1965).

4.5.1 Procedure of Ratio Elimination. This procedure is useful for finding the second mode when ω_1 is known accurately. If a configuration $\mathbf{z}^0$ is assumed and is intended to resemble $\mathbf{z}_2$, it can be written as in Eq. 4.18. After running one cycle of the iteration procedure described in the foregoing section we find $\mathbf{z}^1$ as given by Eq. 4.19. By making $\omega = \omega_1$ in the latter expression we obtain

$$\mathbf{z}^1 \big|_{\omega=\omega_1} = a_1 \mathbf{z}_1 + \sum_{n=2}^{N} a_n \frac{\omega_1^2}{\omega_n^2} \mathbf{z}_n$$

and

$$\mathbf{z}^1 \big|_{\omega=\omega_1} - \mathbf{z}^0 = \sum_{n=2}^{N} a_n \left(\frac{\omega_1^2}{\omega_n^2} - 1 \right) \mathbf{z}_n$$

Taking this as the initial configuration, the interaction procedure will converge to the second natural mode. It is important that in each cycle the initial configuration be taken exactly proportional to the one found in the previous configuration, to minimize roundoff errors that would favor a creeping in of the fundamental mode and make the procedure eventually converge to it. At intermediate cycles of the calculation it may be worth repeating the device through which the fundamental is eliminated.

In principle this procedure could be applied to the calculation of the third and higher modes, once $\omega_2 \ldots$ were known, but the resulting computations are impractical. The fact that it is necessary to subtract quantities of the same order of magnitude causes a loss of accuracy that limits the usefulness of the procedure, especially beyond the second natural mode.

4.5.2 Procedure of Orthogonal Functions. This is a more powerful tool for computing higher modes. It demands an accurate knowledge of the shapes of all modes lower than the one to be computed.

Directly from Eq. 4.20 we get the coefficient of the first, or, more generally, of the nth natural mode in the natural-mode series expansion of the assumed configuration as

$$a_n = \frac{\mathbf{z}_n^T \mathbf{M} \mathbf{z}^0}{\mathbf{z}_n^T \mathbf{M} \mathbf{z}_n} \tag{4.22}$$

This allows us to eliminate any natural-mode component from the assumed configuration. If we wish to compute the ith natural mode, we assume its shape as $\mathbf{z}^0$, compute

$$\mathbf{z}^0 - \sum_{n=1}^{i-1} a_n \mathbf{z}_n$$

Sec. 4.5   ITERATION PROCEDURES FOR HIGHER NATURAL MODES   131

where a_n has been obtained from Eq. 4.22 and from knowledge of the first $i-1$ natural modes, and apply the iteration procedure, which will converge to the desired natural mode.

As with the first procedure it is necessary to minimize roundoff errors, and it may be convenient occasionally to sweep the components of the first $i-1$ modes, that may have crept in, away from the computed configuration at the end of some cycles. These corrections will be quite small normally and will not hinder the accuracy of computation appreciably.

When using equivalent concentrations of inertia forces, Eq. 4.22 cannot be

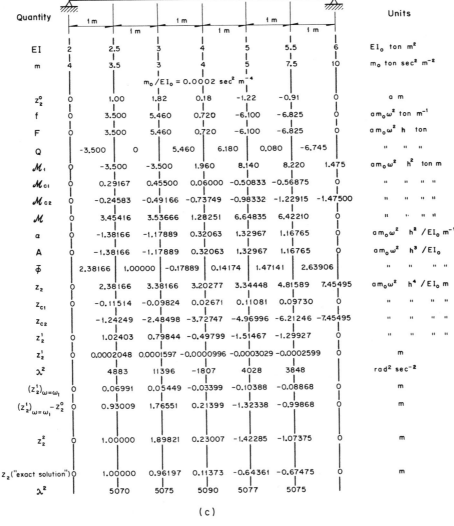

Figure 4.8 (c). Example 4.8, second mode.

132 NUMERICAL COMPUTATION OF STEADY-STATE RESPONSES   Chap. 4

applied directly. One way of using this relation, then, is to replace the components of the inertia matrix with the product of mass per unit length times the spacing between node points, except at free ends, where one would use half this value. This version of Eq. 4.22 is used in Example 4.7.

Example 4.7. Compute the second natural mode of vibration of the system analyzed in Examples 4.3 and 4.5 (Fig. 4.7a).

Solution. We shall use the "procedure of ratio elimination" making use of our knowledge of ω_1. Let us take, arbitrarily

$$\mathbf{z}^0 = \begin{vmatrix} 1 \\ 0 \end{vmatrix}$$

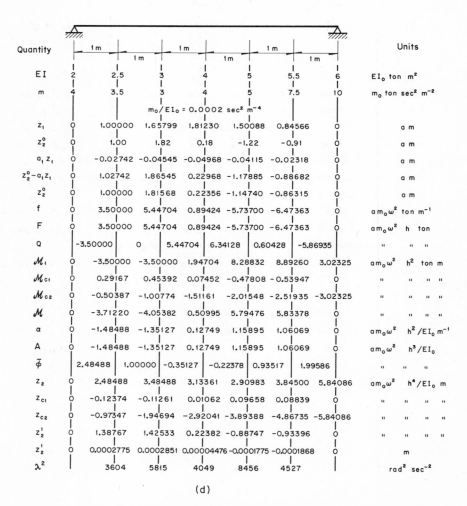

Figure 4.8 (d). Example 4.8, second mode.

Sec. 4.5 ITERATION PROCEDURE FOR THE FUNDAMENTAL MODE 133

The first cycle of the iteration procedure we used for computing z_1 consists of computing

$$\frac{F^0}{\omega^2} = \begin{vmatrix} 4.08 & 0 \\ 0 & 68 \end{vmatrix} \begin{vmatrix} 1 \\ 0 \end{vmatrix} = \begin{vmatrix} 4.08 \\ 0 \end{vmatrix}$$

$$\frac{z^1}{\omega^2} = \begin{vmatrix} 50 & -10 \\ -10 & 6 \end{vmatrix} \begin{vmatrix} 4.08 \\ 0 \end{vmatrix} 10^{-5} = \begin{vmatrix} 2.040 \\ -0.408 \end{vmatrix} 10^{-3}$$

with $\omega = \omega_1 = 14.123$ rad/sec, this becomes

$$z^1 = \begin{vmatrix} 0.4069 \\ -0.0814 \end{vmatrix}$$

whence

$$z^1 - z^0 = \begin{vmatrix} -0.5931 \\ -0.0814 \end{vmatrix}$$

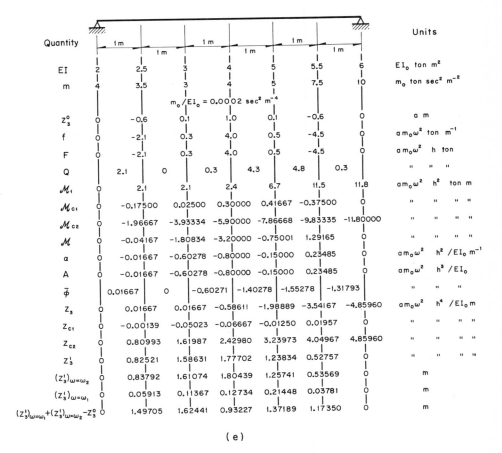

Figure 4.8 (e). Example 4.8, third mode.

134 NUMERICAL COMPUTATION OF STEADY-STATE RESPONSES Chap. 4

One cycle of iteration beginning with the latter configuration as z^0 gives

$$\frac{z^1}{\omega^2} = \begin{vmatrix} 65.628 \\ 9.013 \end{vmatrix} 10^{-5}$$

and

$$\lambda^1 = \begin{vmatrix} 903.59 \\ 903.14 \end{vmatrix}$$

The answer is exact (save for roundoff errors) because the system has only two degrees of freedom; subtracting the component of the fundamental mode from any configuration leaves precisely the second natural mode.

Example 4.8. Compute the second natural mode of vibration of the beam analyzed in Example 4.6 (Fig. 4.8a), using six equal segments.

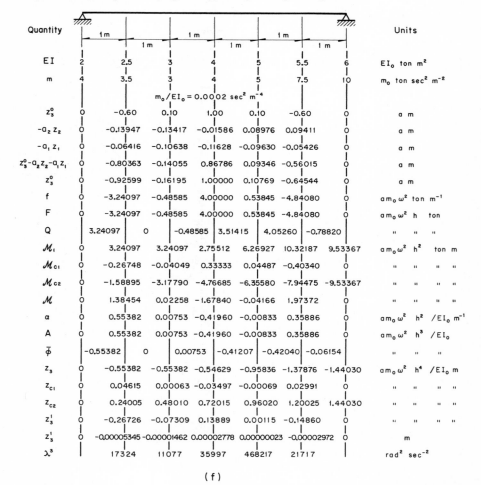

(f)

Figure 4.8 (f). Example 4.8, third mode

Solution. The first cycles of computation with the first two procedures described are shown in Figs. 4.8c and 4.8d, together with the "exact" answer (exact for the discrete system). The first cycles for the third mode are shown in Fig. 4.8e and 4.8f. It is seen that important components of the lower modes creep in with each cycle. This is due to the lack of orthogonality of the three modes because each one actually corresponds to a different discrete approximation to the distributed-parameter system. Moreover, the number of segments would be considered insufficient normally for calculation of the third natural mode. Ordinarily in a problem of this type it would be preferable to use a greater number of segments and resort to the polygonal approximation.

4.6 Rayleigh's Method

If we assume that a conservative system vibrates harmonically with a configuration z, we can compute its maximum potential energy (which it attains when the kinetic energy is zero) from the inertial forces Kz and the deflections z:

$$\max W_p = \frac{1}{2} z^T K z \tag{4.23}$$

The maximum kinetic energy is attained when all deflections are zero and the velocities are maximum. Assuming that the system describes a harmonic motion, the amplitudes of velocity are ωz. Hence, the maximum kinetic energy is

$$\max W_k = \frac{\omega^2}{2} z^T M z \tag{4.24}$$

From the assumption that the system is conservative, it follows that these two energies must be equal. Hence,

$$\omega^2 = \frac{z^T K z}{z^T M z} \tag{4.25}$$

The second member of Eq. 4.25 is known as the *Rayleigh quotient*. We could also have derived it by premultiplying the equation of motion

$$(K - \omega^2 M)z = 0 \tag{4.26}$$

(see Sections 2.1, 2.2, and 2.10) by z^T.

Clearly, if the assumed configuration coincides with the nth natural mode of vibration, the circular frequency obtained from Eq. 4.25 will coincide with the nth natural circular frequency.

The Rayleigh quotient has other useful properties. To derive them, express z in terms of the natural modes:

$$z = \sum_n a_n z_n \tag{4.27}$$

Since the nth natural mode satisfies Eq. 4.26, premultiplication of this equation by z^T gives

$$z_m^T K z_n - \omega_n^2 z_m^T M z_n = 0$$

136 NUMERICAL COMPUTATION OF STEADY-STATE RESPONSES   Chap. 4

so the property of orthogonality,
$$\mathbf{z}_m^T \mathbf{M} \mathbf{z}_n = 0 \quad \text{when } m \neq n$$
can also be written
$$\mathbf{z}_m^T \mathbf{K} \mathbf{z}_n = 0 \quad \text{when } m \neq n$$
Hence, the numerator in Eq. 4.25 is
$$\mathbf{z}_n^T \mathbf{K} \mathbf{z} = \sum_n a_n^2 \mathbf{z}_n^T \mathbf{K} \mathbf{z}_n$$
$$= \sum_n (a_n \omega_n)^2 \mathbf{z}^T \mathbf{M} \mathbf{z}_n$$

In view of Eq. 4.26, dividing both members of Eq. 4.25 by ω^2 gives
$$\frac{\sum_n (a_n \omega_n)^2 \mathbf{z}_n^T \mathbf{M} \mathbf{z}_n}{\sum_n (a_n \omega)^2 \mathbf{z}_n^T \mathbf{M} \mathbf{z}_n} \tag{4.28}$$

If we make $\omega = \omega_1$, every term in the numerator will be at least as great as the corresponding term in the denominator, so the resulting ratio will not be smaller than 1. It follows that ω in Eqs. 4.25 and 4.28 cannot be smaller than ω_1. Replacing ω with the highest natural circular frequency leads to the conclusion that ω cannot exceed this value. Consequently the Rayleigh quotient always lies between the smallest and the greatest natural circular frequency of the system.

The assumed configuration $\mathbf{z}$ may be regarded in Eq. 4.25 as equal to its component of the mth natural mode $a_m \mathbf{z}_m$ plus an error term whose expansion in terms of the other natural modes is $\sum_n a_n \mathbf{z}_n - a_m \mathbf{z}_m$. The substitutions which led to Eq. 4.28 show that in both the numerator and the denominator of Eq. 4.25 every component of the error term is squared. Hence if $\mathbf{z}$ differs from $a_m \mathbf{z}_m$ by a *first-order* error, the resulting ω^2 computed from the Rayleigh quotient will differ from ω_m^2 by a *second-order* error. Therefore, even relatively crude estimates of the natural modes permit rather accurate calculation of the corresponding frequencies.

Now we shall prove that the Schwartz quotient (second member in Eq. 4.17) is never greater than the Rayleigh quotient. The ratio of the Rayleigh quotient to that of Schwartz is
$$\frac{\sum_m (a_m/\omega_m^2)^2 \mathbf{z}_m^T \mathbf{M} \mathbf{z}_m \sum_n (a_n \omega_n)^2 \mathbf{z}_n^T \mathbf{M} \mathbf{z}_n}{\sum_m (a_m/\omega_m)^2 \mathbf{z}_m^T \mathbf{M} \mathbf{z}_m \sum_n a_n^2 \mathbf{z}_n^T \mathbf{M} \mathbf{z}_n}$$
By letting
$$b_n = a_n^2 \mathbf{z}_n^T \mathbf{M} \mathbf{z}_n \geq 0$$
the foregoing ratio may be written
$$\frac{\sum_m \sum_n b_m b_n (\omega_n/\omega_m^2)^2}{\sum_m \sum_n b_m b_n (1/\omega_m)^2}$$
and we aim to show that it is never smaller than 1. Those terms for which

Sec. 4.6 RAYLEIGH'S METHOD 137

$m = n$ are identical in the numerator and denominator. Hence it suffices to compare those terms for which $m \neq n$. These may be grouped in pairs in both numerator and denominator, respectively, of the form $b_m b_n [(\omega_n^2/\omega_m^4) + (\omega_m^2/\omega_n^4)]$ and $b_m b_n [(1/\omega_m^2) + (1/\omega_n^2)]$ and hence it is sufficient to show that

$$\frac{x/y^2 + y/x^2}{1/x + 1/y} \geq 1 \tag{4.29}$$

where x and y stand for ω_m^2 and ω_n^2, respectively, and are both positive. The proof of Eq. 4.29 is equivalent to that of

$$\frac{x^3 + y^3}{x^2 y^2 (1/x + 1/y)} \geq 1$$

or

$$\frac{x^3 + y^3}{xy(x + y)} \geq 1$$

or

$$3 + \frac{x^3 + y^3}{xy(x + y)} = \frac{(x + y)^2}{xy} \geq 4$$

or, since $xy > 0$,

$$(x + y)^2 \geq 4xy$$

which is the same as

$$x^2 + 2xy + y^2 \geq 4xy$$

or

$$(x - y)^2 = 0$$

which is certainly true.

Therefore the Schwartz quotient is surely never greater than the Rayleigh quotient, and consequently it always lies between the Rayleigh quotient and the square of the fundamental circular frequency. It always gives, therefore, a better approximation to this squared circular frequency. Nevertheless Rayleigh's method is advantageous over a single cycle of the iteration procedure in those structures for which the stiffness matrix is known but not the flexibility matrix. The reverse is true when the flexibility matrix is known, whether or not the stiffness matrix is available.

If the Rayleigh quotient is computed for various assumed configurations, the lowest value of the quotient is the best among the corresponding approximations to the squared fundamental circular frequency. The configuration associated with the lowest quotient is the closest approximation to the fundamental mode among the configurations assumed.

The method of Rayleigh–Ritz, which will not be covered here, represents each relevant natural mode as a linear combination of assumed configurations and computes the corresponding coefficients in such a manner as to minimize the sum of squared errors. This criterion can be expressed in terms of energy. Many variations of the method are possible on the basis of other criteria of adjustment.

Example 4.9. Use Rayleigh's method to find the fundamental circular frequency of the system analyzed in Examples 4.3, 4.5, and 4.7 (Fig. 4.7a).

Solution. As in example 4.5 assume

$$\mathbf{z} = \begin{vmatrix} 2 \\ -1 \end{vmatrix}$$

From Example 4.3, the inertia and stiffness matrices, in units of metric tons, meters, and seconds, are

$$\mathbf{M} = \begin{vmatrix} 4.08 & 0 \\ 0 & 68 \end{vmatrix} \qquad \mathbf{K} = \begin{vmatrix} 3000 & 5000 \\ 5000 & 25000 \end{vmatrix}$$

Hence,

$$\mathbf{z}^T\mathbf{Kz} = 17{,}000 \qquad \mathbf{z}^T\mathbf{Mz} = 84.32$$

and Eq. 4.25 gives $\omega = 14.199$ rad/sec, which should be compared with 14.10 rad/sec resulting from the Schwartz quotient applied to the first cycle of the iteration procedure in Example 4.5, and with $\omega_1 = 14.123$ rad/sec, which is the exact answer. The other conclusions of the foregoing article are readily confirmed in this example.

4.7 Two Energy Methods due to Southwell[2]

Assume the given system replaced with a finite number of subsystems, such that the ith subsystem is characterized by the stiffness and inertia matrices $\mathbf{K}_i$ and $\mathbf{M}_i$. Let the subsystem be chosen so that

$$\sum_i \mathbf{K}_i = \mathbf{K}$$

and $\mathbf{M}_i = \mathbf{M}$ for all i. Assume all the subsystems deformed into the configuration of the fundamental mode of the given system. Call this configuration $\mathbf{z}$ and let ω_i denote the fundamental circular frequency of the ith subsystem. Since $\mathbf{z}$ does not necessarily coincide with the fundamental mode of the ith subsystem, the discussion presented in connection with the Rayleigh quotient allows us to write

$$\omega_i^2 \leq \frac{\mathbf{z}^T\mathbf{K}_i\mathbf{z}}{\mathbf{z}^T\mathbf{Mz}} \qquad (4.30)$$

It follows that

$$\sum_i \omega_i^2 \leq \frac{\sum_i \mathbf{z}^T\mathbf{K}_i\mathbf{z}}{\mathbf{z}^T\mathbf{Mz}} = \frac{\mathbf{z}^T\mathbf{Kz}}{\mathbf{z}^T\mathbf{Mz}} = \omega^2$$

where ω is the fundamental frequency of the given system. Under these conditions, then,

$$\omega^2 \geq \sum_i \omega_i^2 \qquad (4.31)$$

---

[2] The present discussion is based on the more elementary presentation by Jacobsen and Ayre (1958, pp. 112–25) where these methods are described and illustrated with many examples.

The equation of motion could also have been written as

$$(\mathbf{M}^{-1}\mathbf{K} - \omega^2\mathbf{I})\mathbf{z} = 0$$

after premultiplication by $\mathbf{M}^{-1}$, where $\mathbf{I}$ is the unit diagonal matrix. In this form $\mathbf{M}^{-1}\mathbf{K}$ plays the role of the stiffness matrix, so that the ith subsystem may be characterized by an inertia matrix that is not necessarily equal to $\mathbf{M}$, provided that

$$\sum_i \mathbf{M}_i^{-1}\mathbf{K}_i = \mathbf{M}^{-1}\mathbf{K}$$

Under these more general conditions, Eq. 4.30 would have been written as

$$\omega_i^2 \leq \frac{\mathbf{z}^T\mathbf{M}^{-1}\mathbf{K}\mathbf{z}}{\mathbf{z}^T\mathbf{z}}$$

leading again to Eq. 4.31.

Now let the subsystems be such that

$$\sum_i \mathbf{K}_i^{-1}\mathbf{M}_i = \mathbf{K}^{-1}\mathbf{M}$$

Premultiplication of the equation of motion by $\mathbf{K}^{-1}$ gives

$$(\mathbf{I} - \omega^2\mathbf{K}^{-1}\mathbf{M})\mathbf{z} = 0$$

so that

$$\omega_i^{-2} \geq \frac{\mathbf{z}^T\mathbf{K}_i^{-1}\mathbf{M}\mathbf{z}}{\mathbf{z}^T\mathbf{z}}$$

and

$$\sum_i \omega_i^{-2} \geq \frac{\mathbf{z}^T\mathbf{K}^{-1}\mathbf{M}\mathbf{z}}{\mathbf{z}^T\mathbf{z}} = \omega^{-2}$$

or

$$\omega^2 \geq \frac{1}{\sum_i \omega_i^{-2}} \qquad (4.32)$$

This expression can also be written in the form

$$T^2 \leq \sum_i T_i^2 \qquad (4.33)$$

where T and T_i are the fundamental periods of the system and of the ith subsystem respectively.

In particular Eqs. 4.32 and 4.33 hold when $\mathbf{M}_i = \mathbf{M}$ and

$$\sum_i \mathbf{K}_i^{-1} = \mathbf{K}^{-1}$$

and the method of analysis implied thereby is known as that of Southwell–Dunkerley. Both Eqs. 4.31 and 4.32 can be derived from energy considerations.

We are now in possession of simple methods that give us two lower bounds to the fundamental circular frequency, in Eqs. 4.31 and 4.32. Either of them can be used together with Ralyeigh's method to bracket this frequency. By judiciously picking the subsystems, either of the two lower bounds can be made to give a close estimate of the fundamental circular frequency.

140 NUMERICAL COMPUTATION OF STEADY-STATE RESPONSES   Chap. 4

Typical applications of Eq. 4.31 include the calculation of the fundamental frequency of flexural beams under the action of axial tension. In one subsystem only the flexural stiffness is taken into account and the axial force is neglected; in the second subsystem the beam is treated as a string under axial tension, and both subsystems are assigned the same masses as the original beam (Jacobsen and Ayre, 1958, pp.112–25).

The Southwell–Dunkerley method has proved even more useful, for example in the analysis of continuous beams in which individual masses are assigned to the various subsystems. It is useful also in the analysis of tall buildings or chimney stacks, in which the flexural and shearing deformations and the effects of rotary inertia and base rotation are treated separately (Jacobsen and Ayre, 1958, pp.112–25 and 502–504).

The methods presented in this section are practical to arrive at rapid estimates of the effects of small changes in structural parameters, as illustrated in Example 4.11.

Example 4.10. Use energy methods to estimate the fundamental frequency of the system in Fig. 4.9(a).

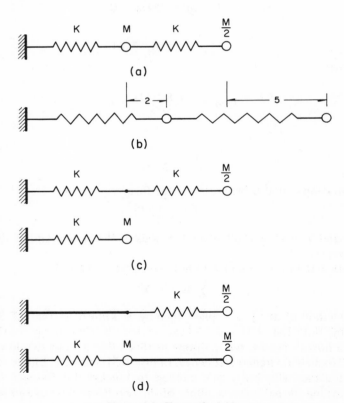

Figure 4.9. Example 4.10.

Sec. 4.7   TWO ENERGY METHODS DUE TO SOUTHWELL   141

Solution. First we shall apply Rayleigh's method to obtain an upper bound. Assume that the displacements are as shown in Fig. 4.9b. The Rayleigh quotient (Eq. 4.25) gives

$$\omega^2 \leq \frac{1 \times 2^2 + 1 \times 1^2}{1 \times 2^2 + (1/2)3^2} \cdot \frac{K}{M} = 0.5882 \frac{K}{M}$$

whence

$$\omega \leq 0.767 \sqrt{\frac{K}{M}}$$

To apply Eq. 4.32, consider the two subsystems shown in Fig. 4.9c. In the first one we have suppressed the center mass, and in the second we have omitted the end mass. Hence, the condition that the sum of the $\mathbf{K}^{-1}\mathbf{M}$ matrices of the subsystems be equal to that of the original system is satisfied. We find

$$\omega_1^2 = \omega_2^2 = \frac{K}{M}$$

so that

$$\omega^2 \geq \frac{1}{2} \frac{K}{M}$$

and

$$\omega \geq 0.707 \sqrt{\frac{K}{M}}$$

We could also have proceeded as in Fig. 4.9d, where we have replaced in each subsystem one spring with a rigid connection. Here,

$$\omega_1^2 = \frac{2K}{M} \qquad \omega^2 = \frac{2}{3}\frac{K}{M}$$

and

$$\omega^2 \geq \frac{K/M}{1/2 + 3/2} = \frac{1}{2} \frac{K}{M}$$

which coincides with the solution based on Fig. 4.9c.

We conclude that ω lies between 0.707 and 0.767 times $\sqrt{K/M}$. Indeed, according to the data in Table 4.1, the exact solution is

$$\omega = \frac{1}{2} 1.530 \sqrt{\frac{K}{M}} = 0.765 \sqrt{\frac{K}{M}}$$

Example 4.11. Estimate the effect, on the fundamental frequency of the structure analyzed in the foregoing example, of reducing the stiffness of the first spring by 20 percent and that of the second spring by 10 percent.

Solution. We consider three subsystems. One is the original structure with $\omega_1 = 0.765\sqrt{K/M}$; the second one has flexibility of $0.250\,K^{-1}$ in its first spring and a rigid second spring; the last system has a rigid first spring and a flexibility of $0.111\,K^{-1}$ in its second spring. All subsystems have the same masses. We find

$$\omega_1^{-2} = 1.706 \frac{M}{K}$$

$$\omega_2^{-2} = 0.250 \times 1.5 \frac{M}{K} = 0.375 \frac{M}{K}$$

$$\omega_3^{-2} = 0.111 \times 0.5 \frac{M}{K} = 0.056 \frac{M}{K}$$

$$\omega^2 \geq \frac{K/M}{1.706 + 0.375 + 0.056} = 0.468 \frac{K}{M}$$

$$\omega \geq 0.684 \sqrt{\frac{K}{M}}$$

This differs from the exact answer by 0.87 percent.

Example 4.12. A tall building has X-braced frames. Its fundamental period was computed neglecting the stiffness of the braces and found to be 3.0 sec. Next the fundamental period was computed taking the braces into account, but assuming pin connections between structural members, and was found to be 2.0 sec. Estimate the fundamental period of the building.

Solution. Since the subsystems analyzed are such that

$$\mathbf{K}_1 + \mathbf{K}_2 \cong \mathbf{K}$$

Eq. 4.31 applies:

$$T^{-2} \geq T_1^{-2} + T_2^{-2}$$

or

$$T^2 \leq \frac{1}{\frac{1}{9} + \frac{1}{4}}$$

from which $T \leq 1.66$ sec. If the shapes of the fundamental modes in the two subsystems are alike, T will be very close indeed to 1.66 sec.

4.8 Effects of Changing a Single Mass or a Stiffness Element

Consider a conservative linear system. Suppose that one element in its inertia matrix and the symmetrical of this element are increased in the same amount leaving all other elements unchanged. Using energy methods (Rayleigh, 1945; Bisplinghoff, Ashley, and Halfmann, 1955, pp.774–79) as well as through other procedures (Courant and Hilbert, 1953, pp.285–86), it has been shown that the natural frequencies of the new system are all equal to or smaller than those of the original system and that, except for the lowest frequency, they are "sandwiched" between those of this system. In other words, if ω_n and ω'_n denote the nth natural circular frequencies of the original and of the modified system, respectively, it has been shown that an increase in a particular member $M_{ij} = M_{ji}$ of the inertia matrix gives

$$\omega_{n-1} \leq \omega'_n \leq \omega_n \quad \text{if } n > 1$$

and

$$0 \leq \omega'_1 \leq \omega_1$$

Evidently the same result is obtained if we decrease a single member of the stiffness matrix or increase one member of the flexibility matrix.

As a corollary, if for some n in the original system we had $\omega_{n-1} = \omega_n$, we obtain $\omega'_n = \omega_n$. These results are most helpful in preliminary studies of design alternatives.

4.9 Step-by-Step Procedures for Steady-State Vibrations

This section is presented essentially as a preamble to Holzer's table and related methods for the calculation of natural modes.

In some dynamic systems it is possible to compute all the pertinent generalized forces, deformations, and velocities at one interior point after specifying the conditions at a boundary. Then it is possible to compute the conditions at a second interior point from those at the first point, and so on, until reaching the remaining boundary or boundaries. If the corresponding end conditions are not satisfied, corrections are introduced that do not violate conditions at the first boundary or conditions of compatibility and equilibrium anywhere in the system. The corrections are multiplied by such factors as to make the corrected solution comply with all the boundary conditions.

The simplest structures to analyze in this manner are conservative, discrete, close-coupled systems subjected to harmonic excitation due to external forces and imposed displacements, all of them in phase. Under steady-state conditions all the masses vibrate with the same frequency and in phase with the disturbance. Hence the maximum inertia forces are readily computed from the displacements, and all that has to be estimated is the amplitude of vibration at a free end or the amplitude of the reaction at a fixed end (or, equivalently, the force in the first interior spring element, or its elongation, or the displacement of the first interior mass).

Consider the simply-coupled system in Fig. 4.10. Its ends are respectively fixed and free and on its masses act the external forces $P_i \sin \omega t$. We shall take external (P) and inertia (F) forces and displacements (x) as positive from left to right, spring forces (Q) as positive when they constitute tension, and spring deformations (Δx) when they constitute elongations. Forces, deformations, and displacements will be computed only at instants when $\sin \omega t = 1$.

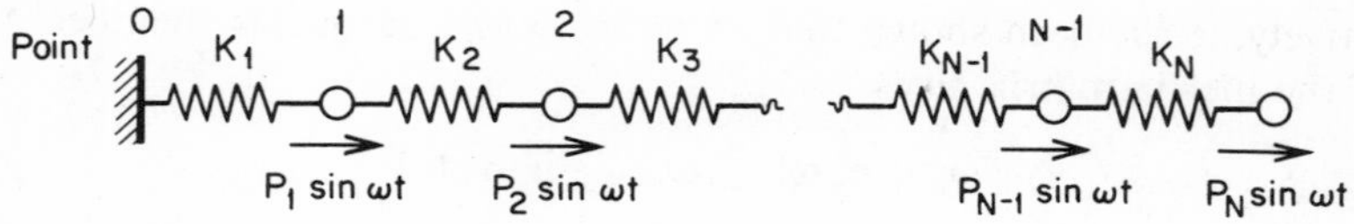

Figure 4.10. Closely-coupled system.

We may begin by specifying zero end displacement at the support and estimate the amplitude of the corresponding reaction. Call it P_0. Then $Q_1 = -P_0$, $\Delta x = Q_1/K_1$, $x_1 = \Delta x_1$, and $F_1 = -\omega^2 x_1 M_1$. (We could equally well have estimated Q_1, Δx_1, or x_1; in the latter case we would have computed $\Delta x_1 = x_1$, $Q_1 = K_1 \Delta x_1$, and $P_0 = -Q_1$.) Equilibrium of the first mass requires the appearance of a spring force $Q_2 = Q_1 - P_1 - F_1$. This is associated with a spring deformation $\Delta x_2 = Q_2/K_2$. The displacement of the second mass is $x_2 = x_1 + \Delta x_2$, and so on.

The inertia force in the last mass will be $F_N = -\omega^2 M_N x_N$. Equilibrium will require the appearance of a force $Q_{N+1} = Q_N - P_N - F_N$ at a nonexistent spring, and in general Q_{N+1} will differ from zero. Next we introduce a correction computed in the same manner as this configuration, beginning with an arbitrary P_0, or an arbitrary Q_1, Δx_1, or x_1, save that we now take $P_i = 0$ at every mass, so that the correction will not alter the dynamic equilibrium attained at all interior points. We will find the need for a force $Q_{N+1,c}$ at the fictitious end spring. Hence, by adding $-Q_{N+1}/Q_{N+1,c}$ times the corrective configuration to the original configuration we satisfy all the boundary, equilibrium, and compatibility conditions. Values at any instant are obtained by multiplying the computed values by $\sin \omega t$.

We could also have started at the free end, estimating x_N and specifying $Q_{N+1} = 0$. Hence, $Q_N = P_N + F_N$, $\Delta x_N = Q_N/K_N$, $x_{N-1} = x_N - \Delta x_N$, and so on. We would generally have found $x_0 \neq 0$ and would have introduced a correction to satisfy this condition.

The second variation is especially useful when all external forces P_i are zero and the boundary condition at the support is $x_0 = a \sin \omega t$, where a is a constant. In this case the solution is found by merely multiplying the initially computed configuration by the appropriate factor. Similarly, if x_0 must be zero and $P_i = 0$ except at $i = N$, the first variation of the method will be convenient, because multiplication of the initially computed configuration by the appropriate factor will yield the solution. Obvious variations can be worked out that are especially suited for disturbances applied at intermediate masses as well as for other end conditions.

Rather than using corrective configurations, the force or the displacement that we have guessed in the foregoing presentation to obtain the initial configuration may be retained as a symbol of unkown value and all quantities expressed in terms of this symbol. Then, on arriving at the end where a boundary condition must be satisfied, we may solve for the unknown quantity and replace it in the computed configuration. For example, we could have left x_1 as unknown. We would have obtained all quantities as first-degree expressions in x_1. By equating Q_{N+1} to zero we can solve for x_1. Alternatively we could have expressed x_1 as a constant plus a factor times x_2, x_2 as a constant plus a factor times x_3, and so on. When reaching the right-hand end we would have solved for x_N. Then, proceeding from right to left we would have successively obtained x_{N-1}, x_{N-2}, ..., x_1.

Systems with viscous damping as well as those in which not all the disturbances are in phase may be dealt with by keeping separate accounts of sine and cosine terms.

In remotely coupled systems it is necessary to estimate two or more forces and/or displacements and introduce as many corrective configurations. Again it is possible to treat one or more of the forces or displacements as parameters of unknown magnitudes and solve for them when arriving at the boundary or boundaries where certain conditions must be fulfilled. In the method due to Goldberg, Bogdanoff, and Moh (1959) for calculation of steady-state and free vibrations of tall buildings with masses concentrated at floor levels, the assumption that all joints in a floor undergo equal rotations and that there is no torsional component of motion, together with neglect of axial and shearing deformations, makes it possible to calculate the configuration of the entire building from two assumed generalized forces or displacements. In the original presentation of the method one generalized displacement is estimated and the other is retained as an unknown parameter. In principle the method may be put in such a form that only numerical computations need be carried out, by assuming two generalized forces or displacements at one end of the building and introducing two correction configurations to cancel errors found in two boundary conditions at the other end of the building. The authors' presentation is preferable from a practical viewpoint because multistory building frames would ordinarily require large values of the corrective quantities, with the ensuing loss of accuracy.

Let subscript i refer to the ith floor level and to the story immediately below it, h denote story height, x lateral displacement, ϕ joint rotation, ψ the drift (story deformations over story height), M the mass assumed concentrated at a floor level, K^c the sum of column stiffnesses in a story, K^g the sum of girder stiffnesses in a floor, M^c the sum of end bending moments in the columns of one story, and M^g the sum of end bending moments in the girders of one floor. M^c is the story moment: it equals the product of story shear and story height. (By "stiffness" we understand EI/L, where $E =$ modulus of elasticity, $I =$ cross-sectional moment of inertia, and $L =$ length of member.) Under the assumptions made, application of, let us say, the slope-deflection equations leads to the following relations.

$$\psi_i = \frac{M_i^c}{12 K_i^c} + \frac{\phi_{i-1} + \phi_i}{2} \qquad (4.34)$$

and

$$-K_i^c \phi_{i-1} + (12 K_i^g + K_i^c + K_{i+1}^c)\phi_i - K_{i+1}^c \phi_{i+1} = \frac{M_i^c + M_{i+1}^c}{2} \qquad (4.35)$$

Also,

$$x_{i-1} = x_i - \psi_i h_i \qquad (4.36)$$

The procedure is applied as follows, again working only at instants when $\sin \omega t = 1$. We estimate the displacement at the top of the building. This is x_N if we say that the structure has N stories. From here we compute the top inertia force,

$$F_N = \omega^2 M_N x_N$$

Adding the external force at that level, say P_N, we may obtain the story shear in the uppermost story. From here,

$$M_N^c = (F_N + P_N) h_N$$

Next we enter Eq. 4.35 with $K_{N+1}^c \phi_{N+1}$ and M_{N+1}^c equal to zero because there are no columns above N. From this expression we get ϕ_N as the sum of a constant and a term proportional to the unknown ϕ_{N-1}. From Eq. 4.34 we obtain ψ_N in terms of ϕ_{N-1}, and Eq. 4.36 gives us x_{N-1} also in terms of ϕ_{N-1}.

The inertia force is computed now at floor $N-1$ from the computed displacement. Adding P_{N-1} and the shear at story N we get the shear in story $N-1$, which allows us to compute M_{N-1}^c. Next, from Eq. 4.35 we can express ϕ_{N-1} in terms of ϕ_{N-2}. The process is repeated down to the foundation. There ϕ_0 is chosen so as to satisfy one boundary condition, for example $\phi_0 = 0$ if the column bases are taken as fixed.

The analysis is repeated to obtain a corrective configuration, taking now $P_i = 0$ at all elevations. This configuration is multiplied by a factor such that its addition to the first configuration will cancel the error in the second boundary condition at the base, for example, $x_0 = 0$. The sum of both configurations solves the problem.

If the structure is subjected solely to harmonic oscillation of the base it suffices to multiply the first configuration by a constant factor. And if the disturbance consists in a single, harmonically oscillating force at the roof, the procedure of analysis may be inverted, going from the base of the building up.

The procedure can be transformed easily into one of successive approximations, with some numerical advantages.[3]

It is possible to improve on the solutions found through application of the present procedure. Once the responses have been computed in this manner, a method, such as moment distribution with floors fixed against sway, permits obtaining story shears that do not, in general, coincide exactly with those found from the forces F and P. These differences may be treated as a new set of disturbances and the analysis repeated to correct the original results.

The methods described for the analysis of close-coupled systems and tall

---

[3] This has been done for calculation of the combined effects of vertical and lateral forces in building frames (Rosenblueth, 1965).

buildings may be regarded as special versions of the very general and versatile method of *transmission matrices*.[4] (The method will be presented here in rather broad lines. We shall use it in a modified form in Chapter 10 for the calculation of earthquake motions at the surface of stratified soil.) Consider a system analyzed as a series of segments and such that specification of certain generalized forces and displacements at any division point, say the $(i-1)$th, between segments allows calculation of the corresponding quantities at the ith point. If these quantities are arranged into a column vector $\mathbf{x}_{i-1}$ for the $(i-1)$th point, linear behavior of the system ensures that $\mathbf{x}_i$ can be obtained by premultiplying $\mathbf{x}_{i-1}$ by a square ("transmission") matrix and adding the effects of external forces that act between points $i-1$ and i:

$$\mathbf{x}_i = \mathbf{T}_i \mathbf{x}_{i-1} + \mathbf{x}_{s,i} \tag{4.37}$$

where $\mathbf{x}_{s,i}$ denotes the effects of these external forces. Proceeding in this way, we can write for the last vector

$$\mathbf{x}_N = \mathbf{S}_1 \mathbf{x}_0 + \sum_{i=1}^{N} \mathbf{S}_i \mathbf{x}_{i,i-1} \tag{4.38}$$

where

$$\mathbf{S}_i = \mathbf{T}_N \mathbf{T}_{N-1} \cdots \mathbf{T}_i \tag{4.39}$$

From an assumed $\mathbf{x}_0$ that satisfies the boundary conditions at point 0 we obtain the entire configuration of the system by applying Eq. 4.37 repeatedly. In general the boundary conditions at point N will not be satisfied. These conditions require application of a correction configuration that complies with Eq. 4.38 with $\mathbf{x}_{s,i} = 0$, everywhere.

There are various ways of arriving at the correction configuration. We may take the corrective $\mathbf{x}_0$ to be a linear combination of as many linearly independent configurations as the boundary conditions at point 0 allow us to choose arbitrarily. Or we may proceed in a manner similar to what is done in the method of Goldberg–Bogdanoff–Moh. Or we may solve for the corrective $\mathbf{x}_0$ thus,

$$\mathbf{x}_0 = \mathbf{S}_1^{-1} \mathbf{x}_N \tag{4.40}$$

writing for $\mathbf{x}_N$ the quantities to be corrected at point N.

For the closely coupled systems treated at the beginning of this article, we may define

$$\mathbf{x}_i = \left| \begin{array}{c} x_i \\ Q_{i+1} \end{array} \right|$$

Then,

$$\mathbf{x}_{s,i} = \left| \begin{array}{c} 0 \\ -P_i \end{array} \right|$$

and

[4] Marguerre (1960). This paper expounds the method in detail for the calculation of natural modes of flexural beams and contains an extensive bibliography on applications to other problems. The reader is referred to Marguerre's work for more sophisticated and expeditious versions of the method than the one described in the present chapter.

$$T_i = \begin{vmatrix} 1 & \dfrac{1}{K_i} \\ \omega^2 M_i & 1 + \omega^2 \dfrac{M_i}{K_i} \end{vmatrix}$$

If 0 is fixed and N is free, the conditions to satisfy are

$$\mathbf{x}_0 = \begin{vmatrix} 0 \\ Q_1 \end{vmatrix} \qquad \mathbf{x}_N = \begin{vmatrix} x_N \\ 0 \end{vmatrix}$$

Advantages of the method of matrices of transmission become manifest in more complicated systems because operations are conveniently systematized and because the method allows analyzing systems having distributed parameters without need to discretize them.

The simplest system having distributed parameters is a uniform shear beam under an oscillating distributed load $p \sin \omega t$ with p constant along the beam axis. At instants of maximum amplitude the differential equation of motion is

$$kx'' - m\ddot{x} = p$$

(see Section 3.2) where k is the stiffness, m the mass per unit length, x the deflection, and the primes denote differentiation with respect to X, the coordinate along the beam axis. This expression can also be written in the form

$$x'' + \left(\frac{\omega}{v}\right)^2 x = \frac{p}{k}$$

where $v^2 = k/m$ and v is the wave velocity in the beam. The general solution of this equation is

$$x = a \cos \mu\xi + b \sin \mu\xi + \frac{p(X_1 - X_0)^2}{k\mu^2}$$

where

$X_0, X_1 =$ abscissas of points 0 and 1, respectively

$$\mu = \frac{\omega(X_1 - X_0)}{v}$$

$$\xi = \frac{(X - X_0)}{(X_1 - X_0)}$$

$a, b =$ constants such as to satisfy the conditions at $X = X_0$

From here,

$$x' = -a\mu \sin \mu\xi + b\mu \cos \mu\xi$$

and the shear is

$$Q = -a\mu k \sin \mu\xi + b\mu k \cos \mu\xi$$

Conditions at $X = X_0$ require

$$a = x(0) \qquad b\mu k = Q(0)$$

Hence

$$x = x(0) \cos \mu\xi + \frac{Q(0)}{\mu K} \sin \mu\xi + \frac{p(X_1 - X_0)^2}{k\mu^2}$$

and

$$\frac{Q}{\mu k} = -x(0) \sin \mu \xi + \frac{Q(0)}{\mu k} \cos \mu \xi$$

so that, if we choose

$$\mathbf{x} = \begin{vmatrix} x \\ \dfrac{Q}{\mu k} \end{vmatrix}$$

we may write

$$\mathbf{x}_1 = \mathbf{T}_1 \mathbf{x}_0 + \mathbf{x}_{s,1}$$

where

$$\mathbf{T}_1 = \begin{vmatrix} \cos \mu & \sin \mu \\ -\sin \mu & \cos \mu \end{vmatrix} \qquad \mathbf{x}_{s,1} = \begin{vmatrix} p(X_1 - X_0)^2 k \mu^2 \\ 0 \end{vmatrix}$$

In this manner it is simple to analyze shear beams whose sections are taken constant for each segment. As will be seen in Chapter 10 the method can be extended easily to systems with internal damping.

An entirely different approach to generalize the step-by-step procedure described for close-coupled systems allows a more systematic treatment of absolutely general discrete structures. The method consists in making the matrix $\mathbf{K} - \omega^2 \mathbf{M}$ tridiagonal. Convenient computer methods are available to accomplish this step (Wilkinson, 1960). Once this has been done the procedure in question applies directly.

4.10 Step-by-Step Procedures for Calculation of Natural Modes

The procedures to be described are, in essence, the same as those presented for the calculation of steady-state responses to harmonic excitation. External forces and imposed displacements will now be zero, and the frequency of vibration will be unknown. The procedures are applied by taking trial values of the frequency, while one generalized force or displacement is assigned an arbitrary amplitude. Those values of the assumed frequency which permit satisfying the boundary conditions are the natural frequencies of the system, and the corresponding computed configurations are its natural modes.

When applied to closely coupled systems, this procedure is known as Holzer's table. Because a single boundary condition need be satisfied at each end, the procedure becomes especially simple. For the system in Fig. 4.10, we begin by assigning ω^2 a tentative value. Next we may assign Q_0 or z_1 an arbitrary magnitude and proceed step by step until reaching point N, where $\mathbf{z}$ denotes a natural mode of vibration. Generally we will find $Q_N \neq 0$ and try a different ω^2, until the right-hand end condition is satisfied. We may also begin at point N, proceed from right to left, and use the displacement of the support as control to decide when we have correctly guessed a natural frequency.

The method may be used without further refinement, merely interpolating

150 NUMERICAL COMPUTATION OF STEADY-STATE RESPONSES Chap. 4

between assumed values of ω^2 in terms of the resulting error in the boundary condition, using graphical or numerical methods for the interpolation. However, the following iteration method is much more convenient. It has been shown (Crandall and Strang, 1957) that, when the boundary condition that is not complied with is one of equilibrium, the following iteration formula

$$\omega'^2 = \omega^2 \frac{\sum\limits_i Q_i \Delta z_i}{\sum\limits_i F_i z_i} \tag{4.41}$$

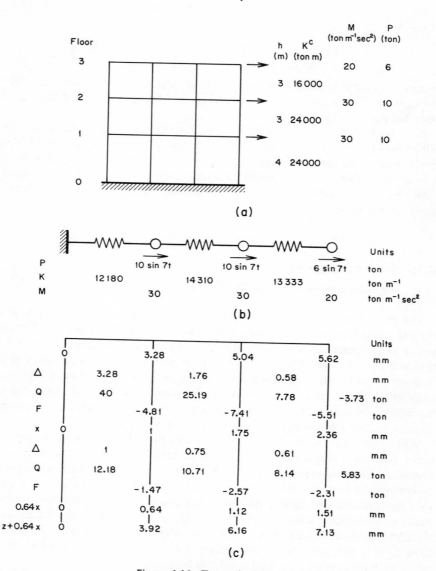

Figure 4.11. Example 4.13.

Sec. 4.10   PROCEDURES FOR CALCULATION OF NATURAL MODES   151

gives a circular frequency ω' that lies closer to one of the natural frequencies of the system, provided ω already lies sufficiently close to that natural frequency. If ω involves an error ϵ, ω' will involve an error of order ϵ^2. Thus, successive application of Eq. 4.41 furnishes a rapidly converging method.

If the boundary condition that was not satisfied is one of displacement at a support, Eq. 4.41 must be replaced with

$$\omega'^2 = \omega^2 \frac{\sum_i F_i z_i}{\sum_i Q_i \Delta z_i} \qquad (4.42)$$

(Damy, 1965).

It is important to verify that the solution satisfies the boundary conditions, for both Eqs. 4.41 and 4.42 give $\omega' = \omega$ (and hence, apparently, indicate that ω is a natural frequency) at certain values of ω that differ from the natural frequencies. Revision of the boundary conditions will, of course, detect such a situation (Damy, 1965).

In the Goldberg–Bogdanoff–Moh procedure, the computations associated with every trial frequency implicitly involve carrying one generalized displacement as an unknown quantity and solving for it at the opposite end, so that only one boundary condition is not satisfied. Equations 4.41 and 4.42 can be adapted to this procedure.

Similar remarks apply to the more general method of transmission matrices. However, when the steady-state procedure is formulated in this manner, there is no need to compute values of z_i at interior points to find

$$z_N = S_1 z_0$$

						Units
P					6 sin 7t	ton
K		12180	14310	13333		ton m⁻¹
M		30	30	20		ton m⁻¹ sec²
x	0	1.64	2.87	3.87		mm
Δx		1.64	1.23	1.00		mm
Q		20	17.59	13.37	9.58	ton
F		−2.41	−4.22	−3.79		ton
x	0	1.03	1.80	2.43		mm
Δx		1.03	0.77	0.63		mm
Q		12.55	11.04	8.39	6	ton
F		−1.51	−2.65	−2.39		ton

(d)

Figure 4.11. Example 4.13.

152 NUMERICAL COMPUTATION OF STEADY-STATE RESPONSES   Chap. 4

which is Eq. 4.38 with $\mathbf{x}_{s,i-1} = 0$, and $\mathbf{S}_1$ is obtained as in Eq. 4.39. Use of Eq. 4.37 can be postponed until the last trial, once a satisfactory value of ω has been attained, unless one wants to use Eq. 4.41 or 4.42 to replace trial and error with successive approximations. In the latter case, Eq. 4.37 must be evaluated for every value of ω.

Example 4.13. Find the steady-state response of the building shown in Fig. 4.11a. The stiffnesses are taken equal to the sum of beam stiffnesses and of column stiffnesses in all frames parallel to the plane of the drawing. Assume column fixity at ground level.

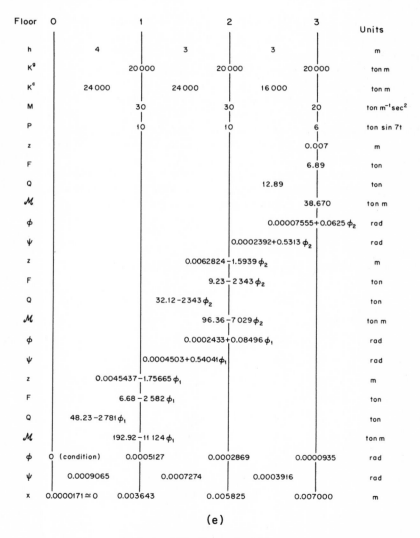

(e)

Figure 4.11. Example 4.13.

Solution. First the building will be analyzed by idealizing it as a shear beam, using Wilbur's formulas (Wilbur, 1935) to compute the story shear stiffnesses. These formulas yield the equivalent close-coupled system in Fig. 4.11b. Its solution is shown in Fig. 4.11c, where we assumed $Q_1 = 40$ ton. This gave us $\Delta z_1 = z_1 = 3.28$ mm. Hence $F_1 = -7^2 \times 3.28 \times 0.03 = -4.81$ ton, $Q_2 = 40 - 4.81 - 10 = 25.19$ ton, and so on. We find $Q_4 = -3.73$ ton. Next we assume, for the corrective configuration, $z_1 = 1$ mm, which gives us $\Delta z_1 = 1$ mm and $Q_1 = 12.18$ ton. Proceeding from left to right with $P_i = 0$ at all levels, we find $Q_4 = 5.83$ ton. Hence we must add $3.73/5.83 = 0.639$ times the corrective configuration to the one we had first computed, to arrive at the solution.

If we had $P_1 = P_2 = 0$ we would have proceeded as in Fig. 4.11d, where we merely multiplied the first configuration by $6/9.58 = 0.626$ to satisfy the condition $P_3 = 6$ ton.

A more accurate solution is computed in Fig. 4.11e using the method of Goldberg, Bogdanoff, and Moh. Here we begin by assuming $z_3 = 7$ mm. Calculations in Fig. 4.11e are self-explanatory.

The difference between the results of idealizing the frames as a shear beam and the more accurate treatment are mostly due in this case to the fact that the building has too small a number of stories to make Wilbur's formulas very accurate. Ordinarily the agreement is far better than found here, provided the frequency of the disturbance does not approach or greatly exceed the fundamental frequence of the structure.

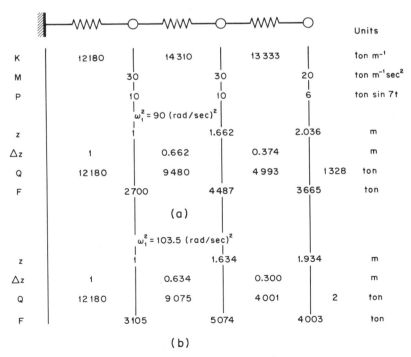

Figure 4.12. Example 4.14.

Example 4.14. Compute the first two natural frequencies of vibration of the building analyzed in the foregoing example.

Solution. For the sake of illustration we shall begin again by idealizing the building as a close-coupled system having the stiffnesses shown in Fig. 4.11b. We shall estimate its fundamental frequency using the Southwell–Dunkerley method. By successively taking $K_1 = K_2 = \infty$, $K_1 = K_3 = \infty$, and $K_2 = K_3 = \infty$, we obtain for ω^2 the values 667, 286, and 152 (rad/sec)^2. Consequently

$$\omega_1^2 \cong \frac{1}{1/667 + 1/286 + 1/152} = 86.4 \text{ (rad/sec)}^2$$

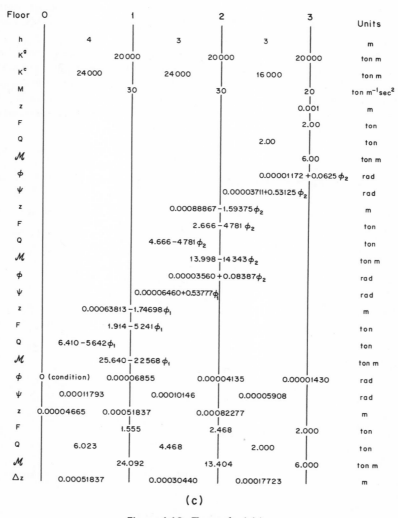

Figure 4.12. Example 4.14.

Sec. 4.10   PROCEDURES FOR CALCULATION OF NATURAL MODES   155

Since this criterion underestimates ω_1^2, we shall assume initially $\omega^2 = 90$ rad/sec². The corresponding computations are shown in Fig. 4.12a, with an assumed $z_1 = 1$ m. This gives an error $Q_4 = 1328$ ton. For the next trial we take $\omega^2 = 104$ (rad/sec)², which is obtained from Eq. 4.41. This gives $Q_4 = -40$ ton. The exact ω_1^2 for the system in Fig. 4.12b is 103.5 (rad/sec)².

To initiate calculation of the second natural frequency we notice that in a uniform shear beam it would equal three times the fundamental frequency. We begin calculations with $\omega^2 = 900$ (rad/sec)². After a first cycle we apply Eq. 4.41 to obtain $\omega^2 = 788$(rad/sec)². The exact answer is $\omega_2^2 = 791.8$ (rad/sec)².

In Fig. 4.12c is shown the first cycle of the Goldberg–Bogdanoff–Moh method for the fundamental frequency, using a trial $\omega^2 = 100$ (rad/sec)². Application of Eq. 4.41 as it stands gives us $\omega^2 = 100$ (rad/sec)² for the second trial. The exact answer is $\omega_1^2 = 106$ (rad/sec)².

The second natural mode computed by this procedure leads after a few cycles to $\omega_2^2 = 921$ (rad/sec)².

As would have been expected, the shear-beam idealization results in a considerably better estimate of the fundamental frequency than of the first harmonic.

Example 4.15. Find the second natural frequency of the stratified soil formation in Fig. 4.13a.

Solution. We shall use the method of matrices of transmission as described in Section 4.8 for shear beams. We shall attempt a first trial with $\omega = 60$ rad/sec, which is a crude estimate of the second natural frequency of a uniform shear beam whose properties are taken as averages of those of the soil formation. From the definition of μ we find, for the three strata, $\mu = 1.549, 1.549,$ and 1.148, respectively. From here,

$$\mathbf{T}_1 = \mathbf{T}_2 = \begin{vmatrix} -0.681 & 0.733 \\ -0.733 & -0.681 \end{vmatrix} \qquad \mathbf{T}_3 = \begin{vmatrix} -0.151 & 0.989 \\ -0.989 & -0.151 \end{vmatrix}$$

Hence,

$$\mathbf{S}_1 = \mathbf{T}_3 \mathbf{T}_2 \mathbf{T}_1 = \begin{vmatrix} S_{11} & S_{12} \\ S_{21} & S_{22} \end{vmatrix}$$

The boundary conditions are

$$\mathbf{z}_0 = \begin{vmatrix} 0 \\ \dfrac{Q_0}{\mu_1 k_1} \end{vmatrix} \qquad \mathbf{z}_3 = \begin{vmatrix} z_3 \\ 0 \end{vmatrix}$$

Therefore, we should have $S_{22} = 0$. This is not the case, so we try new values of ω until the condition is satisfied. In every trial we need only be concerned with calculation of S_{22}. We do not apply Eq. 4.41 because it would require calculation of integrals of continuous functions, and it is not justified to go into this labor in order merely to avoid trial and error.

After several trials and graphical interpolation we find $\omega = 66.5$ rad/sec.

156   NUMERICAL COMPUTATION OF STEADY-STATE RESPONSES   Chap. 4

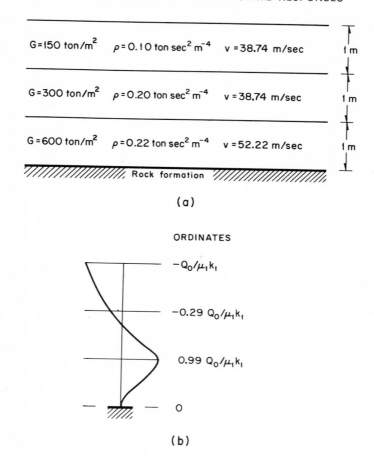

Figure 4.13. Example 4.15.

We substitute into Eq. 4.37 with $\mathbf{x}_{s,i} = 0$ everywhere, to verify that we have computed the *second* natural frequency. This gives us

$$\mathbf{z}_1 = \begin{vmatrix} 0.99 \\ -0.15 \end{vmatrix} \frac{Q_0}{\mu_1 k_1} \quad \mathbf{z}_2 = \begin{vmatrix} -0.29 \\ -0.96 \end{vmatrix} \frac{Q_0}{\mu_1 k_1} \quad \mathbf{z}_3 = \begin{vmatrix} -1.00 \\ 0 \end{vmatrix} \frac{Q_0}{\mu_1 k_1}$$

(Fig. 4.13b), which has one change of sign, indicating that we have indeed computed the second natural circular frequency.

PROBLEMS[5]

4.1. Compute the fundamental period of a cylindrical chimney stack of steel, with circular cross section 6 ft in diameter, whose height is 90 ft, and whose thickness is

---

[5] The solution of problems marked with an asterisk is lengthy.

½ in. Neglect shear deformations, rotary inertia, damping, gravity effects, and base rotation. Perform the computations by taking four segments of equal length and using parabolic approximation to obtain concentrated inertia forces. Compare with Problem 3.1.

Ans. $T_1 = 0.406$ sec.

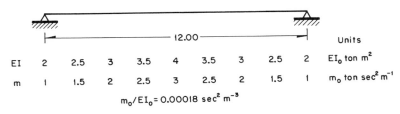

Figure 4.14. Problems 4.2 and 4.3.

4.2. Apply the iteration method described in this chapter to find the first three natural periods of vibration of the beam shown in Fig. 4.14. Use four segments of equal length. Apply (a) the polygonal and (b) the parabolic approximation to find equivalent concentrations of force and of curvature. Verify that (a) satisfies the condition of orthogonality of the natural modes whereas (b) does not.

Ans. (a) $T_{1,2,3} = 1.106, 0.229, 0.089$ sec, (b) $T_{1,2,3} = 1.163, 0.287, 0.147$ sec.

4.3. Solve the foregoing problem for the first two periods with eight segments of equal length and extrapolate to an infinitely small length of segments.

Ans. (a) $T_{1,2} = 1.143, 0.266$ sec, $T_{1,2} = 1.156, 0.283$ sec, (b) $T_{1,2} = 1.158, 0.277$ sec, $T_{1,2} = 1.159, 0.277$ sec.

4.4. Using the parabolic approximation, compute the fundamental period of vibration of the beam solved in Problem 4.2b assuming that (a) there is an axial tension of 30 ton, and (b) there is an axial compression of 30 ton.

Ans. (a) $T_1 = 1.090$ sec, (b) $T_1 = 1.235$ sec.

4.5. Apply Eqs. 2.30 and 4.31 to estimate the fundamental period of this same beam under (a) an axial tension and (b) an axial compression, each equal to one half of the buckling load.

Ans. (a) $T_1 = 1.639$ sec, (b) $T_1 = 0.946$ sec.

4.6. Apply the iteration method of this chapter to find the fundamental period of vibration of the beam analyzed in Problem 4.2 when it carries a concentrated mass weighing 196.2 ton at midspan in addition to the distributed mass. Notice that the derivative of the curvature diagram has a discontinuity at midspan, so that although the parabolic approximation may be used for the combined inertia forces due to the distributed and concentrated masses, Eq. 4.8 does not apply at midspan. Instead, use (a) Eq. 4.5 at that point to find the angle changes of the combined inertia forces and (b) Eq. 4.8 for the effects of the distributed forces and Eq. 4.7 for those of the concentrated force.

Ans. (a) $T_1 = 1.254$ sec, (b) $T_1 = 1.238$ sec.

4.7. Use a step-by-step procedure to find the steady-state displacement amplitude at the end mass in the system of Fig. 4.15. Verify the result using an iteration procedure, beginning with an erroneous assumed configuration.

Ans. 0.546 cm.

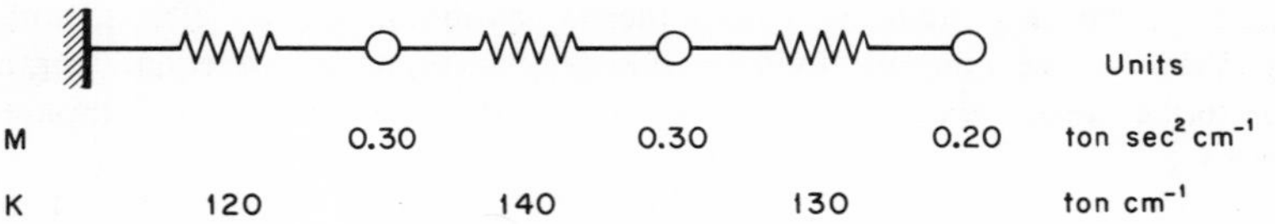

158 NUMERICAL COMPUTATION OF STEADY-STATE RESPONSES Chap. 4

							Units
M		0.30		0.30		0.20	ton sec² cm⁻¹
K	120		140		130		ton cm⁻¹

Figure 4.15. Problem 4.7.

4.8. Use Holzer's table to find the natural periods of the system analyzed in the foregoing problem.

Ans. $T_{1,2,3} = 0.623, 0.226, 0.160$ sec.

4.9. Verify the foregoing results using an iteration procedure.

4.10. Apply Rayleigh's method to estimate the fundamental period of the system analyzed in Problems 4.8 and 4.9. Assume that the deflections of the system are those produced by a statically applied uniform longitudinal acceleration.

Ans. $T_1 \geq 0.621$ sec.

4.11. Use energy methods to estimate the effect of reducing the stiffness of the first spring of the system analyzed in the last problems by 50 percent. Compare with the exact answer.

Ans. $T_1 \leq 0.807$ vs. 0.801 sec.

4.12. Use the Southwell–Dunkerley method to correct the result of Problem 4.1 considering base rotation and shear deformations. Assume that the foundation rigidity in rotation is 50,000 ft-lb/rad, that a hemispherical mass of reinforced concrete 6 ft in diameter is rigidly tied to the structure, and that no other part of the foundation or subsoil participates in the motion. Neglect base translation.

Ans. $T_1 \leq 0.954$ sec.

4.13. Use Holzer's table to find the first two natural periods of the system in Fig. 4.16. Notice that the boundary condition that is not satisfied is one of equilibrium, so that the iteration formula to apply is Eq. 4.42.

Ans. $T_{1,2} = 0.367, 0.199$ sec.

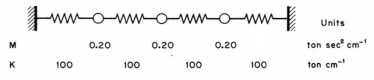

								Units
M		0.20		0.20		0.20		ton sec² cm⁻¹
K	100		100		100		100	ton cm⁻¹

Figure 4.16. Problem 4.13.

4.14.* Apply the Goldberg–Bogdanoff–Moh method to compute the fundamental period of the structure shown in Fig. 4.17. Neglect gravity effects and deformations of the ground. Assume that the ground-story columns are fixed against rotation at their bases.

Ans. $T_1 = 0.436$ sec.

4.15.* Repeat the foregoing problem assuming (a) that the bases of the ground-story columns are hinged and (b) that they are elastically restrained against rotation by a ground-foundation system whose stiffness may be taken equal to twice that of the first-story columns.

Ans. (a) $T_1 = 0.662$ sec, (b) $T_1 = 0.508$ sec.

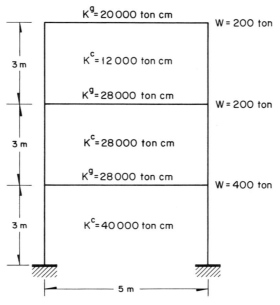

Figure 4.17. Problem 4.14.

4.16. Use the Southwell–Dunkerley method to estimate the fundamental period of the system analyzed in Problems 4.14 and 4.15b when the overall base rotation and horizontal displacement can be represented by a rotational spring with a stiffness of 84,000 ton m and a translational spring with a stiffness of 10,000 ton/m. Assume that the foundation moves as a rigid body and that the mass at each floor is uniformly distributed in plant. Neglect the soil mass.

Ans. $T_1 = 1.367$ sec, $T_1 = 0.762$ sec.

5

NONLINEAR SYSTEMS

5.1 Types of Nonlinearity

Motion of a general type of nonlinear system with a single degree of freedom is governed by the differential equation

$$M\ddot{x} + Q(y, \dot{y}) = P(t) \tag{5.1}$$

which we may derive directly from D'Alembert's principle and where

M = mass
x = absolute displacement of mass
$y = x - x_0$
x_0 = ground displacement
Q = resistance function (restoring force), including effect of damping
P = external force
t = time

M and Q may be functions of time, the applied force, and the history of deformation. Here we shall be concerned only with the case in which M is constant and in which Q is a function of the applied force and history of deformation only. Often the function Q may be broken into one part that depends primarily on y and another part that depends mostly on $\dot{y}$. In that form we arrive at Eq. 1.1 for linear systems.

The resistance function may be single valued and independent of the history of the motion. In this case we say the system is *elastic;* otherwise, it is *inelastic.* Undamped elastic systems are said to be *conservative.*

Examples of conservative, nonlinear systems are structures whose material is strained within the Hookean range but whose geometry changes significantly because of the deformation to which it is subjected. Such is approximately the case of suspension bridges within a wide range of deflections. Such is the case, also, of systems in which deformation is opposed by members that buckle in the elastic range, and in a rough manner, of prestressed concrete frames, whose first major departure from linear behavior is often due to flexural cracks, which may close upon unloading.

We may cite some cases of special interest:

1. $Q = -Q_1$ if $\dot{y} < 0$; $Q = Q_2$ if $\dot{y} > 0$. This is a rigid-plastic system (Fig. 5.1a). The idealization adequately describes the behavior of em-

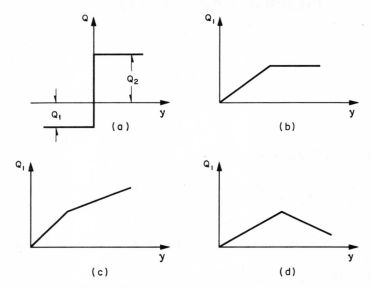

Figure 5.1. Inelastic force-deformation curves.

bankments, rockfill dams, and the like, within a wide range of conditions in which partial slips along "failure" surfaces are considered explicitly. If $Q_1 = Q_2 = Q_0$, let us say, so that $Q = Q_0 \, \text{sgn} \, \dot{y}$, the system is said to be *symmetric*. It is then a special instance of the next type of system.

2. $Q = Q_1(y) + C\dot{y}$, with $Q_1(y)$ as in Fig. 5.1b, and $C =$ constant. This is a linearly damped elastoplastic system. Many steel frames are covered adequately by the idealization.

3. $Q = Q_1(y) + C\dot{y}$, with $Q_1(y)$ as in Fig. 5.1c or d. These systems are often said to be *bilinear* with linear damping. The ascending second line in Fig. 5.1c corresponds to *strain hardening;* the descending branch in Fig. 5.1d corresponds to *strain softening*. These idealizations cover, with sufficient accuracy, most single-degree cases met in practice; Fig. 5.1d describes the behavior of structures whose members undergo inelastic buckling (which may be caused by vertical load) or progressive brittle failure.

A more general load-deformation relationship can be specified (Jennings, 1964) that includes Q_1 in the foregoing examples as special cases and that closely reproduces the load-deformation curves of many types of real structures. Let $Q = Q_1(y)$ denote the load-deformation curve on first loading. We specify that on unloading or reloading,

$$\frac{Q - Q_0}{2} = Q_1\left(\frac{y - y_0}{2}\right) \tag{5.2}$$

where y_0 and Q_0 are the deformation and force at which the loading process was last reversed, except that, if the curve defined by Eq. 5.2 crosses a curve described in a previous load cycle, the load-deformation curve follows that of the previous cycle. Solids and structures that behave in this manner are said to be of the Masing type [Masing (1926); see Fig. 5.2].

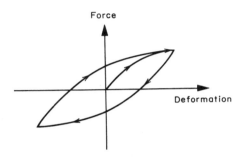

Figure 5.2. Masing-type system.

The stipulation that branches of the hysteretic force-deformation relation should be a two-fold magnification of the first-loading curves seems arbitrary at first glance. Herrera (1965b) has shown that this type of behavior is necessarily associated with all combinations of Coulomb friction and linearly elastic elements (see also Lazan, 1968, pp. 114–15). Thus, the behavior of dry or very permeable granular material is necessarily close to that of a Masing solid provided the stress level is not so high as to cause extensive grain breakage. Experimental evidence confirms this contention (Marsal, 1963a).

It has been shown (Rosenblueth and Herrera, 1964) that a relationship of the type defined by Eq. 5.2, with Q_1 an exponential function of $(y - y_0)/2$, gives rise to systems that behave as though they had an equivalent percentage of critical viscous damping independent of the frequency of vibration, which is a good approximation to the behavior of a wide class of materials over a wide range of conditions. This nonlinear model has the advantage over the representation of multidegree structures with linearly elastic behavior and viscous damping in that the dashpot constants need not be computed in such a way that they furnish a percentage of damping independent of frequency.

Behavior of many nonlinear systems with several degrees of freedom is governed by a matrix differential equation generalized from Eq. 5.1. In order to include the possibility that not all points of the support describe the same motion it is convenient to express the restoring force as a function of a vector obtained by listing both the displacements of the system and the displacements of the support. Behavior of these multidegree systems may be classified in the same groups as that of simple systems and includes also combinations of these groups.

Free vibrations and steady-state, forced vibrations of nonlinear systems have been studied extensively. The methods developed are, however, rarely of use in earthquake engineering because we are usually concerned with responses to transient disturbances. In forced-vibration or sudden-release testing of strongly nonlinear structures, the oscillations are ordinarily so small that structural behavior may be idealized as linear within the testing range when interpreting the test results. Large oscillations of these structures, if imposed, are maintained for only a short number of cycles, so that even in dynamic tests the problem of frankly nonlinear oscillations is nearly always one of transient motions. Accordingly we shall concern ourselves in this chapter with transient disturbances only.[1]

The most powerful methods of analysis available comprise the use of electrical analogs, graphical construction in the phase plane, and numerical procedures. We shall omit description of electrical analogs, although they have been efficiently applied to calculation of responses of elastoplastic systems subjected to earthquakes and to earthquake-like motions.[2] Numerical procedures, used with the aid of digital computers, are completely general and most efficient for the present type of problems.

5.2 Graphical Evaluation of Responses of Simple Systems

The phase plane construction (gyrogram) described in Section 1.6 can be generalized to nonlinear systems with a single degree of freedom (Rojansky, 1948; Jacobsen and Ayre, 1958). The method is based on the following derivation. Equation 5.1 can be written in the form

$$M\ddot{y} + K_i(y + y_{0i}) = a_i M$$

or

$$\ddot{y} + \omega_i^2(y + y_{0i}) = a_i \tag{5.3}$$

where $\omega_i^2 = K_i/M$ and a_i, K_i, and y_{0i} are constants in the interval between t_i and t_{i+1}.

Validity of Eq. 5.3 requires that

$$K_i(y + y_{0i}) - a_i M = Q(y, \dot{y}) - P(t) - M\ddot{x}_0 \tag{5.4}$$

Given $\dot{y}$ we can always express Q as a nonlinear function of y and choose K_i and y_{0i} so that the difference between $K_i(y + y_{0i})$ and Q be kept smaller than

---

[1] A brief survey of methods of analysis for nonlinear systems is found in the work by Klotter (1962).

[2] See for example Bycroft, Murphy, and Brown (1959) and Bycroft (1960).

Sec. 5.2   EVALUATION OF RESPONSES OF SIMPLE SYSTEMS   165

a specified tolerance. We can also approximate $P(t) + M\ddot{x}_0$ by a step function with sufficiently small errors. Hence Eq. 5.4 can always be satisfied with sufficient accuracy.

If we introduce the variable

$$y_1 = y + y_{0i}$$

we can write Eq. 5.3 as

$$\ddot{y}_1 + \omega_i^2 y_1 = a_i \tag{5.5}$$

which is the same as Eq. 1.15 except for the change in coordinates.

Therefore we can draw the gyrogram for nonlinear systems in the same way as for linear systems, provided we change the coordinates every time we enter a new straight segment in the load-deflection curve and provided we compute the necessary step-function approximations. We can expedite the procedure by drawing the load-deflection diagram below the gyrogram using the same scale for y in both graphs.

Horizontal and descending branches in the force-deflection diagram complicate the phase-plane solution somewhat. The horizontal branches lead to parabolas and the descending branches lead to hyperbolas in the $(y, \dot{y}/\omega_1)$ plane. If many problems must be solved having these characteristics it is practical to draw the curves beforehand and trace time markings on them.

Example 5.1. A forcing function, approximated as shown in Fig. 5.3a, acts

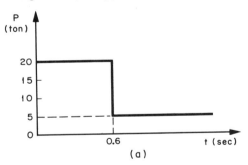

(a)

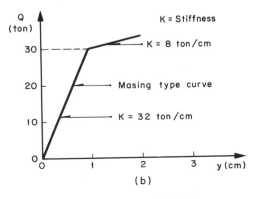

(b)

Figure 5.3. Example 5.1.

166 NONLINEAR SYSTEMS Chap. 5

on an undamped system having a mass of 2 ton sec²/m and a load-deflection curve as shown in Fig. 5.3b. Compute the response of the system.

Solution. First we compute $\omega_1 = \sqrt{32/2} = 4$ rad/sec and $\omega_2 = \sqrt{8/2} = 2$ rad/sec and proceed as shown in Fig. 5.4. The first circular arc O_1B_1 is drawn in the same way as for a linear system, using C_1 as center, O_1 as origin, and $\dot{y}/\omega_1$ as ordinate axis, as the structure has not left the linear range. The time elapsed to the end of the first segment is measured by dividing the angle $O_1C_1B_1$ by ω_1.

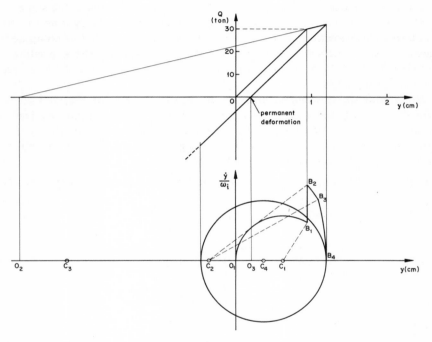

Figure 5.4. Solution of Example 5.1.

The coordinate system is now changed to the one that has O_2 for origin and $\dot{y}/\omega_2$ for ordinates. We find O_2 by projecting, on the y axis of the gyrogram, the trace of the second straight line of the force-deflection diagram. Point B_1 moves to B_2 due to the change in the scale of the ordinates because both B_1 and B_2 correspond to the same value of $\dot{y}$, which is divided by ω_1 in one case and by ω_2 in the other. Using the new coordinate axes we draw an arc with center at C_2. The arc ends at point B_3, which corresponds to the instant t_1 where the forcing function changes magnitude. The construction proceeds again as for a linear system, with an arc whose center is at C_3, until we meet the y axis (point B_4).

At this instant unloading begins, and we must shift the origin to O_3 and use $\dot{y}/\omega_1$ again as ordinates. The structure remains, from then on, in an elastic range (with permanent deformation equal to O_1O_3), so the graphical construction from this instant on is the same as for a linear system.

Addition of damping forces does not complicate the procedure. They can be estimated as the graphical construction goes on and introduced as negative forcing functions. Special types of damping can be dealt with by using special graphical constructions, as we did with logarithmic spirals for viscous damping in systems of linear behavior.

5.3 Numerical Method

The numerical method to be described is a generalization of the one presented in Sections 1.5 and 2.7 (see Newmark, 1959). The generalization is immediate because the relations between acceleration, velocity, and displacement are unaffected by the nonlinear character of the spring and dashpot elements that connect the masses among themselves and the masses to the ground. It suffices to replace the linear relations between displacements and velocities on the one hand and forces on the other with nonlinear relations, and these must have been stipulated if the system has been defined.

Use of stiffness, flexibility, and damping matrices is limited to systems whose load-deformation and load-strain rate relations are idealized as sets of straight segments.

Formulas have not been worked out that would enable computation of appropriate lengths of time intervals to ensure convergence and accuracy of the procedure. These lengths must be chosen by comparison with those which would be adequate for linear systems that would temporarily approximate the behavior of the system analyzed.

Example 5.2. Solve Example 5.1 numerically. Use $\Delta t = 0.25$ sec and $\beta = \frac{1}{6}$.

TABLE 5.1. EXAMPLE 5.2

t (sec)	$\ddot{y}$ (assumed) (cm sec^{-2})	P (ton)	y (cm)	Q (ton)	$\ddot{y}$ (calculated) (cm sec^{-2})	$\dot{y}$ (cm sec^{-1})	Notes
0	—	20	0	0	10	0	
0.25	5.75	20	0.266	8.48	5.75	1.97	
0.50	−3.47	20	0.842	26.95	−3.47	2.25	
0.544	−5.01	20	0.938	30.00	−5.01	2.07	Stiffness change
0.60−	−5.43	20	1.046	30.86	−5.43	1.78	Load change
0.60+	—	5	1.046	30.86	−12.93	1.78	
0.735	−13.41	5	1.167	31.83	−13.41	0	Load reverse
0.75	−13.41	5	1.166	31.82	−13.41	−0.20	
1.00	−6.97	5	0.765	18.97	−6.97	−2.75	
1.25	3.27	5	−0.188	11.53	3.27	−2.29	
1.50	9.73	5	−0.592	24.45	9.73	−0.56	
1.557	9.98	5	−0.608	24.97	9.98	0	Load reverse

168 NONLINEAR SYSTEMS Chap. 5

Solution. The solution is contained in Table 5.1. Only the last cycle for each time interval is given.

5.4 One-Dimensional Wave Transmission

In general, when nonlinear material behavior is idealized in a refined manner, including strain-rate effects, the only practical way of dealing with dynamic problems is to discretize the system in question and use numerical procedures with the aid of a digital computer, even though there is still some difficulty associated with the solution of two-dimensional nonlinear problems, let us say, by using finite-element techniques (Wilson, 1969).

There is often an advantage in dealing with nonlinear continuous media as such in one-dimensional problems when stress is assumed to depend only on strain and on the strain history. Two cases must then be distinguished. In one of them the stress-strain curves for first loading and for reloading are convex, and those for unloading are concave. In the second case the reverse is true.

Consider first the case of a concave first-loading curve. This can be treated by the "pulse method" (Heierli, 1962; Selig, 1964), which is an extension of the method of characteristics that we saw in Chapter 3 in connection with one-dimensional waves in linear media. The method to be presented here is actually a variation of the pulse method, so as to bring it closer to the method of characteristics we presented in connection with wave transmission in media of linear behavior.

We can select a number of points along the stress-strain curve and associate to each point a stress, a strain, a density, and a modulus of elasticity or of rigidity. (If the waves are strictly equivoluminal, the density will be the same for all the points selected.) Suppose that the boundary condition at the free boundary of the medium in question consists in the specification of a stress-time history. First this history is idealized as a series of steps. We can represent this condition, as shown in Fig. 5.5, using as abscissas for the time axis a line that represents the free surface. The stress applied during the ith time interval is denoted by p_i. To each of the stresses p_i selected to idealize the stress-time curve there will correspond a point along the time axis, marking the beginning of a time interval. From each of these points a straight line will depart that represents the path of the characteristic associated with the stress in question if we use as ordinates the depth below the free surface of the medium (X in the figure). The slope of the ith line will be the corresponding velocity of stress-wave transmission:

$$v_i = \sqrt{\frac{G_i}{\rho_i}} \tag{5.6}$$

Here G_i is the tangent modulus of rigidity for point i on the stress-strain curve, and ρ_i is the corresponding mass per unit volume. For normal stresses in a

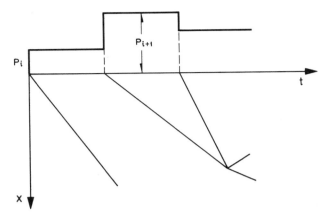

Figure 5.5 One-dimensional wave transmission.

semispace, G must be replaced with the tangent bulk modulus, and for transmission along a slender bar it must be replaced with the tangent modulus of elasticity. Other cases require obvious substitution in place of G.

In Fig. 5.5 the characteristic stemming from the beginning of the $(i+1)$th interval is associated with a smaller wave velocity because p_{i+1} was taken larger than p_i, and the stress-strain curve is convex. The slope of the third characteristic shown in the figure is greater than those of the first two because it corresponds to unloading at the free surface. The rest of the graphical construction in Fig. 5.5 is explained below.

When considering very large strains in the direction of wave propagation, it is necessary to consider finite changes in geometry. This is done neatly by using Eulerian coordinates (coordinates which move with the system) that lead to an expression for the velocity of wave transmission that differs somewhat from Eq. 5.6 (Heierli, 1962; Selig, 1964).

As long as the applied stress increases monotonically, the stress-strain curve is convex, and the waves do not meet their own reflections under conditions such as to decrease their stress, the graphic construction we have just described will apply without change. We find that the only difference with wave transmission in media of linear behavior is that, for the softening material we are now considering, the waves spread out with distance because the characteristics are diverging lines.

Incidence of these waves at the interface with another material is accompanied by refraction and reflection phenomena. The process differs from that taking place in media having linear behavior, in that the reflected wave will change the stresses in the medium through combination with the incoming waves and will therefore be associated with different moduli of elasticity than in the case before reflection. Continuity and equilibrium requirements must be satisfied at the boundary. One way to accomplish this consists of choosing a tentative stress intended to be equal to the one resulting from the sum of the incident and the reflected waves at the interface, on the side of the first medium.

This same stress will exist at the other side of the interface. From the stress chosen we can find the moduli of elasticity for the reflected and refracted waves and proceed as we did in Chapter 3 for linear behavior. The two conditions to satisfy at the interface are those that led to Eqs. 3.16–3.18, but since the medium is now nonlinear, we must proceed by trial and error, as the stiffnesses or moduli k and k_0 are now functions of stress. In this step there is need for care because the reflected wave may be one of unloading, and the corresponding modulus may be many times greater than that of first loading.

A similar situation arises as soon as unloading begins at the disturbance source. Unloading waves ordinarily travel faster than those of first loading. When an unloading wave catches up with one of first loading (and this event occurs at an instant that depends on the discrete points that have been chosen to describe the stress-time curve), refraction and reflection take place as they do at interfaces between different media. This is illustrated in Fig. 5.5 where the characteristics for the beginning and end of the time interval $i + 1$ meet each other.

The method that has been described deals directly with boundary conditions in which the stress is specified. This type of problem is met especially in connection with the effects of blast. In earthquake engineering we are often interested in problems in which it is stipulated that the boundary undergo a certain displacement as a function of time. To translate into the first kind of boundary condition there is need to proceed by trial and error.

Concave stress-strain curves give rise to *shock waves*, rather than the *stress waves* we have just discussed. The reason for this is that the portions of a wave that are associated with high stresses travel faster than those of low stress when the curve is concave. Hence, instead of the wave spreading out as it travels, it compresses itself until it becomes practically a surface that travels with the speed that corresponds to the secant modulus in the stress-strain curve. Usually this situation arises in the transmission of plane or nearly plane P waves in granular soil. Owing to the specialized nature of this phenomenon it will not be dealt with here.[3]

Concave stress-strain curves for first loading often become so steep at high stresses that the material may be reasonably idealized as incompressible beyond a certain stress. The same is true of some curves for unloading, whether the curve for initial stressing was convex or concave. Ideal materials having infinite modulus in some range of stresses are said to be of the *locking type*. Analysis is simplified to some extent by the adoption of these ideal curves; accordingly these materials have received some attention.[4]

The main applications to earthquake engineering of the study of stress waves in nonlinear media lie in the insight it provides to these phenomena and in the possibility it opens of running Monte Carlo analyses to set up bases for design.

[3] The reader is referred to the scanty literature available on shock waves in soils. See, for example, Kompaneets (1956), Duvall (1962), and Hoff (1964).

[4] See, for instance, Prager (1957) and Salvadori, Skalak, and Weidlinger (1960 and 1961).

PROBLEMS[5]

5.1. The frame shown in Fig. 5.6 is subjected to a horizontal force $P = 20$ ton (metric tons) during the time interval $0 \leq t \leq 0.4$ sec and $P = 10$ ton when $0.4 < t \leq 0.6$ sec. Graphically find the maximum roof displacement in numerical value. Assume that the columns have the nonlinear, load-displacement diagram specified

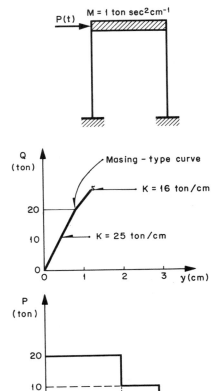

Figure 5.6. Problem 5.1.

in the figure and that the beam is infinitely rigid. Disregard effects of gravity loads, damping, and masses of members.

Ans. max $|x| = 1.5$ cm.

5.2.* Replace the support at the ground floor of the structure of Problem 5.1 with a rigid mass weighing 9.81 ton (metric tons), a linear spring having a stiffness of 10 ton/cm, and a nonlinear dashpot in parallel with it, defined by the relation

---

[5] Solution of problems marked with an asterisk is lengthy.

$F = |c\dot{x}_1|$ and directed opposite to $\dot{x}_1$, where F is the force in the dashpot in tons, c is the dashpot constant in ton sec² cm⁻², and x_1 is the ground-floor displacement in centimeters (Fig. 5.7). Solve the problem numerically.

Ans. max $|x| = 1.73$ cm.

5.3.* The frame shown in Fig. 5.8 is subjected to a horizontal force $P = 50$ ton during the time interval $0 \leq t \leq 0.5$ sec, and $P = 5$ ton for $t > 0.5$ sec. Draw the gyrogram and the graph of displacement and velocity.

Ans. The gyrogram is shown in Fig. 5.9.

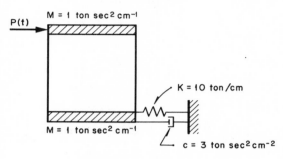

Figure 5.7. Problem 5.2.

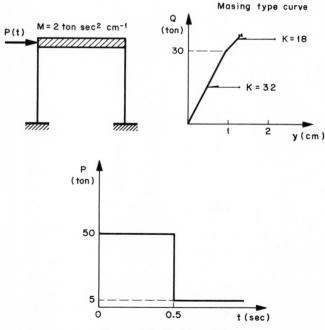

Figure 5.8. Problem 5.3.

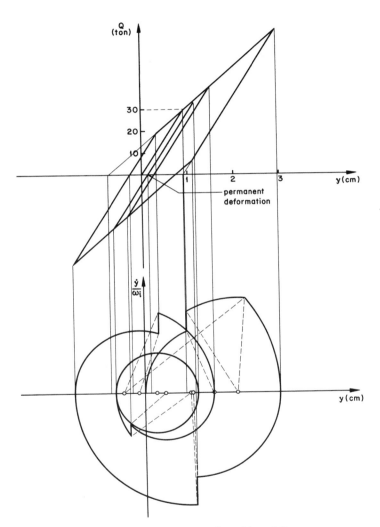

Figure 5.9. Gyrogram of Problem 5.3.

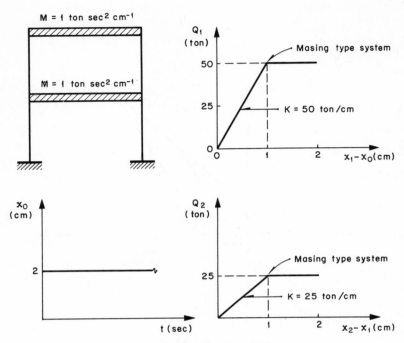

Figure 5.10. Problem 5.4.

5.4.* Each rigid mass of the system in Fig. 5.10 weighs 9.81 ton (metric tons). Both springs are elastoplastic with initial stiffnesses of 50 and 25 ton/cm for the first and second stories, respectively. Numerically compute the maximum absolute displacement, in numerical value, of the first floor as response to a sudden displacement of 2 cm of the support ($x_0 = 0$ for $t \leq 0$ and $x_0 = 2$ cm for $t > 0$).

Ans. max $|x_1| = 1.47$ cm.

Before

After

Damage at Hilo, Hawaii, due to tsunami of May 23, 1960.

6
HYDRODYNAMICS

6.1 General Considerations

The chief hydrodynamic problems of interest in earthquake engineering include dynamic pressures on dams and in tanks, vibration of submerged structures, and wave transmission. A review of the applicable theory of hydrodynamics will be presented first, followed by the pertinent simplifications for each class of problems.

From the beginning we shall neglect water viscosity and assume that we are dealing with small displacements. We shall also omit consideration of air trapping by water and assume that Reynolds' numbers are sufficiently small so that we can neglect all effects of turbulence. Under these conditions, motion of the liquid is governed by the differential equation that governs the dilatation in an elastic solid (Section 3.9):

$$c^2 \nabla^2 \theta = \frac{\partial^2 \theta}{\partial t^2} \tag{6.1}$$

where θ stands for the dilatation $= \sum_{i=1}^{3} \partial x_i / \partial X_i$, x_i is the displacement along the Cartesian coordinate X_i, $c = \sqrt{\lambda/\rho}$ is the velocity of sound, λ is the bulk modulus (Lamé's constant, as $\mu = 0$ in liquids), ρ is the density, ∇^2 is Laplace's operator, and t is time. Applicability of Eq. 6.1 to a liquid under the assumptions made is not surprising because the material cannot transmit S waves.

In hydrodynamics it has been traditional to deal, rather than with Eq. 6.1, with a similar differential equation (Lamb, 1945),

$$c^2 \nabla^2 \phi = \frac{\partial^2 \phi}{\partial t^2} \tag{6.2}$$

where ϕ is a velocity potential such that

$$\frac{\partial x_i}{\partial t} = -\frac{\partial \phi}{\partial X_i} \tag{6.3}$$

We show later that Eqs. 6.1 and 6.2 are equivalent.

The dynamic pressure at any point, taken positive when it is a compression, is
$$p = -\lambda\theta$$
so that
$$\frac{\partial p}{\partial t} = -\lambda \sum_{i=1}^{3} \frac{\partial^2 x_i}{\partial X_i \partial t}$$
and, according to Eq. 6.3,
$$\frac{\partial p}{\partial t} = \lambda \nabla^2 \phi$$
which, from Eq. 6.2 and the definition of c, is
$$\frac{\partial p}{\partial t} = \rho \frac{\partial^2 \phi}{\partial t^2}$$
Since we may choose $\partial \phi / \partial t = 0$ for the liquid at rest, in which we have $p = 0$ (notice that p is additive to the hydrostatic pressure), the last expression is equivalent to
$$p = \rho \frac{\partial \phi}{\partial t} \tag{6.4}$$
Equivalence of Eqs. 6.1 and 6.2 follows from Eq. 6.4. Since ϕ satisfies Eq. 6.2, $\partial \phi / \partial t$ satisfies it too; in view of Eq. 6.4, so does p and, since θ is proportional to p, θ also satisfies Eq. 6.2 and therefore Eq. 6.1.

From the definition of ϕ in Eq. 6.3 we can establish the boundary condition at a surface (bottom or wall) perpendicular to any direction X_i' and which is known to move with a component x_i along this coordinate:
$$-\frac{\partial \phi}{\partial X_i'} = \frac{\partial x_i}{\partial t} \tag{6.5}$$

At the free surface, Poisson's boundary condition for gravity waves is readily established. Indeed, letting $i = 1$ refer to the vertical coordinate, taken positive upwards, the pressure at the level of the original water surface is $g\rho x_1$, where g denotes the acceleration of gravity. From Eq. 6.4 then
$$\frac{\partial \phi}{\partial t} = g x_1$$
Differentiating with respect to time and substituting Eq. 6.3, we obtain
$$\frac{\partial^2 \phi}{\partial t^2} + g \frac{\partial \phi}{\partial X_i} = 0 \tag{6.6}$$

In problems concerning dynamic pressure on dams and vibration of submerged structures it is often permissible to neglect wave motion and assume that the free surface remains at rest. Thus, when undulatory motion is not important, we may replace Eq. 6.6 with the simpler boundary condition
$$\frac{\partial \phi}{\partial t} = 0 \tag{6.7}$$
Later we shall discuss the limits within which Eq. 6.7 is reasonable.

PRESSURES AGAINST DAMS

Every problem of interest will, then, constitute a solution of Eq. 6.2 with the boundary conditions of Eq. 6.5 and either Eq. 6.6 or Eq. 6.7, depending on the importance of wave motion. Once the equation is solved for ϕ, dynamic pressures follow from Eq. 6.4.

6.2 Pressures against Dams

6.2.1 Introductory Note. In the present section we shall be concerned with setting down the theoretical bases that will later allow us to make recommendations for design. The hydrodynamic problems that interest us most directly in connection with dams concern the response of water bodies, as they may affect dams, when the ground undergoes a transient disturbance. It would be desirable to deal directly with the transient problem. Certainly this can be done if we assume that water is incompressible and that the dam and walls of the reservoir do not deform, treat the problem as two dimensional, and neglect the formation of waves in the water surface. But such drastic simplifications hamper the applicability of results seriously. In many problems, it has been possible to do without the most objectionable simplifying assumptions only for steady-state vibrations. Accordingly we shall begin by presenting an elementary solution in which we simplify the problem to the utmost, followed by various steady-state solutions in which we incorporate water compressibility and other refinements. These solutions will serve to guide us in the interpretation of the somewhat simplified transient-disturbance solutions that are presented in Sections 6.2.6-12.

6.2.2 Elementary Solution. The problem of hydrodynamic pressures against a gravity or buttress dam admits idealization as two dimensional. First consider the case of a rigid dam with vertical upstream face and rectangular basin (Fig. 6.1). Let X_1 and X_2 denote the vertical and horizontal coordinates, respectively, in the plane of the drawing, with origin at the base of the dam. Also let H and L denote the height and length of the reservoir at rest. Suppose that the reservoir is so long ($L/H \gg 1$) that it can be taken as semiinfinite. Take the ground motion to be harmonic and purely horizontal. The proper boundary conditions, from Eq. 6.5, are

$$\frac{\partial \phi}{\partial X_1} = 0 \qquad \text{at } X_1 = 0 \qquad (6.8)$$

$$x_2 = a \sin \omega t \qquad \text{at } X_2 = 0$$

where a is the amplitude of ground vibrations. Hence,

$$\frac{\partial \phi}{\partial X_2} = -a\omega \cos \omega t \qquad \text{at } X_2 = 0 \qquad (6.9)$$

We must also have ϕ tending to zero as X_2 tends to infinity. We shall adopt the simplified boundary condition of Eq. 6.7 at $X_1 = H$.

180 HYDRODYNAMICS Chap. 6

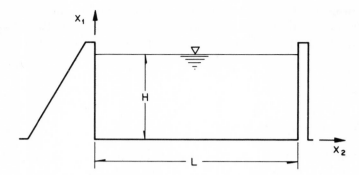

Figure 6.1. Rigid dam with vertical upstream face
and rectangular basin.

Under the assumption of water incompressibility, Eq. 6.2 becomes

$$\frac{\partial^2 \phi}{\partial X_1^2} + \frac{\partial^2 \phi}{\partial X_2^2} = 0 \tag{6.10}$$

Its solution for the boundary conditions adopted and substitution into Eq. 6.4 gives

$$p = -2a\omega^2 \rho H \sin \omega t \sum_{n=1}^{\infty} \frac{(-)^{n+1}}{\mu_n^2} \exp\left(\frac{-\mu_n X_2}{H}\right) \cos\left(\frac{\mu_n X_1}{H}\right) \tag{6.11}$$

where

$$\mu_n = \frac{(2n-1)\pi}{2}$$

Since time and the circular frequency enter only in the common factor $a\omega^2 \sin \omega t$, the solution for incompressible fluid and a rigid dam is not contingent on the assumption of harmonic ground motion. Accordingly, for an arbitrary ground motion $x_0(t)$, Eq. 6.11 becomes

$$p = 2\ddot{x}_0 \rho H \sum_{n=1}^{\infty} \frac{(-)^{n+1}}{\mu_n^2} \exp\left(\frac{-\mu_n X_2}{H}\right) \cos\left(\frac{\mu_n X_1}{H}\right) \tag{6.12}$$

At the upstream face of the dam, $X_2 = 0$.

Substituting in Eq. 6.12 we arrive at a pressure distribution referred to the maximum pressure as shown in Fig. 6.2. The curve lies between a circular arc with center at $X_1 = X_2 = 0$ and a parabola with vertex at the bottom of the reservoir. At $X_1 = X_2 = 0$, p attains its absolute maximum; there it is equal to $0.743\ddot{x}_0 \rho H$ or $0.743(\ddot{x}_0/g)\gamma H$, where γ is the unit weight of water. The total pressure per unit width of the upstream face is $0.543\ddot{x}_0 \rho H^2$, which should be compared with the total hydrostatic pressure, $0.5g\rho H^2$.

Thus, if we neglect the compressibility of water, the hydrodynamic force against a vertical face is only 8.6 percent greater than we would obtain by assuming that it equaled the hydrostatic pressure times the ratio of maximum horizontal acceleration to the acceleration of gravity. The resultant of the

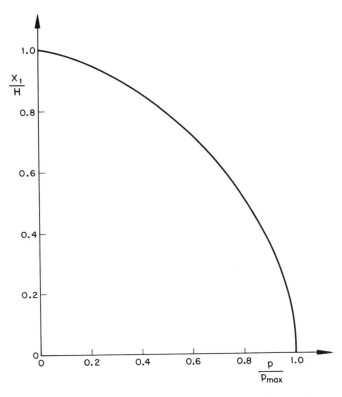

Figure 6.2. Pressure distribution for incompressible liquid.

dynamic pressure on the dam lies at an elevation $X_1/H = 0.401$, as opposed to $X_1/H = 0.333$ for the hydrostatic pressure.

The form of the solution suggests an alternative statement of hydrodynamic pressures as the incorporation of a virtual mass of water fixed to the upstream face of the dam. Dynamic stresses in the dam can be found by replacing the water in the reservoir with a virtual mass, fixed to this face and having the same distribution as the computed hydrodynamic pressures. As in all problems in which the fluid is idealized as incompressible, the virtual mass is independent of the ground motion.

Idealizing the problem as two dimensional and neglecting the compressibility of water allows solution through construction of a flow net (Zangar and Haefeli, 1952; Zangar, 1953). The problem can be solved also by using an electric analog in which the proper boundary conditions are introduced because electric potential is also governed by Eq. 6.10 in two dimensions. The analog consists of a tray containing a uniform-depth, shallow layer of a solution of any salt in water. The condition in Eq. 6.7 requires a grounded conductor, while Eq. 6.8 introduces a conductor at a constant voltage at the bottom of the reservoir. Equation 6.9 implies a linearly varying potential along the face of the dam.

Voltage is probed by means of an electrode at points along this face, and Eq. 6.4 yields the hydrodynamic pressures. The tray can be replaced with a special film-coated paper. The method admits an obvious generalization to three-dimensional problems involving an incompressible fluid.

6.2.3 Inclined Upstream Face.

Under the assumptions that we deal with a two-dimensional problem, that water is incompressible, and that the ground motion is horizontal, pressures acting against an inclined upstream face have been obtained through the use of an electric analog (Zangar and Haefeli, 1952; Zangar, 1953). Results are plotted in Fig. 6.3 and compared with an approximate expression. We see that the distribution of the pressure, normalized in terms of its total magnitude, does not differ widely from that against a vertical face. For many purposes the form of the distribution may be assumed independent of the inclination of the upstream face. The total pressure is always smaller than for a vertical face, and for small slope angles θ with the vertical, it varies approximately as $1 - \theta/100$, where θ is in degrees (Fig. 6.4).

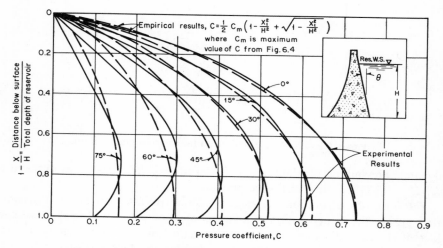

Figure 6.3. Pressures against an inclined upstream face. *After Zangar (1953).*

Results are also available for upstream faces that consist of two planes [Zangar (1953) and Fig. 6.5]. These results show that, when more than half the face is vertical, the pressure may be taken to be the same as against a vertical face without undue error in most cases.

In the light of results shown in Figs. 6.3–6.5, the influence of the inclination and shape of the upstream face on the hydrodynamic pressures exerted against it can be gaged with sufficient accuracy for most applications to the design of gravity dams. Special cases may deserve study by means of an electric analog, model tests, or direct calculation in a high-speed digital computer.

Sec. 6.2 PRESSURES AGAINST DAMS **183**

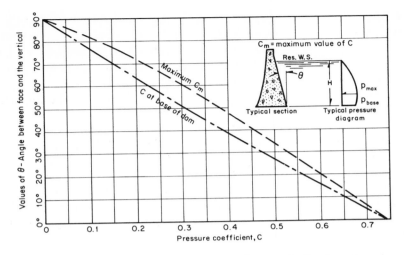

Figure 6.4. Coefficient of pressures against an inclined upstream face. *After Zangar (1953)*.

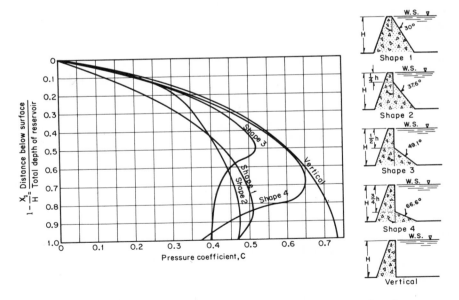

Figure 6.5. Pressure distribution for upstream faces that consist of two planes. *After Zangar (1953)*.

6.2.4 Slope Failures in Reservoirs.

Preserving the simplifying assumptions adopted up to this point, Ambraseys (1957) has used an electric analog to study the dynamic pressures set up by the sudden entrance of a volume of soil into a reservoir. He concludes that these pressures may often be far more significant than those caused by water vibrations induced directly by the ground motion. The study includes finite reservoirs ($L < \infty$) and shows that the importance of the phenomenon increases rapidly as L decreases.

The assumption that the free water surface does not change level, which is essential for the use of the electric analog, precludes the study of the transmission of waves set up by the slope failure. We shall discuss this matter further in connection with tsunamis, in Section 6.5.

Despite the limitations of the available studies, these studies do confirm the general consensus that slope failures are of greater significance in reservoirs than any other cause of hydrodynamic pressures, from the viewpoint of the design of earthquake-resistant dams.

6.2.5 Effects of Water Compressibility.

Explicit consideration of the compressibility of water is important in the analysis of phenomena that involve either the deformation of large bodies of water or the appearance of very high pressures. Usually only the former condition is operative in earthquake engineering. Accordingly, we can anticipate that it will be worth considering water compressibility in the analysis of truly tall dams. For the same reason, model studies of hydrodynamic phenomena with free surface fail to represent adequately the effects of water compressibility.

Refer again to the two-dimensional, semiinfinite reservoir under steady-state vibrations set up by a harmonic, horizontal vibration of the ground, and adopt the boundary condition that neglects change in the free-surface elevation (Eq. 6.7). Under these conditions the solution, first published by Westergaard in 1933 (and indeed the first known application of hydrodynamics to earthquake engineering), has the same form as the solution that neglects water compressibility (Eq. 6.11), except that the denominator μ_n^2 must be replaced with $\mu_n \nu_n$ and the exponent $\mu_n X_2/H$ with $\nu_n X_2/H$, where

$$\nu_n = \sqrt{\mu_n^2 - \frac{\omega^2 H^2}{c^2}}$$

The formula is valid provided $\omega H/c$ does not exceed μ_1. If we take the bulk modulus of water as 2.11×10^5 ton/m^2, which is correct for pure water at about 8°C, we find that the solution is valid up to $H/T = 360$ m/sec, where $T = 2\pi/\omega$ is the period of the disturbance. At $H/T = 0$ the error stemming from neglect of water compressibility is, of course, nil. Referred to the exact answer, this error, in the force that acts against the dam, grows monotonically with H/T. It is less than 4 percent when $H/T = 100$ m/sec and approaches 100 percent as H/T approaches 360 m/sec (Bustamante et al., 1963). At this critical value Westergaard's solution predicts infinite pressures. The pressure distribution, normalized in terms of the total force, is not very sensitive to H/T below the first critical frequency even for a finite reservoir. Beyond the critical value, however, the distribution and the total force change entirely and become very sensitive to the frequency of the disturbance and to the size of the reservoir.

A simple approximation to the dynamic pressures taking into consideration water compressibility consists of multiplying the pressures computed for an incompressible liquid by v_1/μ_1. For water at ordinary temperatures this ratio is

$$\left[1 - \left(\frac{H}{360T}\right)^2\right]^{-1/2}$$

with H/T in m/sec. The error in total hydrodynamic force is insignificant, although this approximation overestimates pressures near the surface and underestimates them slightly near the base of the dam, as will be seen later.

Up to this point we have recognized water compressibility assuming that the reservoir walls and bottom are incompressible. As will be seen in connection with the effects of vertical ground acceleration (Section 6.2.10), the deformability of the material which surrounds the reservoir can reduce hydrodynamic pressures markedly. The reduction is due to partial refraction of sound waves in the liquid back into the rock, so that there is only partial reflection into the liquid.

Surface conditions at the interface may be such as practically to nullify these reflections independently of the deformability of the underlying material. This has been found to be the case in a steel tank subjected to forced vibrations in the laboratory, when its walls and bottom were covered with a layer of sand (Hatano, 1965). To some extent the absence of reflections has been confirmed in forced-vibration tests on prototype dams (Hatano, 1958; Takahashi, 1964). A school of researchers has sprung from this information that maintains that reflections at the interface may be neglected entirely in practice (Hatano, 1966; Okamoto and Kato, 1969). The contention may prove to be tenable even in prototypes. However, it does not follow therefrom, as is also maintained, that hydrodynamic pressures under these conditions are identical to those corresponding to an infinitely rigid liquid. (The difference between the two conditions is illustrated clearly when considering the pressures generated in a rectangular prismatic tank by an almost instantaneous rectangular pulse. When the tank contains a compressible fluid, the pressure pulse travels up and then down; the maximum hydrodynamic force per unit width on any wall is equal to the pulse pressure times the pulse length. When the fluid is incompressible, the maximum force is equal to the pulse pressure times the full height of the wall.) Hence, modeling of reservoirs for the study of hydrodynamic pressures cannot ignore the compressibility of water.

6.2.6 Finite Reservoir. When L is finite and we consider steady-state harmonic vibrations, the question arises of the change of phase between the ground motion at the site of the dam and at the opposite end of the reservoir. A simple way to consider any change of phase consists of assuming the pressures both at $X_2 = 0$ and at $X_2 = L$. Then by superposition we can find the pressures at $X_2 = 0$ when the wall at $X_2 = L$ is out of phase with the dam. The exact expressions for the pressures at any point are available (Bustamante et al., 1963) in the form of series, so that this approach is always accessible. A simpler approach is as follows.

One case of special interest corresponds to the motion $x_0(X_2 = 0) = -x_0$ $(X_2 = L)$. From consideration of symmetry it is obvious that the pressures on the dam are exactly equal to those found when $x_0(X_2 = L/2) = 0$. In this manner, from the pressures on the dam found for the case of a wall at the opposite end of the reservoir remaining at rest, we immediately construct those for a 180° shift of phase between the dam and the wall. Let p_0 denote the pressures for the first case and p_1 for the second condition. The pressures corresponding to in-phase motion of the two ends of the reservoir will be $p_0 - (p_1 - p_0) = 2p_0 - p_1$, as is immediately clear from consideration of superposition of effects because we are dealing with a linear problem. For intermediate phase shifts we can interpolate in terms of the size of the phase angle. Since all the required information derives easily from the solution for pressure against the dam under the condition $x_2(X_2 = L) = 0$, we shall be content mainly with presenting here some results for this case that are found in the literature (Bustamante et al., 1963).

First we notice that below the critical frequency, which corresponds to $H/T = 360$ m/sec, pressures against the dam, normalized in terms of total force, are not especially sensitive to L/H. The total force is a monotonically decreasing function of L/H, but it is practically equal to the value for $L/H = \infty$ if this aspect ratio of the reservoir exceeds about 3. The effects of water compressibility in the range $H/T < 360$ m/sec are more noticeable the smaller the ratio L/H. These comments are illustrated in Fig. 6.6.

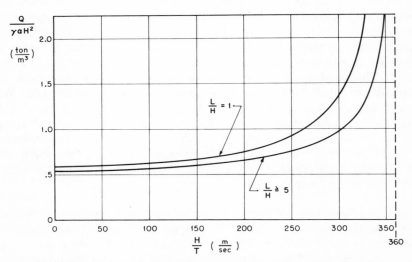

Figure 6.6. Influence of L/H and H/T ratios on total hydrodynamic force. *After Bustamante et al. (1963).*

Beyond the first critical value of ω there is an infinite number of critical or resonant frequencies, each associated with a value of the ratio L/H (Bustamante et al., 1963). The number of critical values within any fixed interval of H/T increases monotonically with L/H. Also, there are important changes in the pressure distribution.

Reflecting the change in pressure distribution, the position of the resultant is affected by frequency. The effect is small below the first critical frequency but cannot be ignored much beyond this value.

A solution for in-phase movement of the dam and the plane at $X_2 = L$, neglecting water compressibility, has produced the curves shown by full lines in Fig. 6.7 (Werner and Sundquist, 1943). Subtracting from twice the pressure that corresponds to $x_0(X_2 = L) = 0$, shown by the dashed curves, gives us the solution for 180° out-of-phase motion of the ends of the reservoir, shown as dot-dash lines.

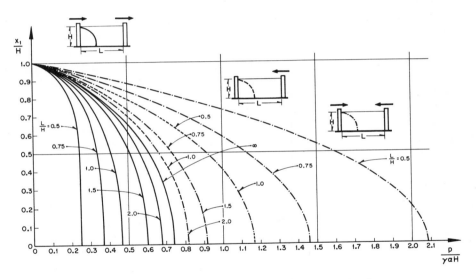

Figure 6.7. Pressure distribution for movement of the dam.

From the solutions available we conclude that the motion of the face of the reservoir opposite the dam is immaterial if its amplitude does not exceed that of the motion at the dam site and provided L/H is not smaller than about 3. In practice, the fact that reservoirs are limited by quite irregular boundaries, except for the dam itself, can be expected to make the motion of the opposite end even less relevant (see remarks concerning the very small magnitude of reflections at fluid-rock boundaries, at the end of Section 6.2.5).

The complete solution for steady-state vibrations contains both a stationary and a transient part. The latter tends rapidly to zero in an infinite reservoir because there is no return of the energy that travels in the liquid away from the dam. In a finite reservoir, however, a rigorous interpretation of the hypotheses adopted would demand preserving the transient part because we have assumed no energy dissipation into the ground and no viscosity in the liquid. Waiving of these simplifying assumptions will not affect the stationary part of the solution seriously but will make the transient portion tend to zero with time. Hence, we are justified in neglecting this part of the solution, but we should recognize that, since energy dissipation may be proportionally smaller than for an ordinary structure, the time required to approach steady-state conditions in a vibrating liquid may far exceed the duration that could be expected from tests on ordinary structures.

6.2.7 Effects of Dam Flexibility and Base Rotation. In addition to the case in which the dam moves as a rigid body describing the ground motion, solutions are available for a dam that deflects, in harmonic motion, according to a prescribed shape (Bustamante et al., 1963). By equating the hydrodynamic force associated with a prescribed deflection to the force required to cause that deflection we obtain an approximate solution applicable to a dam or a tank wall of given stiffness. Results indicate that the flexibility of the dam or wall has a decisive influence on the hydrodynamic pressures and that, depending on the magnitude of the flexibility, the force may increase or decrease by an important fraction of the force associated with a completely rigid dam or wall.

A method of analysis having greater generality is also available (Chopra, 1968; Chopra et al., 1969). However, it is still limited to the fundamental mode of vibration of the dam as specified shape of its deflection curve.

The case of a concrete dam, 300 ft tall, with vertical upstream face, storing a reservoir 300 ft long, subjected to the NS (North-South) component of the 1940 El Centro earthquake, was analyzed by Chopra, Wilson, and Farhoomand (1969). In that particular instance it was found that the maximum hydrodynamic force increased slightly over 100 percent because of dam flexibility. In this study the effects of inertia forces acting in the dam itself are given proper recognition.

It is most likely that explicit consideration of nonlinear behavior of the dam material will give considerably lower maximum forces than have been obtained from analyses that assume linear behavior.

6.2.8 Effects of Motion of the Free Surface. It has been shown that the neglect of gravity waves, which imposes Eq. 6.7 rather than the more accurate Poisson condition in Eq. 6.6, introduces insignificant errors in the total force exerted against a dam with vertical upstream face and semiinfinite basin if H/T is not smaller than about 75 m/sec (Bustamante et al., 1963). The error in the total force can also be shown not to exceed 5 or 20 percent, respectively, if

$$\frac{H}{T} \geq 4.2\sqrt{H} \quad \text{or} \quad \frac{H}{T} = 2.6\sqrt{H} \tag{6.13}$$

where H and T are, respectively, in meters and seconds. For a dam 100 m tall, these limits are $H/T \geq 42$ and 26 m/sec, or $T \leq 2.38$ and 3.85 sec. The upper limits of 5 and 20 percent in the error are probably conservative, and the actual errors are likely to be much smaller. For a very tall dam, applicability of Eq. 6.7 would merely require that H/T exceed 75 m/sec, which is much smaller than the first critical value of 360 m/sec. Hence, the critical frequencies computed with the simplified boundary condition are not excessively in error. But the values of the total force computed, in association with the longer significant periods of earthquake spectra, may be appreciably in error on the unsafe side.

Poisson's boundary condition introduces additional critical frequencies that are quite sensitive to the size of the reservoir. In a circular tank whose height H equals the radius R the first critical frequency occurs when $\omega^2 = 1.75\, g/H$, and in a cubic tank, when $\omega^2 = \pi g/H$ (Lamb, 1945). The corresponding periods for a dam 100 m tall are $T = 15.2$ and 11.3 sec.

These modes are very lightly damped, and the frequencies are not independent of amplitude except for very special dimensions of tanks. These characteristics are even more pronounced for the higher modes of gravity waves, as has been confirmed in laboratory experiments in tanks (Hoskins and Jacobsen, 1934; Jacobsen and Ayre, 1951), which fail to excite the modes in question to any appreciable extent.

We conclude that the critical frequencies in gravity wave motion are not of serious practical concern for the evaluation of the total hydrodynamic force against dams. In very tall structures, there is some doubt in that the contribution of long-period components of earthquakes may be underestimated due to the use of the simplified boundary condition at the water surface, even if the critical wave frequencies are not excited, but the error from this cause is unlikely to exceed 20 percent even in very tall dams.

Although these comparisons reinforce confidence in the calculation of the total hydrodynamic force under the assumption that the free water surface remains horizontal, wave motion will certainly affect the pressures on the dam near the original water surface to a marked extent. Consequently, there is good reason for approximating the pressure distribution for $H/T = 360$ m/sec as a circular arc, rather than a parabola of degree 2.0 to 2.5, even though the latter would give a better approximation to the solution with simplified boundary condition.

The most serious effect that earthquake-excited, hydrodynamic phenomena may have on rockfill dams is the possibility of overtopping. This is particularly dangerous during the construction stage unless special measures are taken to prevent erosion of the crown and of the downstream face.

Overtopping may be caused by a combination of a reduction in the height of the dam (due to compaction and sliding along failure surfaces) and the production of large waves. The latter may in turn originate directly as a consequence of reservoir oscillations, as the result of tilting of the bottom of the reservoir, or as a fresh-water tsunami because of a slope failure anywhere in the reservoir; hence we see the importance of studying the three kinds of phenomena.

Subsidence and tilting of the reservoir, together with the production of seiches, was the cause of water overtopping Hebgen dam by several feet and producing minor damage in 1959 (Sherard, 1967). The fact that approximately

40 million cubic yards of rock slid into Madison Canyon in this instance is also to be reckoned (National Research Council, *National Academy of Engineering, Report*, 1969, Chap. 7).

6.2.9 Natural Modes of Vibration of Water in Reservoirs.

The steady-state solutions we have presented may indeed serve directly as guides for design of dams to resist hydrodynamic pressures, especially when the periods of excitation are sufficiently long as to lie away from the first critical value and not so long as to excite large gravity waves. More realistically, these solutions give an idea of the influence of various factors not considered in the transient solutions to be presented subsequently and permit modifying the latter, even if only in a near qualitative fashion. To arrive at explicit transient solutions first we obtain the natural modes of vibration in the reservoir and the corresponding critical frequencies. And it is precisely at these values that the steady-state solutions break down.

This part and the rest of the present section are essentially based on the works of Kotsubo (1959) and of Flores (1966).

The material presented in the foregoing sections entitles us to assert that most of the problems that concern us in the realm of hydrodynamics with a view of application to the earthquake-resistant design of gravity dams can be based on the study of the two-dimensional vibrations of water contained in a cylindrical, semiinfinite reservoir of rectangular cross section, under the assumptions that the dam is rigid and that there are no surface waves (an exception concerns the possibility of overtopping of rockfill dams, which deserves special treatment). This is the approach we shall follow in connection with the natural modes of the liquid. Such matters as the slope of the upstream face and finiteness of the reservoir can be introduced as refinements to the solutions obtained, and we have already presented results that permit such refinements. We shall also describe results that allow an approximate treatment of cylindrical reservoirs whose section differs markedly from a rectangle. Water compressibility must, of course, be taken into consideration.

The pertinent differential equation of motion (Eq. 6.2, see Fig. 6.1, with $L = \infty$) subject to the boundary conditions

$$\frac{\partial \phi}{\partial X_1} = 0 \quad \text{at } X_1 = 0$$

$$\frac{\partial \phi}{\partial t} = 0 \quad \text{at } X_1 = H$$

and to an arbitrary, prescribed horizontal ground motion $x_2(0, X_2, t) = x_0(t)$, beginning with the liquid at rest at time $t = 0$, admits the following series solution for the hydrodynamic pressure against the dam (Kotsubo, 1959; Ferrandon, 1960),

$$p(X_1, 0, t) = \sum_{n=1}^{\infty} a_n \cos\left(\frac{\omega_n}{c} X_1\right) S_n(t) \qquad (6.14)$$

where

$$a_n = \frac{(-)^{n+1} 8\rho H}{\pi^2 (2n-1)^2} \quad (6.15)$$

$$\omega_n = \frac{(2n-1)\pi c}{2H} \quad (6.16)$$

$$S_n(t) = \omega_n \int_0^t \ddot{x}_0(\tau) J_0[\omega_n(t-\tau)] \, d\tau \quad (6.17)$$

ρ is the mass density of water ($\rho g \cong 1$ ton/m^3), c is the velocity of sound in water ($\cong 1440$ m/sec), and J_0 is Bessel's function of the first class, order zero. (Compare with Eq. 6.12 for an incompressible fluid.)

The form of Eq. 6.14 is similar to that of the corresponding expression for the responses of a multidegree-of-freedom structure that rests on a support having a single degree of freedom and is subjected to an arbitrary ground motion (Eq. 2.17). We may interpret a_n as the participation coefficient for the nth natural mode of vibration, ω_n as the corresponding natural circular frequency, the cosine in Eq. 6.14 as the distribution of the response in the same mode of vibration [indeed, it is the same as the shear distribution in the nth natural mode of a uniform shear beam (Eq. 3.14)], and the function S_n as the absolute acceleration of a single-degree system having a natural circular frequency ω_n, except that the function $\exp(-\zeta_n \omega_n t) \sin \omega_n' t$ of structures has been replaced with $J_0(\omega_n t)$. Because J_0 resembles to some extent a damped sine wave we conclude that hydrodynamic pressures, as responses to an arbitrary ground motion, will have much in common with the responses of damped, multidegree structures to the same ground motion, although we have assumed no internal damping in the water. There are some important differences, though that will become apparent in the following paragraphs.

First if we take $\ddot{x}_0$ to vary as $\sin \omega t$ or $\cos \omega t$ in Eq. 6.17 and make the upper limit of integration go to infinity, we notice that the contribution of the nth natural mode under steady-state harmonic excitation is proportional to $(1 - \omega^2/\omega_n^2)^{-1/2}$. (We had already noticed this result for the fundamental mode when we first considered the effects of water compressibility.) This factor should be compared with $(1 - \omega^2/\omega_n^2)^{-1}$ which is proportional to the contribution of the nth mode in an undamped structure.

Second, $S_n(t)$ as response to a disturbance whose accelerogram is proportional to $\sin \omega_n t$ or $\cos \omega_n t$ (that is, in resonance with the nth natural mode) has an envelope as shown by the full line in Fig. 6.8. For large values of $\omega_n t$ the envelope varies asymptotically as $(2\omega_n t/\pi)^{1/2}$ (Flores, 1966). This variation should be compared with that of the envelopes of damped or undamped structures; the latter envelopes tend, respectively, to a fixed value that is inversely proportional to the percentage of damping, or continue to increase in proportion to $\omega_n t$. The envelopes for structural responses are shown as dashed lines in Fig. 6.8. We conclude that if the nth natural mode of water in a semiinfinite reservoir is to be assimilated to the nth natural mode of a damped structure, both under resonant excitation, the equivalent percentage of damping must be made a decreasing function of $\omega_n t$, tending to zero as $\omega_n t$ tends to infinity. Further, the amplitude of the structural response when there is no damping, in the nth mode, to an impulsive load is twice the static response; in the hydrodynamic case, $S_n(t)$ attains a maximum of 1.47 times its static value (Flores, 1966). And to equate this magnification factor to that in the nth mode of a damped structure, the latter must have a damping coefficient of 27 percent. For any longer-lasting resonant disturbance the equivalent percentage of damping is less than 27 percent if we base the equivalence on the response envelope.

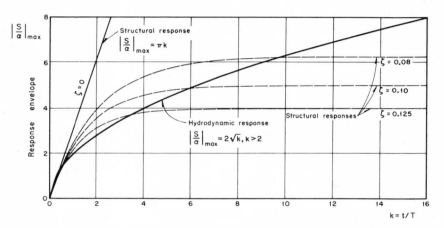

Figure 6.8. Response envelope for harmonic disturbance. *After Flores (1966).*

Once the pressure distributions have been found for all the significant natural modes, the corresponding responses and their combination can be found for any prescribed ground motion, such as is done with multidegree linear systems. However, the numerical procedures we saw in connection with structures subjected to transient disturbances do not apply to hydrodynamic problems, and we must resort to considerably lengthier procedures (Bustamante and Flores, 1966; Chopra, 1967). Some results of these computations are presented in Chapter 12 together with conclusions applicable to seismic design of dams.

When we wish to consider a finite reservoir we must give consideration to additional critical circular frequencies. We saw that the number of these within the interval defined by the fundamental circular frequency ω_1 and an arbitrary, fixed $\omega > \omega_1$ is an increasing function of L/H. As this ratio tends to infinity, all these additional natural circular frequencies tend to ω_1 from above, tending to leave the system with no more than the natural circular frequencies ω_1, $3\omega_1$, $5\omega_1$, ...

To the authors' knowledge, a study of natural modes of the water in reservoirs with an inclined upstream dam face is not available. Until such a study is done it seems warranted to assume that the pressure distribution in the fundamental

Sec. 6.2 PRESSURES AGAINST DAMS **193**

mode changes in proportion to the pressures found when neglecting water compressibility and that neither the natural frequencies nor the higher-mode pressure distributions are affected by the upstream slope.

Thus far we have assumed that the reservoir cross section is rectangular while preserving the assumption that the reservoir is a semiinfinite cylinder and that the upstream face is vertical. Kotsubo (1959) has succeeded in finding the solutions for certain simple nonrectangular reservoir cross sections of practical interest. Some of his results are summarized in Fig. 6.9 and in Table 6.1. Kotsubo concludes that it is conservative and satisfactory for almost any gorge profile to proceed as follows.

1. Assume that the pressure distribution in the fundamental mode is the same as for a rectangular cross section, or modify it in the light of the curves in Fig. 6.9.

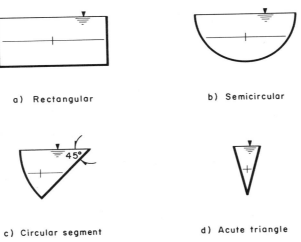

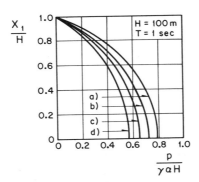

Figure 6.9. Pressure distribution on dam. *After Kotsubo (1959)*.

2. Assume that the pressure distribution in all higher modes is the same as for a rectangular cross section, neglecting the variation of hydrodynamic pressures along horizontal planes.
3. Take the fundamental period equal to that for a rectangular cross section having a depth H, multiplied by $\sqrt{\bar{H}/H}$, where $\bar{H}$ and H are, respectively, the mean and maximum depths for the canyon in question.
4. Take all higher natural periods equal to those for a rectangular section of depth H.

TABLE 6.1. NATURAL PERIODS OF PRISMATIC RESERVOIRS WITH VARIOUS CROSS-SECTIONAL SHAPES (AFTER KOTSUBO, 1959)

Cross section	Exact		Approximate*
	$100T_1/H$	$100T_2/H$	$100T_1/H$
Rectangular	0.278	0.093	0.278
Semicircular	0.237	0.104	0.246
Half quadrant	0.202	0.092	0.208
Acute triangle	0.181	0.079	0.196

*T_1 proportional to $\sqrt{\bar{H}/H}$.

The solutions obtained by Werner and Sundquist (1943) for the longitudinal vibrations of water in cylindrical reservoirs of semicircular and V-shaped cross sections confirm the corresponding results of Kotsubo. Werner and Sundquist include solutions for the transverse vibrations of water in reservoirs of these shapes.

6.2.10 Vertical Ground Motion. For a time several analyses (Napedvaridze, 1959; Chen 'Chzhen'-Chen, 1959) seemed to indicate that ground motions might be idealized ordinarily as horizontal, neglecting the vertical component for the calculation of hydrodynamic pressures, without introducing undue errors. Yet, more recent work (Chopra, 1967) has yielded surprisingly high values of the hydrodynamic pressures due to vertical ground acceleration. This work concerns a rectangular prismatic reservoir. It takes into account water compressibility and assumes that the bottom and walls of the reservoir are altogether rigid and that the disturbance arrives at the entire bottom simultaneously. Under these assumptions it is found that the total maximum hydrodynamic force and overturning moment on a dam 100 ft tall subjected to the vertical component of the El Centro 1940 earthquake exceed three times the hydrostatic values under the action of gravity. Such large values are not credible because of, among other reasons, the performance of existing dams.

Deformability of the rock which surrounds and underlies the reservoir is responsible for part of the overestimate incurred in the simplified analysis. If we continue to take the problem as one dimensional but recognize the deformability of the rock under the reservoir, by idealizing it as a halfspace, we find

Sec. 6.2   PRESSURES AGAINST DAMS   **195**

reductions that ordinarily lie between 20 and 80 percent with respect to the solution for rigid rock (Rosenblueth, 1968a). These reductions depend on the unit weight of rock and on the velocity of P waves in this material as well as on the duration of motion. The method of analysis used is similar to that presented in Section 9.8 for soft soils overlying a semiinfinite homogeneous rock formation.

Other analyses (Herrera, Flores, and Lozano, 1969) indicate a decisive influence of the direction of travel of earth waves relative to the dam. When the ground motion travels toward the structure, pressures may reach several times the values predicted under the assumption that the motion reaches the bottom simultaneously. They are many times smaller when the earth waves travel away from the dam.

As pointed out in Section 6.2.5, there are indications that reflection of sound waves at the interface between water and rock may be negligible in practice, owing to the presence of very deformable material in contact with the contained water. If so, the effects of the vertical acceleration of the ground will be smaller than predicted by any of the existing studies. Further reductions may be expected from the flexibility and nonlinear behavior of the dam itself.

6.2.11 Arch Dams. Kotsubo (1961) has obtained the natural modes of vibration in the plane of symmetry and the corresponding frequencies for the water in a reservoir limited by a vertical, cylindrical arch dam and radial abutments (Fig. 6.10). His solution is too lengthy to present here. The pressure distributions

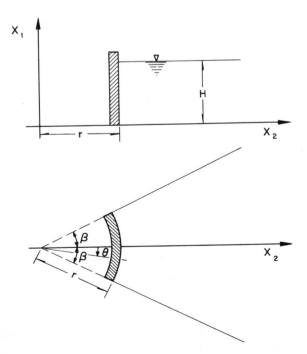

Figure 6.10. Vertical cylindrical arch dam with radial abutments.

computed by him for infinite reservoirs of rectangular cross section, limited by abutments that subtend various angles as first-mode responses to steady-state harmonic excitation, are shown in Fig. 6.11, both for vibrations along the river course and for transverse vibrations. The solid lines represent pressures in phase

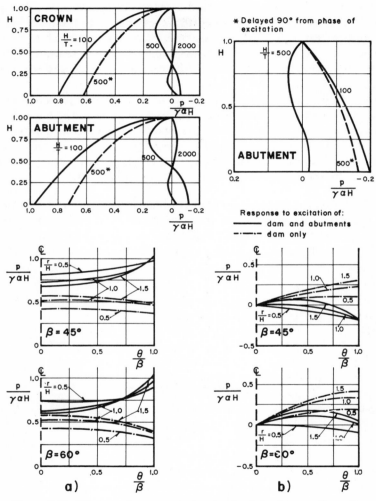

Figure 6.11. First-mode response to steady-state, harmonic excitation both for (a) vibrations along the river course and for (b) transverse vibrations. *After Kotsubo (1961).*

with the ground acceleration, and the dashed lines show pressures 90° out of phase therewith. These results may serve as guides for application to practical problems. Notice that the pressure distributions cannot be regarded any longer as uniform in horizontal planes.

As shown in the figure, the pressures developed as responses to transverse oscillation are considerably smaller than for longitudinal vibrations. Yet, some arch dams are more vulnerable to vibration perpendicular to the plane of symmetry than parallel to it (Ulloa and Prince, 1965).

The foregoing remarks apply to in-phase motions of the abutments. Considerably more critical conditions are conceivable under out-of-phase disturbances (see Okamoto et al., 1964 on instances in which such disturbances have been actually recorded).

The analysis of hydrodynamic pressures in three-dimensional reservoirs subjected to steady-state harmonic excitation can be accomplished in an electrolitic tank using an electric analog (Nath, 1969). The method is especially applicable to the calculation of effects on arch dams. Once the responses to harmonic excitation have been found, we may calculate the effects of an arbitrary disturbance by using Fourier or modal analysis.

6.3 Vibration of Liquids in Tanks

When studying pressures against dams we found that we could ordinarily neglect surface waves but that water compressibility was usually important. For the dimensions that are common in tanks we find that the reverse is true. One problem consists of solving $\nabla^2 \phi = 0$ (see Eq. 6.2) subject to the boundary condition of Eq. 6.5 along the bottom and walls of the tank and to Poisson's condition (Eq. 6.6) at the free surface of the liquid. (For large oscillations the latter condition should be replaced with a nonlinear one.)

In a rigid, completely full tank covered with a rigid lid, the whole mass of liquid moves with the tank as a rigid mass. If, however, there is a small space between the liquid surface and the lid (say 2 percent of the depth of the tank), the pressures exerted by the liquid on the walls and bottom will be practically equal to those for a strictly free surface (Jacobsen and Ayre, 1951). Accordingly, for practical purposes it suffices to study the conditions of a completely full tank and of a truly free surface.

Analytical solutions are available for tanks of various shapes and for such complex problems as interaction between vibrations of the liquid and deformations of the container walls, rotational motion of the liquid induced by translation of the container, and so on (Abramson, 1963).

For small oscillations the velocity potential ϕ and hence the hydrodynamic pressures can be put in the form of a sum of eigenfunctions times the sine or cosine of $\omega_n t$, where ω_n is the nth natural frequency of vibration, and t is time. This form of solution is identical with that for linear, conservative, multi-degree structures and shows that the liquid may be replaced with a number of masses attached to the tank by means of linear springs, one mass being associated with each natural mode. Here we shall be content to present in this manner the solutions that correspond to the first natural mode of rectangular and cylindrical rigid tanks subjected to translatory motion.[1]

With reference to Fig. 6.12a, let R denote the radius of a circular cylindrical tank with flat bottom, H the depth at rest of the liquid contained, and M the mass of this liquid. For calculation of the resultant force exerted by the liquid on the tank and of the corresponding overturning moment, the liquid may be replaced with a mass M_0 rigidly fixed to the tank at an elevation H_0 above the bottom, plus a mass M_1 attached through springs of total stiffness K at elevation H_1 (Fig. 6.12b). These parameters are given by

$$M_0 = \frac{\tanh 1.7R/H}{1.7R/H} M$$

$$M_1 = \frac{0.71 \tanh 1.8H/R}{1.8H/R} M$$

$$H_0 = 0.38H\left[1 + \alpha\left(\frac{M}{M_0} - 1\right)\right] \qquad (6.18)$$

$$H_1 = H\left[1 - 0.21\frac{M}{M_1}\left(\frac{R}{H}\right)^2 + 0.55\beta\frac{R}{H}\sqrt{0.15\left(\frac{RM}{HM_1}\right)^2 - 1}\right]$$

and

$$K = \frac{4.75gM_1^2 H}{MR^2}$$

where g is the acceleration of gravity.

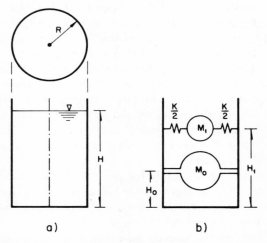

Figure 6.12. Circular cylindrical tank and equivalent masses. *After Housner (1963)*.

---

[1] These solutions as well as Examples 6.1 and 6.2 are taken from Housner (1963), after making some corrections. The corresponding expressions for higher modes in translation as well as in pitching and yawing of rigid, prismatic tanks are found in Graham and Rodriguez (1952). Expressions for the higher natural frequencies of cylindrical containers are found in Jacobsen and Ayre (1951).

The corresponding solution for a rectangular tank that measures $2L$ in the direction of motion is

$$M_0 = \frac{\tanh 1.7L/H}{1.7L/H} M$$

$$M_1 = \frac{0.83 \tanh 1.6H/L}{1.6H/L} M$$

$$H_0 = 0.38H\left[1 + \alpha\left(\frac{M}{M_0} - 1\right)\right] \qquad (6.19)$$

$$H_1 = H\left[1 - 0.33\frac{M}{M_1}\left(\frac{L}{H}\right)^2 + 0.63\beta\frac{L}{H}\sqrt{0.28\left(\frac{ML}{M_1 H}\right)^2 - 1}\right]$$

and

$$K = \frac{3gM_1^2 H}{ML^2}$$

For both shapes of container, $\alpha = 1.33$ and $\beta = 2.0$, if the hydrodynamic moment on the tank bottom is to be included in the computations, while $\alpha = 0$ and $\beta = 1$ if only the effects of hydrodynamic pressures on the container walls are of interest. The solution for a cylindrical tank with a semispherical bottom may be taken to be equal to the one for a tank with flat bottom of same radius and same volume as the tank in question.

The amplitude of the height of waves set up by the vibration may be taken equal to the horizontal displacement amplitude x of mass M_1 times the factor

$$\eta = \frac{0.69KR/M_1 g}{1 - 0.92(x/R)(KR/M_1 g)^2} \qquad (6.20)$$

in cylindrical tanks, and

$$\eta = \frac{0.84KL/M_1 g}{1 - (x/L)(KL/M_1 g)^2} \qquad (6.21)$$

in rectangular tanks. These expressions are satisfactory provided ηx does not exceed about $0.2R$, $0.2L$, or $.02H$. Beyond these limits nonlinear phenomena become important.

Energy dissipation due to viscosity of the liquid can be expressed as an equivalent percentage of critical damping. This quantity decreases rapidly with increasing linear dimensions of the container (Jacobsen and Ayre, 1951) and is only a small fraction of 1 percent for tanks of practical interest. [Graphs are available ("*Nuclear Reactors and Earthquakes*," 1963, pp. 183–209) which facilitate the application of Eqs. 6.18 and 6.19.]

For small values of H/R or H/L, the approximations $T_1 \cong 1.07R/\sqrt{H}$, $T_1 \cong 1.25L/\sqrt{H}$ are useful for estimating the fundamental period of liquids in cylindrical and rectangular tanks, respectively. Here T_1 is in seconds, and H, R, and L are in meters. The error introduced by using these expressions does not exceed 2 percent when H/R is smaller than 0.25. The expressions $T_1 \cong R/\sqrt{H}$, $T_1 \cong 1.25L/\sqrt{H}$ are valid up to H/R or $H/L = 0.7$ with errors smaller than 10 percent.

Example 6.1. Consider a rectangular tank 50 ft long and 10 ft deep ($L = 25$ ft, $H = 10$ ft) that rests on the ground.

Solution. From Eqs. 6.19 for M_1 and K and from $T = 2\pi\sqrt{M_1/K}$ we find the fundamental period to be 5.9 sec. If the undamped displacement spectral ordinate for $T = 5.9$ sec is 1.9 ft, this will be the lateral displacement of M_1 so that, according to Eq. 6.21, the amplitude of sloshing 1.9η is 1.5 ft. This answer refers only to the effects of the ground-motion component of translation in one direction. The translational component at right angles with this, as well as the three rotational components, will induce additional responses.

Example 6.2. Analyze the elevated cylindrical tank shown schematically in Fig. 6.13, having $R = 25$ ft and $H = 15$ ft.

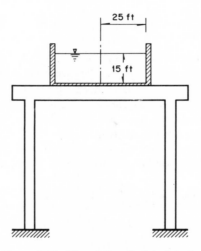

Figure 6.13. Elevated cylindrical tank.

Solution. Applying Eq. 6.18 we find

$$M_1 = 0.52M = 955 \text{ kip}/g$$
$$K = 57 \text{ kip/ft}$$
$$M_0 = 0.35M = 640 \text{ kip}/g$$

If the mass of the platform and of the container is 315 kip/g, the total virtual mass fixed to the supporting structure will be $640 + 315 = 955$ kip/g. Further, if the stiffness of the structure proper is 700 kip/ft, we find the two natural modes of vibration to be

$$\begin{vmatrix} z_0 \\ z_1 \end{vmatrix} = \begin{vmatrix} 0.081 \\ 1.000 \end{vmatrix} \quad \text{and} \quad \begin{vmatrix} 1.000 \\ -0.081 \end{vmatrix}$$

where z_0 is the displacement of M_0 and of the platform, and z_1 is the displacement of mass M_1. The corresponding natural periods are 4.52 and 1.24 sec. There is relatively little coupling between the two degrees of freedom we chose for the analysis. Had we considered the tank as resting on the ground, we would have computed the natural period of M_1 alone, as 4.6 sec. The second natural mode and period approach those of the total mass fixed to the structure disregarding the effects of M_1.

If the pseudovelocity spectral ordinates at both natural periods are 2.0 ft/sec the base shear associated with the first mode will be

$$\frac{4.52 \times 2.0}{2\pi} \frac{1 + 0.081}{1 + 0.0066} \times 0.081 \times 700 = 88 \text{ kip}$$

and that associated with the second mode

$$\frac{1.24 \times 2.0}{2\pi} \frac{1 + 0.081}{1 + 0.0066} \times 1 \times 700 = 253 \text{ kip}$$

so that the maximum base shear is less than $88 + 253 = 341$ kip, but the probable maximum would be considerably smaller. If the tank wsre completely full so there could be no sloshing, its natural period would be 1.95 sec and, with a spectral pseudovelocity of 2.0 ft/sec, the base shear would amount to

$$\frac{1.95 \times 2.0}{2\pi} \times 700 = 434 \text{ kip}$$

6.4 Vibration of Submerged Structures

In most cases of practical interest the earthquake-induced vibrations of submerged structures can be studied satisfactorily under the assumptions that wave action is negligible and that the velocity of the structure relative to the surrounding fluid is sufficiently low that the liquid may be taken as incompressible, inviscid, and irrotational. Under these conditions the phenomenon can be analyzed easily by adding, to the mass of the structure (not considering the buoyant effect of the liquid), the mass of a certain volume of liquid, which gives a total "virtual" mass, and otherwise treating the structure as though it stood in air.

If the structure is a long, rigid prism on flexible supports, moving in a direction perpendicular to its axis, flow of liquid around the structure is essentially two dimensional. Under these conditions, the added mass is that of a circular cylinder of liquid having the same length as the prism and a diameter equal to the width of the projection of the prism on a plane perpendicular to the direction of motion (Fig. 6.14). Experiments (Clough, 1960a) have shown that the additional mass may be of the order of 25 percent greater than this value for rigid prisms on flexible supports, and slightly less than the theoretical value for long, flexible prismatic structures.

Differences between two- and three-dimensional solutions may be important for stocky structures, such as the submerged portions of some bridge piers. For a cylindrical pier the liquid mass that may be assumed to vibrate together with the pier equals the mass of a cylinder identical with the pier times the

202 HYDRODYNAMICS Chap. 6

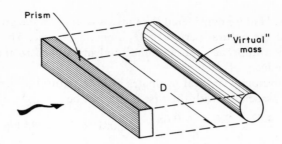

Figure 6.14. Submerged body and its virtual mass.

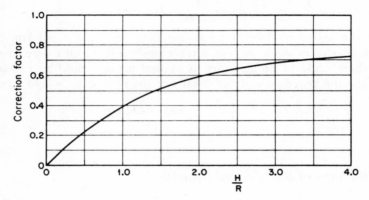

Figure 6.15. Liquid mass correction factor in circular pier.
After Jacobsen (1949).

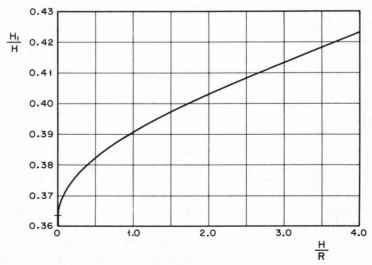

Figure 6.16. Mass centroids of pier virtual mass.
After Jacobsen (1949).

correction factor (Jacobsen, 1959) shown in Fig. 6.15, where H is the submerged height of the pier, and R is its radius. The center of gravity of this mass lies at a height H_1 which varies as shown in Fig. 6.16. The distribution of this mass along the pier is approximately as shown in Fig. 6.17. It seems reasonable to expect that the same solution will hold for the virtual cylinder of liquid associated with piers of other shapes.

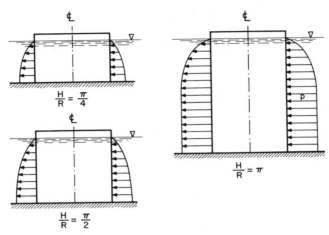

Figure 6.17. Mass distribution along the pier. *After Jacobsen (1949)*.

As with the problems considered earlier in this chapter, damping due to liquid viscosity may be disregarded. Energy dissipation due to radiation into the liquid may be more important, but the model tests to which we have referred indicate that it will not exceed about 2 percent of critical for submerged structures of ordinary dimensions.

6.5 Tsunamis

The Japanese word *tsunami* refers to transient sequences of water waves caused by a local phenomenon. They are also known as *tidal waves* or *seismic sea waves*. Their most common causes are the local change of elevation of the ocean floor, in turn associated with the production of an earthquake and coastal or submarine landslides which may be triggered by earthquakes. As exceptions, volcanic phenomena, such as the 1884 Krakatoa explosion, have given rise to sizeable tsunamis.[2]

Tsunamis are important because of the loss of life and vast property damage

---

[2] There is some discussion concerning the relative importance of the local phenomenon and coupled air-water waves in the cases when volcanic events have produced tsunamis (Van Dorn, 1965; National Research Council, *National Academy of Engineering Report*, 1969, Chap. 7).

that have resulted from the larger ones. One of the worst occurred in Japan on 15 June 1896. The wave rushing onto land was between 75 and 100 ft in height, engulfing entire villages. More than 27,000 persons were killed and over 10,000 houses were destroyed (Leet, 1948).[3]

Landslides in reservoirs have been responsible for one catastrophic event that may be called a fresh-water tsunami—that of Vajont—and certainly deserves attention.

The phenomenon through which a landslide orginates water waves has been simulated in the laboratory by allowing a block to fall vertically as well as by allowing it to slide along a submerged, inclined plane (Wiegel, 1955) and by rotation of rigid blocks (Cruickshank, 1969). When the relationship between the length of the falling body and the initial period of the waves is extrapolated considerably, there is an indication that an underwater slide of a few thousand feet could generate waves with an initial period of the order of 10 or 15 min, that is, of the same order of magnitude as the prevailing periods of large tsunamis. Smaller dimensions would be required if the body slid along a slope.

Coastal landslides have been known to cause small tsunamis. This has been the case repeatedly in Norwegian fjords. And Lituya Bay, in Alaska, affords a classic example of such events. The slide occurred along one side of the T-shaped bay, causing water to surge up the opposite side to an elevation of 1700 ft above the water level. A nearly solitary wave, about 200 ft high, plus two smaller ones moved down the main part of the bay, out to sea. The phenomenon was reproduced in essential features in a model test.

The wave energy liberated in model studies ranges between a fraction of 1 percent and about 2 percent of the net potential energy of the dropping or sliding body. The energy of the wave at Lituya Bay was of the order of 2 percent the potential energy of the mass of soil that slid.

Heaving or subsidence of the ocean floor is the other main cause of tsunamis. The phenomenon has been studied analytically and has been reproduced to scale in the laboratory (Wilson, Webb, and Hendrickson, 1962). Results of these studies are in good agreement with observed characteristics of actual tsunamis. It is found that the details of the leading waves are sensitive to the shape and size of the disturbance and to the velocity with which it takes place, so that these characteristics may be inferred, to some extent, from wave records. The possibilities of such inferences are still limited because of the marked influence of coastal and submarine topography on those details. However, continuing research on the subject holds considerable promise (Van Dorn, 1965).

Former estimates (Iida, 1963b) greatly overestimated the maximum fraction of energy liberated by an earthquake that could be connected into the energy of tsunamis. This was due to an underestimation of the earthquake energies (Wilson, Webb, and Hendrickson, 1962). The revised calculations show that the

---

[3] Remarks in this paragraph and several of those in succeeding portions of this section are taken essentially from Wiegel (1964). The section is also based on Wilson, Webb, and Hendrickson (1962), Van Dorn (1965), and Chap. 7 of the National Research Council, *National Academy of Engineering Report* (1969).

fraction that may be converted into tsunami energy is less than 1 percent on the average and has never exceeded about 2 percent (Wilson, Webb, and Hendrickson, 1962; Van Dorn, 1965).

It has been found (Iida, 1963a) that, for an earthquake to cause an appreciable tsunami, the earthquake magnitude[4] must be greater than approximately $6.42 + 0.017h$, where h is the focal depth in kilometers. Offshore earthquakes with a magnitude greater than $7.75 + 0.008h$ have often caused major tsunamis.

The interval between the first two crests of a tsunami can be taken to define its period. Proceeding in this manner, periods of 6 to 60 min are found to correspond to most of the world's significant tsunamis. There is relatively little correlation between focal distance and wave period for a given tsunami, except at small islands in midocean, where influence of local topography is small. Rather, each coastal location is characterized by some prevailing periods. This leads to the conclusion that such periods are mostly governed by the natural periods of the bays and harbors where they have been measured. The prevailing tsunami periods do increase as a function of magnitude.

The celerity of tsunami waves is given closely by $\sqrt{gH}$, where g is the acceleration of gravity, and H is the depth of water. (Actually, this relationship holds for ratios of wavelength to water depth in excess of, let us say, 10 or 25, and this limit is met by tsunami waves even in deep oceans because of the great lengths of the waves.) In accordance with this expression, the travel times to Honolulu, for example, are 4 to 6 hr from the Aleutian Islands, 5.5 hr from California, 7.5 to 8.5 hr from Japan, and 14 to 15.5 hr from Chile.

The mathematical theory of tsunami waves (Van Dorn, 1965) is far from simple because wave velocity depends on wavelength and on water depth. If the disturbance is axially symmetric and water depth is constant, the waves are circular and spread out much as those caused by dropping a stone into a shallow pond. The wave front moves approximately with the celerity quoted, of $\sqrt{gH}$. Waves behind the front travel at velocities that increase with wavelength, tending as a limit to the celerity of the front. This dependence of velocity on wavelength leads to a difference between group velocity and the velocity of individual waves and to the formation of energy packets similar to what we saw in connection with surface earth waves in Section 3.14.

This picture of the phenomenon is complicated by the fact that most major tsunamis are caused by a disturbance that undergoes displacement as it takes place. For example, slip along a geologic fault begins at one point of the fault and travels at a velocity comparable to that of shear waves in rock (see Chapter 7). The situation is complicated further because differences in water depth distort the shapes of the wave front and of the wave crests.

The Japanese magnitude (Iida, 1963a) for tsunamis is approximately equal to the logarithm to base 2 of the maximum runup, measured in meters. Thus, magnitude 0 corresponds to a maximum runup of 1 to 1.5 m, and magnitude 4 to a maximum runup of 16 to 24 m. The corresponding tsunami energies are

[4] See Chapter 7 for definition of earthquake magnitude.

0.025 × 10²³ and 6.4 × 10²³ ergs and vary approximately as the square of the maximum runup, increasing by a ratio of 2 for every halfunit increase in magnitude.

Techniques are available for extrapolating model test results to predict the maximum runup of tsunamis in the prototype (Van Dorn, 1965). They should be applied with care in view of the difficulties involved in simulating ocean conditions in the laboratory, of the need to use different vertical and horizontal scales, and of the significant effects of free-surface phenomena in very small-scale models. About the smallest scales that are practicable to deal with are 1:600 horizontal and 1:300 vertical, which are the ones of the model of Hilo, Hawaii that has been used effectively in the study of local effects of tsunamis. Such techniques advantageously supplement the calculation of refraction patterns that in turn should include consideration of diffraction patterns. Data are available on the effects of the shape of bays, but these data should be interpreted with care, as the magnification factors for a given shape of bay depend on the open-sea periods of the waves (Wiegel, 1964), and these periods may vary over a wide range.

In view of the immense uncertainties involved in the theoretical aspects of the problem, it is wise to lay much weight on statistical evidence, when this is available. For example, the frequency of occurrence of tsunami maximum runups at Hilo, Hawaii have been computed (Wiegel, 1964) and are shown in Fig. 6.18. The abscissas are the reciprocal of the mean occurrence periods associated with the corresponding ordinates; in other words, if this empirical curve (which for exceptionally high runups may be adequately approximated by an extreme type I distribution) is taken as applicable to the universe of future tsunamis at Hilo, a given maximum runup will be exceeded, on the

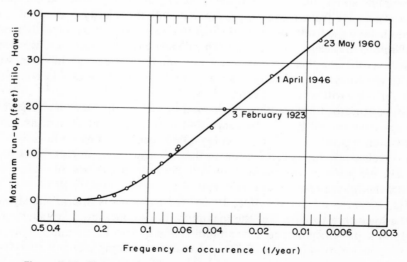

Figure 6.18. Frequency of occurrence of tsunami maximum runups at Hilo, Hawaii. *After Wiegel (1964).*

average, once every T_s years, where T_s is the reciprocal of the corresponding abscissa.

Under the assumption that tsunamis have a Poisson distribution in time, the probability P that a given maximum runup will be exceeded at least once in t years is

$$P = 1 - \exp\left(-\frac{t}{T_s}\right)$$

Thus, there is a probability of 10 percent that a tsunami runup greater than 33 ft will occur at Hilo in any 10-year interval, and a 63 percent probability that such a tsunami will occur in any 100-year interval.

As will be seen in Chapter 8, whether we consider the whole earth or some volume of its crust, the logarithm of the return period of earthquakes whose magnitude exceeds a given value M is related in an approximately linear fashion with M over a wide range of magnitudes. On the other hand, the logarithm of the energy liberated by an earthquake of magnitude M is also a linear function of M (see Chapter 7). Hence, over that range, the logarithm of the return period and the logarithm of the energy liberated must be approximately linearly related. Under the assumption that tsunami energy—and hence the square of maximum runup—is, on the average, proportional to earthquake magnitude, it follows that, over a wide range of runups, the logarithm of the maximum runup and the logarithm of the corresponding T_s must be related in approximately linear manner. For extremely long return periods, however, these relations seriously overestimate the corresponding magnitudes. Hence, they also overestimate the mean rates of occurrence of tsunami runups at any given site, and there must be a dropoff relative to the linear relationship. There is also a dropoff for very small magnitudes, and a similar situation can be anticipated with respect to runups. The situation is illustrated in the logarithmic plot of Fig. 6.19.

When estimating the probable rates of occurrence of tsunami runups at a given site, it is also advisable to compare its conditions of exposure with those of sites for which ampler statistics are available. For example, it has been found that the presence of coral reefs offers some protection. Correlations with coastal slope, however, have not been conclusive (National Research Council, *National Academy of Engineering Report*, 1969, Chap. 7), although there is some indication that steeper slopes are associated with higher runups (Van Dorn, 1965). On flat beaches the linear theory of stationary waves indicates a magnification factor inversely proportional to the $\frac{1}{2}$ power of the beach slope. This theory is inapplicable for the study of tsunami runup at very steep coastlines and only approximately valid for transient phenomena on flat ground. Also, the coastline directly facing the epicenter usually suffers the highest runups, especially at about the point nearest the epicenter (National Research Council, *National Academy of Engineering Report*, 1969, Chap. 7).

It is also worth remembering the relative frequencies of these events in the various oceans. About 62 percent of the known important tsunamis have taken place in the Pacific Ocean, 20 percent in the Indian Ocean, 9 percent in the

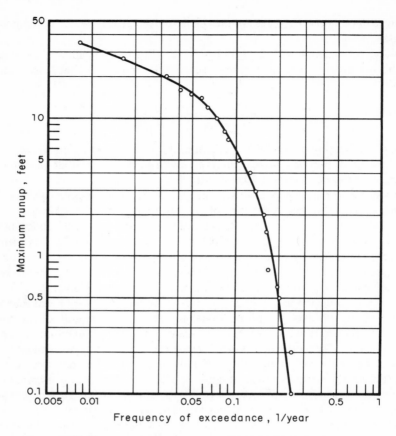

Figure 6.19. Frequency of occurrence of tsunami maximum runups at Hilo, Hawaii. *After Wiegel (1964)*.

Mediterranean Sea, and 9 percent in the northern Atlantic Ocean. Tsunamis are virtually unknown in the southern Altlantic Ocean (Wilson, Webb, and Hendrickson, 1962). The frequency distribution of runups associated with a specific tsunami has been variously described as Gaussian (National Research Council, *National Academy of Engineering Report*, 1969, Chap. 7) or lognormal (Van Dorn, 1965).

The paucity of statistical information on tsunamis leaves the door wide open for uncertainties in extrapolations. Some of the techniques discussed in Chapter 8 for the calculation of the probabilities of occurrence of earthquakes are applicable to tsunamis.

Hydrodynamic pressures associated with tsunamis are often important. Suffice it to mention that estimated hydrodynamic pressures of 400 to 1800 psf were generated at Hilo during the 1960 tsunami.

In order to estimate the pressures generated by runup the phenomenon may be likened to a surge (National Research Council, *National Academy of Engineering*

Report, 1969, Chap. 7). The speed of a surge v may be approximated using the Keulegan equation

$$v = 2\sqrt{gd}$$

where d is the height of the surge. The coefficient in this expression has been found to depend on the degree of saturation and the smoothness of the bottom. Laboratory tests have sometimes given values of the order of 1.8 instead of 2.0.

Once the speed is known, the corresponding pressure can be computed easily. Including the hydrostatic component one finds for the total pressure

$$\frac{1}{2}(\gamma d^2) + c\left(\frac{\gamma}{g}\right)v^2 d$$

where γ is the unit weight of water, and c is a coefficient depending on the slope θ of the water surface relative to the horizontal:

θ	0°	20°	40°	60°
c	1.0	1.3	1.8	3.0

Since few fresh-water tsunamis have occurred in nature in historical times, statistical information for these phenomena is very limited. On the other hand, they can be simulated much more accurately in the laboratory than tsunamis proper, and laboratory studies are undoubtedly justified whenever there is fear that a sizeable volume of rock or soil may slide in a reservoir.

In attempting a preliminary estimate of the size of the waves that may be generated, it is well to recall that most of the studies available concern either two-dimensional or axially symmetric conditions. In practice it is more likely that landslides will take place transversely in an elongated reservoir. The phenomenon may be qualitatively different. For example, there are convincing signs that the slope slides that stopped the flow of the Valdivia River in Chile in 1960 carried the soil up the opposite river bank and allowed it to fall back into the river bed (Duke and Leeds, 1963).

The nature of events to expect following a landslide into a reservoir can be envisaged from this description of the Vajont disaster (Müller, 1964). The Vajont project consists essentially of an arch dam 725 m tall that confines a reservoir the original capacity of which was 169×10^6 m³. The dam was completed in September 1960. Important creep and cracking of the rock were observed, as well as the occurrence of minor slides, especially during and following the first filling of the reservoir.

On 9 October 1963 the largest rock slide in modern times took place. About 270×10^6 m³ of rock slid into the storage lake. The front of the sliding mass was pushed forward 400 m and up 140 m along the opposite slope, after crossing a gorge 80 m wide (Fig. 6.20). The lake contained only 115×10^6 m³ of water at that time. The slide caused a wave that reached 260 m above the lake level and 245 m above the crest of the dam. Falling from there more than 400 m into

210 HYDRODYNAMICS   Chap. 6

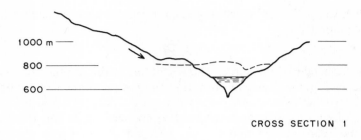

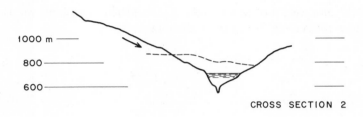

——— Surface before the slide   ——— Surface after the slide

Figure 6.20. Rock slide at Vajont Storage Lake. *After Müller (1964)*.

the valley, the water leveled five villages almost completely. It took about 1900 lives.

The slide occurred in a limestone formation that has signs of being the result of a much earlier slide which had taken place during geologic times. The chief, direct causes of the 1963 event are found in the weakening action of water that saturated the rock mass, in the pore water pressures set up during earlier fillings, in the joint water pressures, and in the buoyance of the submerged portion of the mass.

Apparently, once the slide started, the surface of failure was lubricated by water; this made the phenomenon occur at an extremely high velocity, almost as though the rock had found no resistance to sliding. To this high velocity must be ascribed the violence of the hydrodynamic phenomenon that it set up. Earlier model tests had indicated that, if a major slide were to take place with a duration of 1.5 min, it would cause a wave 25 m in height. The fact that the actual wave was more than 10 times this high points out to a much greater velocity.

The Vajont accident has been quoted not because it bears any direct relation to earthquakes (other than in itself having caused two minor shocks: one during the slide and the other as water fell downstream of the dam), but because it illustrates the type of phenomenon that can be triggered by an earthquake in the banks of a reservoir.

PROBLEMS[5]

6.1. A gravity dam retaining the water of a semiinfinite reservoir of rectangular cross section 100 m deep undergoes steady-state, harmonic oscillations with a frequency of 10 Hz and an amplitude of 2 cm. Find the maximum hydrodynamic pressure exerted against the dam, the maximum force, and the maximum overturning moment.

Ans. $p_{max} = 29.2$ ton/m², $P = 2080$ ton/m, $M = 83{,}100$ ton.

6.2. Repeat the foregoing problem, assuming that the reservoir is bounded by a vertical wall 200 m from the dam, if the wall remains motionless.

Ans. $p_{max} = 29.5$ ton/m², $P = 2110$ ton/m, $M = 83{,}100$ ton.

6.3. Repeat the foregoing problem if (a) the wall moves in phase with the dam and (b) its motion is identically opposite that of the dam.

Ans.

Response	Unit	a	b
p_{max}	ton/m²	24.9	34.1
P	ton/m	1,810	2,410
M	ton	72,400	93,800

6.4. Compute the first three natural periods of the reservoir in Problem 1.

Ans. $T_{1,2,3} = 0.278, 0.093, 0.056$ sec.

6.5. Repeat the foregoing problem under the assumption that the cross section of the reservoir is no longer rectangular, its maximum depth is still 100 m, and its mean depth is 60 m.

Ans. $T_{1,2,3} \cong 0.215, 0.093, 0.056$ sec.

6.6. Solve Problem 6.1 for an upstream face inclined 4 (vertical): 1 (horizontal).

Ans. $p_{max} = 20.6$ ton/m², $P = 1490$ ton/m, $M = 61{,}800$ ton.

6.7.* The reservoir analyzed in Problem 6.1 is subjected to the disturbances $\ddot{x}_0 = (1 \text{ cm/sec}^2) \, \delta(t)$. Compute the corresponding hydrodynamic pressures at midheight in the first and second natural modes as functions of time.

Ans. See Fig. 6.21.

6.8. A prismatic tank that measures 10 by 20 m in plan rests on the ground. It holds water to a depth of 5 m. It is subjected, along one of its diagonals, to the ground accelerogram $\ddot{x}_0(t) = 0.1g$ when $0 < t \leq 0.2$ sec, $\ddot{x}_0(t) = -0.1g$ when $0.2 < t \leq 0.4$ sec, and $\ddot{x}_0(t) = 0$ when $t \leq 0$ and when $t > 0.4$ sec. Compute the maximum height of sloshing.

Ans. 0.049 m.

6.9. Repeat Problem 6.8 for a cylindrical tank holding the same depth and volume of water as the prismatic tank considered.

Ans. (.041 m.

6.10. An elevated tank 5 m square in section contains water to a depth of 5 m. Under steady-state conditions the tank is made to oscillate sinusoidally parallel to one of its faces, with a frequency of 3 Hz and an amplitude of 2 cm. What is the

---

[5] Solution of problems marked with an asterisk is lengthy.

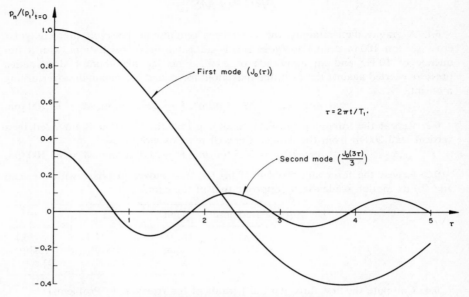

Figure 6.21. Solution of Problem 6.7.

amplitude of the total overturning moment? What fraction of this is taken directly by the bottom?

Ans. 1.75 ton m, 9.2 percent.

6.11. An elevated cylindrical water tank with a hemispherical bottom 3 m in radius holds 100 m³ when completely full. It is covered by a rigid lid. When full, its fundamental period of vibration is 1.5 sec. Use an energy method to compute, approximately, the fundamental period of the system when the tank contains 98 m³ of water.

Ans. 2.57 to 2.95 sec.

6.12. Suppose that the chimney stack analyzed in Problems 3.1 and 4.1 is completely submerged and full of water. Compute its fundamental period assuming that the tributary mass of water outside the stack is 25 percent greater than the value obtained from elementary theory.

Ans. 1.37 sec.

PART 2

Earthquake
Motions
and
Structural
Responses

Fence offset shows 8.5-ft right lateral movement that occurred on San Andreas fault near Woodville in Marin County, north of San Francisco. Note wood frame structures in background are not destroyed. San Francisco earthquake of 1906. (G. K. Gilbert Photo.)

7

CHARACTERISTICS OF EARTHQUAKES

7.1 Causes of Earthquakes

Many phenomena may give rise to earthquakes: volcanic activity, explosions, collapse of cave roofs, and so on. By far the most important earthquakes from an engineering standpoint are of tectonic origin, that is, those associated with large-scale strains in the crust of the earth. This is so because of the frequency of tectonic earthquakes, the energy they liberate, and the extent of areas they affect.

There is still doubt concerning the mechanisms that produce tectonic motions. The most extensively favored theory maintains that tectonic earthquakes are caused by slip along geologic faults. In major earthquakes a chain reaction would take place along the entire length of slip, but at any given instant the earthquake origin would lie in a small volume of crust—practically a point— and the origin would travel along the fault. However, some seismologists hold that earthquakes originate in phase changes of rocks, accompanied by volume changes in relatively small volumes of the crust (Evison, 1963 and 1967).[1] There are inconclusive data to substantiate both theories, and it is conceivable that different tectonic earthquakes are caused by more than one mechanism.

Those who favor the phase-change theory argue that there is little likelihood that geologic faults exist below depths of a few hundred kilometers, because of the high temperatures and confining pressures, and yet data have been interpreted to indicate that earthquakes have originated at depths exceeding 600 km (Gutenberg and Richter, 1954) and up to 800 km. Hence, perhaps some earthquakes cannot be associated with faulting. On the other hand, an evaluation of Southern California earthquakes for the period 1934–1963 strongly sub-

[1] Volume changes of rock to reach equilibrium may be due to important changes in lithostatic compression, caused by migration of the material toward or away from the surface of the earth, or they may be due to the application or removal of large loads such as the weight of glaciers and of water in reservoirs.

stantiates that at least most of these motions originated as slips along geologic faults (Allen et al., 1965). Moreover, the most accurate among the recent analyses of earthquake records have invariably pointed to fault slips as the causative mechanism (see, for example, issues of the *Bull. Seism. Soc. of Am.*, 1964–1969).

Perhaps the most satisfactory conceptual model proposed for the fault-slip mechanism is the one advanced by Burridge and Knopoff (1967). The rock on at least one side of the fault is conceived as a large number of elements having a rigid mass, a linear spring, a linear dashpot, and a Coulomb-friction component. The model explains such matters as foreshocks, aftershocks, and fault creep that are observed in association with a large percentage of major earthquakes.

The question of generating mechanisms is not purely academic. Characteristics of strong motions near the focus are sensitive to the generating mechanism (see, for example, Rascón and Cornell, 1969). Thus, assuming that fault slips cause earthquakes and adopting certain hypotheses about the mechanical properties of rocks, Housner (1965) has concluded that the maximum possible ground acceleration is $0.5g$, where g is the acceleration of gravity.

An upper limit to the ground acceleration has not been established under the assumption that strong ground motions are caused by phase changes in rocks. On the other hand the maximum ground velocity that can be transmitted is limited by the breaking strains of rocks and the velocities of shear waves. Depending on the properties of the rocks in question, maximum horizontal ground velocities of 1 to 3 m/sec are computed (Newmark, 1968). According to Ambraseys (1969), the upper limit lies between 1 and 1.5 m/sec.

These velocities are associated with earthquake intensities greater than those which are considered for most practical purposes. Hence it may be contended, from a practical standpoint, that there is no absolute upper limit to the earthquake intensity. It follows that no matter how conservative a design is, there is always a finite probability of structural failure in any finite interval of time, and there is certainty of eventual failure unless the structure in question is purposely demolished. The conclusion is valid even in so-called nonseismic areas. Seismic regionalization must therefore be taken in the sense of greater or lesser risk, and one acceptable basis for maps of "maximum probable intensity" is the mean recurrence times with which they may be associated.

The simplest quantitative statement about an earthquake consists of grading it by means of a single number—the *intensity* or local destructiveness of the earthquake—in some conventional scale. This is too crude a basis for most engineering purposes. A more adequate description includes the accelerograms or other time histories of three orthogonal translation components of ground motion at a point: the two horizontal and the vertical component. This description is sufficient for the purpose of computing the effects of the earthquake on buildings of small and moderate size. In special instances the space derivatives of ground accelerations become important; this is the case with the rotational components of ground acceleration for slender structures and the soil strains for large civil-engineering works.

7.2 Focus, Magnitude, and Intensity

The *focus, center, hypofocus,* or *hypocenter* of an earthquake is the point in the earth's crust where calculations indicate that the first seismic waves originated.

Epifocus or *epicenter* is the vertical projection of the focus on the earth's surface. In the absence of instrumental data the epicenter is often established, on the basis of damage observed, as the point of most intense shaking. Ordinarily this point does not coincide exactly with the instrumental epicenter. Such terms as *focal distance* and *epicentral* distance refer to distances to a given point of interest, called a *station*.

The *magnitude* of an earthquake is a measure of the energy released. *Intensity* is a measure of the earthquake's local destructiveness. One earthquake will therefore be associated with a single magnitude, while its intensity will vary from station to station.

Richter's magnitude scales are used universally (see Richter, 1958). In its original definition, magnitude (denoted by M) is the common logarithm of the trace amplitude, in microns, of a standard Wood–Anderson seismograph having a magnification of 2800, a natural period of 0.8 sec, and a damping coefficient of 80 percent and is located on firm ground 100 km from the epicenter. Empirical charts and tables are available to correct for epicentral distances different from 100 km and for various conditions of the ground. The correction charts and the definition itself apply strictly only to earthquakes having focal depths smaller than about 30 km. The correction charts are relatively accurate up to epicentral distances of about 600 km.

The teleseismic scale (also denoted by M) and Gutenberg's unified scale (denoted by m) apply, respectively, to focal distances greater than 2000 km and of 600 to 2000 km. They are both determined from the amplitudes and periods of certain earthquake phases. Although intended to coincide with the original definition of magnitude, these two scales, particularly the unified scale, give systematic differences with the original Richter magnitude (Riznichenko, 1962), especially under geologic conditions different from those of California (Jordan, Black, and Bates, 1965). Most of the officially reported magnitudes are determined on the basis of the original or the teleseismic scale (however, the U.S. Coast and Geodetic Survey has retained the unified scale in its reports), and in this book we shall also confine the quantitative use of the term *magnitude* to the M scales. (For very large magnitudes the original Richter definition is always replaced with the teleseismic or the unified scale.)

The early expressions relating magnitudes and energy release have required revision. The following relation enjoys favor among seismologists

$$\log_{10} W = 11.8 + 1.5M \tag{7.1}$$

where M is magnitude and W is energy release in ergs (Gutenberg and Richter, 1956).[2] We shall use this expression in connection with seismicity in Chapter 8.

Nuclear blasts release amounts of energy that compare with those of medium earthquakes. A bomb of one megaton releases about 5×10^{22} ergs. However, only a small fraction of this is converted into seismic waves. It takes about 50 megatons to produce this much seismic energy, which is of the same order as the energy of an earthquake of magnitude 7.3 (Eq. 7.1). Natural seisms of this magnitude or greater occur on the average about seven times a year in the entire world (Gutenberg and Richter, 1954). The most important effects of nuclear explosions are not transmitted normally as earth waves, so that earth motions from this cause deserve a secondary (and specialized) treatment that differs from that of natural earthquakes and will not be dealt with in the present work to any detailed extent.

Almost all intensity scales are subjective and similar in structure to the modified Mercalli (MM) scale (Appendix 2). This scale, widely used in North America, generally resembles the Soviet scale, that of Cancanni–Sieberg, which is used extensively in Western Europe, and the more recent one of Medvedev, Sponheuer, and Karnik, known as MSK-64 (Medvedev and Sponheuer, 1969). Equivalences with other intensity scales have been published (Richter, 1958).

The nature of these subjective scales seems undesirable. Man's reactions to earthquakes depend on numerous factors including previous experience with ground motions. Effects on buildings are contingent on local design and construction practices. Especially objectionable seem clauses in scales that permit assigning an intensity to an earthquake in uninhabited regions in terms of the amplitude of permanent deformations, slope failures, or relative displacements of the ground because, usually, the area of maximum intensity does not follow surface faults at which slip is noticeable, and slope failures often occur in the absence of earthquakes.

Despite their many shortcomings, subjective intensity scales are an important consideration in areas where no strong-motion instruments have been installed and they afford the only means for interpreting historical information.

In order to utilize instrumental data and relate them with the subjective scales, instrumental intensity scales have also been proposed. Those that rest exclusively on the maximum ground acceleration or on the maximum trace of some type of seismograph bear little connection with the destructiveness of the ground motion. Matters improve somewhat if at least the focal distance is crudely taken into account. A rough correlation of this sort is shown in Fig. 7.1 (which is probably applicable to typical earthquakes in California although not necessarily in other regions of the world).

Since the destructiveness is directly related to the energy that the ground motion transmits to human beings and to man-made structures, a much better correlation can be expected and actually has been verified, with the maximum

---

[2] There is a school among seismologists which favors replacing the relatively arbitrary magnitude scales with the physically based "seismic moment," and this quantity can be related directly with energy release (J. Brune and C. Lomnitz, personal communication, 1969).

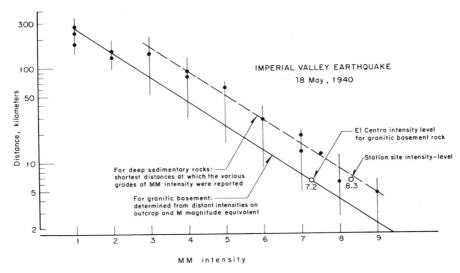

Figure 7.1. Distance-intensity graph for California earthquakes *By Permission from* Earthquake Intensity and Related Ground Motion, *by Frank Neumann, copyright © 1954 by the University of Washington Press.*

ground velocity or, better yet, with the average ordinate of the pseudovelocity spectra, as will be described in Chapters 15 and 17.

The following expression coincides reasonably well with the correlations proposed (Esteva and Rosenblueth, 1964; Rosenblueth, 1964c) and verified between the *MM* intensity I and the maximum ground velocity v (cm/sec);

$$I = \frac{\log 14v}{\log 2} \qquad (7.2)$$

On the average, this correlation holds up to $I = 10$ and overestimates I for higher intensities. It seems advisable to modify the *MM* scale for intensities greater than 10 in accordance with the equation.

There are two other conditions under which Eq. 7.2 overestimates the intensity. One corresponds to ground motions of exceptionally short duration,[3] say with a perceptible duration not exceeding 10 or 15 sec. The second condition corresponds to motions of very soft ground away from heavy structures, since soil-structure interaction will normally decrease the intensity of shaking in the areas of greater interest.[4]

It is not clear which maximum ground velocity one should use in Eq. 7.2. When vertical motions are appreciable compared with the horizontal, the

---

[3] An example is found in the San Salvador earthquake of 3 May 1965 (Rosenblueth and Prince, 1965).
[4] This is the most likely explanation for the high ground velocities associated with some low-intensity earthquakes in Mexico City.

maximum horizontal velocity is probably too small a value, while the maximum absolute (inclined) velocity is probably too high. The difference between these two criteria is not ordinarily important. (We shall discuss the matter of the various components of the earthquake later in this chapter.)

Housner's (1952a) definition of spectral intensity is

$$S = \int_{0.1}^{2.5} V_\zeta dT \qquad (7.3)$$

averaged in two orthogonal directions. Here V_ζ is the pseudovelocity spectral ordinate in ft/sec, for a damping ζ, and T is natural period in seconds. Often ζ is taken equal to 0.2 in Eq. 7.3. Putting S in centimeters rather than feet and dividing by the interval of integration, 2.4 sec, should give values quite close to those of v to be used in Eq. 7.2 to obtain I in the MM scale but not hampered as Eq. 7.2 by the characteristics of earthquakes of very short duration.

7.3 Types of Earth Waves

Whatever the cause of earthquakes, they release two types of body waves: P and S (Section 3.9). Because of the difference in velocity of propagation of these waves one would expect accelerograms some distance away from the focus to consist of two separate trains of oscillations, one for each type of wave. But such accelerograms are always very complicated, and the train of S waves always begins before that of the P waves has subsided. Even the accelerograms of explosion-generated ground motions and those due to a single blow of a forge hammer are quite complicated (Figs. 7.2 and 7.3). Undoubtedly this characteristic of strong ground motions must be attributed to multiple reflection and refraction at irregular and sometimes diffuse geologic interfaces.[5]

A typical earthquake accelerogram (Fig. 7.4) contains three main groups of waves or *phases*: P, or primary; S, or secondary; L, or surface waves, the latter including Love, Rayleigh, and other types of waves. In accelerograms of destructive earthquakes, L waves are usually masked by the tail of the S phase because the accelerations associated with L waves are normally quite small. On the other hand, seismograms of teleseisms are obtained with instruments that magnify long-period components far more than the high-frequency portions; moreover, as we saw in Chapter 3, surface waves decay roughly as cylindrical disturbances and hence much more slowly with distance than body waves, which decay essentially as spherical waves. Consequently, the L phase in teleseismic records ordinarily stands out more prominently than the P and S phases. (In modern seismology the three phases are rarely referred to in such oversimplified terms. A wide variety of waves are recognized within each phase, and it does an earthquake record injustice to speak of less than a half dozen kinds in the L portion alone.)

[5] The matter is well put by Bullen (1953) with basis on Jeffreys' reasoning.

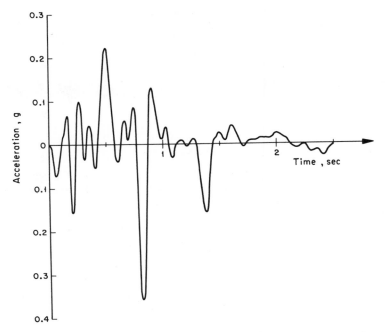

Figure 7.2. Accelerogram of explosion-generated ground motion. *After Carder and Cloud (1959), reproduced by Housner (1961).*

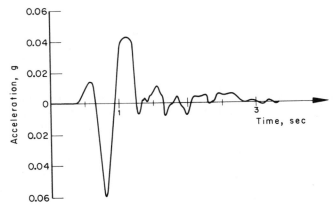

Figure 7.3. Accelerogram of forge hammer-generated ground motions. *After Flygare (1955), reproduced by Housner (1962).*

The times at which signals from the same seismic shock arrive at different stations can be recorded, so that it is possible to determine the travel time of the disturbance as a function of the angular distance from the epicenter. The

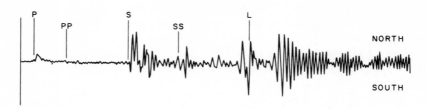

Figure 7.4. Typical earthquake accelerogram. *Adapted from* Elementary Seismology *by Charles F. Richter, W. H. Freeman and Company. Copyright © 1958.*

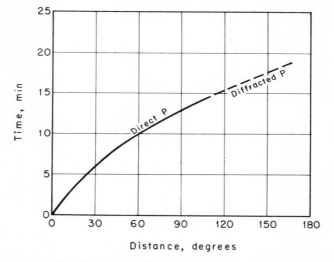

Figure 7.5. Simplified time-distance curve for P-waves. *Adapted from* Elementary Seismology *by Charles F. Richter, W. H. Freeman and Company. Copyright. © 1958.*

distance has been plotted against the corresponding travel time, giving curves such as those shown in Fig. 7.5. The plot of travel time vs. distance for surface waves is essentially a linear relationship, while for *P* and *S* waves the curves are convex.

From graphs of this sort and the time lag between the arrival of *P* and *S* waves it is possible to locate the epicentral distance of various stations for a given earthquake. The intersection of the corresponding circular arcs serves to locate the epicenter. (Actually a study of the seismograms corresponding to the three orthogonal direction at a single station also permits an approximate calculation of the direction at which the epicenter lies and an estimate of focal depth.)

Sec. 7.3 TYPES OF EARTH WAVES 223

Integration of the travel curves permits calculating wave velocities as functions of depth. Results of such computations yield the relationships plotted in Fig. 7.6. The sudden changes in velocities at certain depths mark interfaces between major layers within the earth.

The most conspicuous reflections of waves occur at the earth's surface. Through this type of reflection a *P* wave may give rise to both *P* and *S* waves, called *PP* and *PS* waves, respectively (Fig. 7.7). Similarly, an incident *S* wave, upon reflection, may give both *P* and *S* waves, respectively, called *SP* and *SS*.

Effects of refraction and reflection of seismic waves create a shadow zone (Fig. 7.7) within distances of 105° to 142°. In this region practically no *P* or *S* waves are recorded. The phenomenon, plus a study of reflected waves, has led to the conclusion that the earth contains a liquid core, capable of reflecting *P* and *S* waves and capable of refracting *P* waves.

The symbol *c* is used to indicate a reflection at the core boundary, and the symbol *K* is used to denote that part of a wave that is refracted through the core. Thus a *PcS* wave is a *P* wave that has traveled down to the core boundary and has been reflected as an *S* wave. *PKS* is a *P* wave that has been refracted into

	Depth, km	Velocities, km/sec	
		v_P	v_S
SURFACE	0	5	3
Continental crust		6	3.5
MOHOROVIČIĆ DISCONTINUITY	30*	7? / 8.2	4? / 4.5
Mantle			
	2 900	13.5 / 8	8 / —
Core			
	5 000	10	—
Inner core			
CENTER	6 370	11.5	?

* Depth of the Mohorovičić discontinuity differs widely from region to region and may be double that shown in the table

Figure 7.6. Variation of wave velocities with depth. *Adapted from* Elementary Seismology *by Charles F. Richter, W. H. Freeman and Company. Copyright © 1958.*

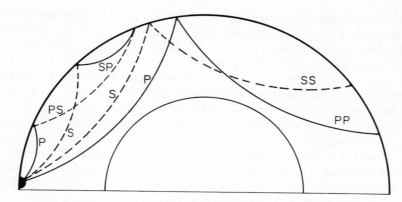

Figure 7.7. Reflections at the earth's surface. *Adapted from* Elementary Seismology *by Charles F. Richter, W. H. Freeman and Company. Copyright © 1958.*

the core and refracted back through the mantle into the S wave in which it finally emerges.

S waves have never been observed passing through the core. This is the main reason for believing that the core is liquid because a fluid cannot transmit shear waves.

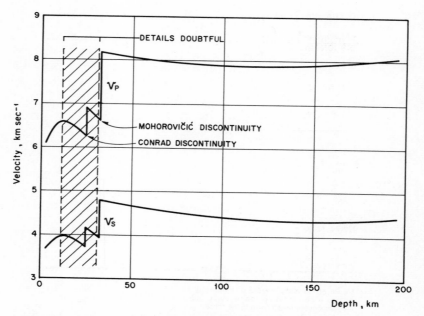

Figure 7.8. Velocity distribution in the upper 200 kilometers of the earth. *After Gutenberg (1955), reproduced by Richter (1958).*

There is evidence that the innermost region of the earth, called the inner core, may have substantial rigidity and be able to transmit S waves. The symbol I is used for P waves in the inner core.

Seismological data have been used to gain information regarding the structure of the earth. A discontinuity has been recognized in this manner at which the P-wave velocity jumps from 6.4 to 6.7 km/sec to 8.1 to 8.2 km/sec (Fig. 7.8). It is named after its discoverer as the Mohorovičić ("Moho") discontinuity, and it defines the lower boundary of the crust proper. The crust defined thus is about 5 km thick beneath oceans and 35 km thick beneath continents.

There are other seismic discontinuities in the earth's crust. The most important is the Conrad discontinuity, where the P-wave velocity changes abruptly from 6.1 to 6.4 to 6.7 km/sec (Fig. 7.8). It exists at varying depths in continental areas and presumably separates a granite from a basalt layer.

7.4 Characteristics of Strong Ground Motions

Because earthquake motions are irregular and each is different from all others, even at a given site, it is important to establish whatever characteristics certain groups of earthquakes may have in common and to base earthquake-resistant design on such generalizations. To this end we shall classify earthquakes into four groups:

1. Practically a single shock. Acceleration, velocity, and displacement records for one such motion are shown in Fig. 7.9. Motions of this type occur only at short distances from the epicenter, only on firm ground, and only for shallow earthquakes. When these conditions are not fulfilled, multiple wave reflections change the nature of the motion. Destructive single-shock earthquakes occurred in Agadir in 1960 (Despeyroux, 1960), Libya in 1963 (Minami, 1965), Skopje in 1963 (Ambraseys, 1964), and San Salvador in 1965 (Rosenblueth and Prince 1965). They have all been associated with moderate magnitudes (5.4 to 6.2), shallow foci (less than 30 km), and effects indicating an almost unidirectional motion, stronger in one than in the opposite sense. If the energy of the motion were broken down in accordance with the frequencies of vibration it excites, a prevalence of short periods of vibration (of the order of 0.2 sec or shorter) would no doubt be found.

2. A moderately long, extremely irregular motion. The record of the earthquake of El Centro, California in 1940, NS component (Fig. 7.10) exemplifies this type of motion. It is associated with moderate distances from the focus and occurs only on firm ground. On such ground almost all the major earthquakes originating along the Circumpacific Belt are of this type. Over a wide range of periods of vibration (on the order of between 0.05 to 0.5 sec and 2.5 to 6 sec) there is, on the average, equipartition of

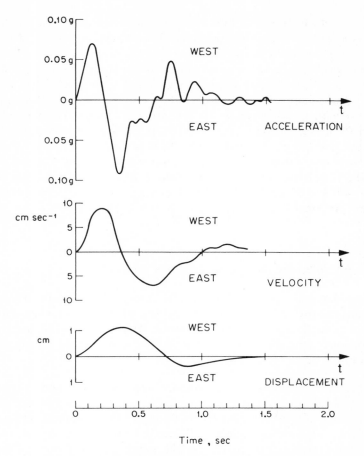

Figure 7.9. Port Hueneme earthquake of 18 March 1957, EW component. *After Housner and Hudson (1958)*.

energy. By analogy with light, we may say that these motions are nearly white noise. They are ordinarily of almost equal severity in all directions.

3. A long ground motion exhibiting pronounced prevailing periods of vibration. A portion of the accelerogram obtained during the earthquake of 6 July 1964 in Mexico City is shown in Fig. 7.11 to illustrate this type. Such motions result from the filtering of earthquakes of the preceding types through layers of soft soil within the range of linear or almost linear soil behavior and from the successive wave reflections at the interfaces of these mantles.

4. A ground motion involving large-scale, permanent deformations of the ground. At the site of interest there may be slides or soil liquefaction. Examples are found in Valdivia and Puerto Montt during the Chilean earthquakes of 1960 (Rosenblueth, 1961), in Anchorage during the 1964

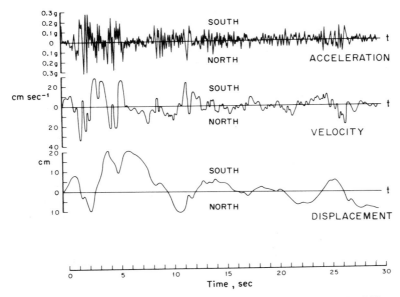

Figure 7.10. El Centro, Cal. earthquake of 18 May 1940, NS component. *After Blume, Newmark, and Corning (1961).*

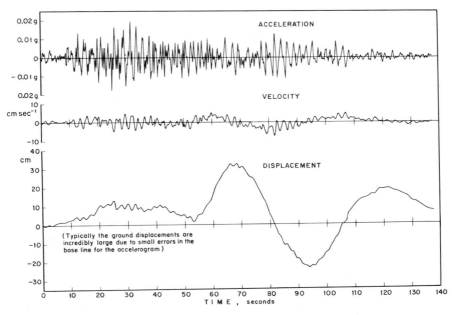

Figure 7.11. Mexico City earthquake of 6 July 1964, NS component. *After Rosenblueth (1966).*

Alaskan earthquake (Shannon and Wilson, 1964), and in Niigata during the Japanese earthquake of that same year (Falconer, 1964; Seed and Idriss, 1966).

There are ground motions of characteristics intermediate between those we have just described. For example, the number of significant, prevailing ground periods, because of complicated stratification, may be so large that a motion of the third group approaches white noise, or soil behavior may be only moderately nonlinear. Still the classification proposed serves the purpose of underlining the possibilities and the limitations of generalizations applicable to design against one but not the other types of ground motion.

The nearly white-noise type of earthquake has received the greatest share of attention. This interest in white noise is due to its relatively high incidence, number of records available, and facility for simulation in analog and digital computers or even for analytical treatment of the responses of simple structures. Because of the chaotic nature of these motions, such analytical studies are based necessarily on the theory of probabilities, while simulation in computers aims at Monte Carlo studies involving statistical interpretation of the results.

Very likely the first kind of earthquake can be dealt with deterministically because of its simplicity. The only serious limitation at present is the meagerness of records available.

The third type of motion obtains from linear filtering of the first or second type. It is, hence, nearly as amenable to analytical treatments as these two and to Monte Carlo studies as the second type.

The fourth kind of earthquake is difficult to study either analytically or through simulation, and it is not especially useful to predict structural responses thereto. Ordinarily it is impractical to attempt a structural design to resist large-scale failure of the ground. Rather, one should normally aim at establishing the conditions under which such phenomena are likely to occur and, where they are likely, either erect the structure in question elsewhere or treat the soil in such a way that the phenomenon becomes unlikely, at least locally.

For the reasons given above and for brevity in presentation, we shall direct the larger share of attention to the second type of motion—nearly white noise.

7.5 Correlation of Ground-Motion Parameters with Magnitude and Focal Distance

For strong earthquakes on firm ground at short and intermediate focal distances, knowledge of the maximum ground acceleration a, velocity v, and displacement d in a given direction permits estimating the shape of response spectra. Roughly, the maximum acceleration A for small damping in a smoothed (locally averaged) spectrum is $4a$, the maximum spectral pseudovelocity V relative to the ground is $3v$, and the maximum spectral deformation D is $2d$

(Blume, Newmark, and Corning, 1961). This information, coupled with the facts that A tends to a as the natural period T tends to zero and that D tends to d as T tends to infinity, permits drawing a fairly accurate free-hand sketch of the smoothed, response spectrum in a logarithmic plot.

Another approach consists in assuming that the lines which mark a, v, and d in the logarithmic plot coincide with the smoothed spectrum for 25 percent damping (Esteva and Rosenblueth, 1964) and estimate the spectra for other percentages of damping from the assumption that ordinates are, on the average, proportional to $(1 + 0.6\zeta\omega s)^{-0.45}$ or to $(1 + 0.5\zeta\omega s)^{-1/2}$ (see Section 9.6), where ζ is the percentage of damping, ω the natural circular frequency, and s the duration of an "equivalent" ground motion having uniform intensity per unit time (about half the duration of perceptible motion for moderate and strong earthquakes; 12.5 sec on the average for strong earthquakes that have been recorded along the western coast of the United States). (Rosenblueth and Bustamante, 1962.)

Figure 7.12 compares the spectra predicted by both of the criteria just described. In Chapter 9 we shall deal to a greater extent with the matter of expected response spectra.

It follows from these remarks that it is sufficient to give means to estimate a, v, d, and perhaps s as functions of magnitude and focal distance. Once these

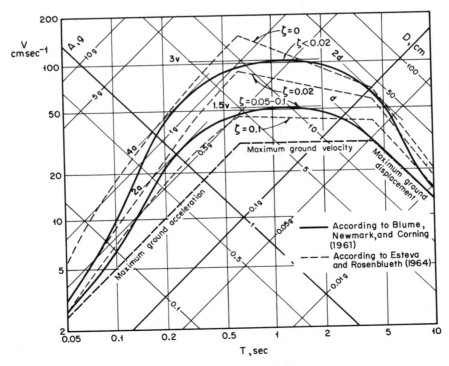

Figure 7.12. Elastic spectra prediction.

parameters are known, the design spectral ordinates can be estimated readily for motions on firm ground. After this has been done, an additional step allows us to estimate the design spectra for motions at the surface of soft soil through which the firm-ground motions may filter.

In response to earthquakes, curves of equal intensity, called *isoseismals*, are often quite elongated in a direction parallel to the main geographic and geologic (especially orographic and tectonic) features of a region [see Figs. 7.13–7.15 for examples (Davison, 1936; Figueroa, 1963)]. This may be due to a greater transmissibility of waves in one direction than at right angles to it, or to a greater extent, it may reflect the manner in which those earthquakes originate.

The assumption must be discarded that systematic departures of isoseismals

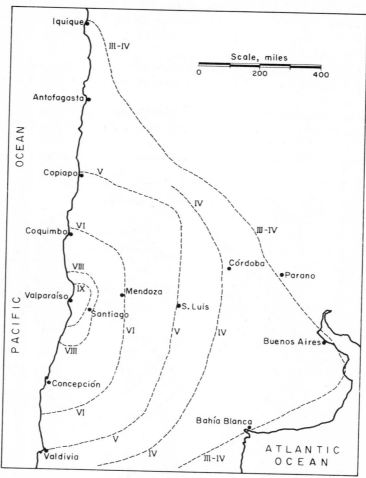

Figure 7.13. Isoseismal lines of the Valparaíso earthquake, 1906. *After Rudolph and Tams (1907), reproduced by Davison (1936).*

Sec. 7.5   CORRELATION OF GROUND-MOTION PARAMETERS   231

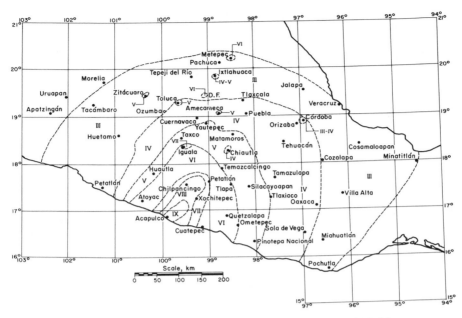

Figure 7.14. Isoseismal lines of the Mexico earthquake of 19 May 1962. *After Figueroa (1963).*

from the circular shape could be ascribed entirely to earthquake characteristics at the focus. If the assumption were tenable, one would find that the characteristics of ground motions set forth by nuclear blasts would necessarily exhibit radial symmetry, and this is certainly not the case. Figure 7.16 shows the results of computations based on records of ground motions during a nuclear detonation (Jordan, Black, and Bates, 1965); the curves depart markedly from circular arcs. In studies of earthquakes for engineering purposes we are rarely interested in epicentral distances in excess of about 600 km. Within such regions the curves in Fig. 7.16 show less anisotropy and irregularity than they do beyond, but the generating mechanism of these motions was quite radially symmetric, and even so the curves depart appreciably from circular arcs.

In many cases, a large dispersion is anticipated in the correlations we seek to establish unless effects of local geologic conditions and the direction of the line joining the epicenter and the station are taken into account, and thus far there seems to be no satisfactory way to do this.

We saw in Chapter 3 that in spherical waves a, v, and d varied with distance as the sum of two terms, one proportional to R^{-1} and the other to R^{-2}, where R is the focal distance. In cylindrical waves all quantities vary as $\ln R$ close to the source, and as $R^{-1/2}$ at long distances therefrom. These results concerned wave transmission in conservative media. There will be further attenuation of the seismic waves due to internal damping. Provided R is not too small or too great, we expect that a, v, and d may be approximated as negative powers of

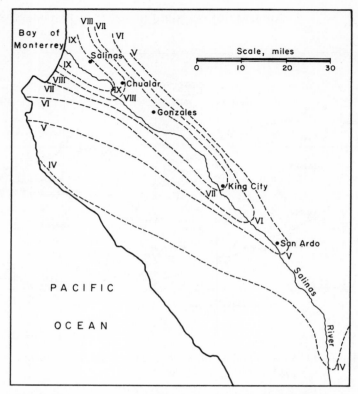

Figure 7.15. Isoseismal lines in the Salinas Valley, California earthquake, 1906. *After Lawson (1908 and 1910), reproduced by Davison (1936)*.

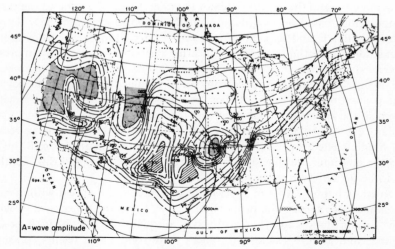

Figure 7.16. Longitudinal wave A/T (mμ/sec) values from short period vertical seismometer for Gnome event of 10 December 1961. *After Jordan, Black, and Bates (1965)*.

R, of the order of R^{-1} to R^{-2}. We also expect that a, v, and d depend linearly on an exponential function of the magnitude because of the way in which magnitude is defined (see Section 7.2).

In establishing empirical relations of maximum acceleration, velocity, and displacement with earthquake magnitude and focal distance, two conditions must be satisfied. First, attenuation due to internal damping in rock is equivalent to multiplication of the wave component that has a frequency ω by a factor $\exp(-\beta\omega R)$, where β does not depend on ω (see Chapter 3). Hence, the ground motion tends with distance to a harmonic oscillation of ever-increasing period. For harmonic vibration, $ad/v^2 = 1$. Hence, as R tends to infinity, this quantity must approach 1. For steady-state, square acceleration waves we find $ad/v^2 = \frac{1}{2}$, and it can be shown that this constitutes the minimum possible value of ad/v^2 (Newmark, 1968).

Second, as R tends to zero, the ground motion differs more and more from harmonic motion. Provided we do not consider areas very close to the focus, the motion will seem to approach white noise for which ad/v^2 is infinite. Hence, we should propose expressions for which ad/v^2 increases markedly as R tends to zero.

The following expressions have been developed by Esteva (1969) for earthquakes on firm ground, with basis on a proposal due to A. J. Hendron, Jr. (Newmark, 1968)

$$a = 1230\, e^{0.8M} (R + 25)^{-2}$$
$$v = 15\, e^{M} (R + 0.17\, e^{0.59M})^{-1.7}$$

(7.4)

and

$$\frac{ad}{v^2} = 1 + \frac{400}{R^{0.6}}$$

Here a, v, and d are in cm/sec^2, cm/sec, and centimeters, respectively, and R is in kilometers. The first two expressions were established by a least-squares adjustment (see Figs. 7.17 and 7.18). The expression for ad/v^2 was chosen so as to satisfy conditions as R tends to zero and to infinity and checked as to order of magnitude with a few data. For most earthquakes of practical interest the third equation gives $ad/v^2 = 5$ to 15.

Combining Eq. 7.2 with the second part of Eq. 7.4 we obtain

$$I = 1.44M + F(R)$$

(7.5)

where F is a decreasing function of R and varies only slowly with M. For moderate R this relation agrees closely with magnitude-intensity correlations that have been proposed by others (Gutenberg and Richter, 1956; Gzovsky, 1962; Seed, Idriss, and Kiefer, 1968).

The proposal in Eq. 7.4 satisfies the conditions we demand of ad/v^2. It is probably adequate at moderate and long focal distances, although the dispersion is high even in this range.

The foregoing expressions imply that an earthquake originates at one focus. In order to recognize more complicated generating mechanisms, it would be

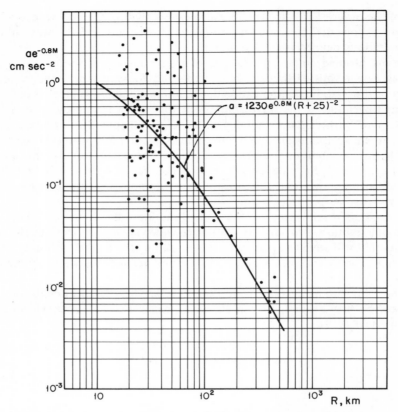

Figure 7.17. Variation of maximum ground acceleration with focal distance. *After Esteva (1969)*.

necessary to adopt functional relationships different from those of Eq. 7.4. Still these expressions have already been adjusted from an earlier version (Esteva and Rosenblueth, 1964) so that they predict finite values of a, v, and d as R tends to zero.

It is difficult to set an upper limit to a at the focus. On the basis of the strength of most rocks, Housner (1965) concludes that the maximum acceleration cannot exceed $0.5g$ ($g = $ acceleration of gravity) after admitting that the earthquake-generating mechanism is always fault slippage. This limit is not acceptable and should be raised to at least $1.0g$, and perhaps even to $1.5g \cong 1500$ cm/sec², if reports of vertical accelerations greater than g during the 1897 Assam earthquake (Richter, 1958) are trustworthy. Or possibly the focal acceleration should vary with M.

The maximum possible ground velocity near the surface of the earth can be established approximately on the following premises. The most violent portion of an earthquake is the S phase. The corresponding waves satisfy the expression

$$x = x(R \pm v_s t) \tag{7.6}$$

Sec. 7.5   CORRELATION OF GROUND-MOTION PARAMETERS   235

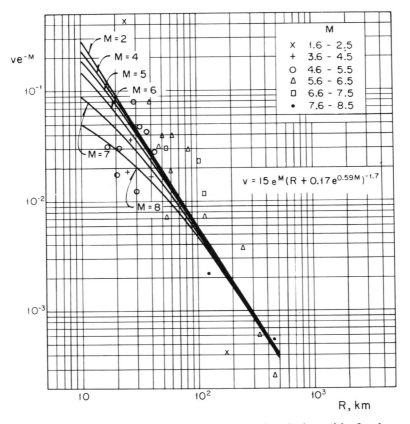

Figure 7.18. Variation of maximum ground velocity with focal distance. *After Esteva (1969)*.

where x denotes transverse displacement, v_s is shear-wave velocity, and t is time. It follows that

$$v_s \frac{\partial x}{\partial R} = \pm \frac{\partial x}{\partial t}$$

Because $\partial x/\partial R$ cannot exceed the maximum shearing strain that the material can withstand, $\partial x/\partial t$ cannot exceed v_s times this value. Most rocks rupture at a shear strain of the order of 0.0005 (in the absence of high lithostatic compression). Hence, with a seismic shear velocity of about 3 km/sec, $v = 1.5$ m/sec, approximately (see Ambraseys, 1969).

This limit on the ground velocity implies a maximum possible *MM* intensity *I* of 11 according to Eq. 7.2. Because of the possibility of the limit on ground velocity being exceeded under high-compressive stresses and in view of the spread of the correlation between v and I, the upper limit in intensity is more properly estimated as 12.

Some data indicate that expressions of the form of Eq. 7.4 systematically overestimate a (and probably v) for very large magnitudes (Housner, 1965). These objections do not altogether invalidate use of such expressions for the purpose of constructing maps of regional seismicity from data on earthquake magnitudes, and we shall use them in this manner in Chapter 8. The presentation will at least be useful for illustrative purposes because the approach that we shall adopt can be applied to more trustworthy correlations when the latter become available.

An expression has also been proposed for expected duration of a white noise of constant intensity per unit time, equivalent to an earthquake of given magnitude and focal distance:[6]

$$s = 0.02\, e^{0.74M} + 0.3R \qquad (7.7)$$

where s is the expected equivalent duration in seconds, and R is in kilometers. Insufficient data have been processed to establish the dispersion of actual equivalent earthquake durations with respect to this formula. It probably involves coefficients of variation not smaller than those of v with respect to the second part of Eq. 7.4.

The procedure described for estimating the expected ordinates of earthquake spectra uses the expected values of a, v, d, and s computed as functions of magnitude and focal distance. Rascón and Cornell (1969) have developed a more direct and satisfactory, even if more time-consuming, approach. It consists of simulating the generation of an earthquake as a series of closely spaced foci along a geologic fault. The earthquake origin is assumed to move from one focus to the next at the velocity of crack propagation (approximately equal to the velocity of shear waves). Thus, successively each focus radiates P and S waves, appropriately randomized (to simulate multiple, irregular reflections and refractions) and modified as they travel through rock. The spectra computed for the simulated earthquakes agree very well with those of the ground motions for which there is information on the causative fault. The same technique can be used for simulating other earthquake-generating mechanisms.

It is true that the approach of Rascón and Cornell does not incorporate the effects of anisotropy and major irregularities in geologic formations, or the triggering of foci that do not lie precisely along the main fault, or even the fact that the different foci usually release widely different amounts of energy during a single earthquake. But the approach does represent a vast improvement over other methods for estimating the spectra of earthquakes associated with given focal characteristics and satisfies most of the present engineering requirements. We shall return to the matter of earthquake simulation in Section 9.10.

7.6 The Three Translational Components of Ground Motions

Suppose that the spectra of a ground motion have been obtained from records in two horizontal, orthogonal directions and that we are interested in a linear

---

[6] Based on Esteva and Rosenblueth, 1964.

system having a single, horizontal degree of freedom oriented arbitrarily with respect to the orthogonal directions. Let $q(t)$, $q_1(t)$, and $q_2(t)$ denote the responses of the system at time t if the degree of freedom of the system were oriented, respectively, at an angle ϕ from direction 1, or parallel to each of the orthogonal directions 1 and 2. Since we have chosen a linear system, we may write

$$q(t) = q_1(t) \cos \phi + q_2(t) \sin \phi$$

Now let Q denote $\max_t |q|$. As Q_1 and Q_2 do not necessarily occur at the same instant, we can surely write

$$Q \leq \sqrt{Q_1^2 + Q_2^2} \tag{7.8}$$

This relationship allows us to compute upper bounds to the responses of simple systems when considering both horizontal components of motion from the bounds that we shall obtain in Chapter 9 for a single component.

Instrumental information on earthquakes of the single-shock type is meager. Ordinarily it is based on approximate evidence of the effects of the motion in the direction in which they were most violent, and there is practically no information about motion at right angles thereto. It is of interest to estimate the expected responses in an arbitrary direction to an earthquake of this type when the maximum is known, because some structures have all their significant degrees of freedom along a single direction. The very meagerness of information does not warrant the development of a theory applicable to earthquakes of this type alone, so we shall assume that the treatment we develop for ground motions of the second type applies also to those consisting essentially of a single shock.

As we shall see, for earthquakes of the second and third types, when the latter arise from filtering of motions of the second type, the expected maximum numerical values of responses to individual earthquakes are of greater interest than their actual maxima. We shall present a method for estimating the expectations of Q and of Q_{max} in term of Q_1 and Q_2, where Q_{max} is the maximum value of Q relative to direction. The approach will be crude but sufficiently accurate for most purposes.

Given Q_1 and Q_2 we must specify the dependence of Q on orientation in such a way as to satisfy Eq. 7.8. Let us assume that the ground motion is the result of two mutually perpendicular and stochastically uncorrelated processes (see Section 9.3), let us say, parallel to directions X and Y. It can be shown (Rascón, 1967) that under these conditions the expected response of a simple, linear system is a maximum or a minimum when its degree of freedom is parallel to either X or Y. Accordingly, let us denote the responses along these two directions by Q_{max} and Q_{min}, respectively. For earthquakes of the second type the maximum responses along opposite senses of the same axis are almost equal to each other. For single-shock earthquakes this is not the case. The difference is irrelevant if the criterion of failure is based on the numerically greatest response. We shall be content to describe events in the first quadrant of the plane. The present results can be extended to more general conditions.

Under the assumptions made, the squares of these two projections along any given direction will be additive. [The reasoning that leads to this conclusion

238 CHARACTERISTICS OF EARTHQUAKES Chap. 7

is similar to the method applied throughout much of Chapters 9 and 10; it rests on the fact that the variances of independent random variables are additive, and that Q is proportional to the variance of $q(t)$.] The resulting expected response Q is given, accordingly, by

$$Q^2 = Q_{min}^2 \cos^2 \theta + Q_{max}^2 \sin^2 \theta \qquad (7.9)$$

where θ is the angle between the directions of Q_{min} and Q (Rascón, 1967).

As shown in Fig. 7.19, there is a strong correlation between $(S_1^2 + S_2^2)^{1/2}$ and S_1/S_2, where S_1 and S_2 are Housner's spectral intensities for zero damping (see Eq. 7.3) corresponding to a number of earthquakes (Housner, 1962), along the directions of the components of recording instruments, and chosen so that $S_1 \leq S_2$. The full curve has been adjusted to the empirical values. Under the assumption that it represents the expected ratio of S_1 to S_2, the corresponding values of S_{min}/S_{max} have been computed. These are shown by the dashed curve. If the relation between S_{min}/S_{max} and $(S_1^2 + S_2^2)^{1/2}$ were deterministic, no values of S_1/S_2 would fall below the dashed line. The fact that only two points fail to meet this criterion supports the working hypothesis of a deterministic rela-

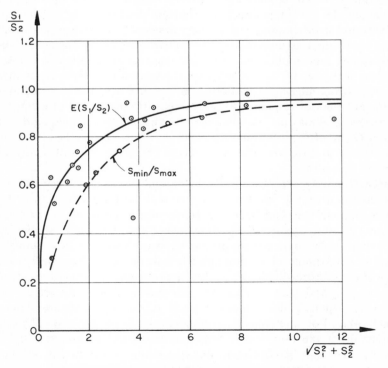

Figure 7.19. Correlation between $(S_1^2 + S_2^2)^{1/2}$ and S_1/S_2. After Rascón (1967).

tion between the ratio of minimum to maximum intensities and the square root of the sum of the squared S_1 and S_2. Errors introduced thereby will surely be overshadowed by the dispersion of future intensities.

The correlation in Fig. 7.19 can be explained on the following basis. A large value of $S_1^2 + S_2^2$ implies a large magnitude and/or a small focal distance. If earthquakes originated as slips along a geologic fault, both conditions would be associated with large subtended angles at the station, from beginning to end of the slip length (see Rascón and Cornell, 1969), except in those cases in which the line joining the center of the slip line and the station is nearly colinear with the fault. This explains why a few points do fall below the dashed line in the figure. It follows that the correlation in Fig. 7.19 is not necessarily applicable to earthquakes caused by other mechanisms.

In accordance with our assumptions, we can replace S with Q in the relations we have obtained. Under the assumption that Q_{min}/Q_{max} is fixed, Rascón has computed the value of $E_\theta(Q)(Q_1^2 + Q_2^2)^{-1/2}$, which is independent of the ratio Q_1/Q_2, where $E_\theta(Q)$ is the expectation of the response with respect to θ, and θ is uniformly distributed in the interval 0, $\pi/2$. This is of interest for structures whose single degree of freedom is oriented randomly with respect to the recording instrument. Results are shown in Fig. 7.20.

Consider next a linear system having two mutually perpendicular degrees of freedom that give rise to two natural modes with identical periods of vibration and percentages of damping. The natural frequency and percentage of damping are independent of the direction of vibration. (This is the case of a massless, circular shaft supporting a concentrated mass.) For a system of this sort the maximum response is always equal to Q_{max} because symmetry makes the orientation irrelevant. Hence, it is of interest to compute a quantity such as Q_{max}/Q_2 given Q_1/Q_2. Results are also shown in Fig. 7.20 for various ratios Q_{min}/Q_{max}.

When Q_{max} has been estimated from evidence of the effects of the earthquake, it is of interest to compute $E_\theta(Q)/Q_{max}$ so as to predict the expected effects on a single-degree structure having an arbitrary orientation relative to the direction of Q_{max}. Figure 7.20 gives values of $E_\theta(Q)/Q_{max}$ as a function of Q_{min}/Q_{max}.

For systems having degrees of freedom that give rise to uncoupled natural modes having different periods of vibration in two mutually perpendicular directions, the criteria of failure are often such that earthquake effects in each direction may be obtained from modal analysis and combination of the modal responses in each direction independently, in accordance with the conclusions presented in Chapter 10. This should be followed by the combination of the effects in the two directions as the square root of the sum of the corresponding squared responses.

The foregoing treatment presumably also applies to earthquakes that result from filtering of motions of the second type through soft mantles. When earthquakes having well-defined, prevailing periods result from filtering of single-shock motions, an intermediate situation apparently exists concerning the distribution of Q_{min}/Q_{max} between the conditions that apply to earthquakes of the first and second types. At least such are the indications derived from one

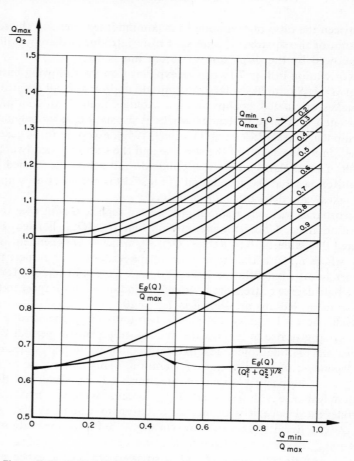

Figure 7.20. Relations between $E(Q)$, Q_1, Q_2, Q_{min}, and Q_{max}. After Rascón (1967).

earthquake whose effects were observed on rock as well as on a sand formation of medium density.[7]

Data are scarce concerning the vertical component of ground motion; nevertheless, some generalizations are warranted. First, the major portion of the energy in an earthquake is carried by S waves, and refraction tends to make these travel close to the vertical direction when the surface material is soft and horizontally stratified. Therefore, the relative importance of the vertical component of ground motion increases with the hardness of the upper ground formations. Second, this relative importance is usually a decreasing function of the ratio of epicentral distance X to focal depth H; the details of the earthquake-generating mechanism must play a decisive role, especially near the epicenter. Under some conditions the ratio of maximum vertical (a_v) to horizontal (a_h)

[7] The San Salvador earthquake of 1965 (Rosenblueth and Prince, 1965).

accelerations probably exceeds 1. The ratio a_v/a_h for the El Centro earthquake of 1940, having $X/H = 2.0$, was 0.85, while a_v/a_h was 0.27 for the earthquake of May 1962 in Mexico City, with $X/H = 0.1$.[8]

Most building codes ignore the vertical component of earthquake motions. There are cases, however, when this component should have great importance in design, as we shall see in Chapter 15.

7.7 Rotational Components and Other Space Derivatives of Earthquake Motions

Over the area covered by a building of moderate dimensions not all particles of soil describe the same motion simultaneously. If the foundation of the building is relatively rigid in horizontal directions, it will tend to average the ground motion, forcing the ground surface to displace as a rigid plate. And if the plan dimension of the building in the direction of wave travel is much smaller than one wavelength (say smaller than one-quarter wavelength), the rotation of the base will be close to the rotation of the ground if we neglect the effects of soil-structure interaction due to the mass polar moment of inertia of the building.

In order to decide whether it is proper to assume that the base rotations of a building may be taken equal to the ground rotations it is worth making the following considerations.

Wavelength equals period times wave velocity. The most significant earth waves from the viewpoint of structural response are those whose periods lie in the neighborhood of the natural periods of the structure. For most buildings the shortest natural periods of interest are of the order of 0.2 sec or longer. On the other hand, the refraction of waves at soil and rock-soil interfaces tends to make shear waves travel upward as they approach the ground surface. Consequently, at the free surface of the soil, they are observed traveling practically at the same velocity as shear waves in the underlying rock, since most of the energy of an earthquake is transmitted as shear waves. It follows that the wave velocities of interest are of the order of 3000 m/sec and that the shortest relevant wavelengths are about $0.2 \times 3000 = 600$ m.

These considerations lead to the conclusion that the assumption of base rotations being equal to those of the ground is not excessively conservative for most buildings, tanks, stacks, and towers, even if it may be inapplicable to much vaster or longer structures, such as dams and some bridges.

Under this assumption the base rotation about a vertical axis, as a function of time, can be computed from the theory of elasticity as

$$\phi = \frac{1}{2}\left(\frac{\partial x_1}{\partial X_2} - \frac{\partial x_2}{\partial X_1}\right) \tag{7.10}$$

---

[8] Here a_h is the maximum of the horizontal accelerations in two arbitrary, orthogonal directions.

where $x_{1,2}$ are ground displacements in the directions $X_{1,2}$. Newmark (1969) has applied this relationship to the calculation of earthquake-excited torsion in symmetric buildings, replacing Eq. 7.10 with

$$\phi = \frac{\partial x_1}{\partial X_2} \qquad (7.11)$$

Assuming that x_1 is caused by a shear wave traveling in the X_2 direction we may write

$$x_1 = x_1\left(t - \frac{X_2}{v_s}\right)$$

where v_s is the velocity of these waves. Hence,

$$\phi = \frac{\dot{x}_1}{v_s}$$

and

$$\max_t |\phi| = \frac{v_1}{v_s} \qquad (7.12)$$

where v_1 is the maximum ground velocity in the direction of X_1.

By a similar reasoning it can be shown that

$$\max_t |\dot{\phi}| = \frac{a_1}{v_s} \qquad (7.13)$$

and

$$\max_t |\ddot{\phi}| = \frac{h_1}{v_s} \qquad (7.14)$$

where a_1 and h_1 are, respectively, the maximum numerical values of $\ddot{x}_1$ and $d^3 x_1/dt^3$.

Consider an infinitely rigid building the side of whose square base measures 30 m and which resists all shear and torque in structural elements located, symmetrically, around its periphery. Suppose that most of the significant earth waves have a period of 0.2 sec and that $v_s = 3000$ m/sec. The maximum derivative of the ground acceleration is approximately equal to $(2\pi/0.2)a_1 = 31.4 a_1$. Hence, according to Eq. 7.14, the maximum angular acceleration is $31.4 a_1/3000 = 0.0105 a_1$. The maximum total shear produced by a_1 is $a_1 M$, where M is the mass of the building. Hence, the maximum shear to be resisted along one side of the periphery, due to base translation alone, is $a_1 M/2$. The torque, on the other hand, is equal to $\ddot{\phi}$ times the building's polar moment of inertia; that is, $\ddot{\phi} M b^2/6$, where b is the side of the building's base, if the mass of the building is uniformly distributed in plan. This torque must be resisted by the periphery, and hence it causes a shear of $\ddot{\phi} M b/12$ on each of the four sides. Substituting the numerical values of the present example we find that, on each side of the periphery, the maximum shear due to torque is 5.25 percent of that caused by translation. The effect would be greater if the resisting structural elements were closer to the center of the base.

In practice the flexibility of a building will modify these results. Using a technique similar to the ones we have described in connection with ordinary response spectra, it is possible to estimate the spectrum of maximum structural rotation, velocity, and acceleration of rotation from given maximum numerical values of ϕ, $\dot{\phi}$, and $\ddot{\phi}$. Proceeding in this manner Newmark (1969) has found that the torques excited by earthquakes in symmetric buildings are ordinarily of the order of 5 percent of the plan dimension (measured perpendicularly to the direction being analyzed) but that they may exceed 10 percent of this dimension in buildings whose fundamental frequency of vibration is high.

If we regard earthquakes as the combination of two orthogonal and stochastically independent ground motions, we conclude from Eq. 7.10 that the expected maximum numerical value of ϕ is equal to one half the square root of the sum of the squared expected maximum numerical values of $\partial x_1/\partial X_2$ and $\partial x_2/\partial X_1$. Hence, under the assumption that the design requirements for transverse vibrations are based on spectral ordinates averaged with respect to θ, the foregoing estimates of dynamic eccentricity in symmetric buildings should be multiplied by one half of the ratio $(Q_1^2 + Q_2^2)^{1/2}/E_\theta(Q)$. This ratio is the reciprocal of the ordinates shown in Fig. 7.20. Now, according to Fig. 7.19, the ratio $S_{\min}/S_{\max}$ (which is the same as $Q_{\min}/Q_{\max}$) lies between 0.80 and 0.95 if $\sqrt{S_1^2 + S_2^2}$ exceeds 4, and it can be shown that this limit corresponds to modified Mercalli intensities greater than about 7. Consequently, for most cases of practical interest, the ratio in question lies between 0.704 and 0.706, so that the eccentricities estimated as described must be affected by a correction factor of about $1/1.41 = 0.71$.

In Section 15.6 we shall return to the matter of seismically-induced torques in symmetric buildings, devoting attention to the relevant structural characteristics and to design implications.

Effects of rotations about a horizontal axis can be estimated through a similar approach. Replacing x_1 with vertical displacements and X_2 with the direction of wave travel in Eq. 7.11, ϕ becomes the rotation about a horizontal axis perpendicular to X_2.

Although the maximum vertical acceleration is usually smaller than the horizontal acceleration, the prevailing periods are also smaller. Hence, it is not unreasonable to assume that the maximum angular acceleration about a horizontal axis would also be of the order of $0.0105 a_1$ for the example considered, where a_1 is the maximum horizontal ground acceleration perpendicularly to the axis of rotation. In a rigid structure 30 m in height this would produce a maximum horizontal acceleration of 31.5 percent of a_1 at the roof. The effect is undoubtedly magnified in practice as a consequence of structural flexibility and of soil-structure interaction, but it is not necessarily magnified more than the effects of translational excitation.

In the foregoing approximate treatment we have assumed that rotation about vertical and horizontal axes is associated with shear waves. This is debatable. It is likely that rotation about a vertical axis is associated largely with Love waves and that rotation about horizontal axes corresponds mostly to Rayleigh waves. The amplitudes of these surface waves are ordinarily much smaller than those of the shear phase in strong-motion earthquakes, but their velocities are also much smaller.

Gravity waves constitute another source of ground rotation about horizontal axes, particularly in soft soil. Their relative importance is a subject of controversy. Indications that they may be significant are the production of fissures in the ground that sometimes release water, sand, and silt (Wilson, Webb, and Hendrickson, 1962), eyewitness accounts of slowly moving visible waves (Lomnitz, 1970), and even the shape of craters on the moon (Van Dorn, 1969). It seems likely that, for gravity waves to be significant in soil, the soil must either liquify or develop pronouncedly softening nonlinear behavior.

On the other hand, it is likely that these phenomena are especially important for structures resting on two formations of very different stiffnesses or on the neighborhood of the interface between such formations, and classical theories of earth waves are not applicable under those conditions. Not until ground rotations are actually recorded will trustworthy estimates of the effects of these phenomena become available.

Most of the foregoing remarks and the approach used for estimating ground rotations are applicable to other space derivatives of ground motion. Thus, the horizontal elongation of the ground in the X_1 direction is given directly by dx_1/dX_1. Longitudinal strains of the ground surface, due to out-of-phase displacements in the direction of wave travel, are important in the design of certain long, civil-engineering works, such as bridges and some dams. The treatment of body and surface waves presented in Chapter 3 permits estimating the amplitudes of these strains, and the methods of analysis in Chapters 2 and 4 allow estimating the ensuing stresses in the structures under the assumption of linear behavior.

One case of special interest concerns long tubular structures, such as tunnels, some penstocks, and transit tubes. This matter is treated in Section 10.4.

Aerial panorama of part of Marin County, California showing the San Andreas rift zone from Bolinas Lagoon (foreground) to Bodega Head. In the great 1906 California earthquake horizontal ground offset along the fault range from 15–20 ft in this region. Olema is at the head of the long finger of water (Tomales Bay in the upper part of the photograph). (Karl V. Steinbrugge collection.)

8

SEISMICITY

8.1 Introductory Note

As we shall see in Chapter 14, any rational attempt at earthquake-resistant design must rest on a probabilistic description of the variables concerned. Among them, the characteristics of future earthquakes involve by far the greatest uncertainties.

If quantitative data were always abundant we could employ traditional statistical methods in describing the probability distributions of the characteristics of future ground motions. Then we would be entitled to take the distribution of the universe identical with the distribution of the sample. This is dramatically not the case when we consider a region of the world where no earthquakes have occurred during the corresponding period of observation. Traditional methods lead to the conclusion that the region is nonseismic. Yet, no sensible engineer would design a major work—a dam or a nuclear power plant—in that region without earthquake provisions.

The scarcity of information requires us to use Bayesian statistics. Bayes' theorem, or the formula of the probabilities of hypotheses, begins with a prior probability distribution of the variable under study, incorporates statistical data, and furnishes a posterior, improved probability distribution.

Because of the difficulty and apparent arbitrariness involved in choosing the prior distribution, the use of Bayes' theorem has been criticized greatly. Alternate methods have been proposed; these actually do no more than disguise an automatic choice of the initial distribution. The fact that we must choose a subjective prior distribution cannot be ignored, and the choice is far less arbitrary than may seem at first sight because every variable we deal with belongs to a phenomenon that bears some similarity to other phenomena with which we are familiar.

Bayes' formula is

$$P(H_j|A) = \frac{P(A|H_j)P(H_j)}{\sum_{i=1}^{n} P(A|H_i)P(H_i)} \qquad (8.1)$$

where H_i, $i = 1, 2, \ldots, n$, denote n exhaustive, mutually exclusive hypotheses, A denotes an event, $P(H_j)$ stands for the prior (absolute) probability that hypothesis H_j be true independently of the occurrence of event A, $P(A|H_j)$ is the (conditional) probability that event A occur when H_j is known to be true, and $P(H_j|A)$ is the posterior (conditional) probability that hypothesis H_j be true when it is known that event A occurred. The denominator in Eq. 8.1 may also be written as $P(A)$.

In the study of seismicity, the information available can be grouped as follows.

1. Similarity with other physical phenomena.
2. Geotectonic features.
3. Statistical data—on the space-time coordinates of seismic foci and amounts of energy released by earthquakes in the entire globe—during the period for which we have information.
4. Qualitative information about the same variables as in the foregoing group, extending back in historic and geologic times, over parts of the earth.
5. Theories and observations on the transmission of seismic waves.
6. Geologic maps and data on the dynamic properties of rock and soil formations.
7. Statistical data on intensities and earthquake records.

The first four groups of data allow for the construction of maps of local seismicity, that is, statements about the probabilities of earthquakes of given magnitudes originating in given portions of the earth's crust. The last three groups of data permit drawing maps of regional seismicity; these are statements about the probabilities of earthquakes of given intensities, or of other given characteristics, shaking a given region of the earth's surface. If one includes much detail derived from geologic information, these maps are known as of microregionalization.

Similarity with other phenomena, such as energy release by radioactive materials and cracking of concrete cylinders in the laboratory, leads us to postulate tentatively that the occurrence of earthquakes is a generalized Poisson process, in the sense that the expected number of earthquakes in a given magnitude range, released per unit time in any given volume of the crust, does not vary with time, and that the probability of an earthquake, in this magnitude range and occurring during a given time interval and volume of the crust, is independent of all previous earthquakes in the entire globe.

In accordance with this postulate we may write

$$P_t(n) = \frac{(\lambda t)^n \exp(-\lambda t)}{n!} \qquad (8.2)$$

where $P_t(n)$ denotes the probability that the number of earthquakes with magnitudes greater than some value M, generated during the time interval t in a given volume of the earth be equal to n, and $\lambda = \lambda(M)$ is the expected value of n per unit time.

Data contradict the assumption of stochastic stationarity [for a definition of these terms see Chapter 9 and Gzovsky (1962)] and of time independence (Aki, 1963; Knopoff, 1964) and space independence (Tsuboi, 1958; Gajardo and Lomnitz, 1960) from past events. There is an evolution in local seismicity, and earthquakes tend to cluster in time and space; this is true even if one disregards fore- and aftershocks in the analysis. However, we shall retain the simplifying assumptions quoted, on the grounds that they are probably adequate for the time intervals in which we are most interested—of the order of expected life spans of civil-engineering works.

Tectonic features, especially geologic faults, have been claimed to be associated quantitatively with local seismicity (Gzovsky, 1962). Yet data from parts of the world, other than those areas from which the proposed correlations were derived, do not bear out the validity of such correlations. One can only state that qualitatively similar tectonic features can be expected to be associated with seismicities of roughly the same order of magnitude. Accordingly, geotectonic information will be used here only for the purpose of dividing the crust into relatively small zones without introducing prior distributions related to local seismicity on the basis of this information alone. We shall also assume that in each of these zones the seismicity per unit volume is a function of depth alone. The manner of incorporating the rest of the information will become apparent in the balance of this chapter.

8.2 Local Seismicity

Available information (Gutenberg and Richter, 1954) indicates that a single parameter—for example, the expected energy release per unit volume and per unit time—is insufficient for engineering purposes, since some regions of the world are noted for the abundance of earthquake swarms, while in others, earthquakes of high magnitude are abnormally prevalent. On the other hand the labor involved in processing these data increases rapidly with the number of parameters chosen, and the amount of data available for each region is so meager that the choice of a large number of parameters is unjustified at present. We shall adopt two parameters to define local seismicity at each site.

Let $\lambda(M)$ denote the expected number of earthquakes that originate in a given volume of the earth's crust per unit time and whose magnitude exceeds M. We assume that local seismicity is constant within a volume V. In accordance with the comments contained in the preceding paragraph we shall take the function $\lambda(M)$ to be determined by two parameters that depend on the coordinates of the centroid of the volume in question. Letting α and β be the parameters, we choose to write

$$\lambda = \alpha V \lambda_0(\beta, M)$$

where V is the volume we are considering, and $\alpha \lambda_0$ is the expected number of earthquakes per unit time and per unit volume.

The following function has been proposed for λ_0:

$$\lambda_0 = e^{-\beta M} \tag{8.3}$$

This expression has been used extensively in the equivalent statement that $\log \lambda$ and M are linearly related (Gutenberg and Richter, 1954).[1] It can be derived from theoretical considerations on the assumption that magnitudes obey the extreme-value theory. Suppose we take a number of independent samples of size n of a variable x and select the largest value of x in each sample; call these maxima x_n. The theory of extreme values (Gumbel, 1958) states that if x has neither an upper nor a lower bound and satisfies other relatively lax conditions, then, for large enough n, the distribution of x_n will asymptotically approach the extreme type I distribution,

$$P(x_n \leq x_m) = \exp\left[-e^{-\beta(x_m - u)}\right] \tag{8.4}$$

where x_m is an arbitrary value of x_n and β and u are parameters of the distribution. Under the assumption that we deal with a generalized Poisson process, Eq. 8.3 can be derived immediately from Eq. 8.4.

Some studies of local seismicity furnish an excellent agreement with the expectations predicted in Eq. 8.3 or with the extreme distribution that it implies (Dick, 1965). Despite this apparent theoretical and empirical substantiation, Eq. 8.3 is untenable for very high magnitudes (Esteva, 1968). According to Eq. 7.1, Eq. 8.3 predicts an infinite amount of energy liberated in earthquake activity per unit time and per unit volume if β is not greater than $1.5 \ln 10 = 3.46$. As we shall see, if we are to attain good agreement with empirical data over intervals of not very high magnitudes, we must adopt values of β between approximately 1.7 and 2.9. Consequently, Eq. 8.3 is indeed inadmissible for very high magnitudes.

It is probable that at every site of the earth there is a maximum possible magnitude determined by the thickness and strength of the crust. For example, magnitudes much greater than 9 in the continents and continental shelves and much greater than 7 under deep oceans (where the crust is thinnest) seem unlikely. Yet there are no good bases for establishing precise upper bounds (J. Brune, personal communication).

This assumption that λ is proportional to $\exp(-\beta M)$ is untenable also for very small magnitudes, as it would predict infinite slip per unit time at geologic

---

[1] Actually the proposal by Gutenberg and Richter relates $\log(\Delta \lambda / \Delta M)$ with M, where ΔM is small compared with one. If we replace the finite increments with differentials, the linear relation amounts to $d\lambda/dM \propto e^{-\beta M}$, where β is a constant; hence, $\lambda \propto e^{-\beta M}$ and hence Eq. 8.3.

faults under the assumption that at least most earthquakes are caused by fault slip (Rosenblueth, 1969a). A significant departure from this exponential relation can be expected at some magnitude smaller than zero with λ tending asymptotically to some finite value when M tends to minus infinity.

On the basis of these considerations a relation of the form

$$\lambda_0 = \exp(\alpha_1 e^{-\alpha_2 M})$$

seems more appropriate than Eq. 8.3, as it predicts a finite amount of energy liberated per unit time and makes λ_0 tend to a finite value when M is made to tend to $-\infty$. Moreover, this expression can give a satisfactory fit to empirical information over a wider range of magnitudes than Eq. 8.3. Finally, it is consistent with the reasoning that there is probably a maximum possible magnitude in each region of the globe but that these maxima are unknown. However, we shall retain Eq. 8.3 throughout the rest of this book for the sake of simplicity. Errors introduced by this simplification are on the safe side and, for purposes of establishing design criteria for the design of most of the structures met in practice, these errors are insignificant. Yet when it comes to the design of special structures for which return periods of failure should be very long, say in excess of 1000 years, as is the case with the most critical parts of nuclear power plants, it is important to refine the estimate of the shape of $\lambda(M)$ curves in the range of very high magnitudes. In such cases, earthquakes caused by extremely rare events (e.g., the impact of large meteorites on earth) may become significant, as their rates of occurrence may be of the same order as those of earthquakes of tectonic origin.

Take now the earth's crust divided into three macrozones, defined by areas that correspond to the Circumpacific Belt, the Alpide Belt, and the low-seismicity macrozone. The boundaries of the macrozones can be derived from the map of epicenters in Fig. 8.1.[2] (The groups of epicenters associated with mid-oceanic ridges should not be taken into account in a crude macrozoning for application in earthquake engineering, as these earthquakes are normally of rather small magnitude.) For illustrative purposes an area will be associated with each macrozone, based on an index figure presented by Gutenberg and Richter (1954); the Trans-Asiatic and Alpide Belts will be treated as a single macrozone. A more accurate definition should be based on a careful consideration of the salient tectonic characteristics of the macrozones. All volumes will be assigned a depth of 800 km, within which all earthquakes have been inferred to originate, as there is reason to believe that they will not originate at greater depths.

To obviate difficulties arising from consideration of joint probabilities, we shall treat the seismicity of each macrozone as independent of that of the

---

[2] The Circumpacific Belt is indicated by the band of closely spaced epicenters surrounding the entire Pacific Ocean in Fig. 8.1. The Alpide Belt extends generally from the middle of the figure toward the left as shown by the slightly more scattered band of epicenters to the west of the Circumpacific Belt in Fig. 8.1. The remainder of the earth can be considered as the low-seismicity macrozone.

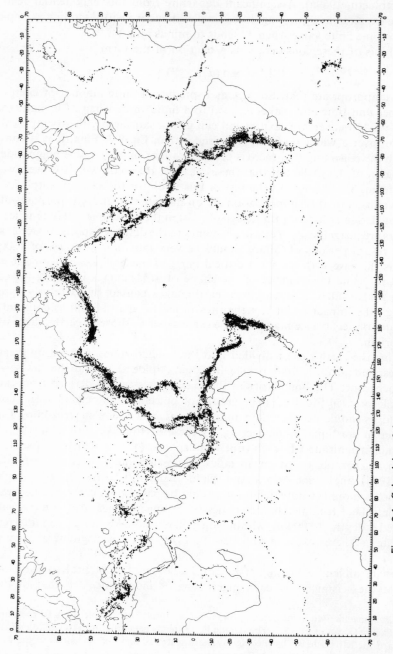

Figure 8.1. Seismicity of the earth, 1961–1967, ESSA, CGS epicenters depths 000–700 km. *After Barazangi and Dorman (1969).*

other two. We shall choose for each λ a gamma-1 distribution, which is the natural conjugate of the Poisson distribution.[3]

When the gamma-1 distribution is chosen, the following procedure applies. Let r' denote the expected number of events during a time interval t' in the prior distribution; let us assign λ (the expected number of events per unit time) in this prior distribution a coefficient of variation equal to $(r')^{-1/2}$; let r denote the number of events observed during a time interval t. Application of Bayes' theorem gives a posterior Poisson distribution with parameter λ, where λ has a gamma-1 distribution with expectation

$$E(\lambda) = \frac{r' + r}{t' + t} \tag{8.5}$$

and coefficient of variation

$$c(\lambda) = (r' + r)^{-1/2} \tag{8.6}$$

We shall use this approach, letting the number of events in question signify the number of earthquakes whose magnitude exceeds a given value, and divide our results by the volume of the zone we are considering. Thus, λ stands for $\lambda(M)$. One extreme approach consists of ignoring all historic evidence, thus assuming complete ignorance in the prior distribution, which amounts to our taking $r' = 0$ and $t' = 0$, so that the expected value of $\lambda(M)$ is equal to the mean number of earthquakes whose magnitude has exceeded M during the 50 years of instrumental records, per unit time.

Neglect of historic evidence is not an asset in our computations, except that it does save considerable searching. Its incorporation would not modify our results substantially, except for lowering the dispersion associated with high magnitudes and changing the corresponding expected λ's slightly, since we deal with large volumes for which $r \gg 0$. It would be desirable to recompute the parameters of the posterior distribution after taking due note of preinstrumental data.

In this treatment we take the λ's for various magnitudes to be independent of each other, which, of course, is objectionable. One way of correcting our results due to this concept consists of requiring that $E(\lambda)$ be a smooth function of M in every macrozone. We may achieve this by adopting Eq. 8.3 and by fitting the experimental data through the use of a traditional statistical method. In this manner we obtain Fig. 8.2, where we have drawn straight lines on semilogarithmic paper to fit the data obtained from Eq. 8.5, using the least-squares method for fitting. The coefficients of variation, in accordance with Eq. 8.6, are equal to $[E(\lambda)(t' + t)]^{-1/2}$, with $t' = 0$ and $t = 3, 35,$ or 49 yr in our case, depending on the magnitude range. Smoothing of the empirical curve introduces changes in the coefficients of variation in the posterior distribution. Esteva (1968) has derived the coefficients of variation shown by the dashed lines in

---

[3] Choice of the natural conjugate for the prior distribution of the parameters of a probability distribution is justified on the basis of mathematical tractability and on the more important ground that then the posterior distribution has the same functional form as the prior (Raiffa and Schlaifer, 1961), which must ordinarily be the case.

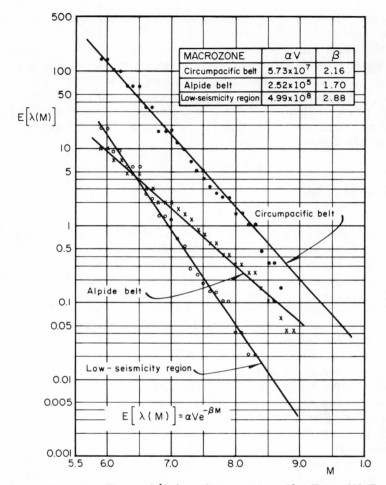

Figure 8.2. Expected λ's in each macrozone. *After Esteva (1968)*.

Fig. 8.3. In so doing he has informally taken into account this effect of smoothing. However, there are numerous data to substantiate the assumption that ln $\lambda(M)$ is practically a straight line over a large range of magnitudes (see, for example, Furumoto, 1966), so that the full lines in the figure are probably justified. Typical gamma-1 distribution curves are shown in Fig. 8.4.

We are ready now to proceed to smaller volumes of the earth's crust. Again we shall adopt gamma-1 distributions for the prior probability, and we shall incorporate the statistical data available.

Consider a volume V' contained in one of the three macrovolumes, say V, we analyzed above. For any given magnitude M the following identity holds:

$$\lambda' \equiv \frac{\lambda'}{\lambda} \lambda$$

Sec. 8.2　　　　　　　　　　　　　　　　　　　　　LOCAL SEISMICITY　**255**

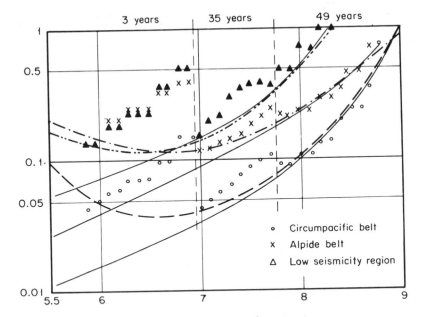

Figure 8.3. Coefficients of variation for λ in the three macrozones. *Data after Esteva (1968).*

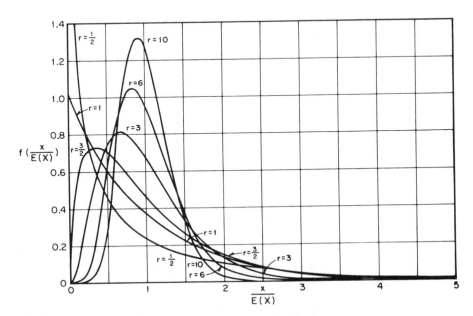

Figure 8.4. Typical gamma-1 distribution curves. *After Schlaifer (1959).*

where λ' is the $\lambda(M)$ that corresponds to V' while λ corresponds to V. The spatial variation of seismicity is incorporated in the ratio λ'/λ, while our ignorance of the seismicity of the macrozone dictates the distribution of λ. We shall assume that λ'/λ and λ are independent variables. This allows us to write

$$E(\lambda') = \frac{V'}{V} E(\lambda) \tag{8.7}$$

$$c^2(\lambda') = c^2\left(\frac{\lambda'}{\lambda}\right) + c^2\left(\frac{\lambda'}{\lambda}\right)c^2(\lambda) + c^2(\lambda) \tag{8.8}$$

In view of Eqs. 8.5 and 8.6,

$$c^2(\lambda) = \frac{1}{E(\lambda)(t' + t)} \tag{8.9}$$

If we assume that the distribution of λ'/λ depends only on V'/V and introduce the restriction that λ' be equal to λ when $V' = V$, we obtain

$$c^2\left(\frac{\lambda'}{\lambda}\right) = \left(\frac{V}{V'}\right)^\gamma - 1 \tag{8.10}$$

where $\gamma(M)$ is a parameter characterizing the macrozone in question.[4]

In order to estimate the value of γ we have computed values of $c^2(n'/n)$ for various groups of zones characterized by the corresponding ratios V'/V and for various earthquake magnitudes. Here n' and n are the numbers of earthquakes known to have originated within portions of a macrozone having volumes V' and V (the entire macrozone), respectively. Results are shown in Fig. 8.5 for the Circumpacific Belt and for the low-seismicity macrozone. Only those values of $c^2(n'/n)$ associated with relatively large n' can be regarded as satisfactory approximations to $c^2(\lambda'/\lambda)$. It can be shown that the ratio of $\ln[c^2(n'/n) + 1]$ to $\ln(V/V')$ increases with the expected value of n' and tends to 1 as n' tends to zero. This indeed happens in Fig. 8.5 as $E(n')$ decreases because either V'/V becomes small, M becomes large, or the seismicity of the macrozone is low. Accordingly, only those data in Fig. 8.5 will be taken into account for which $E(n')$ is considered sufficiently large to yield consistent results. These data correspond to the Circumpacific Belt with $M \leq 8.0$ and $V'/V \geq 0.01$. In these ranges, we find $\gamma = 0.25$. We shall assume that this value of γ applies to the whole earth and to all earthquake magnitudes, as similarity of tectonic features lends weight to this contention.

Substituting Eqs. 8.9 and 8.10 into 8.8 we obtain the coefficient of variation of λ'. Its expectation is given by Eq. 8.7, and we assign it a gamma-1 distribution. In this manner we establish the prior distribution of λ for the smaller zones in which we are interested. Application of Bayes' theorem to incorporate the

---

[4] Equation 8.10 may be proved as follows. Let $x = V'/V$, $y = V''/V$, $c^2(\lambda'/\lambda) + 1 = F(x)$, and $c^2(\lambda') + 1 = G(V')$. Then Eq. 8.8 may be written in the form $G(V') = G(V)F(x)$. Similarly, $G(V'') = G(V)F(y)$. But $G(V') = G(V'')F(x/y)$. Hence, $F(x)/F(y) = F(x/y)$ so that $\ln F(x)$ must be proportional to x. The restriction $F(1) = 1$ yields Eq. 8.10.

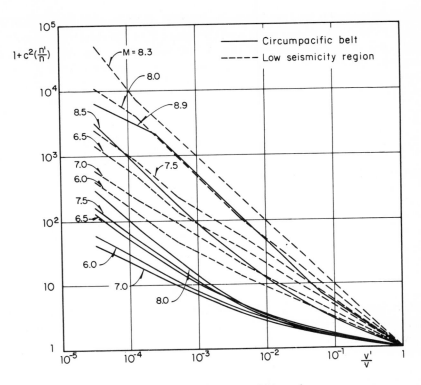

Figure 8.5. Variability of local seismicity within a given macrozone. After Esteva (1968).

available statistical data proceeds in the same manner as for the macrozones, except that we must use fictitious values of r' and t' to enter Eqs. 8.5 and 8.6, consistent with the parameters of the prior distribution. The latter condition requires that we use

$$t' = \frac{r'}{E(\lambda)}$$

and

$$r' = \frac{1}{c^2(\lambda)}$$

First consider a zone having an area of 3.56×10^6 km² within the Circumpacific Belt. Assume that we are interested in earthquakes of magnitude exceeding 7. We find from Fig. 8.2 that $E(\lambda) = 16.0 \; yr^{-1}$. Therefore,

$$c^2(\lambda) = \frac{1}{16.0(0+35)}$$
$$= 0.001786$$

The area of the corresponding macrozone is 93.8×10^6 km². Therefore,

$$E(\lambda') = \frac{3.56}{93.8} 16.0 = 0.607 \, yr^{-1}$$

$$c^2\left(\frac{\lambda'}{\lambda}\right) = \left(\frac{93.8}{3.56}\right)^{0.25} - 1$$
$$= 1.27$$

and

$$c^2(\lambda') = 1.27 + 0.001786 \times 1.27 + 0.001786$$
$$= 1.274$$

The fictitious values for the prior distribution are

$$r' = \frac{1}{1.274}$$
$$= 0.784$$

$$t' = \frac{0.784}{0.607}$$
$$= 1.292 \text{ yr}$$

There is information to the effect that in 35 years, 25 earthquakes of magnitude greater than 7 have originated in this zone, or 0.714 earthquakes per year. It follows from Eq. 8.5 that the posterior expected yearly number of earthquakes having a magnitude in excess of 7 is

$$E(\lambda') = \frac{25 + 0.784}{35 + 1.292}$$
$$= 0.710 \text{ yr}^{-1}$$

with a coefficient of variation, according to Eq. 8.6, equal to

$$c(\lambda') = (25 \times 0.784)^{-1/2}$$
$$= 0.197$$

We would repeat the computations for other magnitudes of interest and draw smooth lines on semilogarithmic plots.

Second, consider a zone whose area is also 3.56×10^6 km² that lies outside both seismic belts and where it is known that no earthquakes of magnitude greater than 7 have originated during a 100-yr period. Computations similar to the ones we have just carried out give, for the posterior distribution,

$$E(\lambda') = 4.3 \times 10^{-6}$$
$$c(\lambda') = 47.7$$

In this presentation we have made no distinction of seismicity with depth, while we know there is a pronounced, systematic variation in terms of it. This should be taken into account first by subdividing each macrozone according to depth. In the absence of this subdivision, the systematic variation may be incorporated by traditional statistical methods, fitting a curve to the pertinent

Sec. 8.2   LOCAL SEISMICITY   259

data. Figure 8.6 shows the variation with depth of the mean annual number of earthquakes whose magnitudes exceed 5.9 for the Circumpacific Belt.

For some zones it is known that most earthquakes originate in the neighborhood of an inclined plane which may or may not coincide with a geologic fault. Such features, and any others that manifest a variation within a zone, should be incorporated in a similar manner.

The foregoing approach may be applied, step by step, independently to the numbers of earthquakes whose magnitudes exceed various values of M. The posterior expectations and coefficients of variation of $\lambda(M)$ will then have a number of discontinuities. The situation may be corrected by fitting a continuous curve to the computed value. Alternatively we may deal with the parameters α and β, which define $\lambda'(M)$ as independent variables. Applications of Bayes' theorem under these conditions presents computational difficulties that may be circumvented by resorting to an approximate procedure due to Esteva (1968), which assigns β a normal distribution or by assigning it a prior gamma-1 distribution. But it often happens that information is very scarce for earthquakes of high magnitudes originating in relatively small zones, and we may lean toward giving more weight to the shape of $\lambda_0(M)$ as derived from a macrozone than as computed *a posteriori* for each zone. If so, it will be proper to compute the parameters of the distribution of λ' for the smallest value of M for which we have ample and trustworthy information and assume that β is the same as for

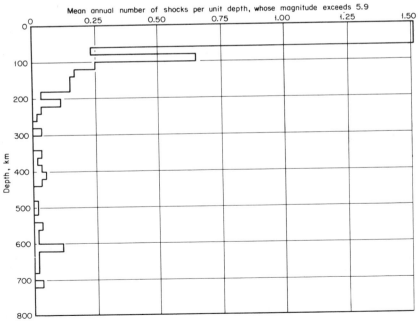

Figure 8.6. Variation of seismicity with depth, Circumpacific Belt. *Data after Gutenberg and Richter (1954).*

the entire macrozone. When using this approach an exception may be made in zones for which we have good reason to assign a locally different value of β and, hence, for which we surely have adequate information concerning the relative rates of occurrence of earthquakes having different magnitudes.

8.3 Regional Seismicity

Our task now is to compute the probability distribution of intensities and other earthquake parameters at a given station. We shall refer to intensities so as to illustrate the manner of obtaining pertinent information about earthquake parameters in general.

Suppose first that there is no direct information about intensities of past earthquakes at the station of interest. From the relation between intensity and magnitude (Eq. 7.5) we can say that, given an intensity I_c and a focal distance R, there is an earthquake magnitude $M(I_c, R)$ such that, if the magnitude of an earthquake exceeds M, its intensity at the station will exceed I_c. Because this correlation has a wide dispersion, we shall refer to I_c as the *computed* intensity; later we shall take into account the differences between the computed and actual intensities.

It follows that the number of earthquakes originating at a distance R from the station and producing, at that point, an intensity greater than I_c in any given span of time is equal to the number of earthquakes originating at that distance and having a magnitude greater than M. The total number of earthquakes whose computed intensity exceeds I_c at the station is the sum of those having magnitudes greater than the corresponding M's at all possible sources, where M is a function of the corresponding R.

We have assumed that earthquake generation, for the magnitudes greater than any given M, is a simple Poisson process. Consequently the earthquakes whose intensity exceeds I_c at the station constitute a simple Poisson process (Parzen, 1964) whose rate is the sum of the rates of generation:

$$\mu(I_c) = \int_V \lambda(M) \, dV \tag{8.11}$$

Here $\mu(I_c)$ is the mean yearly number of earthquakes whose computed intensity is greater than I_c, and V denotes volume.

Because we do not know the actual values of $\lambda(M)$, we must compute the parameters of the distribution of μ, which may be assumed to be gamma-1. It follows from Eq. 8.11 that the expectation and variance of μ are given by

$$E[\mu(I_c)] = \int_V E[\lambda(M)] \, dV \tag{8.12}$$

and

$$\sigma^2[\mu(I_c)] = \iint \operatorname{cov}[\lambda(M_1), \lambda(M_2)] \, dV_1 \, dV_2 \tag{8.13}$$

(Esteva, 1968), where σ^2 denotes variance, and the integrals cover the entire volume of the earth's crust where significant earthquakes may originate.

For the numerical evaluation of Eq. 8.12, one divides each zone into sufficiently small subzones, chooses a value of I_c, finds the corresponding M's and hence λ's, and adds them together. The procedure is repeated for as many values of I_c as desired. If we assume that $\ln E[\lambda(M)]$ is a straight line, $\ln E[\mu(I_c)]$ will also be a straight line, so that two values of I_c will suffice, even if a single value of the slope of this line is obtained analytically.

For Eq. 8.13 essentially the same approach will be applicable, multiplying the dispersion of λ at one subzone by those at all other subzones and by the corresponding coefficient of correlation and computing the sum of these products. It is consistent with our previous assumptions (but nevertheless debatable) that the coefficient of correlation is zero for subzones belonging to tectonically different areas and that it is 1 for subzones of a tectonically homogeneous zone, provided that the shape of the $\lambda_0(M)$ curve is assumed known.

The parameters of the probability distribution of μ must be corrected in order to recognize the dispersion in the correlation between magnitude, focal distance, and intensity. Esteva (1968) has shown that this can be done by means of the expressions

$$E[\mu(I')] = \int_0^\infty \frac{\partial E[\mu(I_c)]}{\partial I_c} P(I > I' | I_c) \, dI_c \tag{8.14}$$

and

$$\sigma^2[\mu(I')] = \int_0^\infty \int_0^\infty \text{cov}\,[K(I_1), K(I_2)] \, dI_1 \, dI_2 \tag{8.15}$$

where

$$K(I_i) = \frac{\partial \mu(I_c)}{\partial I_c}\bigg|_{I_c = I_i} P(I > I' | I_i); \, i = 1, 2$$

I' is a specific value of I, and $P(I > I' | I_c)$ and $P(I > I' | I_i)$ are the probabilities that the actual intensity of an earthquake does not exceed I' given that the computed intensity is I_c or I_i, respectively.

When there is no direct information concerning the intensities of past earthquakes at the site of interest, it is often unnecessary to compute the variance of $\mu(I)$, as design decisions are, in many cases, based solely on the expected values of the rates of occurrence or can be taken with sufficient accuracy as if this were strictly true (see Chapter 14). Both $E(\mu)$ and $\sigma^2(\mu)$ can then be obtained by either of the following procedures.

Procedure 1. Step 1. Compute the prior distributions of $\lambda(M)$. Step 2. Incorporate *statistical* information on magnitudes, by means of Bayes' theorem, to obtain posterior distributions of $\lambda(M)$. Step 3. Use correlations to find the posterior $E(\mu)$ and $\sigma^2(\mu)$.

Procedure 2. Step 1. As in Procedure 1. Step 2. Use correlations to obtain the prior values of $E(\mu)$ and $\sigma^2(\mu)$ at the station of interest. Step 3. Take the statistical information on the occurrences of earthquakes of various magnitudes and translate it into data about intensities at the station by means of correlations.

Step 4. Combine the prior $E(\mu)$ and $\sigma^2(\mu)$ with the statistical data on computed intensities to obtain the posterior values of these parameters.

In the first procedure, $E(\lambda)$ and $\sigma^2(\lambda)$ of the prior distribution permit computing $r' = E^2(\lambda)/\sigma^2(\lambda)$ and $t' = r'/E(\lambda)$. The number of earthquakes with magnitude M (associated with λ) known to have originated in a given zone of the crust is denoted by r and the period of observation by t. These parameters are combined with those of the prior distribution to obtain the parameters of the posterior distribution, $r'' = r' + r$ and $t'' = t' + t$. Similar relations apply to the parameters of the prior and posterior distributions of μ in the second procedure.

When there is no direct statistical information of earthquake intensities at the station, Procedure 1 is preferable, as it saves one step in the computations relative to Procedure 2. When this information is available, however, we must employ the second procedure. The parameters r and t for intensities now include two portions, one derived from information on magnitudes and the other derived from earthquakes of known intensities. In order not to use the same information twice, data derived from magnitudes should not include the periods for which there is information on intensities. (These periods will usually differ for various levels of intensity.) In this case the computation of $\sigma^2(\mu)$ is mandatory both for the prior distribution and for the statistical data derived from earthquakes of known magnitudes, in order to permit computing the corresponding r' and r.

Typical curves showing the rates of occurrence of earthquakes of different intensities are shown in Fig. 8.7. They have been obtained exclusively from statistical information. The curvatures of these lines on a semilogarithmic plot are greater than would be expected from the foregoing description of Bayesian analysis. Indeed, if I_c were a linear function of M and if all curves of $\ln E[\lambda(M)]$ were straight lines, one could show that $\ln E[\mu(I)]$ would necessarily plot a straight line or a concave curve, whereas the curves in Fig. 8.7 are all convex. (The relation between $\ln \mu$ and I would be straight only if β were equal for all the zones where significant earthquakes originate.) Part of the reason for this convexity can be found in the slight convexity of the $\ln \lambda(M)$-curves, part in the minor dependence of $F(R)$ on M in Eq. 7.5, and the rest, very probably, in the nonlinearity of actual soil formations. Curves such as the ones shown in Fig. 8.7 may serve as guides to modify computed curves of $E[\mu(I)]$ before incorporating data on actual intensities.

We may apply essentially the same methods of computation when we wish to arrive at the probability distribution of the rate of exceedance of some given earthquake characteristic, other than intensity, at a station. The characteristics in which we are usually interested are related more directly with structural responses. They include maximum ground accelerations, velocities, displacements, and the duration of ground motion. In the case of linear behavior, we may apply similar methods of analysis even to structural responses themselves. In Eqs. 8.12–8.15 we need only replace I with the appropriate earthquake characteristic and the correlation given by Eq. 7.5 with the appropriate correlation

Sec. 8.3 REGIONAL SEISMICITY 263

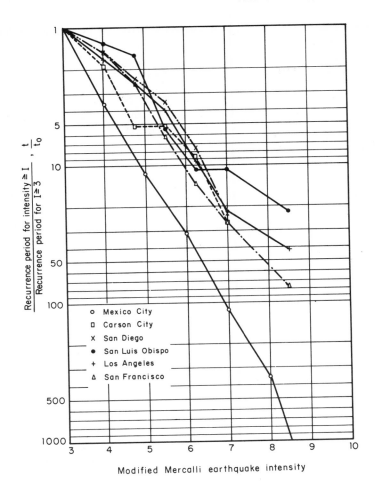

Figure 8.7. Recurrence periods in terms of intensities for several stations.

(Eq. 7.4 or 7.7 or the corresponding relationship from Chapter 9 or 10 in the case of structural responses).

At moderate focal distances Eq. 7.4 may be replaced, without undue error, with the approximate relation

$$y_c = \exp(c_y M) F_y(R) \tag{8.16}$$

where y_c is the computed value of a, v, or d, c_y is a constant, and F_y is a function of R only. If, in addition, we admit that in the range of interest $E[\lambda(M)]$ may be taken proportional to $\exp(-\beta M)$, we find that $E[\mu(y_c)]$ is proportional to y_c^{-b}, where $\mu(y_c)$ is the rate of occurrence of earthquakes whose computed value of a, v, or d exceeds the quantity y_c, and b is a constant proportional to β. Under the circumstances we find that, if the distribution of y/y_c is independent of y_c where y is the actual value of a, v, or d,

$$\frac{y_\mu}{y_{c\mu}} = \left[E\left(\frac{y}{y_c}\right)^{-b} \right]^{1/b} \tag{8.17}$$

where y_μ and $y_{c\mu}$ are, respectively, the actual and computed values of a, v, or d having the same rate of exceedance μ. This result is directly obtained from a relationship derived by Esteva (1968 and 1970), who gives the frequency distributions of y/y_c, required for the evaluation of Eq. 8.17, for maximum ground acceleration and velocity. The resulting ratios $y_\mu/y_{c\mu}$ are shown in Fig. 8.8 as a function of b.

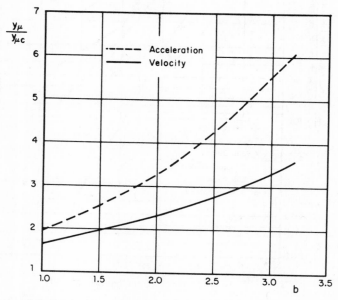

Figure 8.8. Ratio of actual to computed maximum ground characteristics. *After Esteva (1970).*

It follows from Eq. 7.2 that, with the same approximations we have just introduced,

$$I_\mu - I_{c\mu} = 1.44 \ln \frac{y_\mu}{y_{c\mu}} \tag{8.18}$$

where I_μ and $I_{c\mu}$ are, respectively, the actual and computed values of the Modified Mercalli (MM) intensities associated with any given rate of occurrence μ, and y stands for maximum ground velocity.

The foregoing result is consistent with the assumption that $F(R)$ in Eq. 7.5 does not depend on M; this leads to $E[\mu(I_c)]$ proportional to $\exp(-\beta' I_c)$, with $\beta' = \beta \ln 2$. Since β is 2.16, 1.70, and 2.88 for the Circumpacific Belt, the Alpide Belt, and the low-seismicity macrozone, respectively, we find β' equal to 1.50, 1.18, and 2.00. Hence, the mean yearly number of earthquakes whose intensity exceeds $I_c - 1$ will ordinarily be 3.2 to 7.4 times greater than the number whose intensity exceeds I_c.

In regions where the proportion of small earthquakes is relatively high, as where earthquake swarms occur often, larger values of β' should be adopted. This is the case in the central region of the Republic of El Salvador and parts of Japan, for example. Conversely, smaller values of β' will be associated with areas where the ratio of small to large intensities is conspicuously low, as along the Alpide Belt, and in particular at Agadir, Skopje, New Madrid, and other locations where destructive earthquakes have struck most unexpectedly.

In a typical example in which the station is influenced essentially by zones within the Circumpacific Belt, Esteva finds $b = 2.56$ and 2.11 for maximum ground acceleration and velocity, respectively. From Fig. 8.8, then, $a_\mu/a_{c\mu} = 2.4$, and $v_\mu/v_{c\mu} = 1.9$, and from Eq. 8.18, $I_\mu - I_{c\mu} = 1.44 \ln 1.9 = 0.92$. Such discrepancies between computed and actual earthquake characteristics may seem surprisingly high; they are a consequence of the wide dispersions of the correlations used.

When the curve of $\ln E[\mu(y_c)]$ vs. $\ln y_c$ is idealized as a straight line, that of $\ln E[\mu(y)]$ vs. $\ln y$ will be a line parallel thereto. Even when the first line is taken as slightly curved, the present solution furnishes a good approximation by using in Eq. 8.17 and in Fig. 8.8 the value of b associated locally with μ.

In applying these methods to structural responses, one often meets analytical difficulties. One must then resort to simulation procedures[5] or use rough approximations, as will be seen in Chapter 14.

Ordinarily, data such as those presented in Fig. 8.7 as well as overall regional seismicity charts are prepared for conditions on rock of "average" consistency—let us say, soft sandstone or shale, medium limestone, or hard volcanic tuff—and must be corrected when different conditions prevail over a large area before it is possible to draw more detailed seismicity charts. One simple criterion (Gzovsky, 1962) consists in adding one degree to the intensity for rock that is substantially softer or looser than the "average" material, and reducing one degree where the rock is substantially harder. A more elaborate criterion of the same nature has been advanced by Richter (1959).

Partial recognition of the influence of regional geology on seismicity is automatic when statistical data on intensities are incorporated. To be consistent, however, the prior distribution must also recognize this influence.

8.4 Microregionalization

The most elementary way of introducing the influence of detailed local geology consists in locally raising or lowering the intensities of a map of regional seismic-

---

[5] At least one interesting attempt has been made in this direction (Lacer, 1965), although not all the results obtained are directly applicable because they are based on the assumption of linear ground behavior independently of earthquake intensity.

ity as a function of the nature of surficial materials. Richter's (1959) proposal is of this type.

There are several reasons for not using the application of such a simple method of microregionalization. First, not only the geologic classification but the consistency of the surficial material should play a role. Second, the depth of the formation and, in some cases, the characteristics of underlying strata are equally important. Third, the addition or subtraction of a constant term to the regional intensity implies linear behavior of the ground, which is rarely a tenable hypothesis for soft or loose materials when subjected to strong shaking. Finally, the influence of soil characteristics usually involves more than a mere change in intensity; as we have seen, it is often reflected in the appearance of locally prevailing periods and lengthening of the durations of the motions. These matters are discussed in Chapter 9, and some of them are treated quantitatively. Still, in the absence of detailed information on the geology and mechanical properties of the soil in a region, the elementary criteria afford a guide for microregionalizing.

Spectra of microtremors or other results of the analysis of these phenomena are claimed to furnish a quantitative basis for very detailed microregional maps and even for estimating the prevailing ground periods at a site (Kanai, Tanaka, and Osada, 1954). The matter is controversial because of, among other things, soil nonlinearity at high seismic intensities, because only the uppermost soil formations participate appreciably in conditioning such surface vibrations as those due to traffic, and because the characteristics of these minor disturbances are unknown.

9

PROBABILITY DISTRIBUTIONS OF RESPONSE SPECTRAL ORDINATES

9.1 Scope of This Chapter

For any deterministically specified ground motion we can always apply the methods discussed in the first six chapters to compute the responses of any structure whose characteristics have been described deterministically. Yet, future earthquakes can only be described in terms of probability, and there is always uncertainty about structural parameters, especially those pertaining to dynamic behavior. Hence, a more rational approach, more nearly consistent with the problems that concern us, rests in the application of the theory of probabilities. Concepts of probability theory that are applicable in this and the next chapter are presented here as the need for them arises.

Initially we shall take structural parameters as deterministic quantities and introduce probabilities only in the treatment of ground motions. At times we shall simplify the problem further. Since there is far greater uncertainty in the expectation of structural responses at a given site than in the ratio of the responses to their expected values, we shall replace these ratios with their expectations. Indeed, sometimes we shall merely aim at establishing rapid procedures for estimating the expectations of structural responses from such incomplete information about the earthquake motions as the maximum ground acceleration, velocity, and displacement.

Linear systems having a single degree of freedom are interesting as idealizations of a class of structures. Analysis of their responses to earthquakes is, however, of greater interest as a basis for the treatment of more complicated structures.

In Chapter 7 we classified earthquakes into four groups. We shall devote a major portion of our attention to strong motions of firm ground at moderate focal distances. There are several reasons for this preference. First, among destructive earthquakes, these strong motions are the most frequent. Second, more than half the reliable strong-motion records belong to this type. Also, strong motions of firm ground at moderate focal distances lend themselves to convenient mathematical idealization, and a study of their linear filtering leads immediately to a comparable treatment of certain ground motions of soft soils within the linear range of behavior of these materials.

The pronounced irregularity of the strong motion of firm ground sugggests idealizing it as a stochastic process.

9.2 Stochastic Processes[1]

Let $x_n(t)$ be a family of real-valued functions of time and let the index n be a random variable. Such a family of functions constitutes a stochastic process. The process defines a random variable x for each fixed time.

If, for any finite collection of times t_k the random variables $x(t_k)$ belong to a multivariate Gaussian distribution with mean values zero, the random process $x(t) = x_n(t)$ is called *Gaussian*.

A random process $x(t)$ is called *stationary* if, for any finite collection of times t_k and any fixed translation Δt, the random variables $x(t_k)$ have the same joint distribution as the random variables $x(t_k + \Delta t)$. This definition applies when the times t_k are allowed to take any finite value.

A stationary process will be called of *finite duration* if t_k and $t_k + \Delta t$ are restricted to remain within a finite interval, outside which $x(t) = 0$.

The *autocorrelation function* is defined as

$$\phi(t_1, t_2) = E[x(t_1)x(t_2)] \tag{9.1}$$

If $x(t)$ is stationary, the autocorrelation function depends only on the difference $\tau = t_2 - t_1$, in which case

$$\phi(\tau) = E[x(t - \tau)x(t)] \tag{9.2}$$

In stationary processes of finite duration this is true if $t - \tau$ and t fall within the interval in question; if at least one of these instants falls outside the interval, $\phi = 0$. Conversely, if the process is Gaussian and $\phi(t_1, t_2)$ is a function only of $t_2 - t_1$, $x(t)$ is stationary.

Let $x(t)$ be a stationary stochastic process. We wish to make a frequency analysis of this process. We do not expect $x(t)$ to tend to zero as $t \to \pm \infty$. Hence it is not reasonable to take a Fourier transform of this variable. However, if we let $x_s(t)$ be of finite duration s and coincide with $x(t)$ within the interval $-s/2 \leq t \leq s/2$, this signal x_s has a Fourier transform

$$F_s(\omega) = \int_{-\infty}^{\infty} e^{-i\omega t} x_s(t)\, dt \tag{9.3}$$

The *power spectral density* $G^2(\omega)$ of the stationary stochastic process $x(t)$ is defined as

$$G^2(\omega) = \lim_{s \to \infty} \frac{E|F_s(\omega)|^2}{s} \tag{9.4}$$

[1] This section is based partly on the work of Franklin (1963). See also Rosenblueth (1964c).

The expectation is taken over the family of signals $x_n(t)$ in the stochastic process. Thus, $G^2(\omega)$ depends on the whole process and not on any particular sample function.

It can be shown that the power spectral density equals the Fourier transform of the autocorrelation function:

$$G^2(\omega) = \int_{-\infty}^{\infty} \phi(\tau) e^{-i\omega\tau} d\tau \qquad (9.5)$$

The term *shot noise* defines a series of instantaneous pulses

$$x(t) = \sum_i a_i \delta(t - t_i) \qquad (9.6)$$

where both a_i and t_i may be random variables and δ stands for Dirac's delta function. If the probability that one t_i falls within any given time interval of duration dt is a constant of the order of dt and is also independent of the instants at which previous t_i's have fallen, the family of variables $x_n(t)$ constitutes a *Poisson process*.

A shot noise for which $E(a_i) = 0$, and the intervals $t_{i+1} - t_i$ tend to zero as both the duration of motion and the ratio $E(a_i^2)/E(t_{i+1} - t_i)$ remain finite, tends to a Gaussian process. The ratio $E(a_i^2)/E(t_{i+1} - t_i)$ has been called *intensity per unit time* and is generally a time function. When the intensity per unit time is constant, the process is stationary and the shot noise process is called a *white noise*. (Often in the literature, shot and white noise are used as synonyms.) In this case

$$\phi(\tau) = a^2 \delta(\tau)$$

where a is constant. If the white noise is of infinite duration, it has a power spectral density in the usual sense, and Eq. 9.5 gives $G^2(\omega) = a^2$.

It can be shown from Eq. 9.3 that white noise may be regarded as the superposition of infinitely many sinusoidal waves of different frequencies, with a uniform distribution of frequency between zero and infinity. Hence the name of the process.

9.3 Idealization of Earthquakes as Segments of White Noise

Consider a family of ground motions whose accelerograms are white noise of finite duration s. That is, $\ddot{x}(t)$ is zero for $t \leq 0$ and for $t > s$, and it is white within the interval $0 < t \leq s$. (In the formulas of Section 9.2 we must, accordingly, replace x with $\ddot{x}$.) The process is not physically possible because it implies infinite accelerations and does not necessarily have a zero end velocity. It is important that the model predicts the quantities of interest in satisfactory agreement with the prototype. As we shall see, over a considerable range of natural periods of structures subjected to earthquake motion, the prediction is

270 PROBABILITY DISTRIBUTIONS Chap. 9

good assuming that the ground acceleration is white noise. Moreover, a study of these processes sets the bases for more realistic idealizations.

Use of segments of white noise to model earthquake accelerograms was introduced in a paper by Housner (1947). Housner's study showed much similarity between the spectra of earthquakes and those of segments of shot noise in a wide range of natural periods.

Shot noise may be represented as in Fig. 9.1. The sudden changes in ground velocity, u_i, play the role of $a_i \, \delta(t - t_1)$ in Eq. 9.6. If we let max $(t_{i+1} - t_i)$ tend to zero, preserving the ratio $w = E(u_i^2)/E(t_{i+1} - t_i)$ finite and time independent, the process tends to a segment of white noise. From the viewpoint of its effects on structures, this noise may be regarded as Gaussian, since the distribution

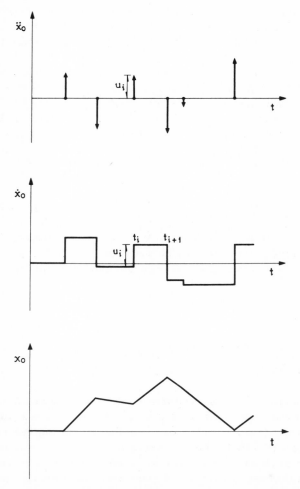

Figure 9.1. Acceleration, velocity, and displacement diagrams of shot noise.

functions of $\int_0^t \ddot{x}_0(\tau)\psi(t-\tau)\,d\tau$ are the same as if the process were Gaussian for any transfer functions having no more than a finite number of finite discontinuities. The ratio w may be called the *intensity per unit time* of the process.

Often the criterion of survival is $Q \leq Q_1$ where $Q = \max|q(t)|$; Q_1 is the structural capacity, and q is the deformation, absolute acceleration, or a response of the structure that is a linear function of either of these quantities (e.g., the force in the spring element, the sum of forces in the spring and dashpot, or any generalized force or deformation). The question arises as to the accuracy and completeness with which we should compute the probability distribution of Q_1. Borges (1956) has shown that the uncertainty in the intensity of future earthquakes is so much greater than that in the ratio of Q_1 to $E(Q)$ corresponding to a given intensity, that little error is introduced by treating Q_1 as a deterministic function of the earthquake characteristics.

We shall drop subscript 1 and speak of the distribution of $Q/E(Q)$, understanding that the expectation $E(Q)$ implies a given set of earthquake characteristics that include intensity (or magnitude and focal distance) and the effects of local geologic conditions. It is clear that $E(Q)$ should be evaluated carefully, but that even a crude description of the distribution of $Q/E(Q)$ will be adequate. For much the same reasons, ordinarily it will be justified to treat structural parameters, including strength, as deterministic quantities.

Matters are simplified if we adopt the assumption that the maximum values of the responses that interest us (absolute acceleration and displacement relative to the base) are proportional to those of the "response" r, defined by

$$r^2 = (\omega_1 y)^2 + (\dot{y} + \zeta\omega_1 y)^2 \tag{9.7}$$

where ω_1 is the undamped natural circular frequency, y is the deformation, and ζ = percentage of damping (see Section 1.4). The simplification stems from the fact that r is the radius vector in a phase-plane representation in which free vibrations are logarithmic spirals (see Section 1.6), and this permits replacing the boundary condition with one that does not depend on the phase angle.

Crandall, Chandirami, and Cook (1966) have shown that there is a significant difference between the expected responses computed from the correct boundary condition and those associated with an absorbing barrier at $R = \max(r)$. However, there is a satisfactory agreement of the distribution of $R/E(R_0)$ derived theoretically with those of the corresponding normalized pseudovelocities of true earthquakes, of segments of white noise, and of Gaussian processes, where subscript 0 identifies undamped responses (Rosenblueth and Bustamante, 1962; Brady and Husid, 1966). This approximation holds for actual earthquakes provided the natural period of the structure is not excessively short, provided the power spectral density, as a function of natural circular frequency, does not have too pronounced a curvature in the neighborhood of the natural frequency of the structure (Caughey and Gray, 1963), and provided $T_1 \ll s$. The first of these limitations is met reasonably well when the natural period is longer, or at least not much shorter, than the period marking the intersection of the lines for a and v in a logarithmic representation of response spectra, where a and v are the maximum ground acceleration and velocity, respectively; in other words, the natural period should not be much shorter than $2\pi v/a$. The second limitation excludes the responses to earthquakes on soft soil in the neighborhood of the prevailing periods of the ground. The limitation that T_1 be much smaller than the earthquake duration is too stringent for practical purposes; it is sufficient that T_1 be not much longer than s, and this covers almost all cases of practical interest.

With these limitations there is little error in assuming that

$$\frac{Q}{E(Q_0)} = \frac{R}{E(R_0)} \qquad (9.8)$$

where Q is a structural response associated with any probability of failure, equal to the probability that R be exceeded.

The probability distributions for various values of the parameter $\zeta\omega_1 s$ are depicted in Fig. 9.2 (Rosenblueth and Bustamante, 1962; Brady, 1966; Brady and Husid, 1966).

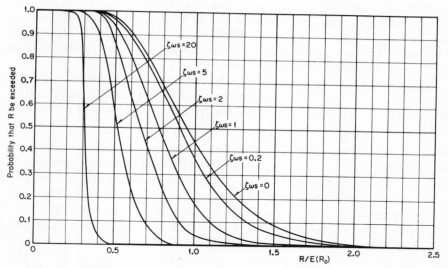

Figure 9.2. Probability distributions of damped responses.

Let us now introduce the factors $\beta = Q/E(Q_0)$ and $\beta_E = E(Q)/E(Q_0)$; these may be regarded as correction coefficients to account for damping. While β_E depends only on $\zeta\omega_1 s$, β is a function also of the probability of failure. These coefficients are displayed in Fig. 9.3. The coefficient β_E may be approximated, with little error, by the expression

$$\beta_E \cong \left(1 + \frac{\zeta\omega_1 s}{2}\right)^{-1/2} \qquad (9.9)$$

(Rosenblueth, 1968c; Rosenblueth and Elorduy, 1969a) or by

$$\beta_E \cong (1 + 0.6\zeta\omega_1 s)^{-0.45} \qquad (9.10)$$

Sec. 9.3   EARTHQUAKES AS SEGMENTS OF WHITE NOISE   273

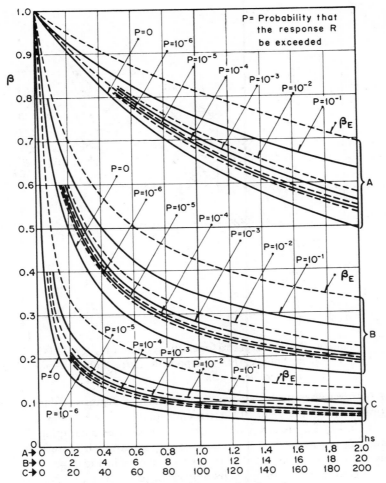

Figure 9.3. Correction factors for damping.

(Rosenblueth, 1964c). The second form is more accurate throughout most of the range of practical interest of $\zeta\omega_1 s$ (with errors smaller than about 3 percent), while Eq. 9.9 tends asymptotically to the right answer as s tends to infinity. In the range of $\zeta\omega_1 s$ between about 8 and 40, β_E varies approximately as $\zeta^{-0.4}$.[2]

Values of β associated with even such relatively large probabilities of failure as 10^{-1} are appreciably smaller than β_E for the same $\zeta\omega_1 s$. This reflects the more pronounced effect that damping has in reducing the peaks than in reducing the mean spectral ordinates. The phenomenon can be appreciated in Fig. 9.4, which

---

[2] This relationship was proposed by Kanai (1957) and has been confirmed adequately in studies of the responses to actual earthquakes, by Arias and Husid (1962) and by Arias and Petit-Laurent (1963).

274 PROBABILITY DISTRIBUTIONS

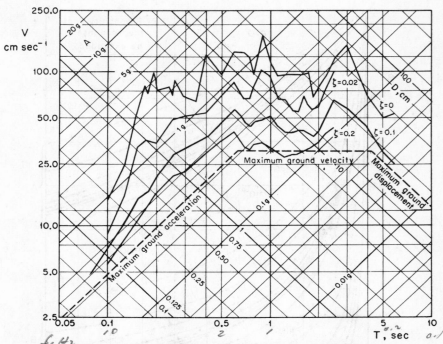

Figure 9.4. Response spectra for elastic systems, May, 1940 El Centro earthquake, NS component. *After Blume, Newmark, and Corning (1960).*

shows the response spectra associated with various degrees of damping for the NS component of the 1940 El Centro earthquake.

Curves in Fig. 9.3 are steep in the neighborhood of zero damping. Going from no damping to any finite ζ decreases the responses much more than doubling ζ. This too can be appreciated in the spectra shown in Fig. 9.4.

The percentage of damping enters only in the product $\zeta \omega_1 s$. Effects of damping therefore vary monotonically with the natural frequency of vibration and with the duration of the ground motion. This conclusion does not hold when ω_1 tends to infinity, since the maximum acceleration in an infinitely rigid structure is necessarily equal to the maximum ground acceleration whatever the degree of damping, so there can be no reduction factor at $T_1 = 0$. The same applies to the pseudoacceleration $\omega_1^2 D$, where D is the spectral deformation, because it can be shown that

$$\lim_{\omega_1 \to \infty} \omega_1^2 D = \lim_{\omega_1 \to \infty} \max |\ddot{x}|$$

[Jenschke, Clough, and Penzien (1965); see Fig. 9.4, for example.]

White noise, as defined above, usually gives a finite end velocity of the ground and tends to underestimate the ratio of expected undamped spectral pseudovelocity to expected maximum ground velocity. This ratio is of interest in some applications. A more satisfactory estimate of its value is obtained from ground motions idealized as a random series of sudden changes in ground velocity, subjected to the restriction that their sum be equal to zero, and then taken to the limit in the same way as we took shot noise to produce white noise. In this way Herrera, Rosenblueth, and Rascón (1966) have found $E(V_0)/E(v) = 2.11$, where V_0 is spectral pseudovelocity for $\zeta = 0$, and v is maximum ground velocity.

It is also more realistic to take the intensity of white noise per unit time, w, to be a function of time. For such processes the distribution of the maximum responses, normalized with respect to their expectation, is practically the same as if w were time independent, provided that $w(t)$ is sufficiently smooth. The same is approximately true of the responses of damped systems if we replace the actual duration of the motion with an "equivalent" duration, which is always smaller than the latter. In principle the equivalent duration should be made a function of ζ but relatively accurate results are obtained by assuming that it is constant.

For the purpose of calculating the distribution of responses normalized with respect to $E(Q_0)$ the idealization of actual earthquakes as segments of white noise, with constant intensity per unit time and an equivalent duration, is satisfactory over a wide range of natural periods. This is illustrated in Fig. 9.5, which compares β_E with the ratio of damped to undamped responses to a large group of earthquakes recorded on relatively firm ground along the west coast of the United States. The coefficient β_E was computed for a constant equivalent $s = 12.5$ sec. The figure also shows a comparison of the theoretical prediction

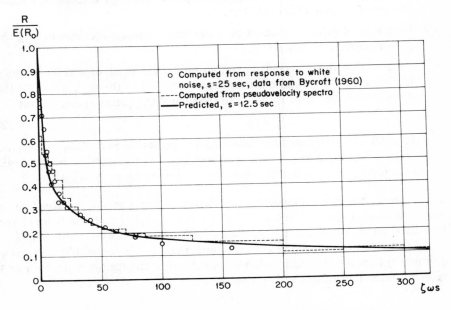

Figure 9.5. Comparison of predicted and computed correction factors for damped average pseudovelocity.

with the responses of simple systems to white noise generated and analyzed in an analog computer.

9.4 Idealization of Earthquakes as Stationary Gaussian Processes

For reasons of simplicity it would be desirable to treat earthquakes as stationary Gaussian processes. This is not directly possible because the probability is 1 that any given response to a motion of this type will be exceeded at least once; hence we would find $E(Q) = \infty$.

In order to be able to use the results available for the mean-squared response to stationary Gaussian disturbances we take the following steps.

1. Compute the functions

$$G^2(\omega) = \lim_{s \to \infty} \frac{E|F(\omega)|^2}{s} \qquad (9.11)$$

where

$$F(\omega) = \int_0^s e^{-i\omega t} \ddot{x}_0(t)\, dt$$

for the family of ground motions in question. The expected squared amplitude of the Fourier transform should include recorded as well as foreseeable ground motions, the latter incorporated through calculations based on considerations of seismic wave transmission.

2. Define the stochastic process of stationary Gaussian motions having the power spectral density G^2.

3. Assume that the expected maximum numerical value of a response to a family of actual ground motions is proportional to the square root of the expected square of the response to the Gaussian process at any instant:

$$E(Q) \propto \sqrt{E(q^2)} \qquad (9.12)$$

Here the left-hand member concerns the earthquakes and the right-hand member concerns the stationary processes. [The instant at which $E(q^2)$ is computed is unspecified because of stationarity.]

The problem reduces now to the calculation of $E(q^2)$ and of the constant of proportionality between the root of this quantity and $E(Q)$. To achieve the first part, let us introduce the *weighting (transfer) functions* or *basic solutions* of the system under study. The transfer function ψ_q associated with the response q of any system having linear behavior is defined as the response $q(t)$ to a unit velocity pulse, that is, to the disturbance

$$\ddot{x}_0(t) = \delta(t)$$

where δ is the Dirac delta. (This disturbance becomes dimensionally consistent when multiplied by the amplitude of an arbitrary velocity pulse.)

Linearity of the system ensures that, given an arbitrary accelerogram $\ddot{x}_0(t)$ starting at $t = 0$, we can write

$$q(t) = \int_0^t \ddot{x}_0(\tau)\psi_q(t - \tau)\, d\tau \tag{9.13}$$

When the disturbance belongs to a stationary Gaussian process characterized by a power spectral density $G^2(\omega)$, we have

$$E(q^2) = \int_{-\infty}^{\infty} |F_q(\omega)|^2 G^2(\omega)\, d\omega \tag{9.14}$$

where $F_q(\omega)$ is the Fourier transform of ψ_q (Crandall, 1958). This relationship is valid whatever the number of degrees of freedom of the structure being analyzed; accordingly we shall make further use of it in Chapter 10.

Simple examples of the foregoing relationship corresponding to the cases when q stands either for y or for $\ddot{x}$, the deformation or the absolute acceleration of a viscously damped simple system, together with the respective Fourier transforms are presented by Rosenblueth (1964c).

It remains to find the constant of proportionality in Eq. 9.12. If $q(\omega_1)$ denotes the absolute displacement, velocity, or acceleration in a simple structure with undamped, natural circular frequency ω_1, then $q(\omega)$ represents the corresponding parameters of the ground motion because an infinitely rigid structure will undergo the same motion as the ground. If we make $\omega_1 = \infty$ in the expressions for $|F_q|^2$ when q stands for x, $\dot{x}$, or $\ddot{x}$ (Rosenblueth, 1964c), we obtain $|F_q(\omega)|^2 = 1$, so that

$$E[Q(\omega_1)] \propto \left[\int_{-\infty}^{\infty} G^2(\omega)\, d\omega\right]^{1/2}$$

Actually, both $E[Q(\omega_1)]$ and $E[Q(\omega)]$ will be infinite if the ground motion is truly Gaussian and stationary, but under the assumption that each response is proportional to the corresponding $\sqrt{E(q^2)}$ we can write[3]

$$\frac{E[Q(\omega_1)]}{E[Q(\omega)]} = \left[\frac{\int_{-\infty}^{\infty} |F_q(\omega)|^2 G^2(\omega)\, d\omega}{\int_{-\infty}^{\infty} G^2(\omega)\, d\omega}\right]^{1/2} \tag{9.15}$$

The approximate validity of the formula breaks down as ζ approaches zero because the integral in the numerator diverges for zero damping. Still, as we shall see, reasonably good results are obtained for even moderately small damping.

A second approach consists of using, as reference for the responses to the disturbances in question, the known solution for stationary white noise of finite duration.[4] This gives

$$\frac{E[Q(\omega_1)]}{E[Q_a(\omega_1)]} = \left[\frac{\int_{-\infty}^{\infty} |F_q(\omega)|^2 G^2(\omega)\, d\omega}{\int_{-\infty}^{\infty} |F_q(\omega)|^2 a^2\, d\omega}\right]^{1/2} \tag{9.16}$$

---

[3] This method is due to Tajimi (1960); see also Rosenblueth (1964c).
[4] The approach is a generalization by Rosenblueth (1964c) of a study by Caughey and Gray (1963).

where Q_a is the maximum numerical value of q as a response to stationary white noise having $G(\omega) = a$ (constant). The accuracy of Eq. 9.16 may be improved by taking the duration of the white noise smaller than that of the actual ground motions because the latter are not stationary. This expression breaks down as ω_1 tends to infinity.

Studies have been made assuming that the power spectral density of the ground motion is the same as if this motion were the stationary Gaussian process resulting from the response of a damped single-degree system to a stationary white disturbance of infinite duration.[5] It can be shown from Eq. 9.16 that the power spectral density of the filtered white noise is proportional to that of the filter $|F_{\ddot{x}}(\omega)|^2$. Acting along these lines, Housner and Jennings (1964) have found an excellent agreement between the mean-power spectral density (defined in Eq. 9.11) of a number of earthquakes recorded on firm ground along the west coast of the United States and the expression

$$G^2(\omega) = B \frac{1 + (2\eta\omega/p)^2}{(1 - \omega^2/p^2)^2 + (2\eta\omega/p)^2} \qquad (9.17)$$

where B is a constant, η the damping ratio of the filter, and p its natural circular frequency. This is of the same form as $|F_{\ddot{x}}|^2$ except for the constant factor.

Expressing G in cm/sec$^{3/2}$ and ω is sec^{-1}, and adjusting for the earthquakes mentioned, Eq. 9.17 becomes

$$G^2(\omega) = \frac{11.5(1 + \omega^2/147.8)}{(1 - \omega^2/242)^2 + \omega^2/147.8}$$

(Fig. 9.6)

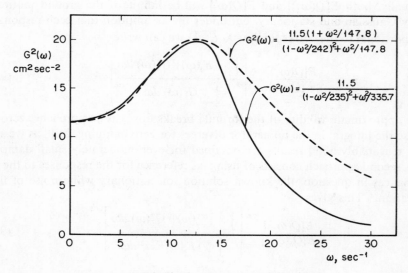

Figure 9.6. Idealized power spectral densities.

[5] This proposal was made by Tajimi (1960) on the basis of work by Kanai (1957).

Sec. 9.4   EARTHQUAKES AS STATIONARY GAUSSIAN PROCESSES   279

Work has been done by Barstein (1960) and Bolotin (1960) on the basis of autocorrelation functions of the form

$$\phi(\tau) = e^{-\alpha|\tau|} \cos \beta\tau \tag{9.18}$$

in which α and β are constants (Fig. 9.7). These autocorrelation functions give rise, through the use of Eq. 9.5, to a power spectral density that can also be put precisely in the form of $|F_{\mathscr{E}}|^2$.

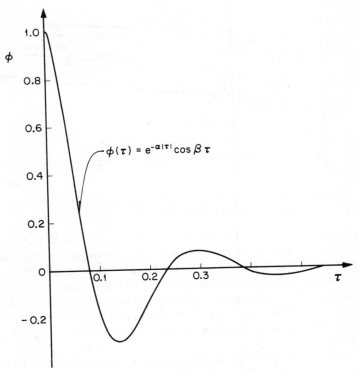

Figure 9.7. Autocorrelation function.

Arias and Petit-Laurent (1962) have used a simplifying assumption about the internal damping in the soil in a study about filtering of plane shear waves due to stationary white noise through a soft mantle of uniform thickness that rests on a semiinfinite formation of rock. They have shown that this process gives rise to a power spectral density that exhibits many peaks as a function of ω and that its envelope follows a curve again of the form defined by Eq. 9.16. This phenomenon of multiple wave reflection and refraction (although not necessarily in a soft mantle) governs the salient characteristics of the motion. To some extent the study in question justifies adoption of the second member in Eq. 9.18 as an approximation to the autocorrelation functions of earthquakes of the type we have been considering.

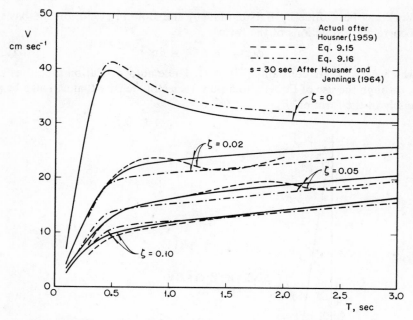

Figure 9.8. Comparison of earthquake pseudovelocity spectra with those from Eqs. 9.15 and 9.16.

A comparison is shown in Fig. 9.8 between the average pseudovelocity spectra that led to the parameters in Eq. 9.17 and the corresponding response spectra derived by substituting Eq. 9.17 into Eqs. 9.15 and 9.16. The last two expressions are complementary in that each gives satisfactory results within its own range of applicability. A pronounced difference between the spectral acceleration calculated by both expressions for short, natural periods can be appreciated in Fig. 9.9, which is based on the same power spectral density as Fig. 9.8; Eq. 9.16 is inadequate in this range.

There is one basic objection to the unconditional adoption of the foregoing treatment. In postulating the model that gives rise to these processes we have taken them to be the motions described by a simple system resting on a ground that undergoes a white noise accelerogram. The end velocity of the ground and hence that of the system are not necessarily zero. This matter can be dealt with in a number of ways, as we shall see under earthquake simulation.

Once the expected responses have been obtained, the assumption that the probability distribution of the responses normalized with respect to their expectation is the same as the distribution of $R/E(R)$ because a segment of white noise furnishes all the information required. Because the expected responses are of much greater interest than this distribution, the accuracy of this assumption ordinarily matters little. However, the assumption leads to entirely erroneous results when ω_1 is extremely large. If the maximum ground acceleration is specified, the distributions of the spectral acceleration and pseudoacceleration tend to become deterministic as ω_1 tends to infinity. The same is true of the

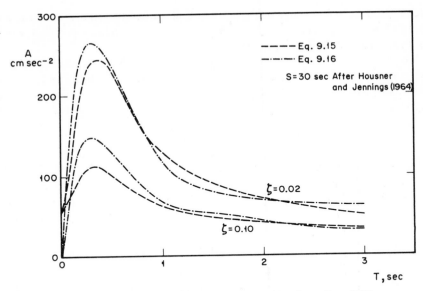

Figure 9.9. Comparison of acceleration spectra from Eqs. 9.15 and 9.16.

distribution of spectral deformation as ω_1 tends to zero if the maximum ground displacement is specified, since the spectral deformation tends toward the maximum ground displacement.

Thus far we have regarded Gaussian processes as superpositions of harmonic motions with an adequate distribution of amplitudes as functions of frequency. We may also regard them as the limits of processes in which we begin with a shot noise accelerogram (Rosenblueth and Bustamante, 1962). We replace each instantaneous pulse with a wave of adequate shape, whose integral with respect to time equals the pulse magnitude and has the same sign as the instantaneous pulse. Finally we let the pulse spacing tend to zero as we keep the appropriate quantities finite. (For the motion to be stationary the shapes of all the waves must be equal to each other.) On this basis we can explain the discrepancy between the results obtained from the Gaussian motion idealization and the responses to true earthquakes in the range of high natural frequencies. In our idealization we have assumed that the ground motion associated with every pulse that began before Q occurred has ceased. The assumption is reasonable, provided the duration of each wave is small compared with the natural period of q; it does not hold when T approaches zero.

We seek an approximate solution applicable in the entire range of natural periods and damping ratios. To this end we recall that Eq. 9.15 holds for $\zeta \gg 0$, and Eq. 9.16 holds when $T \gg 0$. The following expression interpolates between both solutions, approaching each where it is most accurate:

$$E(Q) = \frac{4\zeta\omega_1 E(Q_1) + \eta p E(Q_2)}{4\zeta\omega_1 + \eta p} \tag{9.19}$$

282 PROBABILITY DISTRIBUTIONS　　　　　　　　　　　　　　　　　　Chap. 9

Here ζ and ω_1 correspond to the structure considered; η and p have the same meaning as in Eq. 9.17; $E(Q_1)$ and $E(Q_2)$ are the expected responses furnished by Eqs. 9.15 and 9.16, respectively.

The values $\eta = 0.24$, $p = 25$ sec^{-1}, and $s = 15$ sec correspond closely to the NS component of the 1940 El Centro earthquake. Applying Eq. 9.19 to compute the acceleration spectrum for $\zeta = 0.02$ we can find a satisfactory agreement with the actual spectrum in the entire range of natural periods of interest (Fig. 9.10).

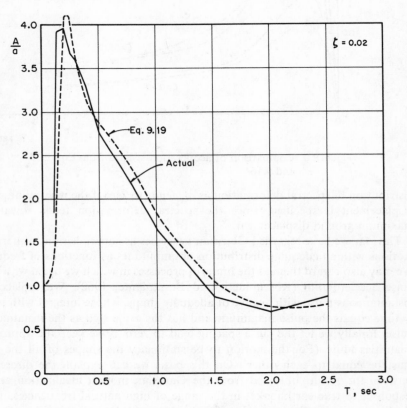

Figure 9.10. Comparison of the smoothed, El Centro 1940 NS component spectrum with that from Eq. 9.19.

For a valid approximation when ω_1 tends to infinity, and when we want to find the distribution of Q, Q_1/a may be taken as a deterministic quantity for any family of earthquakes characterized by given η and p. This waives the objections raised in connection with the range of high natural frequencies.

From the theory of stationary Gaussian processes, Goto and Toki (1969) have developed an expression for the spectral ordinates associated with the fixed expected number of times that they be exceeded per unit time. These responses are approximately proportional to the expected spectral ordinates when the expected number of crossings during the entire duration of the motion is relatively small.

9.5 Idealization of Earthquakes as Transient Gaussian Processes

Shinozuka (1964) has derived expressions for an upper and a lower bound to the cumulative probability distribution of the responses of any linear system to a very general type of transient stochastic processes. It is only required that the processes be Gaussian with zero mean and with the mean squared continuous. Goto and Kameda (1969) have developed an approximation to the probability distribution of earthquake spectral ordinates on the basis of Shinozuka's lower bound and have verified it through computer simulation of earthquakes.

We shall be content with the probability distributions of the normalized responses in terms of their expectations, as derived from the analysis of responses to segments of white noise, because we have seen that this treatment gives a satisfactory approximation in the case of type 2 earthquakes. Accordingly we shall concern ourselves with the calculation of the expectations of these spectral ordinates.

9.6 Expected Spectral Ordinates for Earthquakes of the Second Type

The purpose of this article is to develop a method for estimating the expected spectrum that corresponds to given a, v, and d, where these quantities stand for the ground's maximum acceleration, velocity, and displacement, respectively.

For large magnitudes (say in excess of 6.5 or 7) and moderate focal distances (say 20 to 80 km) the rules given in Chapter 7 are generally adequate. In particular, the assumption that the lines marking a, v, and d in a pseudovelocity spectrum give the expected spectral ordinates for $\zeta = 0.25$ is practical owing to its simplicity.

At longer focal distances these simple criteria no longer apply. One procedure consists of taking the expected, undamped pseudovelocity spectrum proportional to $\omega F_y(\omega)$. This is justified because, if we make $\zeta = 0$ in Eq. 9.16, the expected acceleration spectrum coincides with $\omega G(\omega)$. Hence, if the acceleration spectrum is to be of the same form as $\omega(G)\omega$ in Eq. 9.16, that is, of the form of $\omega F_g(\omega)$, the expected pseudovelocity spectrum must be of the form we have postulated. We shall write now c for ω_1 and η for ζ, since these quantities will refer to the ground motion, not to the structure. Accordingly,

$$E(V) = \frac{b\omega}{\sqrt{(1 - \omega^2/c^2)^2 + (2\eta\omega/c)^2}} \qquad (9.20)$$

where b is a constant.

As $\omega \to \infty$, $\omega E(V)$ must approach a. Consequently, $bc^2 = a$. And as $\omega \to 0$, $E(V)/\omega$ must approach d. Therefore, $b = d$ and $c^2 = a/d$. It remains to stipulate η. We need not, however, deal directly with the equivalent earthquake duration for this purpose because this quantity is implicit in the ratio v^2/ad, as both are related to earthquake magnitude and focal distance.

We notice that for a segment of white noise, $v^2/ad = 0$ and $E(V)/v$ may be taken equal to 2.11. As $v^2/ad \to 1$ and the process approaches harmonic motion, η must tend to zero because Eq. 9.20 has the form of the amplitude of the damped, single-degree velocity response to a harmonic disturbance. For intermediate values of v^2/ad an empirical fit has been made (shown in Fig. 9.11) which yields

$$\eta = 0.237(\sqrt{1 + 150R^{-0.6}} - 1) \tag{9.21}$$

With the values of b, c, and η substituted in Eq. 9.20, it is possible to draw the derived expectation of the undamped spectrum.

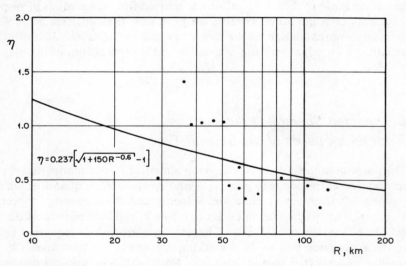

Figure 9.11. Empirical relation between η and focal distance.

Spectra obtained from Eq. 9.20 are not altogether satisfactory because they are symmetrical in a logarithmic plot, while systematically, undamped spectra of earthquakes of the second type have higher ordinates near the intersection of the a and v lines than near that of v and d. From this point of view the rules of Section 7.5 are more adequate for moderate focal distances (see Fig. 9.12).

Also shown in the figure are the expected spectral ordinates obtained by applying the rules due to Newmark and Hall (1969) and Newmark (1969b) that are approximately equal to those found in Blume, Newmark, and Corning (1961) and are supplemented with the amplification factors in Table 9.1 to account for damping. These factors are equal to $E(A)/a$, $E(V)/v$, and $E(D)/d$. They

Sec. 9.6   EXPECTED SPECTRAL ORDINATES

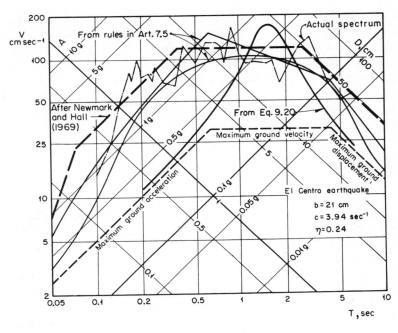

Figure 9.12. Comparison of idealized undamped spectra.

TABLE 9.1 RELATIVE VALUES OF SPECTRUM AMPLIFICATION VALUES [AFTER NEWMARK AND HALL (1969) AND NEWMARK (1969b)]

	Amplification factor		
ζ (%)	Acceleration	Velocity	Displacement
0	6.4	4.0	2.5
0.5	5.8	3.6	2.2
1	5.2	3.2	2.0
2	4.3	2.8	1.8
5	2.6	1.9	1.4
7	1.9	1.5	1.2
10	1.5	1.3	1.1
20	1.2	1.1	1.0

coincide, roughly, with the ones we would obtain for conditions such as those of El Centro 1940 using the rules of Chapter 7.

Notice that at long focal distances the assumption that $Q/E(Q)$ has the same distribution as for a segment of white noise does not apply because the ground motion tends to lose its chaotic character. This is not a serious objection, for only rarely will responses to such motions be of much interest. The dispersions of a, v, and d will then cover up much of this deficiency. A more serious objection to the treatment adopted in the present chapter lies in the fact that at very long distances the effects of damping on the expected responses will differ from those derived on the assumptions applicable to Gaussian processes or to white noise. The difficulty can be surmounted by filtering a Gaussian process through a damped single-degree system so as to give sufficiently peaked mean spectra and resorting either to analytical treatment similar to that in Section 9.7 or to the Monte Carlo techniques of Section 9.9.

9.7 Earthquake Filtering Through Soft Mantles in the Linear Range

In accordance with Chapter 3 we are justified in treating earthquakes at the surface of soft, horizontally stratified soil as the result of filtering of horizontal shear waves going through successive reflections and refractions.

As long as the soil behaves linearly, there are two especially useful ways of studying this filtering process, whether we treat the incident motion as deterministic or as stochastic. We may begin by assuming that the incident motion is an instantaneous unit change in velocity and calculate the ground displacements or accelerations as responses to the incoming pulse. These responses will be the transfer functions of the free soil surface. Through a convolution of this function and the accelerogram of the incident ground motion we obtain the responses of the ground to a deterministic incident disturbance. By using the theory developed in this and the following chapter we can approximate the expectations of responses to a filtered motion that, as an incident disturbance, was idealizable as a Gaussian process.

The second approach involves first finding the Fourier transform of the incoming disturbance. Each incident, steady-state, harmonic motion gives rise to a steady-state harmonic response. The transform of the incident motion is easily analyzed. The antitransform gives the desired response. The Fourier transform technique presented in this chapter permits estimating the expectation of the response when the incoming motion is idealized as a Gaussian process.

From a practical standpoint the second method is more flexible. Its relative simplicity is not impaired by the specification of any dependence of internal damping on frequency of vibration, and this sort of specification is ordinarily chosen to describe the dynamic properties of a material within the linear range. The first method is applied without difficulty when internal damping is a simple function of frequency. Examples of this type of damping include the Sezawa and the constant-Q models of soil behavior. We illustrated these two cases in Chapter 3 up to the point of computing the free-surface accelerations of the soil. Calculation of structural responses thereto and treatment of the resulting transfer functions can be done in straightforward fashion. We shall confine the following presentation to the Fourier transform approach.[6]

[6] This presentation is based on Herrera and Rosenblueth (1965).

Consider N soil strata resting on a semiinfinite rock foundation, as in Fig. 9.13. The rock will be identified by subscript N and the soil strata by subscripts $1, 2, \ldots, n, \ldots, N - 1$ from the top down.

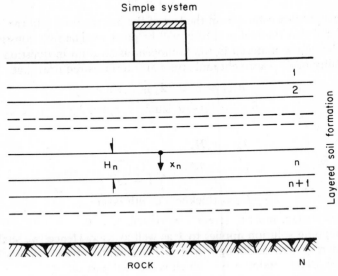

Figure 9.13. Section of subsoil. *After Herrera and Rosenblueth (1965)*.

Equilibrium requires that the differential equation

$$\frac{\partial s}{\partial X} = \rho \frac{\partial^2 x}{\partial t^2} \tag{9.22}$$

be satisfied, where s denotes shearing stress, X is the vertical coordinate, ρ is mass per unit volume, x is horizontal displacement, and t is time (see Eq. 3.23). In every material, s is some functional of the strain $\partial x/\partial X$. If x is of the form $x = \bar{x} \exp(i\omega t)$, with $\bar{x}$ independent of t, and if the stress-strain relation is linear, we may write $s = \bar{s} \exp(i\omega t)$, where

$$\bar{s} = (1 + i\alpha)\mu \frac{\partial \bar{x}}{\partial X}$$

Here μ and α are real functions of ω and independent of t (see Eq. 3.24).

Under these conditions, for a steady-state harmonic disturbance, at each homogeneous stratum Eq. 9.26 reduces to

$$(1 + i\alpha) \frac{\partial^2 \bar{x}}{\partial X^2} + \frac{\omega^2}{v^2} \bar{x} = 0$$

where $v^2 = \mu/\rho$ is again a real function of ω.

With the notation and sign convention of Fig. 9.13, motion in the nth layer has for solution the expression

$$\bar{x}_n(X_n, t) = a_n \cos \eta_n + b_n \sin \eta_n$$

in which
$$\eta_n = v_n x_n$$
$$v_n = \frac{\omega}{v_n \sqrt{1 + i\alpha_n}}$$

X_n is measured downward from the top of the nth stratum, and the square root is taken such that the real part of v_n will be positive. The functions v_n and α_n of ω characterize the material in the homogenous stratum in question.

Continuity of displacement and stress at interfaces requires that
$$a_{n+1} = a_n \cos \lambda_n + b_n \sin \lambda_n$$
$$b_{n+1} = k_n(-a_n \sin \lambda_n + b_n \cos \lambda_n)$$
where
$$\lambda_n = v_n H_n$$
$$k_n = \frac{\rho_n v_n}{\rho_{n+1} v_{n+1}} \sqrt{\frac{1 + i\alpha_n}{1 + i\alpha_{n+1}}}$$
$$H_n = \text{thickness of } n\text{th layer}$$

The ground surface must be stress free. Hence, $b_1 = 0$.

The foregoing solution applies to $\ddot{x}$ as well as to x. Hence, we shall treat the waves and responses as though their displacements were accelerations. At the rock suface, the incoming wave, which is the real part of
$$\frac{a_N + b_N/i}{2} \exp(i\omega t)$$
is stipulated as $\sin \omega t$, that is, the real part of $-ie \exp(i\omega t)$. Therefore,
$$a_N - ib_N = -2i \tag{9.23}$$

Now define the matrix
$$\mathbf{T}_n = \begin{vmatrix} \cos \lambda_n & \sin \lambda_n \\ -k_n \sin \lambda_n & k_n \cos \lambda_n \end{vmatrix}$$
so that
$$\begin{vmatrix} a_N \\ b_N \end{vmatrix} = \mathbf{T}_{N-1} \mathbf{T}_{N-2} \cdots \mathbf{T}_1 \begin{vmatrix} a_1 \\ b_1 \end{vmatrix}$$

With the additional definition
$$\mathbf{U} = \begin{vmatrix} U_1 \\ U_2 \end{vmatrix} = \mathbf{T}_{N-1} \mathbf{T}_{N-2} \cdots \mathbf{T}_1 \begin{vmatrix} 1 \\ 0 \end{vmatrix}$$
we obtain
$$\begin{vmatrix} a_N \\ b_N \end{vmatrix} = a_1 \mathbf{U}$$

because $b_1 = 0$. (Notice that $\mathbf{T}_n$ is a transmission matrix and plays the same role as in Section 4.8.) Therefore Eq. 9.23 reduces to
$$a_1(U_1 - iU_2) = -2i$$

or

$$a_1 = \frac{2}{U_2 + iU_1}$$

The steady-state, ground-surface motion is the real part of $a_1 \exp(i\omega t)$, whose amplitude is the Fourier amplitude spectrum

$$\frac{1}{U_1 + iU_2}$$

Consequently an incoming harmonic vibration of circular frequency ω is magnified in the ratio

$$B(\omega) = \frac{1}{U_1 + iU_2} \qquad (9.24)$$

When we neglect the damping of the soil, λ_n and k_n become real, and Eq. 9.24 may be interpreted in the following way. Vector $\mathbf{U}$ is found by taking the unit vector along an arbitrary U_1 direction, rotating it through an angle λ_1, amplifying (or contracting) it in proportion to k_1 in the direction U_2, and so on, $N - 1$ times (Fig. 9.14). The magnification factor B equals the reciprocal of the amplitude of the vector at the end of the process. This interpretation leads to Takahasi's (1955) graphical solution.

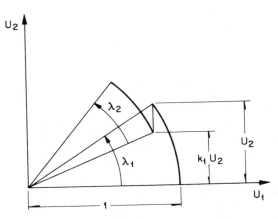

Figure 9.14. Graphical solution. *After Herrera and Rosenblueth (1965) and Takahasi (1955).*

In the case of a single, undamped homogeneous mantle we may write Eq. 9.24 in the form

$$B(\omega) = (k_1^2 \sin^2 \lambda_1 + \cos^2 \lambda_1)^{-1/2} \qquad (9.25)$$

where

$$k_1 = \frac{\rho_1 v_1}{\rho_2 v_2}$$

$$\lambda_1 = \frac{\omega H_1}{v_1}$$

Once $B(\omega)$ has been computed we find the expected spectral ordinates at the surface of the ground by using Eq. 9.25, or Eq. 9.16, or with the magnified power spectral density $B^2(\omega)G^2(\omega)$ in place of $G^2(\omega)$, where $G^2(\omega)$ is the power spectral density of the motion that would take place at the rock surface if the soft soil were absent.

The magnification factors we obtain in this manner are strictly applicable to the Fourier amplitude spectrum. In keeping with the approximate treatment used in connection with responses to earthquakes of the second type, we may idealize the ground motions as stationary Gaussian processes. Accordingly, we are justified in assuming that the magnification factors for the Fourier amplitude spectrum will apply reasonably well directly to the expected response spectra, except for very short periods of vibration.

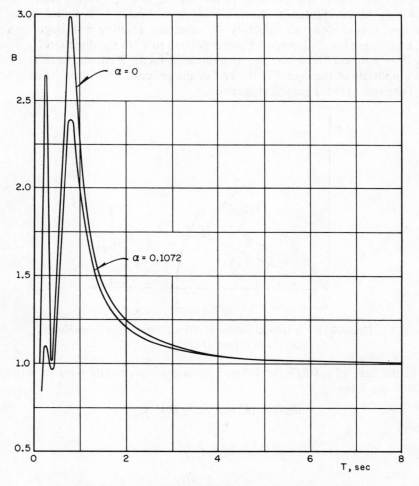

Figure 9.15. Dynamic magnification factor as function of period.

Results of applying Eq. 9.24 to a single mantle that has the following constants, $\alpha = 0.1072$, $\mu = 3270$ ton/m², $v = 152.4$ m/sec, and $H = 30.5$ m, and rests on rock having $\alpha = 0$, $\mu = 29{,}300$ ton/m², and $v = 400$ m/sec, are shown as a function of period in Fig. 9.15. (These are nearly the same data used in drawing Fig. 3.6, as an example in Chapter 3. We could have solved the problem by using the computations performed therein to find the surface motion as response to a Dirac delta incident wave. We preferred to apply the present procedure as an illustration because it permits specifying the percentage of internal damping in the soil as an arbitrary function of frequency.) Also shown are the results obtained when taking $\alpha = 0$ for the mantle which makes Eq. 9.25 applicable. A more convenient way of presenting the results is shown in Fig. 9.16, where the abscissas are frequencies.

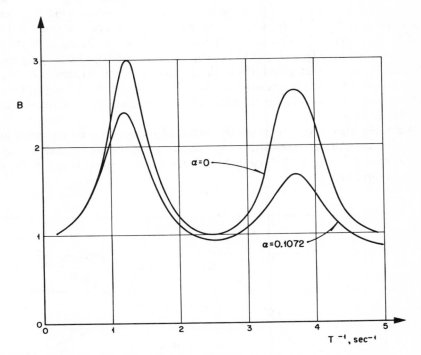

Figure 9.16. Dynamic magnification factor as function of frequency.

Next we illustrate the method proposed, in conjunction with Eq. 9.16, assuming that the power spectral density of the motion that the bare rock would undergo was

$$G^2(\omega) = \frac{25\omega^2/289}{(1 - \omega^2/242)^2 + \omega^2/289} \quad (9.26)$$

in which G is in cm/sec$^{3/2}$ and ω in sec^{-1}. This is of the same form as the average density in Eq. 9.17, but the first term in the parenthesis has been dropped. The reason for this change from Eq. 9.17 to 9.26 lies in that we are using Eq. 9.16, which gives the expected undamped acceleration spectral ordinates on rock proportional to $G(\omega)$; through this modification in the shape of $G(\omega)$ we satisfy the obvious requirement that $1/\omega$ times $G(\omega)$ remains finite as ω tends to zero. (It should not be construed that Eq. 9.26 has a form more closely representing the power spectral density of actual earthquakes on firm ground than does the form of Eq. 9.17. The opposite is the case. But the approximations involved in Eq. 9.16 lead to the necessity of using a fictitious spectral density that will give the correct expected acceleration spectrum for zero damping.)

Using Eq. 9.26, the expected, undamped spectral pseudovelocity on rock, call it $E[V_0(\omega_1, 0)]$, was computed and is shown in Fig. 9.17. Also shown is $\beta_E(0.05)E[V_0(\omega_1, 0)]$. Here $\beta_E(\zeta)$ is the correction factor to account for a damping ratio ζ of critical in a simple system according to Fig. 9.4, which is based on the idealization of the motion as a segment of white noise. The resulting approximation is compared in Fig. 9.17 with the more accurate $E[V(\omega_1, 0.05)]$—the expected pseudovelocity spectrum for 5 percent damping at the rock surface—computed from Eq. 9.16 through numerical integration, using for F_q, $1/\omega_1$ times the Fourier transform of the acceleration transfer function. We see that the two approximate methods of computing the expected, damped pseudovelocity spectrum on rock do not differ markedly in results except for very short natural periods.

We could also have computed the expected pseudovelocity spectrum using ω_1 times the Fourier transform of the deformation transfer function. Results are compared in Fig. 9.18. Again the differences are negligible except for very short natural periods.

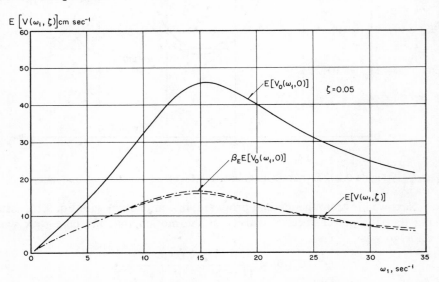

Frgure 9.17. Spectral pseudovelocity on rock. *After Herrera and Rosenblueth (1965).*

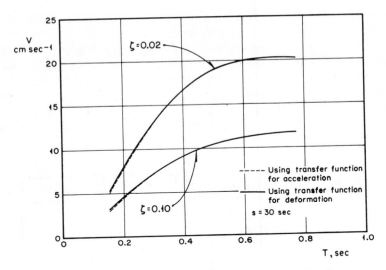

Figure 9.18. Comparison of spectral pseudovelocities from acceleration and deformation spectra.

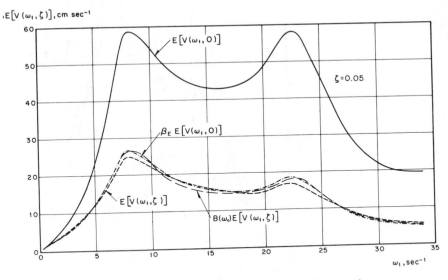

Figure 9.19. Spectral pseudovelocity on soft ground.

Next, Eq. 9.16 was used in conjunction with Eq. 9.26, replacing $G(\omega)$ with $B(\omega)G(\omega)$, where $B(\omega)$ was obtained from Eq. 9.24, to obtain the expected pseudovelocity spectra at the surface of the soft mantle. Results are shown in Fig. 9.19 and compared with the less accurate values of $B(\omega)E[V(\omega, \zeta)]$ and $\beta_E E[V(\omega_1, 0)]$ as approximation to $E[V(\omega_1, \zeta)]$, where V denotes spectral pseudovelocity at the ground surface. Some error is introduced through these devices in the neighborhood of natural periods for which the resulting expected pseudovelocity spectrum has large curvature.

We conclude from the comparisons that the correction factors to account for damping in the structure and the magnification factors to account for filtering through the soft soil may be computed once and for all, independently of the power spectral density of the incident motion. These factors are then applied to the undamped spectra at the surface of bare rock. Important errors appear only in the neighborhood of natural periods for which the resulting spectra have pronounced curvatures.

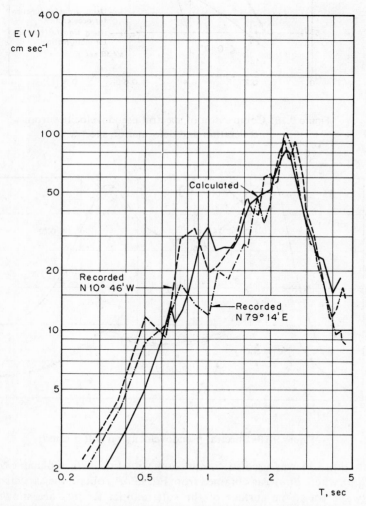

Figure 9.20(a). Comparison for theoretical and actual spectra, earthquake of 11 May 1962 for $\zeta = 0.02$. *After Herrera, Rosenblueth, and Rascón (1965).*

Herrera, Rosenblueth, and Rascón (1965) have applied the most accurate method of those presented in conjunction with Eq. 9.16 to predict the expected spectra at the surface of the Valley of Mexico using dynamic properties of the ground derived from field and laboratory measurements. For the power spectral density of the incident motion, empirical equations (Esteva and Rosenblueth, 1964) were used to correlate known magnitudes and focal distances of actual earthquakes with spectral ordinates on firm ground. The comparison between the predicted mean spectra and those computed from the records of these earthquakes, obtained at the surface of the soft clay, shows reasonable agreement (Fig. 9.20a) once all the calculated values of a, v, and d are boosted, multiplying them by 2.0. This 100 percent increase is not surprising in view of the large dispersions involved in the correlations available.

Because of the peculiar conditions of the Valley of Mexico, the magnification of spectral ordinates due to the soft soil is extremely high in the neighborhood of certain periods; it exceeds 50 and 10 for undamped and damped soil, re-

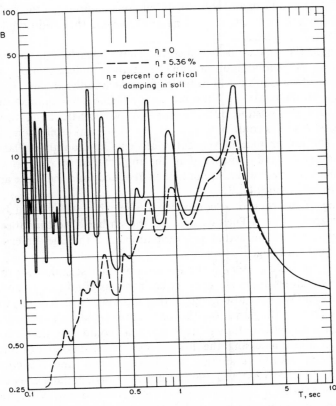

Figure 9.20(b). Dynamic magnification factor for the Valley of Mexico. *After Herrera, Rosenblueth, and Rascón (1965).*

spectively (Fig. 9.20b). The resulting expected, undamped spectra are very peaked at these periods, as can be appreciated in the figure. Consequently the mere multiplication of undamped spectra by the correction factors β_E that hold for the responses to white noise underestimates the damped spectral ordinates at and near the prevailing ground periods and overestimates these ordinates away from the periods in question. We know that for a white or nearly white disturbance the expected spectral responses for damping ratios greater than about 2 percent vary as $\zeta^{-0.4}$ for moderate ζ and as $\zeta^{-0.5}$ for a very high percentage of damping. Instead, responses to earthquakes having pronounced prevailing periods vary, in the neighborhood of these periods, roughly as ζ^{-1} if ζ exceeds about 2 percent, just as do the responses to harmonic, steady-state ground motions. Away from these points of near resonance the responses are relatively insensitive to damping.

A treatment of the one-dimensional problem, very similar to the one developed here, has been published independently by Roesset and Whitman (1969). They show that the deformability of the rock halfspace can be taken into account by introducing an additional, "equivalent" viscous damping in the soil and assuming that the soil rests on a rigid base. However, this additional damping is a function of frequency.

As long as the soil can be assumed to behave linearly, filtering of the various components of ground motion can be studied independently. The vertical component of translation will ordinarily be much reduced when traversing soft, horizontal strata because of the refraction phenomena quoted earlier. These tend to turn the seismic disturbances into vertical-traveling waves, so that only filtered P waves contribute appreciably to vertical oscillation of the ground surface. The importance of vertical relative to horizontal oscillations on soft ground is usually even smaller than the importance of P relative to S waves on rock because the P waves are of higher frequency and therefore are ordinarily damped more drastically.

Filtering of the rotational components of earthquakes cannot be treated as plane horizontal waves. Their existence at the surface of soft mantles stems from the differences between actual soil formations and the one-dimensional model we have been studying.

Thus far we have preserved the assumption that seismic disturbance enters the filtering mantles and undergoes the entire series of reflections and refractions as plane waves. We have also assumed that the waves which return to the underlying rock are lost altogether. Neither assumption is fulfilled in nature. There are no adequate analytical studies available that would lift these restrictions or that would allow assessing the magnitude of the errors introduced by the simplifications. Available studies of wave refraction and reflection at wavy boundaries (Asano, 1960 and 1961; Herrera, 1964), finite-element analysis of sloping boundaries in two dimensions (Idriss and Seed, 1967; Idriss, 1968), and studies about the effects of soil irregularities (Knopoff and Hudson, 1964; Karal and Keller, 1964; Herrera, 1965) still require work before they can be applied to improve the prediction of earthquake spectra. Adequacy of the simplifications introduced must be judged solely in the light of empirical evidence.

As stated in Chapter 2, the accelerograms of real earthquakes on soft alluvial valleys have been reconstructed successfully from the motions recorded at the bare surface of rock a short distance away using an approximate one-dimensional theory that in essence constitutes the basis of the theoretical treatment in the present section. Kanai (1959) has reported excellent agreement between spectra computed from actual records of small and moderate earthquakes and the results of a theory for undamped spectra that practically coincides with the one in this section for the conditions in which he experimented.

A severe test for the present treatment is implied in Fig. 9.20. Here the largest magnification corresponds to a natural period of about 2.5 sec. A few kilometers northwest of the site where the earthquakes were recorded and the dynamic properties of the soil were determined that led to the spectra in Fig. 9.20, the prevailing period of the ground is 2.0 sec (Bustamante, 1965; Esteva, Rascón, and Gutiérrez, 1969); less than 0.5 km east of the first site the clay is seriously preconsolidated, while at the site in question it is practically virgin (Zeevaert, 1964). Hence, the assumption that the soil is horizontally stratified over a large extension is far from correct in this instance. This accounts, in all likelihood, for the small discrepancies between predicted and actual earthquake spectra in Fig. 9.20.

Additional confirmation of the present one-dimensional theory comes from Kanai's semiempirical formulas (Kanai, 1957 and 1961) for the spectra at the surface of soft soil. These formulas are based on a simplified analysis that treats the filtering mantles as a linear, single-degree system. After adjustments and incorporation of an empirical relation between the prevailing ground period and the internal damping of the soil, the most satisfactory of those expressions for the expected pseudovelocity spectrum reads

$$E(V) = 10^{0.61M - 1.73 \log_{10} R - 0.67}$$
$$\times \left(1 + \frac{1}{\sqrt{[(1 + \alpha)/(1 - \alpha)]^2 [1 - (T/T_g)^2]^2 + [0.3/\sqrt{T_g}]^2 (T/T_g)^2}}\right) \quad (9.27)$$

where

M = magnitude
R = epicentral distance
$\alpha = \rho_1 v_1 / \rho_2 v_2$
ρ_1, ρ_2 = densities of the materials of the layer and of the subjacent medium, respectively
v_1, v_2 = wave velocities in the layer and in the subjacent medium, respectively
T = natural period
T_g = predominant period of waves

By adjusting parameters, Eq. 9.27 has been found to stand in good agreement with the spectra of a number of actual earthquakes recorded in Japan and in the western United States (see Fig. 9.21). We state that this fact also confirms the theory developed in the present section because Eq. 9.27 comes close to the results of applying Eq. 9.24 to a single, strongly damped stratum; under this

298 PROBABILITY DISTRIBUTIONS Chap. 9

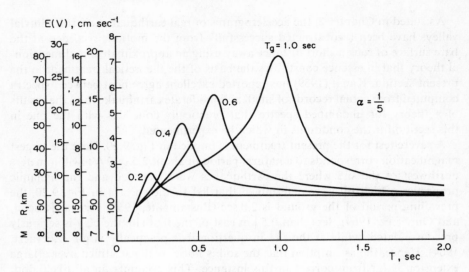

Figure 9.21. Velocity spectra obtained by using Eq. 9.27 adjusting parameters. *After Kanai et al. (1966).*

condition the higher prevailing frequencies lose importance, especially when the fundamental one is not excessively low. Of course Kanai's formula cannot be expected to furnish satisfactory results in extreme conditions, such as those of the Valley of Mexico, where the fundamental prevailing period is long and the first few harmonics are important.

A word of caution is necessary when applying the deterministic one-dimensional theory of multiple wave reflection to earthquakes whose two horizontal components have very different characteristics. As pointed out earlier, there are indications that a unidirectional motion in rock may develop, through filtering, an important horizontal component at right angles with this direction. The second component must be taken into account in design even though calculations fail to indicate as much. The matter is not especially important for strong earthquakes of the originally chaotic type because both horizontal components are likely to be important for them.

It is dangerous to extrapolate conclusions based on the assumption of linear behavior of the filtering medium to conditions in which the soil behaves in even a mildly nonlinear manner. For example, magnification factors due to multiple wave reflection, derived from measurement of minor ground shaking, have greatly overestimated the severity of strong earthquakes in many, commonly met types of soil formations (Wiggins, 1961).

9.8 Earthquake Filtering through Nonlinear Materials

Practical methods of analysis of wave transmission through materials outside the range of linear behavior have not yielded to being used with an analytical stochastic-process treatment. Their usefulness is limited to the study of filtering of individual earthquakes and hence to earthquake simulation.

In Section 5.4 we presented a graphical procedure that preserves continuity of the medium neglecting damping in the analysis of wave transmission. The procedure is adequate for the study of earthquake transmission through strongly nonlinear media in which time dependence of the stress-strain relations is unimportant. Again the method is limited to analysis of individual earthquakes. Furthermore, it is excessively time consuming and inaccurate for the study of long, complicated earthquakes. For simulation studies, one should use computers applied to finite-element or other lumped-parameter models of the soil formations.

Remarks concerning the three-dimensional character of the phenomenon in linear media apply equally in the nonlinear range. Moreover, the corresponding effects cannot be treated rigorously as independent, and the vertical displacements of ground surface due to compaction of granular materials under shaking merit special attention.

9.9 Computer Simulation of Earthquakes

Many problems concerning structural response to earthquakes of the second and third types are too complex for analytical treatment. Even the study of the distribution of maximum responses of linear, single-degree systems requires drastic simplifications. In order to assess the adequacy of these approximations and to study earthquake effects on more complicated structures, it would be desirable to compute the responses of the structures in question to a large number of actual ground motions, as recorded in a trustworthy manner. To accomplish this, one would have to possess a large number of records for every set of conditions of interest. Differentiation between such sets involves many variables: magnitude, focal distance, local geologic conditions, and so on. The computed responses would then be processed statistically. But the vast number of records required for such an approach are not available, nor are they expected to be for many years. It is necessary to simulate them in a computer and verify that the records produced in this fashion are indeed representative of records of true earthquakes.

The first characteristic we would like to incorporate into the simulation is the randomness of earthquake motions. A simple, acceptable way to do this is to generate white noise. Real processes, even in a computer, cannot be white over all frequencies from zero to infinity. Actually what is done is to produce a noise having a flat Fourier spectrum over a wide band, from zero frequency up to frequencies many times or thousands of times higher than the highest natural frequency of interest. This can be accomplished in an analog computer by using commercially available peripheral equipment.

In digital computers white noise can be generated by using algorithms that produce a sequence of random numbers with Gaussian distribution and zero mean. Once this is done we specify that these numbers define the ground acceleration at closely spaced instants. Between these instants one may interpolate the ground acceleration linearly for ease in subsequent computations (Franklin, 1963).

White noise adequately simulates families of ground motions for the study of their effects on structures whose natural periods fall within a nearly flat portion of the expected velocity spectrum of the earthquake and whose periods do not change appreciably with amplitude of oscillation or with time. Accordingly these processes may be used for the analysis of linear or mildly nonlinear systems whose significant natural periods fall between the periods which correspond to the intersections of a, v, and d in a logarithmic representation of spectra.

White noise has been used as earthquake simulation to study the responses of both elastic and elastoplastic simple systems as well as shear beams intended to idealize tier buildings (Bycroft, Murphy, and Brown, 1959; Bycroft, 1960). The usefulness of results obtained in this manner is debatable for buildings having very high ductility factors. Even within the elastic range it is debatable for the uppermost stories of buildings, in which the higher modes play an important role, as the corresponding natural frequencies are associated with a portion of true earthquake-expected velocity spectra that is far from flat.

As a step in refining simulation to extend its range of applicability, white noise has been passed through a single-degree, linear filter (Housner and Jennings, 1964; Penzien and Liu, 1969). The filter behaves as a simple system whose base undergoes the input disturbance, and the filtered motion is that of the mass of the system. Consequently the power density of the filtered process is given by expressions of the type of Eq. 9.17 with coefficients that depend on the parameters of the filter.

The end velocity of these motions can be forced to be zero by ending the motions at an instant when they meet this condition or modifying the base line appropriately. This tends to leave excessively large velocities at intermediate stages of the history of the motion and to produce excessively large ground displacements and fails to simulate the dying out of actual ground motions. The first two objections have been dealt with by introducing corrections to the base line, such as to minimize the mean-squared ground velocity. The third objection has been dispensed with by making the intensity of the motion per unit time a time function. This has been accomplished in some cases by multiplying a segment of a stationary Gaussian process by a time function (the multiplication can be done at the white noise stage or after filtering of the white process) and in others by allowing the oscillations of the filter to build up, remain almost stationary, and die out by themselves. These methods have produced reasonably realistic simulated earthquakes (Amin, Ts'Ao, and Ang, 1969), in some cases corresponding to specific combinations of magnitude and focal distance (Jennings, Housner, and Tsai, 1969). (Small families of the latter simulated motions are available from their authors upon request.)

Incorporation of the effects of magnitude and focal distance in earthquake simulation can be accomplished by using the correlations given in Chapter 7 between these variables and the ground's maximum acceleration, velocity, and displacement as well as the duration of the motion. In turn the parameters of the filter and the duration or rate of decay and intensity of the segment of white noise to be filtered can be chosen so as to produce the desired characteristics of the simulating processes. Bustamante and González have shown how this can be accomplished using a method based, in part, on the one developed by Housner and Jennings (1964).

Other methods of generating simulated earthquakes in computers have been used. For example, the accelerogram

$$\ddot{x}_0(t) = \sum_i (t - t_i) a_i H(t - t_i) \exp\left[\alpha(t - t_i)\right] \sin \omega(t - t_i) \qquad (9.28)$$

has received considerable attention (Bogdanoff, Goldberg, and Bernard, 1961; Goldberg, Bogdanoff, and Sharpe, 1964). Here t_i marks the beginning of the ith function of the type $t \exp(-\alpha t) \sin \omega t$, a_i is a random number, H is the Heaviside step function, and α and ω are constants. Inspection shows that these acceration diagrams closely resemble damped, filtered, white noise. The factor $t - t_i$ accounts for an increase in amplitudes similar to the buildup of the response of the filter, while the exponential factor accounts for a decrease similar to damping of the vibrations of the filter.

The adequacy of a simulation criterion depends on how the resulting motions are to be used. Thus, Amin, Ts'Ao, and Ang (1969) have shown that the expected response spectral ordinates of transient Gaussian processes are practically insensitive to the time dependence of the intensity per unit time. However, they have also found that this is not the case for the responses of nonlinear systems to the same motions. Accordingly, from the viewpoint of effects on some types of systems a segment of stationary Gaussian process is a satisfactory idealization of earthquakes (we have seen that for some purposes even a segment of white noise will do), while a study of the responses of other systems requires an appropriate consideration of the variation of intensity per unit time. And we shall see in Chapter 11, certain nonlinear systems are sensitive to the time dependence of the energy content of the motions. Thus, the expected responses of softening systems with low over all damping (time dependent plus hysteretic) to motions in which the prevailing period lengthens with time may appreciably exceed their responses to motions having a time-independent frequency content.

The most satisfactory process of earthquake simulation, from the viewpoint of its generality, is the one developed by Rascón and Cornell (1969), described in Section 7.5.

Choice of one device over others depends on the characteristics of the structures to be analyzed, on time available for analysis, on the computer hard- and software available, and on the number of computations to be performed.

So much for earthquakes of the second type. Once they have been simulated we must filter them to produce those of the third type. This can be accomplished conveniently, as discussed previously, by idealizing the soil as a discrete, closely coupled system, having dashpots in parallel with spring elements, or by using finite-element techniques.

In order to gain as much information as possible from earthquake simulation in computers and in order to make the conclusions as dependable as possible, it is advisable to supplement them with approximate analytical treatments of the problems under study. The analytical treatment, leading to results adjusted on the basis of the Monte Carlo studies, also permits extrapolating the results. This is important in problems involving so many variables that the unaided Monte Carlo methods become impractical. Additional confirmation of the orders of magnitude of Monte Carlo results should in general be obtained from spotchecks using records of actual earthquakes.

Remarks made in connection with earthquake simulation in computers apply with appropriate changes to physical model testing under random loading. The matter of physical models is discussed in Chapter 17.

9.10 Distribution of Responses to Earthquakes

Consider a family of earthquakes of the second type defined by the expectations of the corresponding spectral responses and assume that all these earthquakes have the same intensity. One would anticipate that the responses of a given simple system to these earthquakes would approach the extreme type 2 distribution for large responses because we would be selecting the maximum response to each earthquake from a series of maxima, roughly independent of each other.

It can be shown (Rosenblueth, 1952a) that, as the design response tends to infinity in an undamped simple system, the probability that it be exceeded tends asymptotically to $\exp(-\alpha^2)$, where $\alpha = Q/E(Q)$. This is precisely the exponential type 2 distribution,

$$P\left[\frac{Q}{E(Q)} \leq \alpha\right] = 1 - \exp(-\alpha^k) \tag{9.29}$$

with $k = 2$.

The ratio of the response of a damped, simple system to that of the corresponding undamped system, both associated with the same probability that they be exceeded, tends to a finite ratio as this probability tends to zero, given by the curve $P = 0$ in Fig. 9.3. It follows that, asymptotically, the distribution of responses of damped systems also tends to the extreme type 2 (Eq. 9.29) with $k = 2$ for large responses, although convergence may be quite slow for strongly damped structures.

The applicability of the extreme distribution in question is supported by the Monte Carlo study of Bogdanoff, Goldberg, and Bernard (1961), who, however, arrive at somewhat different values of k. This discrepancy may be attributed to the criteria used for earthquake simulation in that study.

Penzien and Liu (1969), in a different simulation study, found that the distribution of spectral ordinates can be approximated by the extreme type 1 distribution over a wide range of the normalized responses. Over this range there is not a pronounced difference between extreme types 1 and 2.

Our interest should be directed, actually, not toward the distribution of responses to a single ground motion but toward that of the maximum response over a long period of time. We saw in Chapter 8 that the distribution of the maximum intensity at any given site during a given time interval rapidly approaches the extreme type 2 distribution as the interval is made to increase. Consequently, if we consider the expected responses of any simple system to each earthquake during this time interval and select the maximum expectation, its distribution will also approach the extreme type 2. The k value will differ, however, from that of the maximum intensity. Convergence will be relatively rapid because we already begin with a function that approximates this distribution.

10

RESPONSES OF LINEAR MULTIDEGREE SYSTEMS

10.1 Introductory Note

In conjunction with Chapter 9 we shall successively study here the response to earthquake distributions of types 2 (approximately white noise) and 3 (with prevailing periods). Then we shall apply our conclusions to representative structures.

For the most part our approach will use the transfer functions, $\psi_q(t)$, defined as in the foregoing chapter as the response $q(t)$ to the excitation $\ddot{x}_0(t) = \delta(t)$. We shall be content to present the theory for systems whose base has a single degree of freedom. The expression we shall give can be extended to structures whose base has any number of degrees of freedom; the functions ψ_q will then become vectors, each of whose components represents the response to a Dirac delta disturbance in one of the degrees of freedom of the base. In a more synthetic presentation the functions $\psi(t)$ would be matrices whose terms would represent the ith response to the Dirac delta disturbance in the base's kth degree of freedom.

Whenever convenient we shall convert expressions in terms of the functions ψ into expressions in terms of natural modes. The step will be limited to structures that can be idealized as though they had classical modes; other systems will require retention of the ψ function expressions.

Once the basic solution for a response q is known and the ground motion is specified, we can obtain the response to this motion from Eq. 9.13:

$$q(t) = \int_0^t \ddot{x}_0(\tau)\psi(t - \tau)\,d\tau \qquad (10.1)$$

Since we shall make so much use of transfer functions, it is worth presenting some simple examples. When the structure in question has classical modes we may write

$$\psi(t) = \sum_i \psi_i(t) \qquad (10.2)$$

where each ψ_i corresponds to one natural mode of vibration (see Chapter 2). In a system of this sort, every ψ_i can be written as a sine function with an appropriate phase shift because each natural mode behaves as a simple system. In an undamped system if all the natural frequencies are multiples of the fundamental frequency, the resulting ψ functions for the combination of natural modes will be periodic with a frequency equal to the fundamental frequency of the system; this is the case with a shear beam in cantilever, whose naural frequencies are in the ratios 1:3:5... (see Section 3.2). If the natural frequencies are all multiples of some other finite quantity but not of the fundamental frequency, the transfer function will again be periodic, but its period will exceed the structure's fundamental; otherwise ψ will be aperiodic. Examples of these situations are depicted in Fig. 10.1a.

In a damped system the conditions for damped periodicity of the ψ functions include the same as those for periodicity in the corresponding conservative system, and besides, the modal damping ratios ζ_i must be inversely proportional to the damped natural circular frequencies ω_i'.

When the period of the functions ψ in an undamped system is very long, the addition of even a small amount of damping will greatly affect the shape of successive waves of the transfer functions.

In a system with two degrees of freedom having classical modes, when the two natural frequencies are close to each other, that is, when $(\omega_2 - \omega_1)/$

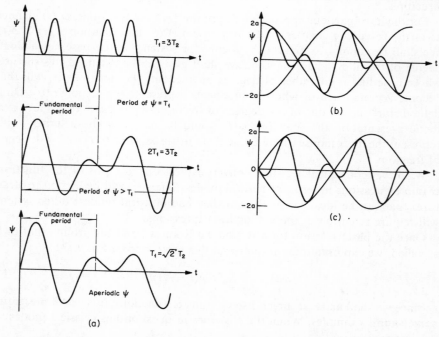

Figure 10.1. Transfer functions in conservative systems having two degrees of freedom.

$(\omega_2 + \omega_1) \ll 1$, the transfer functions exhibit beating and behave as a sine or cosine wave of frequency $\omega_2 + \omega_1$ modulated by a trigonometric function of frequency $\omega_2 - \omega_1$. (The phenomenon can be described in a manner similar to the approximate treatment of the group velocity of seismic waves we saw in Section 3.14.) This situation is illustrated in Fig. 10.1b for the case $\psi = a(\sin \omega_1 t + \sin \omega_2 t)$ and in Fig. 10.1c for $\psi = a(\sin \omega_1 t - \sin \omega_2 t)$. If both natural modes have the same degree of damping, the modulating function ζ will also be a damped sine or cosine wave, and its degree of damping will equal $(\omega_2 + \omega_1)\zeta/2(\omega_2 - \omega_1)$, which may be many times greater than ζ.

If the response we wish to calculate is a deformation, a velocity of deformation, or an absolute acceleration, according to Chapters 1 and 2 we can write, respectively,

$$\psi_y(t) = -\sum_i \frac{a_i}{\omega'_i} \exp(-\zeta_i \omega_i t) \sin \omega'_i t$$

$$\psi_{\dot{y}}(t) = -\sum_i \frac{a_i}{\omega'_i} \exp(-\zeta_i \omega_i t) \left(\cos \omega'_i t - \frac{\zeta_i}{\sqrt{1-\zeta_i^2}} \sin \omega'_i t \right) \quad (10.3)$$

$$\psi_{\ddot{x}}(t) = \sum_i a_i \omega_i \exp(-\zeta_i \omega_i t) \left[\frac{(1-2\zeta_i^2)}{\sqrt{1-\zeta_i^2}} \sin \omega'_i t + 2\zeta_i \cos \omega'_i t \right]$$

where y denotes the displacement of a point in the structure, relative to the ground; $\ddot{x}$ is the absolute acceleration of the same point; a_i is the participation coefficient for the ith mode and depends on the mode shape and on the point of interest; ζ_i is the degree of damping in the ith mode; $\omega'_i = \omega_i(1-\zeta_i^2)^{1/2}$ is the corresponding damped frequency.

The transfer functions for forces and stresses can always be expressed as linear combinations of the $\psi_{\ddot{x}}$'s, and in undamped systems they can also be expressed as linear combinations of the ψ_y's. For example, the transfer function for shear S at any elevation in a vertical, undamped shear beam is $k\partial \psi_y/\partial X$, where k is the shear stiffness of a unit length and X is the elevation of the section considered. Using the results in Section 3.2, we immediately find, for a uniform shear beam fixed at its base and free at the top,

$$\psi_y = -\frac{4}{\pi} \sum_{n=1}^{\infty} \frac{1}{(2n-1)\omega_n} \sin \frac{\omega_n}{c} X \sin \omega_n t$$

where

$$c = \sqrt{\frac{k}{m}}$$

$m =$ mass per unit length

$$\omega_n = \frac{(2n-1)\pi c}{2H} \quad n = 1, 2, 3, \ldots$$

$H =$ total height of beam

(see Eqs. 3.10–3.14). It follows that

$$\psi_S = -\frac{4k}{\pi c} \sum_{n=1}^{\infty} \frac{1}{2n-1} \cos \frac{\omega_n}{c} \sin \omega_n t \quad (10.4)$$

Summation of this series gives the transfer functions shown in Fig. 10.2a. Each consists of a succession of rectangular waves of period $4H/c$, equal to the fundamental period of the shear beam. The functions alternate in value between $\sqrt{km}$, 0, $-\sqrt{km}$, and 0. The duration of each interval within which $\psi_S = 0$ is $2X/c$.

The transfer functions for overturning moment $\mathcal{M}$ are also of interest. Using the result we have for shears and the relation

$$\psi_\mathcal{M} = \int_X^H \psi_S \, dX$$

we find the triangles shown in Fig. 10.2b. At elevation X, the legs of the triangles are $(H - X)\sqrt{km}$ vertical and $2(H - X)/c$ horizontal.

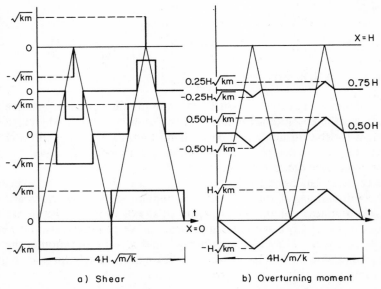

a) Shear   b) Overturning moment

Figure 10.2. Transfer functions in conservative shear beams.

We shall use these results in the analysis of two-degree systems and shear beams to illustrate several points in the present chapter.

10.2 Responses to White Noise

Consider the class of transient disturbances whose accelerogram is the product of white noise and a deterministic function of time, $f(t)$, which is zero for negative t. The distribution of responses of every linear system at any specified instant is Gaussian, with expectation zero. Consider now the class of systems with damped periodic transfer functions. If, for a given f, the damped period of ψ_q, T' is sufficiently short, the ratio of successive maxima of $|q|$ will have a distribution practically independent of the shape of ψ_q. Therefore, the distribution of $Q/E(Q_0)$ and, hence, the ratio $E(Q)/E(Q_0)$ for single-degree systems will be directly applicable if we replace $\zeta\omega_1$ with $\zeta_i\omega_i$, which is a constant for each system, because we have assumed that ψ_q is damped periodic. (Here, as in Chapter 9, Q is $\max_t |q(t)|$, $E(\cdot)$ denotes expectation, and subscript 0 refers to the undamped system.)

It follows from Eq. 10.1 that, for the systems at hand,

$$Q^2 \propto \int_{t_1}^{t_1+T'} \psi_q^2 \, dt \tag{10.5}$$

where Q is the response associated with any given probability of failure and t_1 is such that $\psi_q(t)$ is damped periodic for $t \geq t_1$.

When q is the pseudovelocity (ω_1 times the displacement relative to the ground) of a conservative, single-degree system, ψ_q is $\sin \omega_1 t$. The integral in Eq. 10.5 is then $T'/2$. Hence,

$$Q = V_0 \left(\frac{2}{T'} \int_{t_1}^{t_1+T'} \psi_q^2 \, dt \right)^{1/2} \tag{10.6}$$

Now, if ψ_q is periodic the ψ_{qi}'s are orthogonal in every interval equal to the period of ψ_q. We conclude from Eq. 10.5 that, for systems having periodic transfer functions of sufficiently short period,

$$Q^2 = \sum_i Q_i^2 \tag{10.7}$$

where Q_i is the design response in the ith natural mode; that is, Q_i is chosen so that the probability that it be exceeded at least once in the ith mode be the same as the probability that the structure's response Q be exceeded at least once.

The validity of Eq. 10.7 is not affected if ψ_q is damped periodic. It is only required that f be a sufficiently smooth function of time and that the integral of f^2 over every interval of duration T' be small compared with its integral up to an instant beyond which most of the responses attain their maxima.

So as to widen the type of linear system that we can treat, look now at white noise proper ($f = 1$) of infinite duration. It can be shown that, because the distribution of $q(t)$ is Gaussian with zero mean, the square of $|q(t)|$ associated with any fixed probability that it be exceeded is proportional to $E[q^2(t)]$. Hence, the square of $E|q(t)|$ is also proportional thereto. This suggests assuming that the squared maximum response to a transient accelerogram of the form $f(t)$ times white noise is proportional to $E[q'^2(t)]$, where the prime now identifies responses to a segment of white noise and t is arbitrary. From the fact that the variance of a sum (integral) of independent variables is equal to the sum of the corresponding variances, and $E[q^2(t)]$ is var $q(t)$, we conclude that

$$E[q'^2(t)] \propto \int_{-\infty}^{t} \psi_q^2(t - \tau) \, d\tau$$

$$= \int_0^{\infty} \psi_q^2(t) \, dt$$

and, because we have assumed that $Q^2 \propto E[q^2(t)]$ and $E[q^2(t)] \propto E[q'^2(t)]$,

$$Q^2 \propto \int_0^\infty \psi_q^2 \, dt \tag{10.8}$$

When q is the pseudovelocity of a single-degree system, the second member in Eq. 10.8 gives $\frac{1}{2}\zeta\omega_1$. The result differs from that for the square of the maximum response to a segment of white noise, which depends on the duration of motion and on the probability of failure. Let us adjust the percentage of damping so that the result of applying Eq. 10.8 will coincide with the expected response. To this end we shall employ Eq. 9.9. We seek, then, the "equivalent" degree of damping ζ_i' which will make $2\zeta_i'\omega_i \propto 1 + \zeta_i\omega_i s/2$, where s denotes the duration of the segment of white noise. The answer must be such that $\zeta_i' \doteq \zeta_i$ when s tends to infinity, since the time at which Q occurs tends to infinity with s. We find

$$\zeta_i' = \zeta_i + \frac{2}{\omega_i s} \tag{10.9}$$

Thus, we may use Eq. 10.8 with the increased damping ratios given by Eq. 10.9 in the system's natural modes of vibration. This will be approximately correct for single-degree systems and can be expected to be satisfactory for a wide class of multidegree systems.

If we assume that all the modal damping ratios are small compared with unity, Eqs. 10.2 and 10.8 lead to the approximate relation

$$Q^2 = \sum_i Q_i^2 + \sum_{i \neq j} \sum \frac{Q_i Q_j}{1 + \epsilon_{ij}^2} \tag{10.10}$$

where

$$\epsilon_{ij} = \frac{\omega_i' - \omega_j'}{\zeta_i'\omega_i + \zeta_j'\omega_j}$$

and Q_i is to be taken with the sign that ψ_{qi} has when it attains its maximum numerical value.

Equation 10.10 improves over Eq. 10.7 when ψ is not damped periodic, and usually it is not. The difference between these expressions tends to zero when the natural frequencies tend to become well differentiated, when the damping ratios tend to zero, or when the ratios s/T_i' tend to infinity, where T_i' is the ith damped natural period. To illustrate the influence of the second term in Eq. 10.10, Fig. 10.3 has been prepared for a two-degree system having $\omega_1' = \omega_2'$ and $\zeta_1' = \zeta_2'$.

When ψ_q is damped periodic, the probability distribution of $Q/E(Q_0)$ is roughly equal to that for a single-degree system having $\zeta\omega_1 = \zeta_i\omega_i$. When it is not damped periodic, the distribution of $Q/E(Q_0)$ may differ appreciably from that in the corresponding single-degree system even if ζ_i is inversely proportional to ω_i. Independently of the relation beween ζ_i and ω_i, use of Eq. 10.10 implies a probability distribution for Q because the modal responses in the second member correspond theoretically to the same probability of exceedance as Q. In some cases this assumption may be quite in error. The most extreme shape of ψ consists of a series of spikes. The distribution of Q is then extreme type

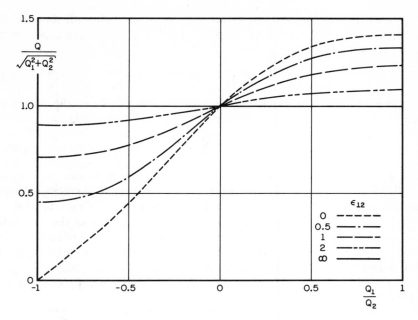

Figure 10.3. Ratio of responses computed from Eqs. 10.10 and 10.7.

1. More accurate statements on the distributions of structural responses will require analysis of results obtained from earthquake simulation.

Although derivation of Eq. 10.10 was quite heuristic, we notice that it cannot give very large errors in two-degree systems when $\epsilon_{12} \ll 1$, when $\epsilon_{12} \gg 1$ and $|Q_1| \cong |Q_2|$, or when $|Q_1| \gg |Q_2|$, whatever the value of ϵ_{12}. Also it satisfies the obvious restrictions $Q \geq |Q_1 - Q_2|$ in these systems, and $Q \leq \sum_i |Q_i|$ in all cases.

The last of these restrictions has served as basis for estimating the design response from model analysis, by taking $Q = \sum_i |Q_i|$. Although in some instances this expression may be adequate for design purposes, it often provides a gross overestimate of Q.

Equation 10.10 is limited to systems that have classical modes of vibration. Besides, we can conceive ψ functions of very long period and small degrees of damping, for which this expression would give poor results. Further to widen the range of applicability of our results we shall take as basis for estimating Q the responses at a specified instant to a transient accelerogram of the type $f(t)$ times white noise. Proceeding as for stationary disturbances we expect that, approximately,

$$Q^2 \propto \int_{-\infty}^{t} f^2(t - \tau) \psi_q^2(t - \tau) \, d\tau$$

If we take $t = 0$ and $f = \exp(-t/s)$ this expression can be written in the form

$$Q^2 \propto \int_0^\infty \exp\left(\frac{-2t}{s}\right) \psi_q^2(t) \, dt$$

In this manner we obtain answers that practically coincide with those of Eq. 9.10 for single-degree systems having $T_1 \ll s$ and $\zeta \ll 1$. We can expect better results by assuming

$$Q^2 \propto \max_t \int_0^t f^2(\tau)\psi_q^2(t-\tau)\,d\tau \tag{10.11}$$

with $f = \exp(-ct)$. For single-degree systems this produces the correction factor

$$\beta_E \doteq \sqrt{\frac{\xi - \xi^\theta}{\theta - 1}} \tag{10.12}$$

where $\theta = \zeta\omega/c$ and $\xi = \theta^{1/(1-\theta)}$. Equation 10.12 approximates the "exact" β_E derived in Chapter 9 from the analysis of responses to segments of white noise, provided we take $c \cong 2.5/s$ (see Fig. 10.4). In fact, it may be that Eq. 10.12 is more satisfactory than the "exact" solution in view of the variability of intensity per unit time in actual earthquakes, but the matter is of rather academic interest because, rigorously, β_E should be a function of natural frequency and of damping, not only of their product. For this reason and because Eq. 10.11 does not lend itself to simple modal analysis, we shall not explore the matter further.

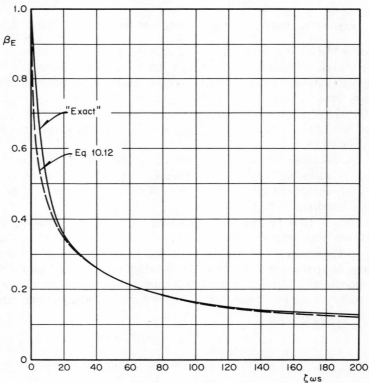

Figure 10.4. Correction factor for viscous damping.

10.3 Responses to Gaussian Processes and to Earthquakes

Consider a single-degree system of period T', subjected to a family of disturbances which are segments of duration s of a stationary Gaussian process. If in the neighborhood of T' the power spectral density of the motions varies smoothly with period, the distribution of the maximum responses of the system to this family of disturbances lies close to the distribution of its maximum responses to a segment of white noise of duration s and whose power spectral density equals that which the segment of Gaussian process has at T' (Caughey and Gray, 1963). Therefore, the expressions we derived for combining modal responses apply when the disturbances are a segment of stationary Gaussian process with essentially the same limitations, provided the power spectral density or the expected spectral pseudovelocity is a sufficiently smooth function of period in the neighborhood of the periods of the significant natural modes.

Particularly for systems without classical modes we may use an approach parallel to the one which assumes that maximum squared responses to the disturbance are proportional to the expected squared responses at a specified instant to the product of $f(t)$ and white noise, replacing white noise with the stationary Gaussian process. The use of Fourier transforms in the manner of Tajimi (1960) is then indicated.

The foregoing approaches cannot be expected to apply far outside the range of periods defined by the intersections of the a and v lines and the v and d lines in a logarithmic plot of the spectra of the motion. For extremely rigid structures the maximum acceleration is approximately equal to that of the ground. For extremely flexible structures Q tends to $\max_t |\psi(t)| \int_0^\infty \ddot{x}_0(t)\, dt$. For structures subjected to earthquakes of the third type, the treatment presented in Chapter 9 is directly applicable.

The theory developed by Shinozuka (1964) is of greater generality than the foregoing results, even if it only furnishes upper and lower bounds to the probability distributions of the responses.

Very simple ground motions, such as earthquakes of type 1, do not lend themselves to idealization as Gaussian processes, but their importance in earthquake engineering is not so great, and their effects can be analyzed deterministically.

10.4 Application to Uniform Shear Beams

Consider now the transverse oscillations of a uniform shear beam. In Section 10.1 we derived expressions for the ψ functions for shear and overturning moment in undamped, uniform shear beams. If the ground motion is idealized as white noise, undamped spectral accelerations turn out to be inversely proportional to the square root of the mean ψ^2 over a long time interval (see Eq. 10.5).

314 RESPONSES OF LINEAR MULTIDEGREE SYSTEMS   Chap. 10

It follows that design shears are distributed according to a parabola with the vertex at the top of the shear beam, and overturning moments vary as a parabola of degree $\frac{3}{2}$, also with the vertex at the top of the beam. The base shear and base overturning moment are readily found from the consideration that the integral of the sine squared of ωt over a long time interval is nearly half the time interval. Because the transfer function of the base shear alternates between $+$ and $-\sqrt{km}$, the corresponding design value is $\sqrt{2km}$ times the design spectral pseudovelocity and the design shears are

$$S = V\sqrt{2km\left(1 - \frac{X}{H}\right)} \qquad (10.13)$$

where V is the spectral pseudovelocity, k is the shear stiffness of the unit of length, m is the mass per unit length, X is the elevation, and H is the height of the beam. Similarly we find for the design overturning moments

$$\mathcal{M} = HV\sqrt{\frac{2}{3}km\left(1 - \frac{X}{H}\right)^3} \qquad (10.14)$$

These results are shown in Figs. 10.5 and 10.6. Throughout the beam the design shears and overturning moments are greater than the values that would be obtained from a static analysis based on a uniform horizontal acceleration that would give the same maximum shear as Eq. 10.13 or the same maximum

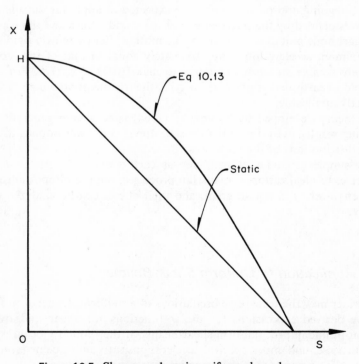

Figure 10.5. Shear envelope in uniform shear–beam.

moment as Eq. 10.14 (also shown in the figures). The differences are most pronounced near the top of the beam.

The shears and overturning moments are not proportional to m, as a purely static analysis would indicate. All other variables being fixed, S and $\mathcal{M}$ are proportional to the square root of the mass per unit length. Recalling that the fundamental period T_1 is $4H/c$, where $c = \sqrt{k/m}$ is the speed of shear waves in the beam (see Eq. 3.11), we can write Eq. 10.13 in the form

$$S = \frac{4VM}{T_1} \sqrt{2\left(1 - \frac{X}{H}\right)} \tag{10.15}$$

$$\mathcal{M} = \frac{4A_1 M}{2\pi} \sqrt{2\left(1 - \frac{X}{H}\right)}$$

where $A_1 = (2\pi/T_1)V$ is the spectral acceleration corresponding to the fundamental period, and $M = mH$ is the total mass of the shear beam. Equation 10.15 can also be written as

$$S = 0.90\, A_1 M \sqrt{1 - \frac{X}{H}} \tag{10.16}$$

Thus, the base shear is about 10 percent smaller than the mass of the beam times the spectral acceleration associated with the fundamental period of vibration.

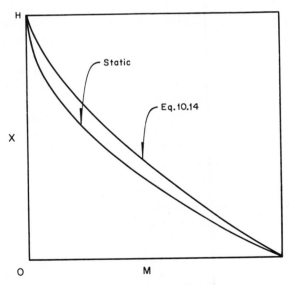

Figure 10.6. Overturning moment envelope in uniform shear–beam.

Overturning moments furnished by Eq. 10.14 are smaller than the values that would be obtained statically from the design shears in Eq. 10.13; that is, $\int_{H-x}^{H} S\, dX$. The ratio of $\mathcal{M}$ to this integral is 0.866 throughout the structure. The reason for this difference is found in the fact that maximum shears are not attained simultaneously at all elevations; indeed, S changes sign along the beam in the second mode and in all higher modes. The question will be discussed again in Chapter 15.

Thus far we have assumed that the spectral pseudovelocity is independent of the natural period and that the shear beam is not damped. We can lift both restrictions and derive approximate solutions by breaking up the transfer functions for shear and moment into their damped harmonic components and combining the corresponding modal responses in accordance with Eq. 10.7 or with Eq. 10.10. These and other applications, to shear beams and to other structures, of the theory developed in the foregoing ariticle are presented by Rosenblueth and Elorduy (1969a) and will be discussed in Chapters 15 and 16.

10.5 Responses to the Simultaneous Action of Several Components of Ground Motion

In structures having linear behavior, any response we wish to consider is a linear function of the responses to the disturbances in the various degrees of freedom of the base. If each disturbance belongs to a stationary Gaussian process, the response in question will also be stationary Gaussian. It may be treated as though it occurred in a structure having a single degree of freedom and the disturbance in this degree belonged to a stationary Gaussian process. If all the components are multiplied by the same deterministic function of time, so will the equivalent single process.

The fact that the resulting motion may be Gaussian is not sufficient to allow us to compute the expectation of the corresponding responses or other parameters of their distributions. We must also specify the correlation between the various components of the ground motion. When these components may be taken as completely uncorrelated, the expected maximum numerical value of any response will be the square root of the sum of squared expectations associated with the various components. Even when there is considerable correlation between the three components of translation and between these and the rotational components of motion, we may treat their combined effects as though the various components were uncorrelated provided the corresponding ψ functions are well differentiated. Under these conditions, if

$$q(t) = \sum_k q_k(t)$$

denotes the total response, where q_k is the response due to the kth degree of freedom of the base, such that

$$q_k(t) = \int_0^t \ddot{x}_k(\tau)\psi_k(t-\tau)\,d\tau$$

where x_k denotes the ground motion along this component and ψ_k is the corresponding transfer function, we may write

$$Q^2 = \sum_k Q_k^2$$

and hence,
$$E(Q) = \sqrt{\sum_k [E(Q_k)]^2}$$
where $Q_k = \max_t |q_k(t)|$, and Q and Q_k are associated with the same probability of exceedance.

However, if the shapes of the ψ_k functions are such that strong correlation between the $\ddot{x}_k$'s results in an appreciable correlation between the quantities $\ddot{x}_k(\tau)\psi_k(t - \tau)$, this conclusion does not apply. Particularly when the ψ functions are similar to each other in two orthogonal directions, there may be coupling between the corresponding natural "modes" of vibration.

Actually the natural modes are not purely displacements contained in vertical planes when there is coupling between degrees of freedom in orthogonal vertical planes; this situation may operate in the actual structure, while the ordinary idealization we would make of it, based on its nominal geometrical and mechanical characteristics, would not disclose such coupling.

Coupling may stem from a minor departure from radial symmetry in an inverted pendulum or a vertical shaft. Expected maximum responses will then usually be much larger than we would predict when ignoring coupling. A specific example of this nature, which involves a nominally radially symmetric chimney, has been analyzed by Yamamoto and Suzuki (1965). In principle we may deal probabilistically with these cases by appropriately modifying the transfer functions.

In many practical problems there is a strong correlation between the various components of motion. This is especially true when the components correspond to parallel displacements of various points of support or to rotation and translation of long bases. Indeed, the opposite sides of a gorge across which lies an arch dam have been recorded to move in such a way that their displacements may be roughly idealized as equal to each other except for phase shifts that may be expected to vary slowly throughout the earthquake (Okamoto et al., 1961).

When we idealize the disturbance along one of the degrees of freedom of the base as being the same as the one along another degree of freedom with a constant time shift, we can use a single equivalent transfer function to treat both disturbances as in a structure whose base has a single degree of freedom. Thus, the expression
$$q(t) = \int_0^t [\ddot{x}_1(\tau)\psi_1(t - \tau) + \ddot{x}_1(\tau - t_{12})\psi_2(t - \tau)] \, d\tau$$
where t_{12} is the time shift, may be put in the form
$$q(t) = \int_{-t_{12}}^t \ddot{x}_1(\tau)\psi_{12}(t - \tau) \, d\tau$$
where
$$\psi_{12}(t) = \begin{cases} \psi_1(t) & \text{when } t < t_{12} \\ \psi_1(t) + \psi_2(t - t_{12}) & \text{when } t \geq t_{12} \end{cases}$$

Any other type of correlation between the various x_k's may be treated in a similar manner. The main source of difficulty lies in the scarcity of information about correlations between the various components. Important civil-engineering works may justify carrying out analyses for a variety of assumed correlations between the components, assigning a probability to every assumption, and computing probability distributions of the responses.

10.6 Flexible Pipes and Tunnels

Very long structures supported continuously throughout their length constitute a special case of systems whose bases have many degrees of freedom—an infinite number of degrees, in fact. Examples of these conditions are pipes, tunnels, and roadbeds. If the structure is sufficiently flexible, relative to the surrounding or underlying soil, it will follow essentially the displacements and deformations that the soil would have if the structure were absent. Under these circumstances it is possible to advance a particularly simple approximate solution for the stresses set up in the structure. If the structure lies underground, the displacements of the soil can usually be computed with sufficient accuracy from the ground displacements taking into account phenomena of wave travel and wave reflection.

The soil displacements can be broken into components associated with the various types of waves: shear, longitudinal, and different kinds of surface waves. In turn the displacement in each type of wave will have components along the longitudinal axis of the structure and the two principal axes of inertia of its cross section. Consider any one of these components and suppose it is associated with a wave that travels with the veloicty c in the structure's longitudinal direction. Let x denote this component of displacement and X a coordinate axis measured longitudinally along the structure. If the soil is linearly elastic and homogeneous, the displacement will satisfy the differential equation

$$\frac{\partial^2 x}{\partial t^2} = c^2 \frac{\partial^2 x}{\partial X^2}$$

(see Chapter 3). It follows that the curvature of the structure is equal to $\ddot{x}/c^2$ where $\ddot{x}$ is the acceleration of the soil, perpendicular to the structure's axis, in the component of motion that travels parallel to the axis. From here we can immediately compute the bending moments about the principal axis of inertia in question and find the corresponding stresses in the structure.

The axial strain of the structure is $-\dot{x}/c$, where $\dot{x}$ is now the velocity of the soil in the longitudinal direction for the component of motion that travels parallel thereto. The resulting stresses are also found immediately. From the stresses in the structure an estimate can be made of the soil-structure contact pressures and shears, and the importance of the stiffness of the structure can be gaged.

Ordinarily it is difficult to distinguish between the various component waves that form an earthquake record. We can always make the conservative assumption that the acceleration in which we are interested is equal to the entire acceleration in the direction we are considering. We may also make various assumptions concerning the breaking of the motion into component waves and assign each assumption a probability that it be valid, as suggested for structures that rest on a finite number of supports.

The most important effects of earthquakes on structures of this kind are associated with strongly nonlinear deformations of the soil or sliding along geologic faults. Such phenomena are treated in Sections 16.3 and 16.5.

11

RESPONSES OF NONLINEAR SYSTEMS

11.1 Introductory Note

Nonlinear behavior of a structure may be due to stress-strain relations, large deformations, conditions of support, conditions of failure, or a combination of these causes. Usually nonlinearity due to material behavior or to conditions of support involves inelasticity; geometric nonlinearity (due to changes in geometry usually caused by large deformations) is associated with reversible processes, and nonlinear criteria of failure may well accompany linear force-deformation relations. Whatever the nature of nonlinearity, no entirely satisfactory analytical treatment is available, except for very simple conditions. In cases of practical interest one must use Monte Carlo methods or resort to relatively crude upper bounds on responses. Still, these are often adequate for design purposes.

In this chapter we shall devote most of our attention to criteria for establishing pertinent bounds and describe results of simulation studies for some ideal systems; these results may serve to guide judgment in the analysis and design of real structures.

11.2 Nonlinear Criteria of Failure

The simplest criteria of fracture do not involve stress history but rely only on the maximum value of some function of the principal stresses. Nonlinearity may be due to the fact that (1) the coordinates at which the critical combination of stresses takes place depend on the conditions of loading, (2) the directions of the principal stresses at the critical point are a function of the pattern of loading, (3) a criterion of failure exists depending on a nonlinear function of the principal stresses, or (4) there is a combination of these factors. Later on we shall discuss structures made up of materials having nonlinear stress-strain relations.

Consider first the case in which the position of the critical point depends on the force distribution. We cannot apply the results of the preceding chapters directly, for we based our reasoning on the assumption that the magnitude of any one response—for example, the bending moment at a given section of a structural member, would determine failure, whereas now we face a condition in which the structure may fail because the response at a different place of the structure—the bending moment or the shear at a different section or in a different member—reaches *its* critical value, and the response on which we had fixed our attention is still short of its own critical value. In a continuous beam, for example, the critical section is clearly a function of the end bending moments. Thus we meet a structure that behaves linearly up to failure but cannot be designed rigorously on the premises that apply to linear structures. The situation is due entirely to the transient and irregular nature of earthquakes, since it does not arise in a structure under steady-state harmonic excitation or in one that vibrates in a single natural mode. In both these cases the critical section is deterministically fixed.

We can visualize the problem better by considering discrete structures as examples. Suppose that a system having linear behavior will suffer collapse if the numerical value of at least one of the responses $q_i(t)$, $i = 1, 2, \ldots, n$, exceeds the corresponding critical value, say Q'_i at least once. Stated in this manner the problem is linear and differs from the ones we have been considering in that now there are many possible modes of failure; in other words, we have a more complicated boundary condition. However, if we form the variable $q(t) = \min_i [Q'_i - |q_i(t)|]$, which is a nonlinear function of the disturbance, failure will occur when $q(t)$ becomes negative. Hence, we show the nonlinear nature of the problem.

Let P_i denote the probability that Q'_i be exceeded at least once independently of the responses in the other $n - 1$ generalized coordinates. If we assume independence of the n events "Q'_i be exceeded," the probability of collapse will be

$$P = 1 - \prod_{i=1}^{n} (1 - P_i) \max{}_i P_i \qquad (11.1)$$

The criteria we have given for evaluating design responses are implicitly associated with a constant probability of failure. Hence, they lead to unsafe design of multidegree systems of this kind if the designs are deemed adequate for single-degree systems.

To illustrate this matter, consider a system having 10 modes of failure, all with the same probability of failure and extreme type 2 distribution with $k = 2$. If we design for a probability of failure of 10^{-4}, application of Eq. 11.1 tells that the design responses should be 11 percent greater than we would derive from individual consideration of each mode of failure. If $P = 10^{-2}$, the required percentage increase in design responses is 22 percent.

The assumption that all P_i's are equal overestimates the required increase in design responses. For example, with $P = 10^{-2}$ if the ratios of actual to required design responses are spaced uniformly between 1.00 and 1.09, the percentage increase for the response having the smallest of these ratios is no longer 22 but 10 percent.

Bounds and simple approximations to the probability of failure of statically loaded plastic structures have been developed (Cornell, 1967; Ang and Amin, 1968; Jorgenson and Goldberg, 1969). Equation 11.1 is based on assuming the independence of the various modes of failure in what concerns the responses as well as the critical values of responses. It often happens that, owing to strong correlations between the modes of failure, P can be approximated by $\max_i P_i$. However, when a plastic system is loaded dynamically, the correlations among the different modes of failure may operate in a very unfavorable manner, making Eq. 11.1 most unsafe, as we shall see when we treat structures made of nonlinear materials.

A problem similar to the one we have considered exists when the criterion of failure is conditioned by the principal stresses at a given location and the principal directions differ in the various natural modes of vibration. This situation arises in structures such as arch dams when analysis concerns crack formation. The criterion of failure may be as simple as that of maximum tensile stress, but the direction of this stress will be a function of time, although the location of the maximum tension may not change greatly with time or not change at all. Analytical intractability of the problem is evident from the fact that principal stresses are nonlinear functions of the normal and shearing stresses parallel to a set of orthogonal axes, and these stresses in turn are correlated random variables. Monte Carlo studies are indicated to cover problems of this nature.

Whether the location, the direction, or both may change, either discretely or continuously, we obtain an unsafe bound to the design responses by fixing our attention on a number of directions and locations of the potentially critical stress and analyzing individually for these, which we can do by applying the methods of modal analysis expounded in Chapter 10. When we can establish a ratio of the most unfavorable response to the one at a fixed position and fixed orientation, we can immediately establish a safe bound for the design responses. If the upper and lower bounds are sufficiently close to each other, we omit further analysis.

All the foregoing remarks apply to design against earthquakes of the nearly white noise type and against ground motions that result from filtering of these seisms. For design against earthquakes that consist of a single shock or are the result of filtering these through soft soil, a deterministic treatment suffices.

11.3 Single-Degree Systems with Symmetric Nonlinear Force-Deformation Relation

11.3.1 General Approximate Method of Analysis. Studies abound in which the responses of special types of single-degree systems to actual or simulated earthquakes have been computed and in some cases have been analyzed. Elastoplastic systems stand out as an example. Such studies cannot be generalized easily to systems having other characteristics. A more general method is desirable even if it is not very precise.

Here we shall present a method of analysis applicable to all single-degree systems with force-deformation curves symmetric about the origin, subject only to the condition that the system does not deteriorate. All elastic systems having a symmetric force-deformation relation fall within this limitation, whether linear or nonlinear, if their damping is any univocal function of the rate of deformation; so do all hysteretic systems. For the method to be applicable, we must have computed first the linear-system spectra of the motion of interest.

The assumption that a system does not deteriorate is often debatable. It would be desirable to possess a general method capable of dealing with structures that deteriorate. Apparently, further study is needed before this becomes possible. Therefore, we shall be concerned with earthquake types 2 and 3.

Consider a structure of the kind in question, defined by its *skeleton curve*—the static force-deformation curve on first loading, by the shape of its static force-deformation cycles at all pertinent levels of maximum deformation, and by the relation between force and rate of deformation. Figure 11.1 illustrates the static part of these relations.

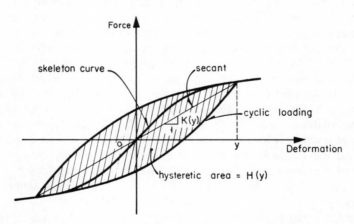

Figure 11.1. Static force-deformation curves of nondeteriorating structure.

We may join each point on the skeleton curve with the origin by means of a straight line. Every point that we choose on the skeleton curve defines an equivalent linear system whose mass is the mass of the original structure. Its stiffness is the slope of the straight line through the origin, that is, the corresponding secant stiffness of the actual structure, and its percentage of damping is

$$\zeta'(y) = \frac{H_\zeta(y) + H(y)}{2\pi K(y) y^2} \qquad (11.2)$$

where $H_\zeta(y)$ denotes the energy dissipated in damping per cycle under steady-state harmonic oscillations of amplitude y, $H(y)$ is the energy dissipated in hysteresis per cycle of amplitude y, and $K(y)$ is the secant stiffness at this same deformation.

When damping is viscous, Eq. 11.2 can be put in the form

$$\zeta'(y) = \zeta(y) + \frac{H(y)}{2\pi K(y)y^2} \tag{11.3}$$

where $\zeta(y)$ denotes the percentage of critical damping of viscous nature corresponding to the mass of the system and to a stiffness $K(y)$.

In the foregoing formulation we have assumed that the energy dissipation due to hysteresis and the energy dissipation that results from the rate of deformation are independent of each other. This does not limit the present treatment. When the assumption does not hold, it is sufficient to incorporate the total energy dissipation per cycle into a single term that replaces $H_\zeta + H$.

Now we postulate that the response of the actual structure is equal to a weighted average of the responses of all equivalent linear systems defined in the foregoing manner. Taking maximum numerical values of deformation for responses, we write

$$D = \frac{1}{D} \int_0^D \alpha(y) D'(y)\, dy \tag{11.4}$$

where D is the maximum numerical value of the deformation of the structure as response to the ground motion in question; $D'(y)$ is D for an equivalent linear system having the same mass as the structure considered, a stiffness $K(y)$, and a percentage of critical damping $\zeta'(y)$; $\alpha(y)$ is a positive weighting factor such that

$$\int_0^D \alpha(y)\, dy = D$$

We arrive at slightly different answers if we choose responses other than deformation for this purpose and treat them as we did D and D' in Eq. 11.4. The pseudovelocity and the square root of the maximum strain energy lend themselves especially for the purpose when the range of equivalent natural periods is associated with a roughly constant pseudovelocity spectrum. Absolute acceleration is especially well suited in the range of very short natural periods.

In Eq. 11.4, D is unknown. Because it appears in the second member we must begin by assuming it and proceed by trial and error. Ordinarily this may be replaced with a converging iterative process.

When the skeleton curve has a descending branch, more than one answer may be obtained for D through this procedure. The correct answer is the smallest D furnished by Eq. 11.4. For example, if the skeleton curve has one or more maximums we begin by assigning D tentative values between zero and the first maximum. If the procedure furnishes an answer, this is the correct value of D.

If it does not, we try values of D between the first and second maximum of the skeleton curve, and so on.[1]

We shall assume that Eq. 11.4 is valid with $\alpha(y) = 1$. We recognize that the weighting factor should be a decreasing function of y when the initial natural period of the structure is relatively short, while it should vary in the opposite manner for relatively flexible structures.

We know that the prevailing periods of an earthquake increase monotonically from beginning to end of the ground motion. This fact tends to invalidate Eq. 11.4, making it err on the unsafe side, even with α's different from 1, in the case of systems whose equivalent periods also increase monotonically with the amplitude of the response. The effect is not too serious for most structures of practical interest, as we shall verify in subsequent examples.

As a further simplification we postulate that the nonlinear response is approximately equal to that of an equivalent linear system having the same mass as the given structure and a period and damping equal to the mean period and mean damping of the equivalent systems defined previously. In this version of the method we begin by assuming a maximum deformation. Compute the equivalent undamped period

$$T' = \frac{2\pi M^{1/2}}{D} \int_0^D K^{-1/2}(y)\,dy \qquad (11.5)$$

where M is the given mass. Compute the equivalent percentage of damping,

$$\zeta' = \frac{1}{2\pi D} \int_0^D \frac{H_\zeta(y) + H(y)}{K(y)y^2}\,dy \qquad (11.6)$$

and take D equal to the maximum deformation of a linear system having this natural period and percentage of damping.

The simplified approach furnishes information on the probability distribution of D, not only on its expectation. This version of the method is relatively accurate when the expected linear spectra are smooth. Existence of a pronouncedly prevailing period within or near the range of the natural periods of the equivalent linear systems may influence the nonlinear responses to a significant extent; Eqs. 11.5 and 11.6 are then more likely to be appreciably in error than Eq. 11.4.

Veletsos (1969a) has shown that the responses of nonlinear systems to a "half-cycle displacement input" provide a qualitative description of their responses to earthquakes. The accelerogram of this input has three symmetrical triangular portions chosen so as to give zero end velocity and displacement. The first and third portions have a sign opposite that of the second triangle. This scheme gives linear and nonlinear response spectra that resemble those of

---

[1] This deterministic treatment ignores the probability that some earthquakes belonging to the family in question produce deformations exceeding the value associated with the next maximum. To take this probability into account we may proceed first in the manner described, take the computed D as a first approximation to $E(D)$, assign $D/E(D)$ the probability distribution corresponding to a linear system, and repeat the procedure for a number of probabilities of exceedance.

actual earthquakes of the second type, although the ratio ad/v^2 is much smaller than is usual for such earthquakes.

After establishing approximate upper bounds on the responses of nonlinear systems, we shall illustrate the present method in applications to certain elastic and hysteretic systems. In all the examples we shall use the simplified version of this general method.

11.3.2 Approximate Upper Bounds. Responses of nondeteriorating, single-degree systems having symmetric force-deformation relations can be bounded approximately from above on the basis of the response spectra for linear systems. We begin by establishing an equivalent percentage of damping for the nonlinear system. This can be done by evaluating Eq. 11.6 in an approximate manner. A cruder value is obtained by estimating D and evaluating the equivalent percentage of damping for a linear system whose mass equals the given mass and whose stiffness is equal to the secant stiffness of the nonlinear system at a deformation of $D/2$. For softening systems that have damping of the viscous type it is conservative to take the equivalent damping equal to that in a linear system whose stiffness equals the initial tangent stiffness of the given system. Three approximate upper bounds result from taking the maximum deformation of the nonlinear system equal to each of the following quantities, all of them evaluated for the linear spectrum associated with the equivalent percentage of damping found as in the foregoing paragraph.

1. The maximum deformation in the linear spectrum.
2. The deformation required to give, under the skeleton curve of the nonlinear system, an area equal to $\frac{1}{2}M$ max V^2, where M is the mass of the nonlinear system and max V is the maximum spectral velocity or pseudovelocity of all linear systems.
3. The deformation required to give, in the skeleton curve, an acceleration equal to the maximum spectral acceleration.

Suppose that the skeleton curve is as shown in Fig. 11.2, that the mass of the system is 300 lb in.$^{-1}$ sec^2, that we have estimated an equivalent viscous damping of 10 percent critical, and that we want to find an upper bound to the response of this system to the NS component of the 1940 El Centro earthquake. We use the spectral curve for 10 percent damping in Fig. 9.4. We find that the maximum deformation in this spectrum is 11.3 in., that the maximum pseudovelocity is 26 in./sec, and that the maximum spectral acceleration is 295 in./sec^2. We conclude that $D \leq 11.3$ in. Since a deformation of 3.7 in. gives an area under the curve in Fig. 11.2 equal to $(\frac{1}{2})$ 300 $\times$ 26^2 = 101,400 in. lb, $D \leq 3.7$ in. And since a deformation of 6.8 in. gives a force of 88,500 lb and hence an acceleration of 88,500/300 = 295 in./sec^2, $D \leq 6.8$ in. Consequently, $D \leq 3.7$ in.

The upper bounds obtained in this manner often err greatly on the safe side and occasionally may err slightly on the unsafe side. The latter situation is likely

328 RESPONSES OF NONLINEAR SYSTEMS Chap. 11

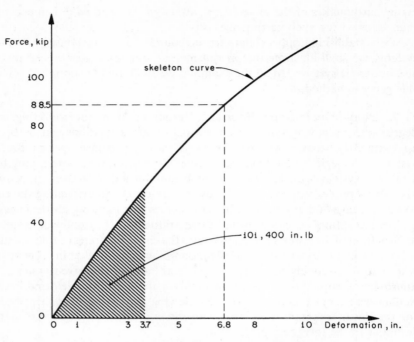

Figure 11.2. Example of skeleton curve of nonlinear system.

to arise in practically undamped, nonlinear elastic systems whose secant stiffness decreases rapidly with amplitude.

Justification for the approximate upper bounds described is found in Eq. 11.4 after we write two other expressions similar to it—one for the square root of the strain energy and one for absolute acceleration. The first member is not greater than the result of replacing D' with max D in the second member, and the same is true for the two other similar expressions.

In some of the following examples, we shall evaluate the adequacy of this upper bound as well as that of the simplified method described above.

11.3.3 Elastic Systems. We shall consider two examples of undamped nonlinear elastic systems. The first type corresponds to hardening systems whose force-deformation curve is a third-degree polynomial:

$$q = (1 + k^2 y^2) K_0 y \tag{11.7}$$

Here q is the nonlinear spring force, K_0 the stiffness for small deformations, y the deformation, and k a constant (Fig. 11.3).

This type of behavior is usually associated in practice with geometric nonlinearity. It describes quite accurately the force-deformation relation of a tense cable the distance between whose supports does not change when the system is subjected to a transverse force at midspan. The relation bears characteristics of

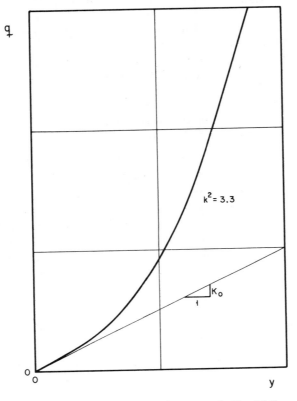

Figure 11.3. Force-deformation curve in Eq. 11.7.

the behavior of suspension bridges. Acceleration and velocity spectra of such systems are available (Murphy and Bycroft, 1956).

First we shall calibrate the general method against these results. As with all other damped elastic systems, Eq. 11.6 gives $\zeta' = 0$. In the present case, Eq. 11.7 gives

$$K = \frac{q}{y}$$
$$= (1 + k^2 y^2) K_0$$

The natural period for small oscillations is

$$T_0 = 2\pi \sqrt{\frac{M}{K_0}} \qquad (11.8)$$

It follows from Eq. 11.5 that

$$\frac{T}{T_0} = \frac{1}{D} \int_0^D (1 + k^2 y^2)^{-1/2}\, dy$$
$$= \frac{1}{kD} \ln\left(kD + \sqrt{1 + k^2 D^2}\right) \qquad (11.9)$$

330 RESPONSES OF NONLINEAR SYSTEMS Chap. 11

For purposes of comparison we choose $k^2 = 3.3$, which is the largest value for which the nonlinear responses are reported. Smoothing the linear spectrum and applying Eq. 11.9 we obtain the curves in Fig. 11.4. We conclude that the agreement is satisfactory. The figure also shows the upper bounds obtained in the manner we have described. In some ranges of initial natural periods the bound is much on the conservative side.

The second group of elastic systems that we choose to consider are bilinear, with zero stiffness in the second branch:

$$q = K_0 y \quad \text{when} \quad |y| \leq y_y$$
$$ = K_0 y_y \quad \text{when} \quad |y| \geq y_y$$

Here y_y is the yield-point deformation (Fig. 11.5).

We meet conditions that produce quite closely this type of behavior in certain prestressed-concrete frames failing in tension due to flexure. In these structures the main departure from linear behavior stems from the opening of flexural cracks that close again on unloading. Actual frames of this kind have thin

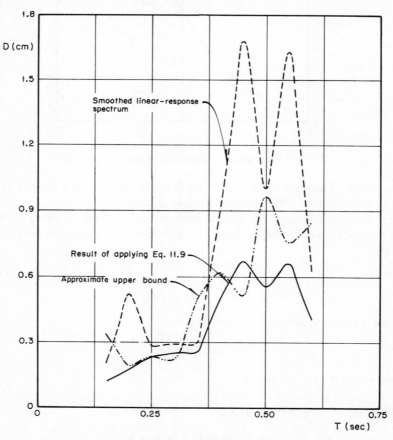

Figure 11.4. Comparison of spectra for elastic nonlinear systems.

Sec. 11.3   SINGLE-DEGREE SYSTEMS   331

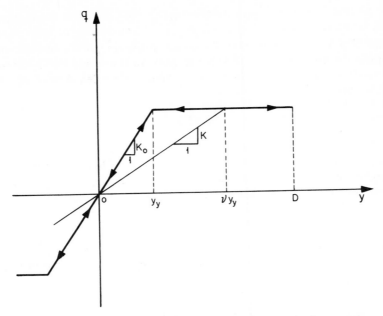

Figure 11.5. Force-deformation curve of bilinear elastic system
with zero stiffness in the second branch.

hysteretic loops (see Spencer, 1968 and 1969, for example), but the study of the present, highly idealized behavior sheds light on that of real structures.

When $|y| \leq y_y$, the natural period is again given by Eq. 11.8. For larger deformations,

$$\frac{K}{K_0} = \frac{1}{\nu}$$

where $\nu = y/y_y$. Letting $\mu = D/y_y$, where μ is known as the *ductility factor*, Eq. 11.5 becomes

$$\frac{T'}{T_0} = \frac{1}{\mu}\left(1 + \int_1^\mu \nu^{1/2}\, d\nu\right)$$
$$= \frac{1}{3\mu}(1 + 2\mu^{3/2}) \qquad (11.10)$$

This expression suffices to estimate the responses of the nonlinear systems in question, when defined by their ductility factor, from the spectrum of undamped linear systems. We can also use it to get a general picture of these nonlinear spectra for earthquakes of the second type.

Let T_{av} mark the period at which the straight lines representing the maximum acceleration of the ground, a, and velocity v intersect in a logarithmic plot of pseudovelocity spectra, and let T_{vd} mark the period at which v and d intersect, where d is the maximum displacement of the ground:

$$T_{av} = \frac{2\pi v}{a} \qquad T_{vd} = \frac{2\pi d}{v}$$

When $T \ll T_{av}$, we know that the linear spectrum gives a nearly constant acceleration approaching a as T tends to zero. In this range the linear spectral deformation is roughly proportional to T^2. Hence, if we denote this deformation by D_0, we may write

$$\frac{D}{D_0} \cong \left(\frac{T'}{T_0}\right)^2 \qquad (11.11)$$

where T'/T_0 is given by Eq. 11.10.

When $T_{av} < T < T_{vd}$, the spectral pseudovelocity of linear systems is roughly independent of natural period. Therefore, if T_0 lies between T_{av} and T_{vd} but $T_0\sqrt{\mu}$ is not much greater than T_{vd},

$$\frac{D}{D_0} \cong \frac{T'}{T_0} \qquad (11.12)$$

When $T_{vd} \ll T$, the spectral deformation of linear systems is nearly independent of period and approaches d as T tends to infinity. Hence, when $T_{vd} \leq T_0$,

$$\frac{D}{D_0} \cong 1 \qquad (11.13)$$

If μ is relatively large and T_0 lies between T_{av} and T_{vd} but not much closer to T_{av} than to T_{vd}, the nonlinear spectral deformation will be appreciably influenced by the responses of equivalent linear systems in the range $T > T_{vd}$. Hence D/D_0 will lie between the values given by Eqs. 11.12 and 11.13. Similarly, if T_0 is slightly smaller than T_{av}, D/D_0 will lie between the values in Eqs. 11.11 and 11.12.

For the case $T_{av} \leq T_0 \ll T_{vd}$, we could have chosen pseudovelocities rather than deformations for averaging the responses of the equivalent linear systems. Since the expected maximum strain energy is roughly independent of T in the range $T_{av} < T < T_{vd}$, we would have obtained for D that value which makes the area under the skeleton curve equal to the area under the force-deformation line of a linear system. Following this criterion we obtain

$$\frac{D}{D_0} \cong \frac{\mu}{\sqrt{2\mu - 1}} \qquad (11.14)$$

which differs less than 7 percent from Eq. 11.12 over the entire range of possible ductility factors.

For $T_0 \ll T_{av}$ we could have taken the spectral acceleration instead of the deformation. We would have arrived at the conclusion that, for any value of μ, the spectral acceleration equals that of a linear system having a natural period T_0. Hence,

$$\frac{D}{D_0} \cong \mu \qquad (11.15)$$

This result differs appreciably from Eq. 11.11 for large ductility factors. According to Eq. 11.11, as μ tends to infinity, D/D_0 tends asymptotically to $4\mu/9$, while Eq. 11.15 states that the ratio of spectral deformations remains equal to μ.

An approximate upper limit of D is the smallest of the following deformations,

Sec. 11.3   SINGLE-DEGREE SYSTEMS   333

$$\max D_0(T) \qquad \frac{T_0 \mu}{2\pi\sqrt{2\mu-1}} \max V_0(T) \qquad \left(\frac{T_0}{2\pi}\right)^2 \mu \max A_0(T)$$

where subscript 0 identifies the undamped linear spectra.

For purposes of comparison we shall take the responses of structures of this type to the initial 6.29 sec of the record of the 1940 El Centro earthquake NS component.[2] For this segment, $T_{av} = 0.70$ sec and $T_{ad} = 3.84$ sec. Consequently, when $T_0 = 2.86$ sec, we expect Eq. 11.13 to apply for large μ, erring slightly on the unsafe side for small μ. Results are shown in Fig. 11.6a. The agreement is excellent.

When $T_0 = 0.57$ sec, D/D_0 should approximate the results of Eq. 11.12 or

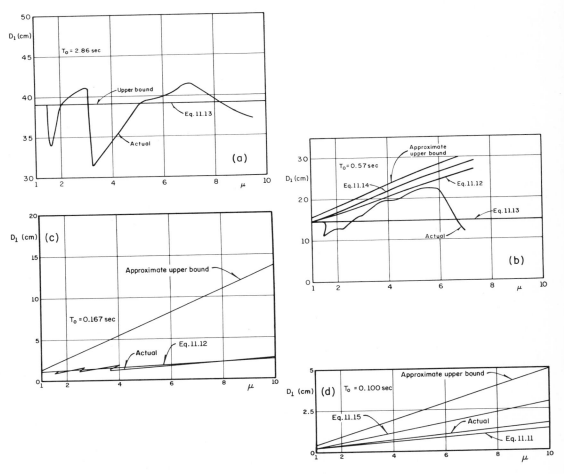

Figure 11.6. Response spectra of bilinear systems.
Actual spectra after Bielak (1966).

[2] The computed responses are due to Bielak (1966).

11.14 for small μ and tend to those of Eq. 11.13 for large ductility factors. This is roughly the case in Fig. 11.16b. In the intermediate to short-period range, results are available for $T_0 = 0.167$ sec. Comparison with Eq. 11.12 again shows a satisfactory agreement in Fig. 11.6c.

Finally, for the relatively short natural period $T_0 = 0.10$ sec, results lie between those of Eqs. 11.11 and 11.15 but are considerably closer to the former (Fig. 11.6d). The figures also depict the approximate upper bounds. In general we find a substantial overestimation of the nonlinear system's deformations, especially for very long and for very short initial, natural periods.

The approximate upper limit in Fig. 11.6a is exceeded by structural responses in some ranges of the ductility factor. This is due to the near synchronous lengthening of the prevailing periods of the earthquake and an increase in the secant flexibility of the structure.

The same reasoning explains the computed behavior of elastic bilinear,

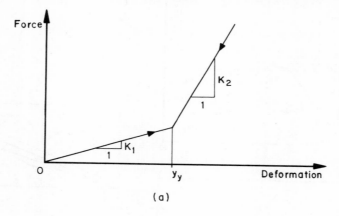

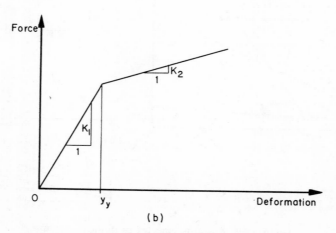

Figure 11.7. Hardening and softening bilinear systems.

single-degree systems subjected to actual and to simulated earthquakes. Let K_1 and K_2 denote stiffnesses associated with small and large deformations, respectively. $K_1 < K_2$ defines a hardening system (Fig. 11.7a), and $K_1 > K_2$ defines a softening one (Fig. 11.7b). Let y_y denote the deformation associated with the break in the force-deformation curve. As y_y tends to the spectral deformation D_1, corresponding to stiffness K_1, the bilinear system's maximum deformation, D also tends to D_1. As y_y tends to zero, D tends to D_2, the spectral deformation corresponding to K_2. In the intermediate range $(0 < y_y < D_1)$, it is found that, on the average, when the system is subjected to records of actual earthquakes, D is appreciably smaller than min (D_1, D_2) in hardening systems, while it appreciably exceeds max (D_1, D_2) in softening systems (Yeh and Yao, 1969; Yao, 1969). Yet, when the disturbance is a simulated earthquake whose relative frequency content is not time dependent (transient shot noise), D varies gradually between D_1 and D_2 as y_y goes from D_2 to zero (Yeh, 1969; Yao, 1969).

11.3.4 Elastoplastic Systems. Let us turn our attention to the classical, hysteretic, elastoplastic system. The shape of the force-deformation curve is depicted in Fig. 11.8.

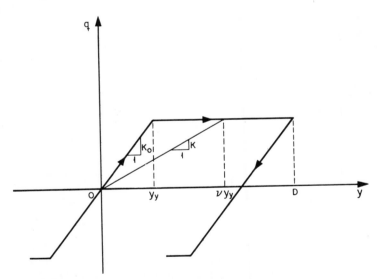

Figure 11.8. Force-deformation curve of elastoplastic system.

In order to apply the simplified method of analysis, we begin by computing the equivalent period from Eq. 11.5. The result is expressed by Eq. 11.10 because the skeleton curve is the same as for the bilinear elastic systems we have just analyzed. If there is no energy dissipation dependent on strain rate, Eq. 11.6 gives the equivalent percentage of critical damping as

$$\zeta' = \frac{1}{2\pi D} \int_{y_y}^{D} \frac{4(y-y_y)K_0 y_y}{(K_0 y_y/y)y^2} dy$$
$$= \frac{2}{\pi\mu} \int_1^{\mu} \left(1 - \frac{1}{v}\right) dv$$
$$= \frac{2(\mu - 1 - \ln \mu)}{\pi\mu} \tag{11.16}$$

This expression gives an equivalent percentage of damping that increases monotonically with the ductility factor and tends to 0.637 as μ tends to infinity.

To calibrate the method proposed we shall compare it first with the responses to the initial 6.29 sec of the El Centro 1940 record we used previously. We must begin by estimating the equivalent white noise duration so as to be in a position to evaluate the effects of damping. From the averaged viscously damped linear-system spectra reported by Bielak (1966), excluding the responses of very rigid systems, we construct Fig. 11.9. We compare the correlation factor that accounts for damping in linear systems with the curve for the reduction in expected response, as presented in Fig. 9.3. The adjustment shows that the equivalent white noise duration 6.29 sec gives satisfactory results.

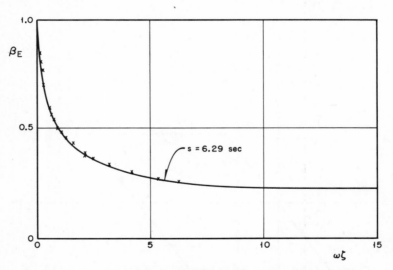

Figure 11.9. Correction factor due to damping for initial 6.29 sec of El Centro record compared with factor for equivalent segment of white noise having a duration of 6.29 sec.

Using this curve and the theory in Chapter 9 to compute correction coefficients, together with the equivalent damping ratio from Eq. 11.16, we obtain the predicted elastoplastic responses in Fig. 11.10. These were arrived at by reducing the predicted bilinear responses in Fig. 11.6. They are compared in the figure with the actual elastoplastic maximum deformations, computed from the data obtained by Bielak (1966). All the deformations are referred to the

Sec. 11.3 SINGLE-DEGREE SYSTEMS 337

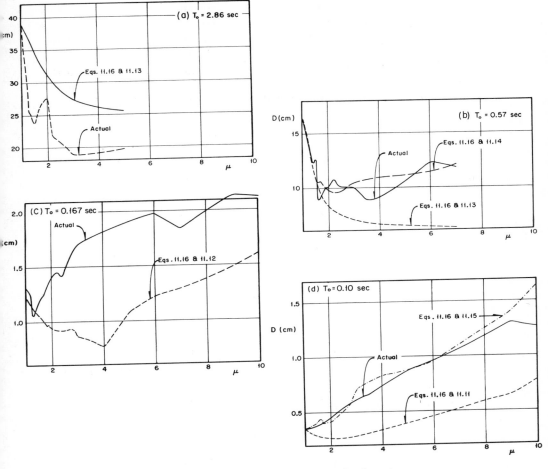

Figure 11.10. Response spectra of elastoplastic systems.
Actual spectra after Bielak (1966).

undamped, linear spectral value associated with the natural period for small oscillations. These natural periods are the same as the ones we chose for the bilinear elastic systems. Agreement between predicted and actual elastoplastic responses is good.

For long initial natural periods the effects of damping are small. Hence in this range there is little difference between the predicted responses of bilinear and elastoplastic systems having the same initial natural period. The difference should actually be smaller in accordance with the remarks made in connection with the weighting factors in Eq. 11.4. Indeed, very flexible structures will reach their maximum response at the end of a monotonic increase of deformation as a function of time. For these structures there should be no equivalent damping due to hysteresis, as only their skeleton curve is significant.

338 RESPONSES OF NONLINEAR SYSTEMS Chap. 11

Next we take the complete record of the same component of the El Centro 1940 earthquake, again with no actual viscous damping. Proceeding as we did in the foregoing example, we find an equivalent white-noise duration of 15 sec. We use available linear-system spectra from Veletsos and Newmark (1960), smooth them to dispose of pronounced peaks and valleys, and supplement this information with calculations carried out in accordance with Chapter 9. Comparison between the computed and actual, smoothed, elastoplastic spectra is shown in Fig. 11.11.

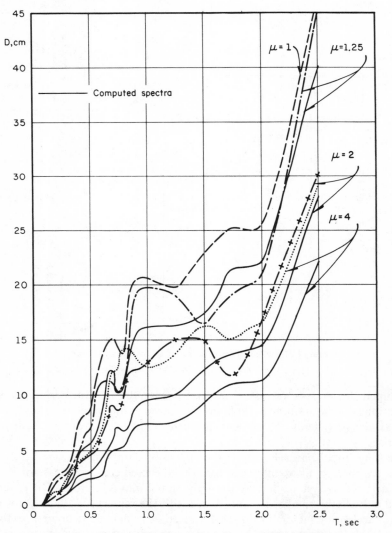

Figure 11.11. Computed and actual undamped elastoplastic spectra. Complete El Centro 1940 NS record. *Actual spectra after Veletsos and Newmark (1960).*

Sec. 11.3 SINGLE-DEGREE SYSTEMS **339**

Finally, let us apply the same simplified method to the spectra of elastoplastic systems having a dashpot element in parallel with the nonlinear spring element. According to Eq. 11.16 we must add, to the equivalent damping in Eq. 11.16, the average percentage of critical damping due to strain rate effects. At any level of deformation this damping is inversely proportional to the square root of the secant stiffness

$$\zeta' = \frac{\zeta_0}{\mu}\left(1 + \int_1^\mu \sqrt{v}\, dv\right)$$
$$= \frac{\zeta_0}{3\mu}(1 + 2\mu^{3/2}) \qquad (11.17)$$

where ζ_0 is the percentage of damping in the linear range.

Using the sum of equivalent degrees of damping in Eqs. 11.16 and 11.17 and proceeding as we did when $\zeta_0 = 0$, we obtain the curves in Fig. 11.12, corresponding to $\zeta_0 = 0.1$. Agreement in Figs. 11.11 and 11.12 between predicted and actual elastoplastic spectra is fair, although the present criterion is rather conservative for undamped systems.

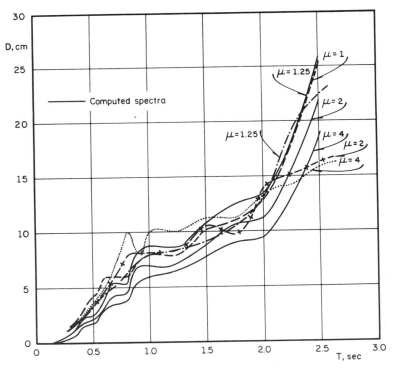

Figure 11.12. Computed and actual damped elastoplastic spectra. Complete El Centro 1940 NS record with 10 percent critical damping. *Actual spectra after Veletsos and Newmark (1960).*

340 RESPONSES OF NONLINEAR SYSTEMS   Chap. 11

For elastoplastic structures subjected to earthquakes of the second type, an even simpler approach is adequate, provided μ does not exceed a value of the order of 10 and provided T_0 is not too short. Given an equivalent white noise duration s and values of T_{av} and T_{vd}, there is a wide range of natural periods over which the reduction in response, due to the damping that is equivalent to hysteresis (Eq. 11.16), is slightly more pronounced than the increase due to the lengthening of eqivalent period relative to the natural period for the linear range (Eq. 11.10). When $s \gg T_{av}$, this range covers all natural periods longer than about T_{av}. If we take for the linear system equivalent to the given structure a system whose natural period and damping are equal to those of the elastoplastic system in its linear interval, we find that in this range of natural periods it is slightly conservative to take, for the elastoplastic structure, expected maximum deformations equal to those of this equivalent linear system. This contention is confirmed in Fig. 11.13, which compares the corresponding spectra. (Notice that the figure directly gives the maximum deformations; spectral accelerations are obtained through division by μ of those in the graph; spectral velocities do not admit a simple interpretation for elastoplastic systems.[3])

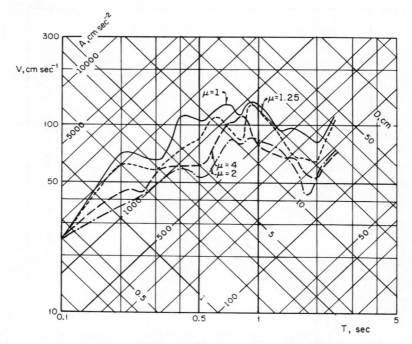

Figure 11.13. Comparison of undamped elastoplastic spectra.
Data after Veletsos and Newmark (1960).

---

[3] The corresponding curves by Veletsos and Newmark (1960) directly furnish the spectral accelerations; spectral deformations are obtained through multiplication by μ of the spectral elastic components of displacement given in that reference.

When the condition $s \gg T_{av}$ is not fulfilled, the range of natural periods over which the approximation quoted in the foregoing paragraph is conservative moves to longer values of T_0. Such is the case with responses to the first 6.29 sec of the El Centro record, as may be appreciated in Fig. 11.10.

When a moderate, actually viscous damping is superimposed on hysteretic energy dissipation, the agreement improves between the responses of elastoplastic systems and their linear equivalents having the same characteristics as the linear branch of the former. This is brought out clearly in Fig. 11.14. The reason

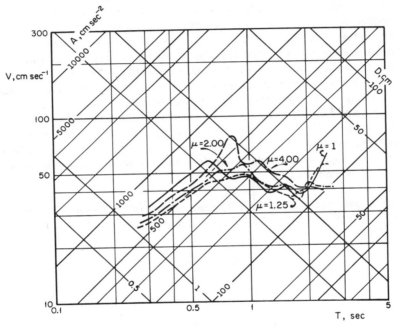

Figure 11.14. Comparison of damped elastoplastic spectra having 10 per cent critical damping. *Data after Veletsos and Newmark (1960).*

for the improvement is found in that the combined equivalent damping brings about a correction factor larger than the product of the individual viscous- and hysteretic-damping correction factors. The combined damping is less effective in reducing responses than the product of the individual effects, and this compensates for the conservativeness of the criterion expounded for systems deprived of viscous damping.

Studies abound in the literature which show the adequacy of this criterion (see Veletsos, Newmark, and Chelapati, 1965, for example). Nevertheless, it should be taken with care, as it errs on the unsafe side for a wide range of periods when the earthquake duration is not sufficiently long, and it always errs in this direction for very rigid structures.

Veletsos (1969) explains the behavior of conventional elastoplastic systems as follows. The main difference between these and linear systems having the same initial natural period and percentage of damping lies in the lower effective stiffnesses of the former structures. Second, they have hysteretic damping. In the range of very long, initial, natural periods the responses are insensitive to both period and damping. Hence, linear and nonlinear spectra practically coincide in this range.

As we focus on initial natural periods in the range T_{av} to T_{vd}, the lengthening of the effective natural period tends to produce larger responses of elastoplastic systems, but this is more than compensated by the effect of hysteretic damping. Consequently, elastoplastic responses are on the average, slightly smaller than the corresponding linear responses. In the case of certain earthquakes of short duration, such as those of type 1, inelastic responses may be much smaller than the corresponding linear responses in this range (Poceski, 1969).

Finally, for very short initial periods the effect of natural periods is decisive, and so the inelastic spectral ordinates tend to exceed those of linear systems.

Penzien and Liu (1969) have found that the distributions of responses of elastoplastic systems subjected to segments of stationary Gaussian disturbances can be approximated closely by extreme type 1 distributions. The coefficient of variation is appreciably higher than for linear systems having the same initial percentage of damping.

11.3.5 Rigid-Plastic Systems. The special case of elastoplastic systems defined by $K_0 = \infty$ (Fig. 11.15) deserves particular attention. The ductility factor is

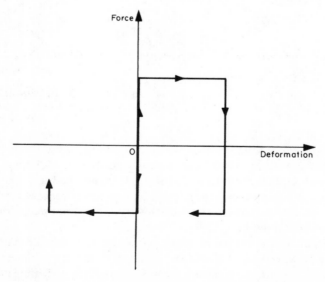

Figure 11.15. Force-deformation curve for rigid-plastic system.

either 1 or infinity depending on whether the maximum ground acceleration is smaller or greater than the value required to make the system yield.

For these structures we cannot use an equivalent linear system having the characteristics of the small amplitude range of the nonlinear structure. We shall employ the more general method we presented earlier, in its simplified version.

We can write the yield force as AM, where A is the acceleration required to make the structure yield, and M is the mass of the structure. Hence in an elastoplastic system the initial stiffness can be written as AM/y_y and the initial natural period as

$$T_0 = 2\pi\sqrt{\frac{y_y}{A}}$$

$$= 2\pi\sqrt{\frac{D}{A\mu}}$$

where D is the maximum deformation. Substituting T_0 in Eq. 11.10 and letting μ tend to infinity we get for the equivalent natural period of a rigid-plastic structure

$$T' = \frac{4\pi}{3}\sqrt{\frac{D}{A}} \qquad (11.18)$$

From Eq. 11.16, $\zeta' = 0.637$ for structures of this type.

When $A > a$ the structure does not deform under the earthquake's action. When $A \leq a$ and $T_{av} \leq T' < T_{vd}$, $V \cong v$ for $\zeta = 0.25$ (Chapter 7). Hence, for $\zeta' = 0.637$, $V \cong (0.25/0.637)^{0.4} v = 0.69v$. But $D = (T'/2\pi)V$. Consequently, $D \cong 0.11vT'$. Substituting in Eq. 11.18 and solving for T' we obtain $T' \cong 1.93v/A$. Hence,

$$D \cong 0.21\frac{v^2}{A}$$

On the other hand, by equating strain energies we get

$$D \cong \frac{0.69^2 v^2}{2A}$$

$$= 0.24\frac{v^2}{A}$$

When $A \leq a$, the maximum deformation can also be estimated in the following manner due to Newmark (1965a, b, and 1970). Consider the effect of a rectangular acceleration pulse of amplitude a and duration v/a (Fig. 11.16a). The ground velocity diagram is as shown by the solid line in Fig. 11.16b. From this we must subtract the deceleration effect due to A, shown by the dashed line. When the two lines meet, the structure has the same absolute velocity as the ground, and hence, it ceases to deform. The instant at which this happens is given by $t = v/A$. It follows that the maximum deformation, given by the shaded area in the figure is

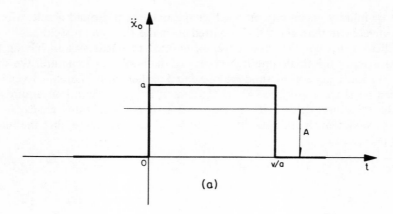

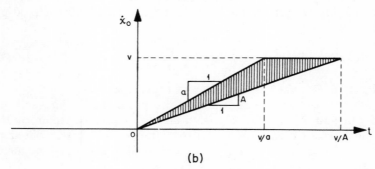

Figure 11.16. Rectangular pulse:
(a) Accelerogram
(b) Velocity diagram.

$$D = \frac{v}{2}\frac{v}{A} - \frac{v}{2}\frac{v}{a}$$
$$= \frac{v^2}{2A}\left(1 - \frac{A}{a}\right) \tag{11.19}$$

This expression may be used for estimating the maximum deformation by an earthquake having a maximum acceleration a and maximum velocity v. However, if the computed D exceeds the maximum ground displacement d, Eq. 11.19 should be replaced with $D = d$.

Figure 11.17 shows a comparison between the deformations obtained by this criterion and those produced by four earthquakes of the second type, after normalizing the records by adjusting their acceleration and time scales to give $a = 0.5g$ (g = gravity acceleration) and $v = 30$ in./sec. The criterion is seen to be slightly conservative for the estimate of the expected deformation. The figure also shows the relation $D = 0.24v^2/A$ which, over a wide range of the ratio A/a, provides a satisfactory estimate of $E(D)$.

Sec. 11.3   SINGLE-DEGREE SYSTEMS   **345**

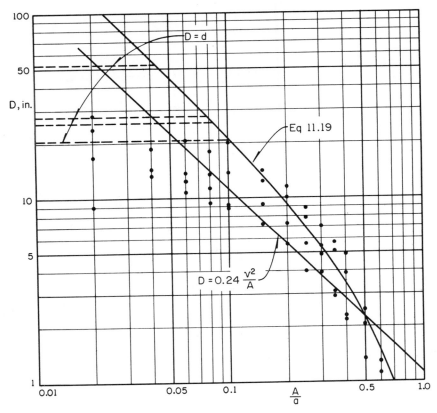

Figure 11.17. Comparison between computed and actual deformations of rigid-plastic systems. *Data after Newmark (1965a).*

11.3.6 Masing-Type Systems. In Chapter 5 we described the broad class of hysteretic systems of the Masing type. It will be recalled that if the skeleton curve of a system of this type is given by

$$q = q_0(y) \tag{11.20}$$

the unloading and reloading curves that begin at (q_1, y_1) follow the expression

$$\frac{|q - q_1|}{2} = q_0\left(\frac{|y - y_1|}{2}\right) \tag{11.21}$$

for every pair of values y_1, q_1, except that, where the curve defined by Eq. 11.13 intersects the force-deformation curve of an earlier cycle, it follows this curve.

A conceptual model of Masing-type structures is shown in Fig. 11.18a (Herrera, 1965; Lazan, 1968; see also Iwan, 1969). In the simple case shown in Fig. 11.18b, this model behaves as an elastoplastic system. Many classical types of force-deformation relations may be regarded as special cases of Masing behavior.

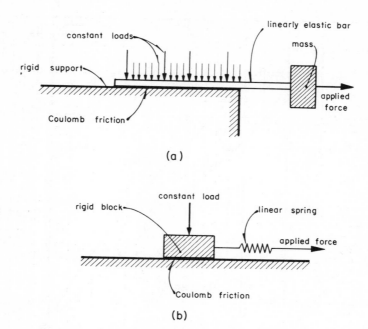

Figure 11.18. Conceptual models of Masing-type systems.

Consider a Masing structure whose skeleton curve is of the type proposed by Ramberg and Osgood (1943):

$$\frac{y}{y_1} = \frac{q}{q_1} + \alpha \left(\frac{q}{q_1}\right)^r \tag{11.22}$$

where q_1, y_1, α, and r are positive constants. The first two of these may be regarded as defining the force and deformation scales, while the other two are related to the degree and type of nonlinearity (Jennings, 1963). We agree that Eq. 11.21 defines the curve in the first quadrant and that the curve is symmetric about the origin. A wide range of force-deformation curves can be produced by suitably choosing the parameters, as shown in Fig. 11.19. In particular, $r = \infty$ defines elastoplastic systems.

Masing-type structures whose skeleton curve follows Eq. 11.22 have been analyzed under simulated earthquakes using Monte Carlo techniques (Jennings, 1963). Results have not been presented, however, in a manner that admits ready interpretation. Owing to the similarity between these structures and those of elastoplastic behavior, it seems certain that the conclusions at which we arrived for the latter will hold at least as well for Masing systems governed by Eq. 11.22 when we define the equivalent linear system as one having the properties to which the Masing-type structure tends as the amplitude of oscillation is allowed to tend to zero.

With this definition of equivalent linear system, we may advance the following conservative criterion applicable to all single-degree, Masing-type systems whose

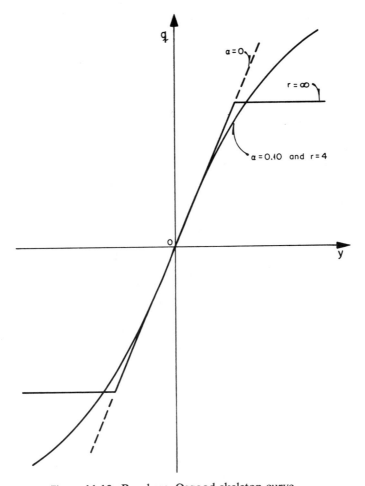

Figure 11.19. Ramberg–Osgood skeleton curve.

skeleton curve is intermediate between the one defined in Eq. 11.22 and an elastoplastic curve.

$$\text{If } T \ll T_{av}, \text{ then } A \cong A_0.$$
$$\text{If } T_{av} \leq T < T_{vd}, \text{ then } W \cong W_0.$$
$$\text{If } T_{vd} \leq T, \text{ then } D \cong D_0.$$

In these expressions, T is the natural period of the equivalent system; $T_{av} = 2\pi v/a$ and $T_{vd} = 2\pi d/a$ mark the abscissas at which the lines corresponding, respectively, to a and v and to v and d intersect in a logarithmic representation of spectra, where a, v, and d are the ground's maximum acceleration, velocity, and displacement; A, W, and D stand for expected maximum acceleration, strain energy, and deformation of the Masing-type system; subscript 0 refers to the equivalent linear system.

348 RESPONSES OF NONLINEAR SYSTEMS Chap. 11

There are indications that the criterion is adequate for design purposes in much more general cases, provided only that the skeleton curve is monotonic. Still, for some systems the initial tangent stiffness may be so high, leading to such a stiff equivalent system, that this criterion becomes inadequate. Take for example the force-deformation skeleton relation described in previous chapters,

$$q = ky^{1-\beta} \tag{11.23}$$

where k and β are constants, with $\beta \ll 1$ leading to a quasilinear system. Since $dq/dy = \infty$ at $y = 0$, the equivalent linear system whose properties coincide with the small amplitude characteristics of the given structure cannot be used. We must use the general method of analysis. When $\beta \ll 1$ and D is reasonably large, this gives us an equivalent natural period practically independent of D; in all cases it gives us an equivalent percentage of damping $\zeta' = \beta/\pi$, strictly independent of D.

11.3.7 Stiffness-Degrading Systems. Penzien and Liu (1969) have analyzed the responses of "stiffness-degrading" structures to segments of stationary Gaussian processes. The force-deformation relations of these systems are as shown in Fig. 11.20. On first loading they behave like elastoplastic systems, but on unloading or reloading, their initial stiffness is decreased as a function of the maximum deformation imposed in previous cycles.

The probability distributions of the maximum deformation of these structures

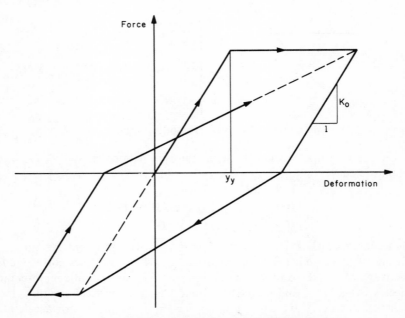

Figure 11.20. Stiffness degrading system. *After Penzien and Liu (1969).*

can be approximated closely by extreme type 1 distributions. The means and coefficients of variation are intermediate between those for the corresponding linear and elastoplastic systems.

11.3.8 A Common Type of Braced Structure.
Consider a single-degree system whose resistance to lateral forces is provided by an X-brace consisting of very slender bars of elastoplastic material (Fig. 11.21). If the bars take no ap-

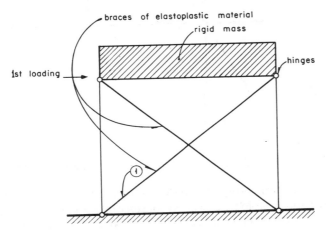

Figure 11.21. X-braced structure.

preciable compression, loading as shown in the figure makes bar 1 resist the entire force. Eventually this bar will yield, giving a skeleton curve similar to that of an elastoplastic structure (Fig. 11.22). On unloading, when the deformation becomes such that bar 1 takes no tension, horizontal displacement will meet no resistance until the system passes through its original configuration. From then on, loading in an opposite sense will generate a curve that is the image of the first loading curve, about the origin. In subsequent loading cycles the braces begin taking the load as soon as the deformation reaches the maximum value it has attained in that sense minus the elastic recovery. (Apparently this type of behavior was first analyzed by Tanabashi and Kaneta, 1962; see also Veletsos, 1969.)

A similar situation arises in structures such as bolt-anchored chimneys. A bolt that yields in tension takes no compression and indeed does not contribute to resist overturning until the base rotation again attains the maximum value of previous cycles, minus its elastic recovery, and does so with the same sign.

Applying the general method of analysis we find for these structures the same equivalent natural period as for elastoplastic systems and for bilinear elastic systems, both having the same skeleton curve (Eq. 11.10). The equivalent percentage of damping is equal to or smaller than half the value in Eq. 11.11.

For relatively long, equivalent natural periods, the structure will behave much as an elastoplastic one with half as much equivalent damping because essentially

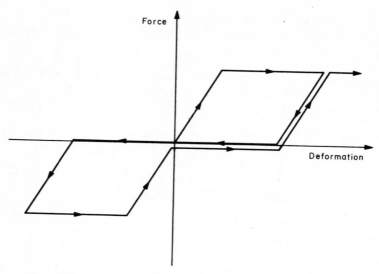

Figure 11.22. Force-deformation curve for structure in Figure 11.21.

both structures will undergo only one hysteretic cycle or one half a cycle. For short, equivalent natural periods, ζ' will be considerably smaller than the value in Eq. 11.11 because the number of cycles will be large, and hysteretic energy dissipation occurs only once for every amplitude of deformation in this type of structure. Hence its behavior will approximate that of a bilinear elastic system having the same skeleton curve. Independently of the initial natural period, the average spectral responses of structures of this type lie between those of conventional elastoplastic systems and those of bilinear systems, all with the same skeleton curve (see Veletsos, 1969).

The response spectra of bilinear elastic systems can be predicted satisfactorily over a wide range of natural periods by taking their maximum strain energy equal to that for an equivalent linear system that has the same characteristics as the linear range of the given structure. Therefore this criterion will also be adequate for the braced structures in question, provided their initial natural period is not so long as to make their spectral deformations approach the maximum displacement of the ground. Essentially this conclusion has been confirmed by measuring the maximum deformations of chimney anchor bolts (Rosenblueth, 1964b).

11.3.9 Effects of Gravity. Consider an elastoplastic inverted pendulum or one-story single bent subjected to both gravity loads and earthquake motion (Fig. 11.23). Neglecting the effect of gravity and assuming that the mass is entirely concentrated at the column tops and that all plastic hinges in the bent develop simultaneously, the force-deformation curve of the system is as shown by line OAB in Fig. 11.24.

Sec. 11.3   SINGLE-DEGREE SYSTEMS   351

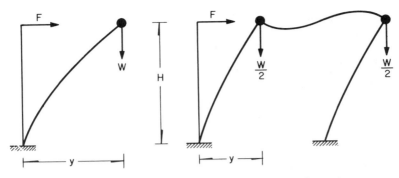

Figure 11.23. Inverted pendulum and one-story frame bent subjected to lateral and gravity loads.

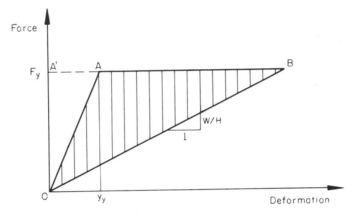

Figure 11.24. Force-deformation relation for elastoplastic structure with gravity effects.

Segments OA and AB may be idealized as straight lines if the maximum rotation of the system is sufficiently small that there is a negligible error involved in taking this angle equal to its sine. Husid (1967) has shown that, for the entire range of practical dimensions, the maximum angle that may be of interest is sufficiently small from this point of view. Under the same conditions the effects of vertical accelerations due to horizontal ground motion may be neglected.

Gravity may be taken into account by subtracting Wy/H (line OB in the figure) from the ordinates, F, of OAB, where W is the weight supported by the structure, y is its lateral deformation, and H is its height.

Whatever the value of the yield force F_y and the slope W/H of the line representing gravity effects, every elastoplastic structure subjected to a Gaussian disturbance of sufficiently long duration eventually collapses. Let t_0 denote the time to collapse. The expected value of t_0 is always finite.

Husid (1967 and 1969) has analyzed the responses of a large number of structures of the type illustrated in Fig. 11.23 under the action of simulated and real earthquakes of type 2. He finds that the average time to collapse can be approximated closely by an expression that can be put in the form

$$E(t_0) = \frac{CH}{(BW/F_y)^2} \tag{11.24}$$

where C is a constant and B is a quantity proportional to the root mean square of the ground acceleration.

The ratio F_y/W is the fraction of the weight at which the structure would yield statically, neglecting the effect of gravity. It may be regarded as a seismic design coefficient.

Equation 11.24 may be somewhat rationalized as follows. The structures analyzed by Husid have initial natural periods, neglecting gravity effects, of 0.5 to 2.0 sec (which cover the range of greatest practical interest) and a damping ratio of 0.02. In this range of natural periods the corresponding pseudovelocity spectra are, on the average, almost independent of natural period. Hence, the strain energy required for failure is also practically independent of natural period. It may be conjectured that an almost constant amount of energy is fed to an elastoplastic system in a given time. The energy fed per unit mass during the time interval t_0 is proportional to $B^2 t_0$. On the other hand, if we do not subtract the area of the triangle OAA' in Fig. 11.24, the maximum strain energy in the system per unit mass is proportional to $(F_y/W)^2$ and to H. Collapse is practically certain to occur soon after the structure attains a deformation equal to the abscissa of point B in the figure. Therefore, $B^2 t_0$ must be proportional to $(F_y/W)^2 H$. Hence, we arrive at Eq. 11.24.

The foregoing heuristic derivation rests on the assumption that the length $A'A$ is much smaller than $A'B$. The assumption is justified in most practical instances, for as shown by Husid (1967), $A'A$ is usually smaller than 1/30 and rarely as large as 1/10 of $A'B$, since the factor of safety against elastic buckling is usually very large. Still, when this assumption is not met, Eq. 11.24 may err seriously on the unsafe side. In the extreme case when point B approaches point A, the structure approaches a situation in which it buckles without an earthquake; that is, $E(t_0)$ tends to zero.

Husid has found that, if B is taken as 2.9 for the NS component of the 1940 El Centro earthquake and 2.1 for the Taft 1952 accelerogram, and H is expressed in feet, C turns out equal to 2000 in Eq. 11.24. The effect of the vertical component of ground motion on the expected time to collapse has been found to be negligible (Husid, 1967).

For bilinear, hysteretic, softening structures with a stiffness K_2 of the second straight segment in the force-deformation curve, $E(t_0)$ is an increasing function of K_2. When K_2 exceeds the slope W/H of the gravity-effect line in a force-deformation diagram (OB in Fig. 11.24), there can be no collapse. Husid (1967 and 1969) has found the following approximate, empirical relation,

$$E(t_0) = \frac{CH}{(BW/F_y)^2[1 - (K_2 H/W)^{0.8}]} \qquad (11.25)$$

The problem of computing the probability distribution of t_0 or of the maximum response of the present type of structure to stochastic disturbances has apparently not been solved. Nevertheless a roughly approximate solution has been obtained by Husid (1967).

Consider now a bent whose columns are hinged top and bottom and is braced by two diagonal bars as in Fig. 11.21, such as to give a lateral force-deformation of the type shown in Fig. 11.22. If the system is designed for the same seismic coefficient F_y/W, as the corresponding flexural frame, and has the same initial stiffness, its strain energy on first loading will also be the same as for the flexural frame. However, in successive loading cycles, it will dissipate less hysteretic energy. For very flexible structures (inital period greater than T_{vd} we may expect approximately the same values of $E(t_0)$ independently of the type of structure. However, collapse will, on the average, occur appreciably earlier in the braced rather than in the flexural frame if the initial period is shorter than approximately T_{av}, and only slightly earlier for intermediate values of the initial period.

It is also to be expected that braced flexural frames will behave in a fashion that is intermediate between those of unbraced, purely flexural, and hinged, braced frames.

11.3.10 Remarks on the Responses of Symmetric One-Degree Systems.

We have presented a general approximate method of analysis applicable to systems that do not deteriorate. The closest we have come to considering deterioration refers to stiffness-degrading systems and to the special type of braced structures in the last few paragraphs; for these, the method only allowed establishing limits between the responses of nondeteriorating systems.

The reserve energy technique developed by Blume (1960) attempts, among other things, to incorporate more general manners of failure than the ones defined by excessive deformation. In principle it can deal with deteriorating systems. However, there is apparently no quantitative way to measure the rate or extent of deterioration in terms of the factor that the technique introduces to account for this phenomenon. Indeed, the number of load cycles does not enter the formulation.

For certain Masing-type structures, including classical elastoplastic systems with small or moderate ductility factors, we found that the equivalent linear system could be chosen such that its stiffness coincided with the initial stiffness of the given structure, obviating the need for cycles of trial and error or iteration to improve the estimated response. This simpler approach is limited in scope, and the dynamic constants of the corresponding equivalent linear systems differ from those in the more general method of analysis. In fact, given a small family of earthquakes and the set of responses thereto, we could define equivalent linear systems in an infinite number of ways, so that their responses coincide approximately with those of the nonlinear structure in question.[4] For every equivalent stiffness we would find a different equivalent percentage of damping.

The simplified version of the method, which we applied systematically in the foregoing examples, uses one equivalent linear system to estimate the response of a nonlinear one. Adequacy of this procedure requires that the expected linear spectrum be relatively smooth. To find the responses to earthquakes of the third type with comparable accuracy, it is necessary to average the responses of linear systems over a range of natural periods and damping ratios, through use of Eq. 11.4.

11.4 Single-Degree Systems with Asymmetric Force-Deformation Curves

Thus far we have only considered systems whose force-deformation curve is symmetric about the origin. This need not be the case even for systems of linear behavior. But if behavior is linear, the only difference between the force-deformation relation in the first and third quadrants lies in the breaking strengths for positive and negative deformations. This difference is not especially significant when the system is subjected to earthquakes of the second or third types. This applies even to the extreme condition in which the strength for deformations of one sign is infinite; it is found that the responses associated with a given probability of failure are nearly the same whether we consider the expectation of $\max_t |q|$ or of $\max_t (q)$, the former exceeding the latter by a small fraction.

The situation is essentially the same for elastic, nonlinear systems but is quite different for inelastic ones. Yielding is cumulative in the latter. Raising the yield-point stress of one sign will often make the responses of the opposite sign increase, and the effect may be very pronounced.

To illustrate the effects of asymmetric yielding, consider undamped, rigid-plastic systems of this sort. Even a small amount of asymmetry elicits most of the effect associated with the maximum possible asymmetry—infinite yield force for deformation of one sign. Force-deformation curves then take the shape shown in Fig. 11.25. We shall confine our attention to this type of rigid-plastic system.[5]

---

[4] Choice of the stiffness of the equivalent linear system is restricted by the fact that its expected responses without damping exceed those of the actual structure, so that the equivalent percentage of damping turns out to be positive. This still leaves a wide range of possible equivalent stiffnesses. The restriction is not strictly fulfilled when we choose the stiffness of a linear system, equivalent to an elastoplastic one without viscous damping, equal to the latter's initial stiffness, since over a finite range of initial natural periods, the expected undamped responses are smaller than those of the elastoplastic structure (Veletsos, 1969).

[5] The present section is based partly on Esteva, Sánchez-Trejo, and Rosenblueth (1961) and on Newmark (1965a, b, and 1970).

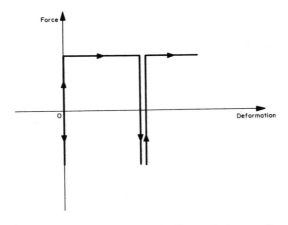

Figure 11.25. Rigid-plastic system yielding only in one direction.

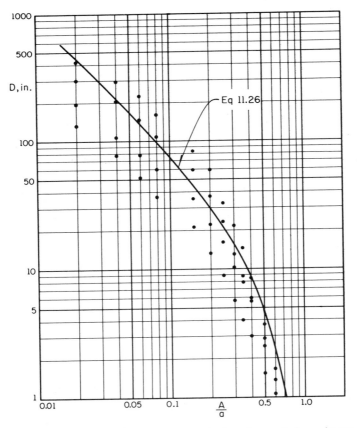

Figure 11.26. Comparison between actual and expected maximum deformation. *Data after Newmark (1965a)*.

The expected maximum deformation can be approximated by the effect of a single pulse, as given by Eq. 11.19, times the equivalent number of pulses in the earthquake under consideration. This number is an increasing function of a/A. By adjusting to the responses in Fig. 11.26, computed by Newmark (1965a, b, and 1970) to a group of earthquake records, normalized as for Fig. 11.17, it is found that the equivalent number of pulses may be taken proportional to $1 - A/a$, and the expression

$$D = \frac{2v^2}{A}\left(1 - \frac{A}{a}\right)^2 \tag{11.26}$$

furnishes a good approximation to the expected maximum deformation. However, the mean equivalent number of pulses corresponding to a given ratio A/a varies considerably from one earthquake to another. It seems likely that a more elaborate formula, involving the earthquake duration as well as its maximum acceleration and velocity, would give more satisfactory results.

The increase in expected response due to asymmetry in yielding force is apparent in Eq. 11.26. If we keep a constant as we let A tend to zero, D will tend to infinity. In a rigid-plastic system with symmetrical yield forces, on the other hand, D tends to the maximum displacement of the ground.

11.5 Multidegree Structures

The number of variables involved in the responses of nonlinear, multidegree structures to earthquakes is so high that it allows not more than qualitative statements to be advanced that are based on most of the practically unrelated calculations that have been made on the effects of actual (Penzien, 1960; Hisada, Nakagawa, and Izumi, 1965; Kabayashi, 1965) or simulated (Bycroft, 1960; Saul, Fleming, and Lee, 1965; Wen and Janssen, 1965) earthquakes of the second type. On the other hand, elementary considerations show that very general methods for the design of structures of this sort should not be established by extrapolating directly from conclusions that apply to the seismic responses of nonlinear, single-degree systems or from design criteria applicable to multidegree structures subjected to static loading.

To elaborate on the latter point, consider an elastoplastic, multidegree system that would give a certain ductility factor, equal in all the degrees of freedom of the structure, when subjected to a certain ground motion. Suppose the structure were redesigned by raising the generalized forces required to make it yield in all but one of its degrees of freedom, and that the strength in this degree of freedom were left unaltered. (The degrees of freedom in question could be the story deformations in a multistory building allowed to deform only in translation; all but one of its stories would be strengthened in redesign.) Strengthening can be made such that all inelastic, strain energy dissipation take place through deformation of the unstrengthened degree of freedom. This degree would act in a way as a fuse, the rest of the degrees of freedom remaining in the linear range. Deformation in the one degree of freedom in which strength was not raised would have to be greater, in some cases many times greater, than was required in the original design. This situation easily brings about severe failure.

A specific example was solved, albeit in a crude manner, to illustrate this (Rosenblueth, 1964b). The example concerns a 20-story building designed originally to resist a certain earthquake, developing a ductility factor of 5 in all stories. The building is redesigned hypothetically so that no story leaves the elastic range except the one between the first and second floors. Under the assumptions that this story is twice as rigid as the rest and that the strain energy must be preserved, it is found that the first story must develop a ductility factor of 21 and withstand extremely large deformations. Such demands cannot be met ordinarily, so the situation would imply collapse.

Qualitatively the same undesirable situation arises in any inelastic—not necessarily elastoplastic—structure that is weaker than required in several degrees of freedom and/or stronger than necessary in the rest. This may be the result of partial underdesign and/or partial overdesign.

We see that one classical proposition applicable to plastic structures under static loading (partial strengthening never decreases structural capacity) does not apply to inelastic structures under dynamic loading. Furthermore, the conclusion we had derived for single-degree inelastic systems—that under a wide set of conditions their expected maximum deformation is approximately equal to that of an equivalent linear structure having the initial characteristics of the given system—may err seriously on the unsafe side for multidegree systems.

The foregoing discussion has been based on the assumption that the structures in question are to be designed against a single earthquake. Hence our conclusions apply with greater generality than this argument implies. Even if a structure is designed to develop a uniform ductility factor under the action of a certain ground motion, it may have to develop locally very large ductility factors when subjected to an entirely different ground motion. Still, the assumption of strictly elastoplastic behavior is unrealistically severe.

Analyses have been made of multistory, framed buildings having bilinear, strain-hardening, Masing-type, moment-rotation relations (Clough, Benuska, and Wilson, 1965). These analyses have led to the conclusion that, for frames designed for a given fraction of elastic analysis requirements to resist a family of earthquakes reasonably similar to each other, the ductility factors developed are of the same order as in single-degree systems having the same shape of their force-deformation diagrams and designed for the same fraction of elastic analysis requirements. The conclusion holds if we define the ductility factors in the multistory structures as the ratios of maximum to yield-point story deformations. The ductility factors that correspond to rotations at elastoplastic hinges of the structural members are often about 100 percent greater than these values. Finally, in frames of this type that are partially underdesigned or overdesigned, the required ductility factors at the weaker stories and weaker hinges may be many times greater than for reasonably well-designed frames.

Borges and Ravara (1969) report on Ravara's (1968) analysis of several six-story buildings having reinforced concrete frames subjected to actual earthquake records. Ravara idealizes the story shear-deformation curves as approximately that of the Masing type, with a bilinear, strain-hardening skeleton curve and parabolic transitions between straight segments in unloading and reloading. He defines the "structural ductility factor" as the ratio of maximum structural displacement relative to the ground to the maximum that the structure can undergo in the range of elastic behavior. He finds that this ratio is always smaller than the story ductility factor (assumed equal at all stories). When the latter lies between 2.1 and 5.1, the structural ductility factor varies between 1.6 and 2.1, depending chiefly on the relative yield-point shears in the various stories. The most unfavorable situations correspond to partial over- or underdesign.

Borges and Ravara note that the ratio of the structural to the story ductility factor is, in general, a decreasing function of the number of stories. Essentially the same conclusions are derived by Veletsos (1969) from the analysis of a family of three-story, elastoplastic buildings. Analyses by Clough and Benuska (1966) indicate that elastic analysis of multistory buildings generally overestimates ductility requirements in columns and underestimates them in girders.

A series of 10-story frames were analyzed by Hanson and Fan (1969) under the action of a magnified initial portion of the El Centro 1940 earthquake, NS component. Some frames were purely flexural, while others were either totally or partially braced, with force-deformation curves of the type illustrated in Fig. 11.22, using light, diagonal bracing. Axial deformations of members were neglected except in the braces. All the structures were designed for the same set of lateral seismic coefficients. It was found that some of the braced structures having exceptionally flexible members became unstable under earthquake action.

The result is in apparent contradiction with those obtained by Goel (1969) for typical 10- and 25-story frames. Goel concludes that the effects of gravity loads (the so-called $P\text{-}\Delta$ effects) are insignificant. It should be noted, however, that Hanson and Fan found no indication of impending collapse by instability a few seconds or even a fraction of a second before this took place. This result agrees with some of the time histories obtained by Husid (1967 and 1969) for one-story frames.

The simultaneous action of several components of ground motion on nonlinear systems has received scanty attention in the literature. A rigorous analysis of the probability of yielding or failure would require the simultaneous consideration of all the components of base motion even in the range of linear behavior. As we saw in Chapter 10, this can usually be obviated by using approximate methods of modal analysis. But in nonlinear systems the fact that effects of the various components are not additive makes their simultaneous consideration necessary, at least until satisfactory, approximate criteria become available.

In a study on the effects of the two orthogonal components of one real earthquake and several simulated ground motions on an elastoplastic single-story frame, Nigam and Housner (1969) found that explicit recognition of interaction between responses to the two components usually lowers the spectral velocity of the system but may give appreciably higher or lower ductility factors. Still, in the range $\mu < 8$, there is, in most cases, little difference between the ductility factors found when neglecting and when recognizing interaction. If these conclusions can be generalized on the basis of additional research, interaction of the horizontal components of translation will be negligible in most practical problems. However, this will surely not be the case with interaction between components of base translation and base rotation.

11.6 Rigid-Plastic Systems with Distributed Parameters

The behavior of certain structures subjected to earthquakes can be studied satisfactorily by idealizing them through distributed parameters and their stress-strain or generalized force-deformation relation as rigid-plastic.

When the positions of potential plastic hinges or surfaces of sliding in these structures form a discrete set, it is irrelevant whether we take parameters as discrete or distributed. The remarks on analysis made in the previous section apply without need for adjustment. If the possible locations of the hinges or surfaces in question are infinite and cannot be predicted at the outset of the analysis, a different approach is needed.

The following examples will illustrate both situations.

1. A framed structure.
   a. The structure is subjected to earthquake and to a finite set of static, concentrated loads. Potential sections for plastic hinges reduce to the joints, the points of static load application, and the sections at which yielding-moment changes (such as sections of bar cutoffs in reinforced concrete frames or of cover-plate cutoffs in steel frames).
   b. The structure is under combined earthquake and static, distributed loads. Plastic hinges may develop at any section within a segment near the center of every beam span.
2. A rock-fill dam having an inclined, weak, thin, impervious, cohesive core.
   a. For accelerations acting in one direction, most of the surface of sliding will lie in the core, and so the actual shear strength of the rest of the dam will be practically of no consequence.
   b. For accelerations in the opposite direction the same dam may develop a surface of sliding in any location within a fairly large volume.

We cope with both the framed structure and the dam having an infinite num-

360 RESPONSES OF NONLINEAR SYSTEMS Chap. 11

ber of possible locations of plastic hinges or surfaces of sliding by discretizing these locations into a manageable number of them. The artifice is tantamount to replacing the distributed loads with a sufficiently large number of concentrated forces and replacing the rockfill—implicitly idealized as a continuous medium—with a number of sufficiently thin wedges, at the interfaces of which we assume Coulomb friction. More ambitious treatments are possible and sometimes desirable. For rigid-plastic frames, a method that has been developed for static loading (Clyde, 1966) can be extended to study earthquake effects, but we shall not deal with it in this text. We shall direct our attention to a relatively simple problem in soil or rockfill stability that will give insight for the design of structures made of these materials.

Consider a mass of uniform, cohesionless material limited by a horizontal free surface, a horizontal interface with rock, and a plane slope (Fig. 11.27).

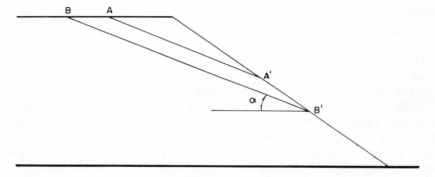

Figure 11.27. Slope of rigid-plastic material.

Assume that the angle of internal friction is independent of pressure. Our first concern is with the shape of the sliding surfaces. An application of the calculus of variations shows that, of all possible surfaces passing through a pair of points on the boundary (A and A' in the figure), the surface requiring the smallest horizontal acceleration to start sliding is the plane. As soon as sliding begins along a set of planes, sufficiently high shears will not develop in other surfaces to cause sliding. Consequently, all sliding surfaces must be plane.

Now let us examine the elevation of the sliding planes. Suppose that the horizontal acceleration is such as to cause a state of impending sliding along the parallel planes AA' and BB'. The second plane lies at a lower elevation than the first plane. The horizontal acceleration required to initiate relative motion along either plane is

$$a = \frac{\tan \phi - \tan \alpha}{1 + \tan \phi \tan \alpha} g \qquad (11.27)$$

where ϕ is the angle of internal friction, α is the angle that planes AA' and BB' form with the horizontal, and g denotes the acceleration of gravity. If sliding should begin along AA', the horizontal accelerations at all lower elevations must be sufficiently large to cause sliding along a lower parallel plane. If sliding begins along BB', horizontal accelerations in the wedge BOB' must be smaller than a, and there can be no relative motion at a plane such as AA'. Consequently, if angle α is critical for a given base acceleration, the surface of sliding must be the lowest possible plane forming an angle α with the horizontal.

Purely on the basis of equilibrium considerations (D'Alembert's principle), we have shown that every surface of sliding must be a plane passing through the toe of the slope. (The conclusion applies with appropriate adaptations to slopes of arbitrary shape.) Convergence of all wedges at the toe introduces there a singularity that cannot be disposed of so long as we retain the assumption of small displacements.

On the basis of experimental evidence in models, surfaces of sliding consisting of circular arcs (Seed, 1966) or pairs of planes (Goodman, 1963; Seed and Goodman, 1964) have been used in the analysis of slope stability under earthquake motions. Effects of the singularity mentioned in the foregoing paragraph are probably insufficiently significant to explain the disparity between our conclusions and the results of model testing. The most likely explanation lies in the fact that the models in question did not exactly fulfill the assumptions we adopted concerning uniformity of the material and independence of ϕ relative to the normal pressure. Triaxial tests run to supplement the model tests show that there is a slight curvature associated with the negative second derivative in the Mohr envelopes at low values of the confining pressure, or perhaps a small real or apparent cohesion. Convex Mohr envelopes or the existence of cohesion would indeed be associated with concave surfaces of sliding. For the cases in question the small curvature of the surfaces of sliding has a negligible effect on the motion of the crest.

In order to determine which surfaces actually become active, consider the dynamic equilibrium of n wedges limited by potential planes of sliding (Fig. 11.28; see Esteva, Sánchez-Trejo, and Rosenblueth, 1961). Let $B_i g$ denote the acceleration of the ith wedge relative to the one below, W_i the weight of the ith

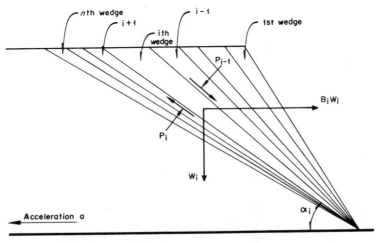

Figure 11.28. Conditions of dynamic equilibrium in wedges.

362 RESPONSES OF NONLINEAR SYSTEMS Chap. 11

wedge, α_i the inclination of its lower face, and P_i the normal force on that plane. From the conditions of the dynamic equilibrium of the ith wedge and the assumption that there is relative motion along all the interfaces, it follows that

$$P_n = P_{n+1} \frac{\cos(\phi + \alpha_n - \alpha_{n+1})}{\cos \phi} + W_n[(\cos \alpha_n - a \sin \alpha_n) - \sum_{i=1}^{n} B_i \sin(\alpha_n - \alpha_i)]$$

$$P_n \tan \phi = P_{n+1} \tan \phi \,[\cos(\alpha_{n+1} - \alpha_n) - \sin(\alpha_{n+1} - \alpha_n)]$$
$$+ W_n[(\sin \alpha_i + a \cos \alpha_i) - \sum_{i=1}^{n} B_i \cos(\alpha_n - \alpha_i)]$$

Solution of the resulting system of equations permits computing the relative accelerations $B_i g$. If any one of them turns out negative, there is no relative

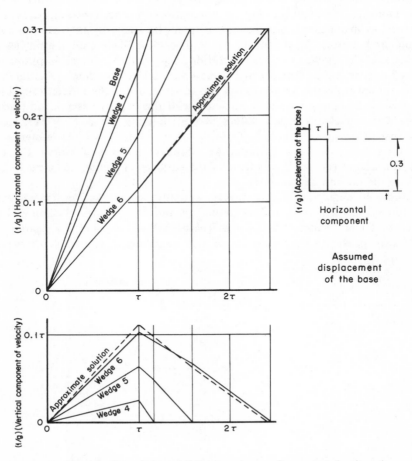

Figure 11.29. Comparison of the responses of a system having six potential wedges and a rigid body on an inclined plane. *After Esteva, Sánchez-Trejo, and Rosenblueth (1961).*

displacement along the corresponding interface, and the analysis must be repeated with a different number of wedges. At any instant two adjoining wedges may reach a nil relative velocity; from then on they will displace as a single body.

Replacing the base accelerogram with a series of constant acceleration steps permits the analysis of any noncohesive, rigid-plastic slope subjected to any base motion. For each step, one must determine the planes at which sliding takes place and the magnitudes of relative motion. This process has been applied successfully by Bustamante (1964 and 1965a) with the aid of an analog computer.

A cruder but more expedient approach consists of assuming that the displacements of the crest of the slope will be the same as those of a rigid block resting on an inclined plane parallel to the slope and having a contact angle of friction between this plane and the block equal to the angle of internal friction of the soil (Newmark, 1965a, b, and 1970). Figure 11.29 shows a comparison of the vertical displacements in question for the specific example depicted in the figure when the analysis is based on six potential wedges. It is seen that the error involved in this assumption is not serious, at least for conditions similar to those analyzed here.

The acceleration required to make the block slide upward is ordinarily too high to merit serious consideration. Hence, the block behaves practically as a rigid-plastic system that can yield in one direction only, and the results we quoted in Section 11.5 apply directly to it. The yield acceleration is found from Eq. 11.26 after replacing the slope angle for α.

A comprehensive study of the response of structures having a trapezoidal cross section is more complex because there will be interaction between displacements of the material near the crest along directions roughly parallel to each slope. Slightly conservative results are obtained by assuming independence of the effects of ground accelerations in opposite directions.

The approach presented here can be extended to more general conditions such as those of slopes in which soil properties vary as a function of position and in which the angle of internal friction depends on the normal pressure.

Damage to concrete walls and gunite panels in reservoir in Sacramento due to sloshing water. Dixie Valley (Churchill County) earthquake of December 1954.

12

EARTHQUAKE EFFECTS ON RESERVOIRS

12.1 Introductory Note

Using the principles of hydrodynamics set forth in Chapter 6, introducing the characteristics of strong-motion earthquakes as described in Chapter 7, and adapting the theory of linear multidegree systems developed in Chapter 10, we shall attempt here to derive the most significant features of hydrodynamic response to earthquakes. In principle we should also make use of the treatment presented in Chapter 11 for nonlinear systems, since some phenomena of interest in the seismic response of liquids are markedly nonlinear. This will not be done in view of the very empirical stage of knowledge of these phenomena, which does not justify their refined treatment.

Disturbances cause essentially two types of hydrodynamic response: compression waves and surface waves. We shall approach both phenomena separately. Owing to the ranges of frequencies involved, we find that in most cases one or the other type of phenomena is significant, and the other can be ignored.

12.2 Hydrodynamic Pressures on Dams

Given a ground motion, the geometry of a reservoir, and the dynamic characteristics of a dam, a solution can, in principle, always be found for the hydrodynamic pressures using the theory in Section 6.2. Dams are usually such important structures that it often is justified to analyze in this manner the effects of entire families of actual and simulated earthquakes for which the structure is to be designed. Still it is desirable from many points of view—chiefly to perform preliminary designs of the more important dams and final designs of lesser structures—to possess typical results applicable to representative conditions.

The essential difference between the action of the various components of ground motion lies in the nature of the main energy-dissipating mechanisms. If we idealize the reservoir as a semiinfinite cylinder and assume that the ground motion arrives simultaneously to the entire bottom and walls, energy due to the longitudinal component of ground motion is dissipated in the form of waves that travel away to infinity. This mechanism does not operate for the transverse and vertical components of ground motion. Hence, when analyzing the effects of the latter two types of disturbance, it is more important explicitly to recognize other mechanisms of energy dissipation, especially the deformability of the bottom.

First consider a semiinfinite prismatic reservoir with a rectangular cross section under longitudinal excitation. Let the upstream face of the dam be a vertical plane. Assume that the dam and the reservoir bottom and sides do not deform. Let us recognize water compressibility but take the simplest of the usual boundary conditions at the free surface (atmospheric pressure at the original elevation of this surface). Using the approximate theory in Chapter 10 we can write

$$Q^2 = \sum_n Q_n^2 \tag{12.1}$$

where Q is a response (maximum numerical value associated with a given probability of exceedance), and Q_n is its corresponding value in the nth natural mode. Also,[1]

$$Q_n = \max_t \left| \int_0^t \ddot{x}_0(\tau) \psi_n(t - \tau) \, d\tau \right| \tag{12.2}$$

where t is time, $\ddot{x}_0$ the ground acceleration, and ψ_n the transfer function for Q_n. When $Q = p$ = hydrodynamic pressure on the dam, Eqs. 6.14–6.17 allow us to write

$$\psi_n = a_n \omega_n \cos\left(\frac{\omega_n}{c} X_1\right) J_0(\omega_n t) \tag{12.3}$$

where X_1 is the elevation, a_n and ω_n are given by Eqs. 6.15 and 6.16, and J_0 is Bessel's function of the first class, order zero.

Equation 12.1 assumed that

$$\sum_{m \neq n} \sum \int_0^s \psi_m \psi_n \, dt \ll \sum_n \int_0^s \psi_n^2 \, dt$$

where s is the duration of a segment of white noise, equivalent (in the sense of Chapters 9 and 10) to the family of earthquakes of interest. This assumption is almost always justified (Rosenblueth, 1968b).

When $Q = p$, Eq. 12.2 can be written in the form

$$p_n = a_n \tilde{A}_n \cos\left(\frac{\omega_n}{c} X_1\right) \tag{12.4}$$

---

[1] Responses that are actually of interest are those acting in the downstream direction. Their maximums are, on the average, smaller than the corresponding maximum numerical values. However, due to the oscillatory character of the responses, this difference amounts to no more than a few percent (Rosenblueth, 1968a) and will be disregarded.

Sec. 12.2 HYDRODYNAMIC PRESSURES ON DAMS

where

$$\tilde{A}_n = \omega_n \max_t \left| \int_0^t \ddot{x}_0(\tau) J_0[\omega_n(t - \tau)] \, d\tau \right| \qquad (12.5)$$

is formally similar to spectral acceleration, with J_0 replacing the damped trigonometric function of acceleration. Hence we are justified in calling the curve $\tilde{A}(T)$ the *hydrodynamic spectrum* of the given motion, where $T = 2\pi/\omega$ represents the reservoir's natural periods of vibration.

Because of the similarity between the initial portion of the function J_0 and a damped cosine curve having a damping ratio of about 15 percent, we expect $\tilde{A}$ to fall close to the spectral acceleration for this percentage of damping.

Using the methods described in Section 6.2, several hydrodynamic spectra have been computed (Figs. 12.1 and 12.2). In the ranges of periods covered by the figures, $\tilde{A}$ coincides, on the average, with the spectral acceleration A for a damping ratio $\zeta = 0.15$. On the average, $\tilde{A}/a \cong 1.3$, where a is the maximum ground acceleration, and indeed, according to Table 9.1, this ratio corresponds approximately to $\zeta = 0.15$.

This result applies to natural periods much smaller than $T_{av} = 2\pi v/a$, where v is the maximum ground velocity. It has been shown (Rosenblueth, 1968b) that for longer periods, $\tilde{A}$ is approximately equal to the spectral acceleration for a damping coefficient that varies as shown by the solid line in Fig. 12.3. The dashed line indicates how this equivalent percentage of damping may vary for intermediate natural periods.

Given the hydrodynamic spectral ordinate associated with the fundamental mode $\tilde{A}_1$, we can find the corresponding pressure distribution in the first mode by substituting in Eq. 12.4 with $n = 1$. The result is shown in Fig. 12.4 in terms

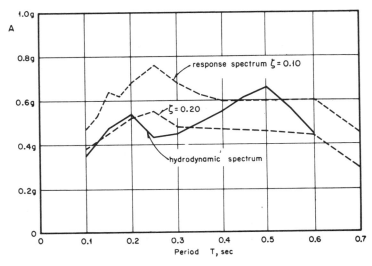

Figure 12.1 Hydrodynamic spectrum for NS component of El Centro 1940 earthquake. *After Flores (1966).*

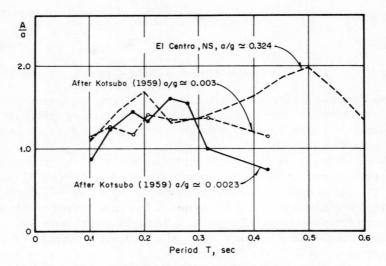

Figure 12.2. Hydrodynamic magnification factors for three earthquakes. *After Flores (1966)*.

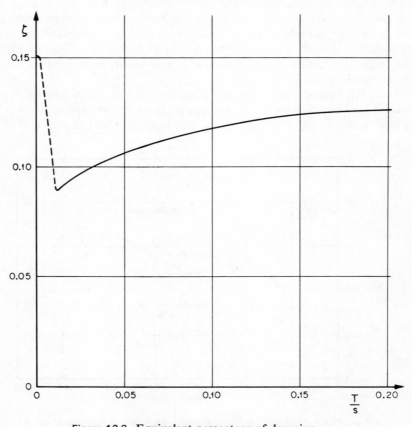

Figure 12.3. Equivalent percentage of damping.

Sec. 12.2 HYDRODYNAMIC PRESSURES ON DAMS 369

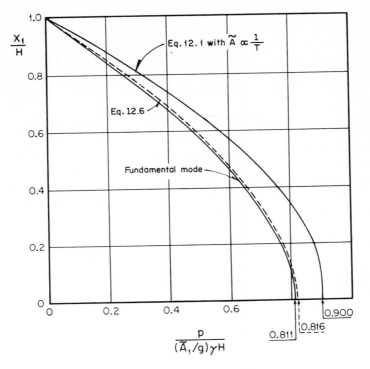

Figure 12.4. Hydrodynamic pressure distributions.

of $(\tilde{A}_1/g)\gamma H$, where g is the acceleration of gravity and γ is the unit weight of water. The abscissa at the base is $8/\pi^2 = 0.811$.

The fundamental period of the liquid is $T_1 = 4H/c$ that, for water at ordinary temperatures, is $H/360$ (T_1 in seconds, H in meters). Hence in most applications T_1 is not much longer than T_{av}. In the range of natural periods, the hydrodynamic spectrum may be assumed, then, to be independent of T. Under the assumption that $\tilde{A}_n = \tilde{A}_1$, Flores (1966) has found

$$\frac{p}{(\tilde{A}_1/g)\gamma H} = \sqrt{\frac{2}{3}\left[1 - 3\left(\frac{X_1}{H}\right)^2 + 2\left(\frac{X_1}{H}\right)^3\right]} \tag{12.6}$$

(Fig. 12.4). The abscissa at the base is $\sqrt{2/3} = 0.816$. In this case most of the hydrodynamic pressure is due to the fundamental mode.

For very deep reservoirs subjected principally to the effects of nearby earthquakes (such as the ones caused or triggered by filling of the reservoir) for which T_{av} may be relatively short, the first few natural periods may fall in the approximately hyperbolic branch of the acceleration spectrum. Assuming that $\tilde{A}$ is inversely proportional to T and applying Eq. 12.1, we find the corresponding curve in Fig. 12.4, which is a second-degree parabola with the vertex at the free surface. The abscissa at the base is 0.900. In this case $\tilde{A}_1$ may be taken equal to the spectral acceleration associated with the percentage of damping given by Fig. 12.3; it is conservative to take this as 9 percent.

This curve is proportional to the distribution of design shears that we found for a uniform shear beam subjected to white noise, in Section 10.4. The reason for this is that the modal shear distributions in the beam are proportional to the modal hydrodynamic pressures, and we are taking the same design spectrum in both cases.

As we shall see in Chapter 15, the parabola with vertex at the top is too conservative in the uppermost stories for design of relatively uniform buildings because it overestimates the importance of the higher modes of vibration. Often, a more realistic shear distribution is a parabola with the vertex at the bottom. The reasoning above justifies the adoption of this type of curve, with $p = 0.816 (\tilde{A}_1/g)\gamma H$ at the bottom, for the hydrodynamic pressure distribution when T_1 is expected not to exceed T_{av} greatly:

$$\frac{p}{(\tilde{A}_1/g)\gamma H} = 0.816\left(1 - \frac{X_1^2}{H^2}\right) \tag{12.7}$$

For reservoirs that are not too deep, we may take $\tilde{A}_1 \cong 1.3a$, which refines Eq. 12.7 into

$$\frac{p}{(a/g)\gamma H} = 1 - \frac{X_1^2}{H^2} \tag{12.8}$$

This curve may be compared with Westergaard's (1933) parabolic approximation,

$$\frac{p}{(a/g)\gamma H} = 0.816\left[\frac{1 - X_1/H}{1 - 2.36(H/1000T)}\right]^{1/2} \tag{12.9}$$

where T is the period of a supposedly harmonic excitation. It is customary to apply Eq. 12.9 with $T = 1$ sec. Hence, for $H = 100$ m we obtain the curve shown in Fig. 12.5, which also shows the result of using Eq. 12.8.

The approximation proposed for shallow and moderately deep reservoirs (Eq. 12.7) is less conservative than Westergaard's near the top of the reservoir. This may seem objectionable, since the theories that serve as bases for all the foregoing expressions neglect the effects of surface waves. However, Eq. 12.7 already gives pressures slightly greater than Eq. 12.6 near the top of the reservoir. Moreover, radiation damping due to deformability of the bottom and walls of the reservoir can be expected to affect the pressures due to higher modes to a greater extent than those of the fundamental mode and therefore compensate approximately for the effects of surface waves.

In principle the shears or total hydrodynamic force and overturning moments due to the hydrodynamic pressure should not be computed by integrating the pressure distribution but by combining the corresponding modal responses. However, for flat or nearly flat hydrodynamic spectra the influence of higher modes is so small that the error introduced by directly integrating the design pressures may be disregarded.

Chopra (1967) has computed the total hydrodynamic force and the overturn-

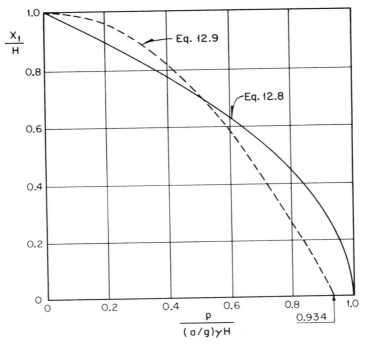

Figure 12.5. Design shear distributions.

ing moment at the base of three dams subjected to the NS component of the 1940 El Centro earthquake. He used the same assumptions which led to the foregoing analysis, but he computed the responses as functions of time, so he has not introduced the simplifications involved in the modal analysis. This permits, therefore, calibrating these simplifications.

The dams analyzed were 100, 300, and 600 ft tall. The corresponding fundamental periods are 0.085, 0.254, and 0.509 sec. The equivalent damping ratios are 15, 15, and 9 percent. And $\tilde{A}_1$ is approximately 0.40g, 0.55g, and 0.64g.

By integrating Eq. 12.7 we find the total hydrodynamic force equal to $\frac{2}{3}p_0 H = 0.554(\tilde{A}_1/g)\gamma H$, where p_0 is p at $X_1 = 0$. A second integration yields the basal overturning moment $\frac{1}{4} p_0 H^2 = 0.204(\tilde{A}_1/g)\gamma H^3$. In terms of the hydrostatic force and moment these values are $1.088(\tilde{A}_1/g)$ and $1.224(\tilde{A}_1/g)$, respectively.

Table 12.1 compares Chopra's results with those obtained from Eq. 12.7. Agreement is satisfactory. The table includes a comparison with Eq. 12.8, which shows that the hydrodynamic force and moment are, respectively, $1.33a/g$ and $1.5\ a/g$ times the hydrostatic values. The error is excessively large for the second and third dams. However, the hydrodynamic spectral ordinate is exceptionally high for these dams (Figs. 12.1 and 12.2). Under representative conditions, results of applying Eq. 12.8 can be expected to be satisfactory.

The influence of the inclination of the upstream face and the cross-sectional

TABLE 12.1. COMPARISON OF RESPONSES TO THE EL CENTRO 1940 EARTHQUAKE, NS COMPONENT [AFTER ROSENBLUETH (1968b)]

H (ft)	Total hydrodynamic force*			Overturning moment at the base*		
	Exact†	Eq. 12.7	Eq. 12.8	Exact‡	Eq. 12.7	Eq. 12.8
100	0.44	0.44	0.43	0.50	0.49	0.48
300	0.57	0.60	0.43	0.64	0.67	0.48
600	0.71	0.70	0.43	0.79	0.79	0.48

*In terms of the hydrostatic force or moment.
†Results obtained by Copra (1967).
‡Taken from the graphs of Do.

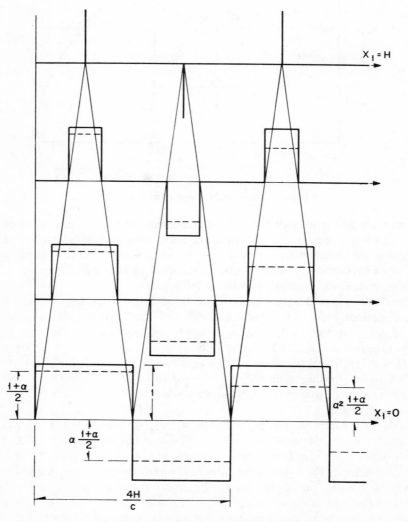

Figure 12.6. Transfer function for hydrodynamic pressure.

Sec. 12.2   HYDRODYNAMIC PRESSURES ON DAMS   **373**

shape of the reservoir may be taken into account in an approximate fashion using the results in Section 6.2. Other matters such as the deformability of the dam, the reservoir bottom and walls, surface waves, and the influence of an irregularly shaped reservoir in plan require more refined analyses.

Now consider a prismatic, semiinfinite reservoir of rectangular cross section, subjected to a vertical disturbance. Again let the upstream face of the dam be a vertical plane. If we assume that the dam and the walls and bottom of the reservoir do not deform, the transfer functions for hydrodynamic pressure, shear, and overturning moment are like the solid lines in Figs. 12.6–12.8. The figures show the wave-travel lines used in applying the method of characteristics to find the transfer functions at various elevations, as we did in Chapter 3 when studying wave travel in horizontally stratified soil.

Transfer functions for pressure and shear are proportional to those for shear and overturning moment in a uniform shear beam (Fig. 10.2). These functions are strictly periodic, as there is no energy dissipation. Hence, when the dis-

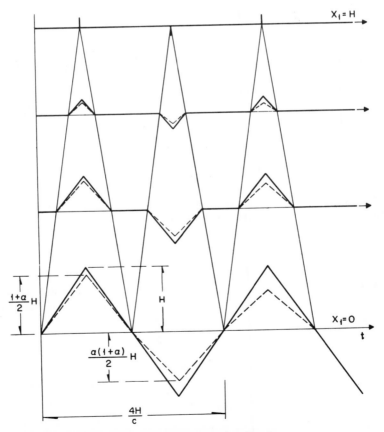

Figure 12.7. Transfer function for shear.

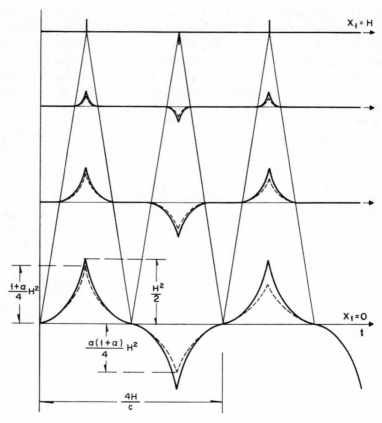

Figure 12.8. Transfer function for overturning moment.

turbance is white noise (hyperbolic acceleration spectrum), it is convenient to use Eq. 10.8, which can be put in the form

$$Q^2 \propto \int_0^{T'} \psi^2 \, dt \qquad (12.10)$$

where T' is the period of ψ. This way we find the distributions displayed in Fig. 12.9. They are proportional to powers of the depth. Pressures are proportional to $(H - X_1)^{1/2}$; shears to the $\frac{3}{2}$ power, and overturning moments to the $\frac{5}{2}$ power. Using subscript 0 to denote basal values we find

$$S_0 = p_0 H / \sqrt{3} = 0.577 p_0 H \qquad (12.11)$$

$$M_0 = p_0 H^2 / \sqrt{20} = 0.244 p_0 H \qquad (12.12)$$

and

$$p_0 = \rho c V \sqrt{2} \qquad (12.13)$$

(Rosenblueth, 1968a), where ρ is the mass per unit volume of water, S are shears, M are moments, and V is the spectral pseudovelocity. Since $V = A_1/\omega_1$,

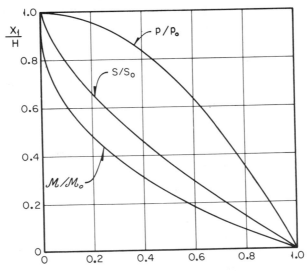

Figure 12.9. Relative distributions of pressures, shears, and moments.

and $\omega_1 = 2\pi/T_1 = 4H/c$, we may also write Eq. 12.13 as $p_0 = 0.900(A_1/g)\gamma$ where A_1 is the spectral acceleration associated with the fundamental period.

The result coincides with the result we found when the disturbance was horizontal white noise if we replace A with $\tilde{A}$. The reason for this correlation is that the natural modes are the same in both cases; only the trigonometric functions of time are replaced with Bessel functions for a longitudinal disturbance.

The base shear is 0.866 times the integral of the design pressures, and the base moment is 0.968 times the integral of the shears. These coefficients are constant throughout the height of the dam. The reason for their differing from 1 is, as in shear beams, that the design pressures do not act simultaneously at all elevations, nor do the design shears. Since design in practice is usually not based directly on pressures but on shears, moments, or a combination of these two generalized forces, the computed shear distribution may be used directly, because the reduction required, relative to the shear integral to obtain moments, in only 3.2 percent.

The effect of rock compressibility may be gaged by the same analytical method used in Chapter 3 for the study of one-dimensional wave reflection and refraction in horizontal soil layers. Let

$$\alpha = \frac{c_0\gamma_0 - c\gamma}{c_0\gamma_0 + c\gamma}$$

where subscript 0 refers to the rock. Refraction from rock into water amplifies waves in the ratio $(1 + \alpha)/2$, relative to the situation we had with infinitely rigid rock. Successive refracted waves are reduced in the ratio α in every half cycle and α^2 in every cycle. Hence the transfer functions shown by the dashed lines in Figs. 12.6–12.8. Successive multiplication of the transfer function ordinates by α is approximately equivalent to the introduction of a damping ratio ζ_1 in the first mode, such that

$$\exp(-\zeta_1 \omega_1 T_1) = \alpha^2$$

Hence, $\zeta_1 = (1/\pi) \ln(1/\alpha)$ and, in the highest natural modes, $\zeta_i \omega_i = \zeta_1 \omega_1$. According to Eq. 9.9, the effect of damping is to multiply modal responses by

$$\beta_E = \left(1 + \frac{\zeta_i \omega_i s}{2}\right)^{-1/2}$$

approximately. Substituting the values of ζ_i and introducing the magnification factor $(1 + \alpha)/2$, we find that the effect of rock compressibility is to multiply all the responses approximately by

$$B = \frac{1 + \alpha}{2}\left(1 + \frac{s}{T_1}\ln\frac{1}{\alpha}\right)^{-1/2} \tag{12.14}$$

This factor is displayed in Fig. 12.10. The reduction due to rock deformability is more pronounced for the longer earthquake durations and shallower reservoirs.

The foregoing treatment is based on a white noise idealization of the disturbance. For acceleration spectra that are not flat, including the case of hyperbolic spectra, the results we obtained for a longitudinal excitation are directly applicable to vertical excitations if only the hydrodynamic spectral ordinates $\tilde{A}$ are replaced with the spectral accelerations A associated with a damping ratio $(1/\pi) \ln(1/\alpha)$.

The pressures obtained in this manner are usually considerably smaller than those computed by Chopra (1967) for vertical excitations neglecting rock

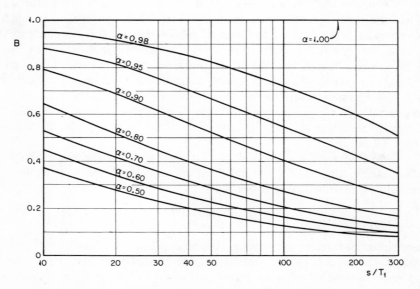

Figure 12.10. Reduction coefficient for rock compressibility.

compressibility. Chopra finds excessively large pressures (see Section 6.2.10). While recognizing that this compressibility brings hydrodynamic effects into a more credible range, there may be other, even more significant phenomena to justify further reductions of the effects of wave reflections (Sections 6.2.5 and 6.2.10), but more experimental confirmation is required before such criteria may be safely incorporated in practice.

12.3 Submerged Structures

Most submerged works of interest in civil engineering have horizontal dimensions so much smaller than the lengths of significant compression waves that water compressibility may be ignored when dealing with such structures. Indeed, the wavelength associated with a period T is equal to the velocity of sound in water multiplied by T. We are rarely interested in values of T smaller than about 0.1 sec, both because of the range of the most significant natural periods of vibration of structures and because of the range of prevailing periods in acceleration spectra. Hence we are rarely interested in waves shorter than about 144 m. Water compressibility cannot be expected to play an important role in the seismic responses of structures whose dimensions in plan do not exceed about $\frac{1}{6}$ of the shortest waves of interest, or about 24 m. Most submerged structures, such as intake towers and bridge piers, comply with this limitation.

It follows that for most submerged structures seismic analysis may be carried out in the same way as for structures in air, by merely assuming the addition of a virtual mass. We can find the magnitude of the virtual mass from hydrodynamic analyses that idealize water as incompressible. Damping due to submersion may also be disregarded. These analyses often involve additional approximations due to a simplified treatment of structural characteristics. The matter can be corrected by extrapolating results of small-scale model tests, such as those described in Section 6.4, since the assumed water incompressibility validates elementary extrapolation to the prototype.

An exception to some of the above statements is found in the effects of surface waves for structures partially submerged or for structures that nearly reach the surface. Additional research is needed to elucidate these phenomena in areas concerning damping and local pressure distribution due to sloshing.

Further research is also required to permit a satisfactory treatment of large, submerged structures for which hydrodynamic phenomena will be appreciably influenced by water compressibility.

12.4 Tanks

Treatment of liquid containers is similar to that of submerged structures; their dimensions are ordinarily small enough so that water compressibility may be ignored in the analysis. Again the results of laboratory tests on small-scale models may be extrapolated directly to prototypes. And once more, damping due to the presence of water in the tank is ordinarily too small to merit consideration.

Practically the entire hydrodynamic phenomenon in tanks is associated with surface waves. The phenomenon may be idealized as linear so long as the waves remain low—that is, provided the surface slope does not become so large as to introduce a significant difference between itself and the sine of the angle in question. With this limitation the kernel for each natural mode is a sine function; hence, the methods developed in Chapter 10 for linear multidegree structures apply directly. The formation of tall waves leads to a condition similar to that of nonlinear elastic systems.

Matters that deserve additional research in connection with seismic analysis of containers include the effects of water compressibility in tanks of large dimensions (say in excess of 24 m), the development of methods to calculate or to extrapolate from small-scale model tests, the behavior of liquids in tanks having bafflers sufficiently effective to introduce appreciable damping due to turbulence, and better-suited methods for predicting the height of waves without the need for model testing, since presently available methods are too crude to be satisfactory.

Failure of concrete column in Caracas in earthquake of July 29, 1967. *American Iron and Steel Institute.*

13

BEHAVIOR OF MATERIALS AND STRUCTURAL COMPONENTS UNDER EARTHQUAKE LOADING

13.1 Introductory Note

The present chapter describes behavior. The ranges of parameters in which we are mainly interested are determined by both the prevailing characteristics of ground motions and the most common dimensions of structures and of their components. The disturbance may range from a relatively sharp shock, through chaotic disturbances lasting a few dozen seconds, to nearly steady-state harmonic oscillations that last up to 4 or 5 min. Because of the usual range of structural natural periods, the most significant excitations have periods of 0.1 to 20 sec, the shorter values associated with masonry and concrete and the longer ones with steel.

The number of load applications lies somewhat below what is called ordinary fatigue. We are essentially concerned with "low-cycle fatigue" at medium to low rates of loading, and data in this range are still scarce.

In fatigue proper the stress levels with which one is usually concerned are low enough that the stress-strain curves often remain practically straight, and it suffices to report the effect of the number of cycles on strength. In low-cycle fatigue, effects on the entire stress-strain or force-deformation diagram are usually significant because in most cases we deal with pronouncedly nonlinear behavior. This does not mean that such information is always available.

It has been customary to test in low-cycle fatigue by subjecting specimens to cycles of stress or load that oscillate between fixed levels and are applied until failure occurs or until the test is discontinued. With materials and structural members that exhibit a wide scatter in their capacity—such as reinforced concrete members in diagonal tension—this procedure requires an enormous number of tests. More expedient loading programs can be devised, although it cannot be claimed that they reproduce earthquake effects much more realistically than the constant level procedure, and interpretation does become more involved.

Earthquake excitation usually produces stress histories that differ greatly from steady-state harmonic oscillations. This further complicates the interpretation of experimental data. There is little evidence to support the applicability of criteria that, with certain limitations, are known to afford satisfactory results in predicting damage due to fatigue proper under random loading.[1]

One other item complicates the issue in relation to the usual type of fatigue testing. In response to earthquake motions, descending branches in stress-strain and force-deformation diagrams are usually of interest because their onset does not usually imply collapse. Information in this range is particularly scanty.

13.2 Damping

Energy dissipation may take place through feedback into the ground, losses into the surrounding medium or contained fluids, internal damping, and friction at connections.

The energy that is fed back into the ground is, in part, lost to the structure because it is sent out as waves that travel long distances; in part it is directly transformed into heat due to internal damping in the ground. The first type of energy dissipation can be incorporated into seismic analysis by considering soil-structure interaction. The second type will be dealt with in Section 13.9; we have already encountered it in connection with the filtering of earth waves through soft mantles.

In all cases of practical interest energy transfer into the atmosphere is negligible. The same holds for the energy lost by small and moderate, completely submerged structures into the surrounding water and for internal damping in liquids inside containers of small and moderate capacity (see Sections 6.3 and 6.4). In contrast we have seen that dams subjected to hydrodynamic pressures behave as though the energy transferred into the reservoir liquid and carried away were equivalent to 9 to 15 percent of critical viscous damping. There is need for additional studies and information on partially submerged structures of small and moderate sizes and on the equivalent damping for even completely submerged structures of dimensions intermediate between those of an intake tower and those of a large dam.

Internal damping may be due to a variety of mechanisms (Lazan and Goodman, 1961; Lazan, 1968) including magnetoelasticity in certain metals (to which we shall make no further reference), and friction at grain boundaries in

[1] Even in high-cycle fatigue the usual criteria disregard the influence of the order of load application. The matter is probably not too important for large numbers of random load applications, but it may become decisive in low-cycle fatigue. In masonry walls, for example, effects of applying one high load followed by a number of smaller ones are probably more severe than those of the process in reverse order because of the chafing that can be anticipated at cracks.

most structural materials. The latter phenomenon is, in essence, of the same type as friction at connections but occurs in a microscopic scale.

Phenomenologically we may classify internal damping according to its dependence on frequency. Frequency-independent damping is associated with Coulomb friction at grain boundaries and is often known as hysteretic. Necessarily it involves nonlinear stress-strain relations under static loading. The corresponding resistance to motion in steady-state vibrations occurs in phase with elastic strains. The conventional differentiation between this type of damping and inelasticity is arbitrary, merely a matter of degree, and not an especially happy one. Some materials, such as structural grade steel, are almost perfectly elastic up to the yield point when they are stress relieved and carefully machined. In them, however, the present type of energy dissipation may become apparent at nominal stresses far below yielding. This must be attributed to stress concentrations and to residual stresses, which may make the actual local stresses far exceed their nominal values computed on elementary bases.

The remaining energy dissipation is frequency dependent. At low stress levels in some materials, such as polymers, this type of damping can be idealized accurately as linear with strain rate. Hence it may properly be called viscous damping. It constitutes practically the only significant type of internal energy dissipation in these materials.

In any structure having linear behavior the percentage of critical damping is amplitude independent. This is nearly true in polymers up to about the fatigue endurance limit. For higher stresses the percentage of critical damping grows with increasing maximum stress. In most metals the percentage of critical damping is an increasing function of maximum stress even at stresses smaller than the endurance limit and grows more rapidly with stress beyond this limit.

The concept of equivalent viscous damping for structures that do not deteriorate is a useful although dangerous one. Given the stiffness and mass of an arbitrary linear system, which we choose to call *equivalent* to a given nonlinear, single-degree structure, we can always compute the percentage of critical viscous damping that will make the linear system respond with the same amplitude as the nonlinear structure to a given periodic disturbance (see Chapter 11). Usually it is agreed that the mass in the equivalent system should be equal to that in the structure in question, so that the equivalent stiffness remains arbitrary. The range of possible choice is large, as it is ordinarily required only that the resulting viscous damping be positive and perhaps that the equivalent stiffness lie between the minimum and maximum tangent or secant stiffnesses of the actual structure for all deformations equal to or smaller than the maximum experienced under the steady state in question.

By equating the energy dissipation per cycle in the nonlinear structure and in its linear equivalent, we find the degree of viscous damping required in the latter as given by Eq. 11.3. At resonance, the percentage of equivalent damping is inversely proportional to the equivalent stiffness and hence may vary over a wide range depending on our choice of the latter.

As an example consider a single-degree elastoplastic structure. If we choose as equivalent linear spring element one having the elastoplastic system's initial stiffness, we find that the maximum possible equivalent damping at resonance is 15.9 percent of critical (Jennings, 1964). If we define the stiffness of the equivalent linear system as the secant stiffness of the actual structure at maximum deformation, the ratio of hysteretic to strain energy approaches 8 as the ductility factor tends to infinity, so that the equivalent coefficient of damping at resonance can approach $2/\pi$, or 63.7 percent of critical.

Futile discussions have arisen from differences in the choice of the equivalent linear stiffness. In the analysis of responses to earthquakes this choice is necessarily arbitrary. We shall favor selection of the secant stiffness at maximum deformation in view of its serving as the basis for the relatively general method of analysis of nonlinear single-degree systems that we presented in Section 11.3. At any rate, the decision should be made explicit in every instance.

Representative values of material damping are found in the literature; see, for example, Lazan and Goodman (1961) and Lazan (1968).

13.3 Effects of Rate of Loading

In any given cycle of loading the stress-strain relation is a function of the rate of loading. Usually only a portion of this effect can be ascribed to viscous damping. We shall begin by referring to specimens of various materials under a uniform, uniaxial state of stress applied monotonically.

With most metals there is only a small difference between the stress-strain curves obtained in an ordinary static test and those associated with the highest rates of loading that are significant during an earthquake. An idea of the orders of magnitude involved can be obtained from the data in Figs. 13.1 and 13.2. A more pronounced effect is noticeable in specimens of structural-grade steel free from stress concentrations and from residual stresses. These specimens exhibit upper and lower yield points when loaded at a sufficiently low rate. The specimens retain only the upper yield point up to the stress-hardening range at the highest rates of loading, of an order that is of interest in earthquake engineering (Fig. 13.3). The phenomenon is, however, mostly of academic concern, since stress concentrations and residual stresses ordinarily met in practice are sufficiently high that the upper yield point entirely disappears and only the lower value remains significant at relevant rates of loading.

Concrete undergoes a somewhat greater increase in modulus of elasticity and in strength, as illustrated in Figs. 13.4 and 13.5. (Data for concrete in tension were recorded in tests on notched cylinders.) In most clays the effects are even more pronounced (Figs. 13.6 and 13.7), while with sands there is hardly a noticeable influence of the loading rate in tests lasting between tenths of seconds and several minutes (Fig. 13.8), although the effect may be important at very slow rates and high stresses.

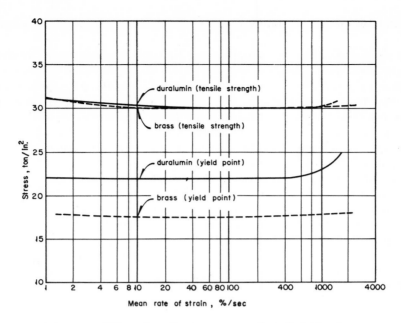

Figure 13.1. Effect of strain rate on strength and yield point for certain metals. *After Evans (1942).*

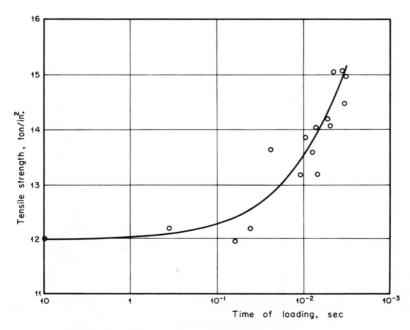

Figure 13.2. Effect of strain rate on strength of cast iron. *After Evans (1942).*

386 BEHAVIOR OF MATERIALS AND STRUCTURAL COMPONENTS   Chap. 13

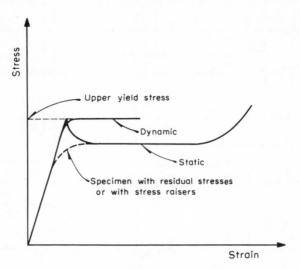

Figure 13.3. Static and dynamic stress-strain curves for structural-grade steel.

The modulus of elasticity and the strength of timber and of most polymers are quite sensitive to the loading rate (Figs. 13.9–13.12).

When our concern is with only the range of small strains, the viscoelastic formulation presented in Section 3.5 is well suited for describing behavior under a single cycle of loading as well as under a large number of cycles. Many attempts have been made at developing elaborate rheological models to describe behavior (Freudenthal and Roll, 1957–1958; Hult, 1966) whether this involves practically linear relations or a strongly nonlinear component. Any such model is usually adequate over a limited range of loading rates and, ordinarily, inapplicable to the description of behavior under a large number of loading cycles.

In specimens subjected to a nonuniform uniaxial state of stress, the rate of loading is also nonuniform. Because the stress-strain relation depends on the rate of loading, the stress distribution will differ from the one that is obtained from the assumption that the stress-strain relation that holds in a static test of a specimen under uniform stress will apply throughout the specimen that has a nonuniform state of stress. A considerably improved solution is obtained by assuming that every point in the specimen in question has the same stress-strain time relation as a specimen under a uniform, uniaxial state of stress. An early study of this type for reinforced concrete beams was made by Glanville (1930). He showed that, under the assumption of a linear distribution of longitudinal normal strains, the stress distribution in the compression zone of concrete was practically linear up to relatively high strains. This result is due to the increase in strain rate with distance from the neutral axis. A more complete and accurate study of the question was done later by Rüsch (1960).

This type of analysis rests on the assumption that the stress-strain time relations are independent of the strain gradient. This is not truly the case. For

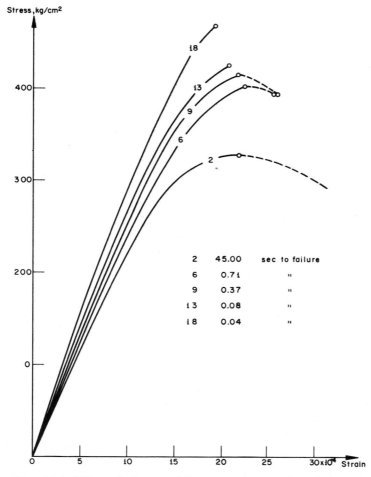

Figure 13.4. Effect of time to failure on stress–strain relation of concrete in compression. *After Hatano and Tsutsumi (1959).*

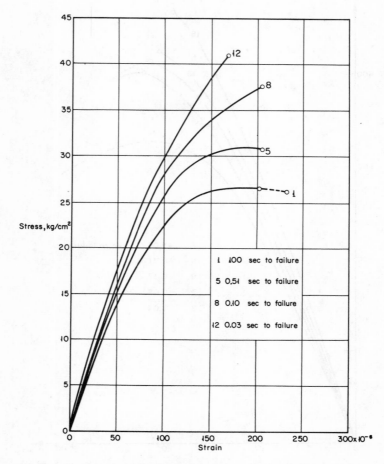

Figure 13.5. Effect of time to failure on stress–strain relation of concrete in tension. *After Hatano (1960)*.

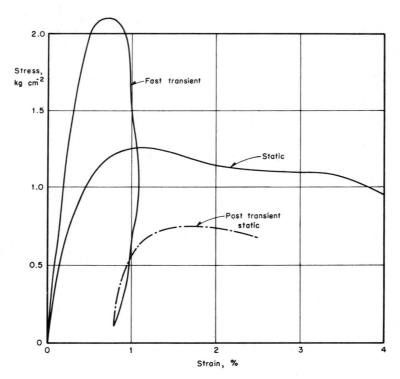

Figure 13.6. Effect of time of loading on stress–strain relation of clays. *After Casagrande and Shannon (1948).*

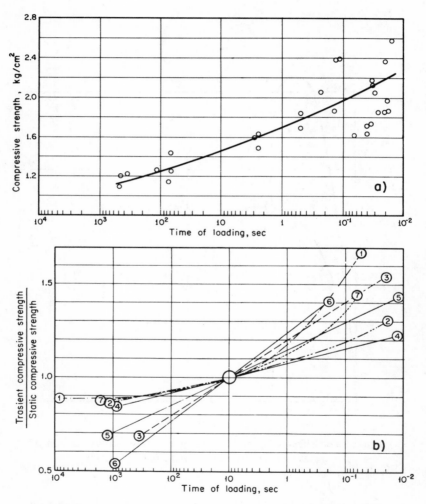

Figure 13.7. Effects of rate of loading on the strength of some saturated clays. *After Casagrande and Shannon (1948)*.

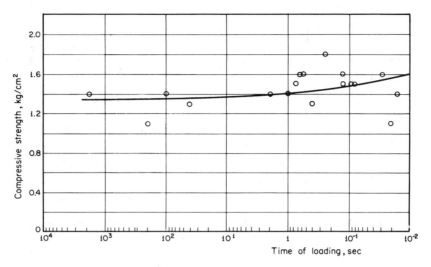

Figure 13.8. Effect of time of loading on triaxial compressive strength of sands. *After Casagrande and Shannon (1948).*

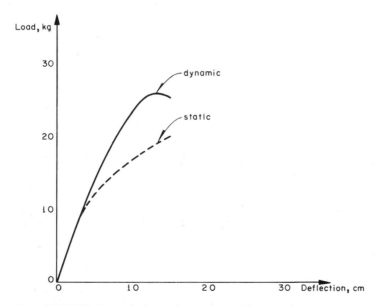

Figure 13.9. Static and dynamic stress-strain curves for Japanese cedar cantilever beams. *After Hisada and Sugiyama (1966).*

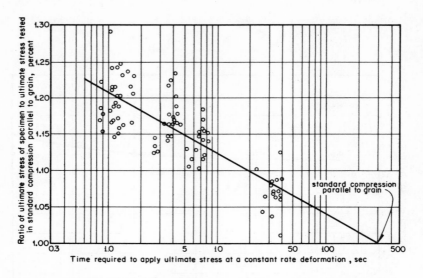

Figure 13.10. Effect of loading rate on ultimate strength of timber. *After Brokaw and Foster (1945)*.

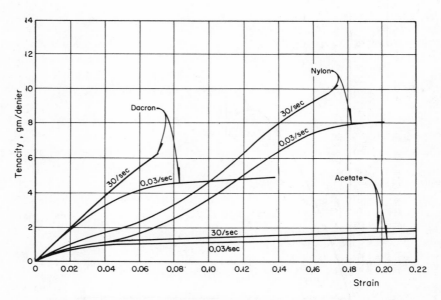

Figure 13.11. Effect of loading rate on the tenacity-strain relation of different polymers. *After Backer (1966)*.

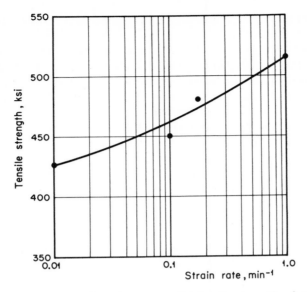

Figure 13.12. Effect of strain rate on the tensile strength of resin. *After Machlan and Edmunds (1962). Reproduced from Baer:* Engineering Design for Plastics, 1962, by Litton Educational Publishing, Inc., *with permission.*

example, Mindlin (1965) has shown that under conditions in which the time variable does not play a significant role and within the range of linear behavior, the strain gradient may significantly affect the stress-strain relations, which can be explained in terms of couple stresses. The matter apparently deserves additional study when time is significantly involved and when behavior is frankly nonlinear, especially in relatively coarse-grained materials.

In triaxial states of stress the effects of rate of loading may differ significantly from conditions under a uniaxial state. For example, when concrete is adequately confined, it behaves as a practically plastic material at high stresses and becomes appreciably more time sensitive than under uniaxial stresses (Richart, Brandtzaeg, and Brown, 1928).

Steel may fail in a brittle manner when it is subjected to either certain bi- or triaxial states of stress, low temperature, rapid loading, or where it has certain welding or metallurgical defects (Hall et al., 1967).

13.4 High-Cycle Fatigue of Simple Specimens[2]

In the question of high-cycle fatigue, emphasis is laid on strength, not on deformations. This is because, throughout most of the range of load repetitions,

---

[2] Comprehensive references on the subject include a report from the Battelle Memorial Institute (1949) and Yokobori (1964).

the stresses applied are quite small, so that materials behave in an almost linear manner up to fracture.

Typical *S-N* curves ($S =$ stress, $N =$ number of applications) in logarithmic plots for many materials can be idealized as two lines, the first of which is often practically straight and the second is parallel to the N axis. The stress associated with this second line is called the *endurance limit*. Below this stress the material withstands any number of applications.

Not all materials exhibit an endurance limit. With those that do not, any stress causes failure if applied a sufficient number of times.

In many fatigue tests the stress applied varies between two limits. To represent these results for any given material on a single chart, a *Goodman diagram* is conventionally used (Salmon, 1953). Similar diagrams can be drawn for specific values of N rather than for the endurance limits and are used for materials with or without an endurance limit.

Fatigue behavior depends on the size and shape of the specimens tested and on the distribution of applied stresses. These phenomena are partly due to nonlinear stress-strain relations, partly to statistical effects, and partly to the effects of stress gradient. The latter can be taken into account by using the theory of couple stresses (Mindlin, 1965). These stresses affect the stress distribution as a function of a characteristic length that depends on the coarseness of the structure of the material.

To take into account the variables mentioned in the foregoing paragraph, empirical "stress-concentration factors" have been tabulated as functions of the material, size, and shape of the specimen and the conditions of loading (Battelle Memorial Institute, 1949).

Results of fatigue tests are characterized by a wide spread. Hence, their description simply in terms of mean values is rarely satisfactory. Accordingly the formal theory of *reliability* (probability of survival) has found increasingly wide application in fatigue. The scarcity of experimental data in the range of small probabilities of failure makes it important that the theory be supported on sound conceptual (or stochastic) models. A model that has led to satisfactory results of rather general applicability is based on the assumption that the expected increase in the extent of damage is proportional to the damage already wrought in the specimen (Rascón, 1967).

Up to this point we have referred to loading cycles in which the minimum and maximum applied loads remain constant throughout the test. For many practical problems it is important to waive this restriction and obtain information on the effects of more general programs of loading. The results are adequately approximated under some conditions by the theory developed by Palmgren and Miner (Miner, 1945). The theory is synthesized assuming that failure occurs when

$$\sum_i \frac{n_i}{N_i} = 1 \qquad (13.1)$$

where n_i is the number of cycles applied in which the stress varies between the limits S_i' and S_i'', and N_i is the number of cycles that would be required to cause failure if the specimen were subjected only to a periodic stress history in which the stress would vary monotonically from S_i' to S_i'' and back to S_i'. In one sense, n_i/N_i may be viewed as the percentage of damage caused by those stress cycles which oscillate between S_i' and S_i''. The specimen fails when the accumulated damage reaches 100 percent.

This theory is based on the assumption that the order in which the various cycles are applied is irrelevant. The assumption stands in contradiction with vast experimental data on many different materials (Newmark, 1952), so that the theory is inadequate for predicting the effects of fatigue in completely general circumstances. Yet, when the loading process is stochastic with a poor correlation between successive cycles, Eq. 13.1 can be expected to afford a reasonably accurate criterion of failure in high-cycle fatigue.

Under the assumption that the N_i's are known deterministically for given sets of the stress limits S_i' and S_i'', Miles (1954) has calculated the expectation of $\sum n_i/N_i$ that corresponds to a white noise disturbance. Crandall, Mark, and Khabbaz (1962) have calculated the variance of this ratio.

13.5 Behavior of Simple Specimens under a Small Number of Loading Cycles

A distinction is traditionally made between the fatigue strength of materials and that of structural members. This does not stand on rigorous bases, since properties of materials are necessarily determined from tests on specimens, and these are characterized by size, shape, and surface conditions. Nevertheless there is room for distinguishing between the properties of simple specimens and those of structural components, as the latter are influenced by a number of additional conditions. Particularly in relation with low-cycle fatigue it is pertinent to retain this distinction.

A theory especially aimed at describing behavior of mild steel under low-cycle fatigue under uniaxial states of stress was developed by Yao and Munse (1962). It postulates that failure occurs when

$$\sum_i \left(\frac{\Delta_{pi}}{\epsilon_{ui}}\right)^z = 1 \qquad (13.2)$$

where

$$z = 1 - 0.86 \frac{\Delta_{ni}}{\Delta_{pi}}$$

The term Δ_{pi} is the difference between the maximum positive plastic strain and maximum negative plastic strain during the ith reversal of plastic strain (Fig. 13.13); ϵ_{ui} is the failure tension plastic strain for the particular precompression plastic strain ratio of the cycle (for mild steel an average $\epsilon_{ui} = 0.85$ may

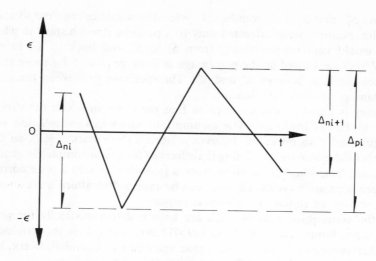

Figure 13.13. One cycle in the history of plastic strain.
After Yao and Munse (1962).

be assumed); Δ_{ni} is the precompression plastic strain induced before tension plastic strain is started in the cycle.

Kasiraj and Yao (1969) have applied the criterion of Yao and Munse to a group of single-degree frames, analyzing their responses to the records of two earthquakes. They find that, for a given ductility factor, the damage inflicted by an earthquake is, in general terms, an increasing function of natural frequency. The situation can be ascribed to the fact that the number of alternating cycles and hence the left-hand member in Eq. 13.2 are increasing functions of natural frequency.

The fact that Yao and Munse deal with plastic strains rather than stresses makes their approach much more satisfactory in low-cycle fatigue than that of Palmgren and Miner. However, the limitations stemming from neglect of the order of application of cycles, which we pointed out in connection with the criterion of Palmgren–Miner, characterize as well the Yao–Munse and other usual criteria used for the study of low-cycle fatigue.

For some materials at least, low-cycle fatigue resistance in triaxial states of stress can be predicted satisfactorily using a criterion developed by Mizuhata (1969), which uses the octohedral shear strain as the significant variable.

Much of what we said of high-cycle fatigue applies directly to behavior under a small number of load applications. However, we must now emphasize the changes that the stress-strain curve undergoes in successive cycles. When these changes involve reductions in stress for given strains or when they are associated with appearance of visible signs of damage, we speak of deterioration.

When the applied stresses are sufficiently small so that there is little departure from a linear stress-strain relation, it is usually adequate to assume that there

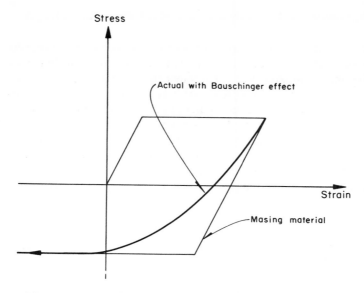

Figure 13.14. Stress-strain relation for structural grade steel.

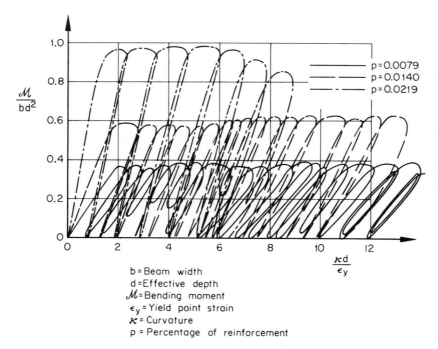

b = Beam width
d = Effective depth
$\mathcal{M}$ = Bending moment
ϵ_y = Yield point strain
κ = Curvature
p = Percentage of reinforcement

Figure 13.15. Repeated loading of singly-reinforced beam in bending. *Data after Sinha, Gerstle, and Tulin (1964b).*

398 BEHAVIOR OF MATERIALS AND STRUCTURAL COMPONENTS Chap. 13

is no deterioration. If the maximum stress exceeds the endurance limit, applicability of this assumption is also limited by the condition that the number of cycles be small compared with the number that would cause failure. Under these conditions, the idealization of material behavior as hysteretic without deterioration—for example as a Masing material (Section 11.3.6)—is satisfactory.

At the higher stress levels there is usually some "softening" of the material within the first cycle. Structural-grade steel, for example, exhibits the Bauschinger effect, through which the reversed-loading, stress-strain curve is associated with smaller stresses than the Masing type of idealization would predict (Fig. 13.14). Plain concrete and masonry develop microcracks that reduce the stiffness

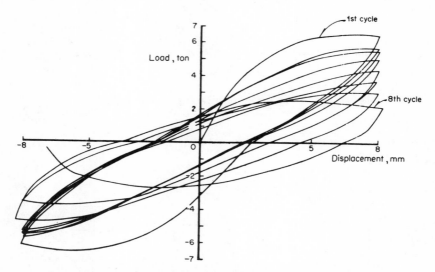

Figure 13.16. Load-deformation relations for reinforced-concrete columns. *After Yamada, Kawamura, and Furui (1966)*.

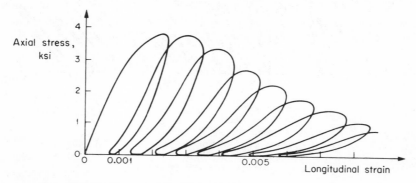

Figure 13.17. Repeated loading of plain concrete in compression. *After Sinha, Gerstle, and Tulin (1964a)*.

of these materials when they are subjected to additional cycles. Reinforced concrete and masonry develop macroscopic cracks, so that in additional cycles, only the steel is available for resisting tension. Depending on the percentage of reinforcement the reduction in stiffness may be substantial (Fig. 13.15). This reduction may even take place before the first cycle, as the material may develop cracks due to volume changes brought about by temperature variations or shrinkage.

A moderate number of loading cycles reaching high stresses may modify the shape of the stress-strain curve to a large extent. Figures 13.16–13.18 illustrate this point for different conditions.

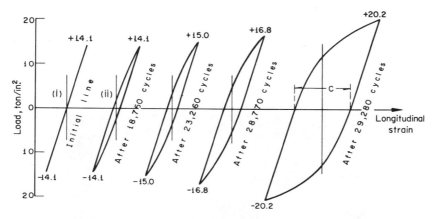

Figure 13.18. Repeated loading of axle steel. *After Salmon (1953)*.

13.6 Behavior of Structural Components

Arbitrarily we shall separate the discussion of the behavior of structural components into sections that pertain to flexural members, joints, frames and continuous beams, and diaphragms including shear walls. There are few data for other types of structural components.

13.6.1 Flexural Members.
We shall confine our discussion to the behavior of beams of reinforced concrete with rectangular cross section and to that of steel I-beams. This information may serve as basis for extrapolation to the behavior of other flexural members on which data are meager.

In earthquake-resistant design it is often important to check the shear strength of reinforced concrete columns of circular cross section. Indeed, probably the largest structural members to have failed are the columns of circular section of the Macuto–Sheraton Hotel in Caracas (Fig. 13.19), which suffered diagonal tension cracks during the 1967 earthquake. This question

400 BEHAVIOR OF MATERIALS AND STRUCTURAL COMPONENTS   Chap. 13

rarely occurs in design for gravity loads, so most building codes ignore it. Faradji-Capón and Díaz de Cossío (1965) have performed static tests on columns of this type, subjected to various axial loads. They conclude that the shear strength may be computed as for rectangular sections, using as effective concrete area the gross cross-sectional area. The effectiveness of transverse reinforcement may be taken equal to that in rectangular sections having the same area and spacing of this reinforcement.

In reinforced concrete beams behavior under a short number of load applications can be predicted fairly well from information on the behavior of simple specimens, with the following exceptions.

1. As soon as a beam develops an important diagonal tension crack, it deteriorates rapidly (Meli and Díaz de Cossío, 1964) as evidenced by the curve in Fig. 13.20. Moreover, the length of the beam increases steadily as successive loading cycles produce an increasing number of ever-widening, diagonal tension cracks (Pauley, 1969).

Figure 13.19. Column in the Macuto-Sheraton Hotel, diagonal tension failure. *Photo Courtesy American Iron and Steel Institute, New York.*

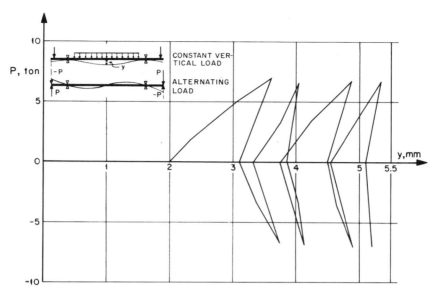

Figure 13.20. Repeated loading of reinforced-concrete beam failing in diagonal tension. *After Meli and Díaz de Cossío (1964).*

2. A similar situation probably exists as soon as an appreciable slip of the reinforcing bars takes place relative to the surrounding concrete or as soon as incipient split failure becomes manifest in a region of anchorage or in the neighborhood of a bar bend. Figure 13.21 illustrates a split failure caused by a particularly poor anchorage detail. The importance of stress concentrations in regions of bar cutoffs is such that, even under static loading, the *addition* of short tensile bars may lower the diagonal tension capacity of beams (Baron, 1966).

3. Bars having hooks or bends develop an appreciably smaller strength under high-cycle fatigue than similar straight bars (Pfister and Hognestad, 1964). As illustrated in Fig. 13.22 this phenomenon is noticeable after a moderate number of loading cycles.

4. Under static or cyclic loading that does not involve alternating flexure, the compression steel in straight members does not ordinarily buckle out of the concrete even at high strains and even in the absence of restraining ties or stirrups. In fact, the concrete cover is normally sufficient for the purpose; moreover, the curvatures induced where there is important bending make the longitudinal steel tend to move inwards.

The latter phenomenon does not apply to corner reinforcement, but a moderate amount of ties or stirrups suffices to keep this steel in place. It does not apply to curved members either, and in this case it is necessary to provide closely spaced ties or stirrups to protect the extrados steel

Figure 13.21. Split failure at a defective anchor detail.
Courtesy L. Esteva, México.

when it is in compression and the intrados steel in tension. Another exception is found in the longitudinal steel that undergoes the smaller compressive stress in columns under combined longitudinal force and bending, especially if the loads are sustained over long periods of time, so that the concrete creeps and gives place to large compressive strains.

But where the sign of the bending moment alternates and reaches such values that the steel that yields in tension during part of a cycle acts in compression during another part of that cycle, the tendency of these bars to buckle out of the member is so great that there is definite need for transverse reinforcement that will effectively confine them. Approximate methods of analysis are available[3] for determining the required spacing of transverse reinforcement in these conditions.

In properly detailed beams of reinforced concrete the conservative criteria advanced by Blume, Newmark, and Corning (1961) are applicable even under a moderate number of load repetitions. The criteria consist of idealizing the moment-curvature relation as elastoplastic with a limiting maximum curvature. Its parameters are chosen so as to give approximately the same strain energy per unit length as obtained from tests carried up to the ultimate bending moment.

[3] Bresler and Gilbert (1961) have derived one such criterion assuming that the member is subjected to static loading. Since that study neglects the contribution of the concrete cover, it actually applies to alternating flexure of the type mentioned herein. The principal simplifying approximation that it introduces is based on neglecting the curvature of the member.

Sec. 13.6   BEHAVIOR OF STRUCTURAL COMPONENTS

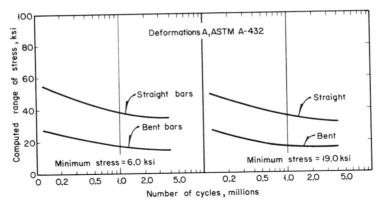

Figure 13.22. Repeated loading of reinforced concrete beams showing the effects of bar bents or hooks. *After Pfister and Hognestad (1964).*

The curvature and moment at yield of the tensile reinforcement mark the break in the idealized moment-curvature relation. They can be derived in a straightforward manner and will not be given here. Usually the ultimate moment may be taken equal to the yield moment without undue error. Accordingly all we need now is the ratio of curvature at ultimate to curvature at yield in order to define the entire moment-curvature relationship. According to the criteria in question, in a moderately reinforced rectangular section without compression reinforcement, this ratio is given by

$$\frac{\alpha_u}{\alpha_y} = \frac{\epsilon_{cu}}{\epsilon_y} \frac{1-k}{1-k_u} \tag{13.3}$$

where α denotes curvature, u refers to conditions at ultimate, y refers to conditions at yield, ϵ_{cu} is the ultimate strain of the concrete, ϵ_y is the yield strain of the tensile steel

$$k = \sqrt{(np)^2 + 2np} - np$$

and

$$k_u = \frac{pf_y}{f_{cu}}$$

Here n is the modular ratio (modulus of elasticity of steel divided by that of concrete), p is the percentage of reinforcement, f_y is the yield stress of the steel, and f_{cu} is the average compressive stress in the concrete at ultimate.

In unconfined plain, ordinary concrete cylinders under axial load ϵ_{cu} varies between 2 and 3.5×10^{-3}. However, under bending it is conservative to take $\epsilon_{cu} = 4 \times 10^{-3}$.

The average concrete stress f_{cu} may be taken as $0.7 f'_c$ for $f'_c \leq 5$ ksi, where f'_c is the standard cylinder strength of concrete, and as

$$f_{cu} = 1.5 \text{ ksi} + 0.4 f'_c$$

for higher strengths.

The addition of compression reinforcement increases the ductility because, while it affects k slightly, it reduces k_u significantly, as it can be usually assumed that

$$k_u = \frac{(p - p')f_y}{f_{cu}}$$

when $p' < p$.

In any case, the cutoff $\alpha_u/\alpha_y \leq 20$ is indicated because it would be difficult to develop a much higher ductility under alternating loads, and there would be little advantage in practice even if it could be developed.

Instead of Eq. 13.3, the cruder estimate

$$\frac{\alpha_u}{\alpha_y} = \min\left(\frac{10}{p - p'}, 20\right)$$

has also been used (Newmark and Hall, 1968), as it usually gives conservative results.

For a given moment diagram the actual curvature diagram in a beam is a complicated function of many variables. Blume, Newmark, and Corning recommend that it be idealized as illustrated in Fig. 13.23. To the bending moment diagram in Fig. 13.23a (taken as a straight line with point of inflection at A and support at B for purposes of illustration) there corresponds an elastic curvature diagram, 1 in Fig. 13.23b. The actual curvature diagram may be as line 2 in the figure. This exceeds the elastic diagram by the shaded area, which is approximately equivalent to a plastic hinge rotation at point B. The simplification proposed consists of replacing the actual diagram 2 with the elastic line

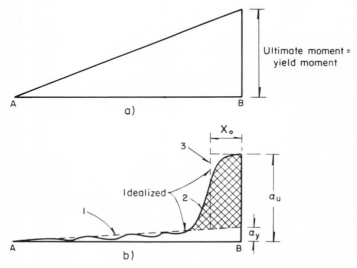

Figure 13.23. Probable and idealized curvature diagrams in reinforced concrete. *After Blume, Newmark, and Corning (1961).*

Sec. 13.6 BEHAVIOR OF STRUCTURAL COMPONENTS **405**

1 and the plastic diagram 3, assumed to extend over a length X_0. The authors suggest that X_0 be assumed equal to the effective depth of the member but not to exceed one half the distance to the point of contraflexure.

Approximately, then, the beam may be replaced with one that behaves linearly up to the curvature α_y, plus one that behaves as a mechanism having rigid members connected by hinges. The maximum rotation at a hinge is $X_0(\alpha_u - \alpha_y)$.

The ductility factor developed by a structural member is, of course, smaller than the ratio α_u/α_y and usually much smaller. And as we saw in Chapter 11, the ductility factor of a structure is smaller than that of its individual components.

Roy and Sozen (1964) have developed empirical formulas for estimating the ductility factors that members of reinforced concrete can develop.[4] The entire load-deformation history in pure flexure under successive cycles can be predicted with fair accuracy using a method developed by Bertero and Bresler (1969).

Repeated yielding of the longitudinal reinforcement and diagonal cracking often lead to concrete spalling (Fig. 13.24). This phenomenon causes a partial loss of bond (referred to in the literature as bond degradation). In turn this may produce failure in the zones of anchorage through progressive slip between concrete and reinforcement or split of the concrete.

In beams having compression reinforcement the widening of top and bottom

Figure 13.24. Spalling of concrete cover. *Photo Courtesy of L. Esteva, México.*

---

[4] The formulas do not provide for explicit consideration of the number of loading cycles but are conservative for most practical purposes, in which only a moderate number of load reversal is of interest.

flexural cracks, associated with cumulative plastic elongation of the longitudinal steel, makes this steel carry all the shear. This causes the steel to develop kinks and may lead to splitting failure of the concrete.

Spalling of the concrete, bond degradation, widening of cracks, and in some cases, the Bauschinger effect in the steel, together with a decrease of the modulus of elasticity of concrete under repeated cycles all contribute to reducing the stiffness of reinforced concrete frames (Bertero and Bresler, 1969). As a consequence their energy absorption capacity may be reduced quite drastically even if their strength is practically unimpaired, and even if their ductility factor, as it is usually defined, is increased in successive loading cycles. Most of the deleterious effects of repeated loading can be obviated by careful detailing and the use of closely spaced, closed ties ("tie-stirrups") in zones of plastic hinging.

Orthodox criteria for design of reinforced concrete members are almost exclusively concerned with strength, while ductility and energy absorption receive little consideration. This seems to follow from the fact that such considerations would revoke some of the fundamental oversimplifications of conventional ultimate strength design (Sozen and Nielsen, 1966).

Among the many questions easily overlooked in traditional design is the fact that very high ductility can develop only if the zones of plastic hinges are spread over long segments of the structural members; this requires that the reinforcement have an appreciable strain-hardening range beyond yielding. Also, splices of tensile reinforcement must be detailed with more conservative criteria than when designing against static loads, so that the concrete does not fail at these localized zones under seismic forces smaller than those which cause extensive yielding and strain hardening of the reinforcement. Generous use of confining ties in regions of splicing is especially advisable to prevent this manner of premature splitting of the concrete. Congestion of reinforcement, particularly at splices, has often been found to cause failure during earthquakes, whether these splices have been regions of tension, compression, or flexure.

Some experimental data are available on low-cycle fatigue of reinforced concrete members of intermediate slenderness ratios under combined axial and lateral forces (Yamada, Kawamura, and Furui, 1966; Newmark and Hall, 1968; Yamada, 1969). Yamada (1969) includes test results for concrete-filled steel pipes that exhibit exceedingly high energy absorption in the inelastic range. The test results reported by Newmark and Hall (1968) indicate that, as the axial load is increased, the ductility factor α_u/α_y decreases gradually to values of the order of 6 or somewhat greater when the axial force is sufficiently large to cause balanced failure (simultaneous yielding of the tension steel and crushing of the concrete in the compression face). The reduction is less pronounced with steel having a well-defined yield plateau and strain-hardening range than with other types of reinforcement. For larger axial forces the moment-curvature relation must be idealized differently, as the steel no longer yields in tension.

Prestressed concrete beams usually develop much smaller hysteretic areas than comparable beams of reinforced concrete. When their major departure from linear behavior is due to the opening of flexural cracks in tension, their hysteretic loops are quite thin (Spencer, 1968 and 1969), and their behavior may be idealized as elastic bilinear with an equivalent damping that is not more than a few percent of critical. At the same time these members are far less prone to deterioration under repeated loads than their reinforced concrete counterparts.

In steel I-beams and WF-beams the "softening" of material in the flanges, due to the Bauschinger effect, reduces its tangent modulus of elasticity causing the flanges to buckle under a few cycles. The phenomenon is sometimes accompanied by tearing of the web. Its onset is very sensitive to the level of straining and to the number of load reversals. Several tests have been performed on cantilever beams to study this phenomenon. The usual mode of failure is illustrated schematically in Fig. 13.25. Figure 13.26 gives the relation between number of load reversals and maximum strain required to cause buckling in these specimens.

One might expect the maximum strain and the number of cycles associated with flange buckling to be increasing functions of the thickness-width ratio of the flange. However, tests with width-thickness ratios between 14 and 23 show an insignificant effect of this variable (Torvi, Olson, and Davenport, 1966). Even tests on mild steel beams of square section exhibit a pronounced effect of the number of cycles of alternating yielding (Royles, 1966) much in agreement

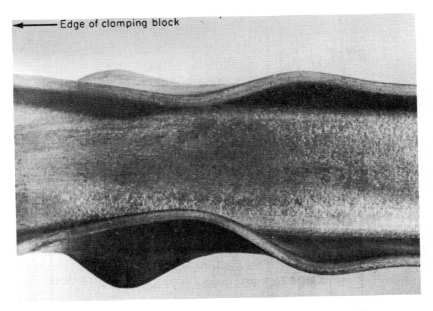

Figure 13.25. Typical local buckling of flanges in steel I-beam subjected to alternating flexure. *After Bertero and Popov (1965).*

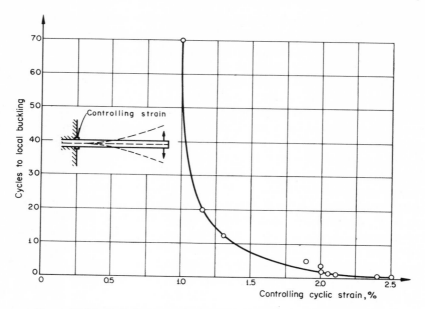

Figure 13.26. Relation between number of load reversals to failure and maximum strain. *After Bertero and Popov (1965).*

with what can be derived from low-endurance tests under uniform uniaxial strain.[5]

The tests on I-beams do indicate an appreciable effect due to stiffeners. Indeed, the maximum strains associated with failure at any given number of load reversals below 200 are about 300 percent lower than those in Fig. 13.26, apparently because the latter had the column stub acting as stiffener at the section of maximum strain. Thus, it is to be expected that closely spaced stiffeners will somewhat extend the safe domain under repeated, alternating loads. And it seems safe to expect a more favorable behavior of box sections under such loading. Although data are lacking on the behavior of stiffened or closed sections under alternating loads, the information contained in Fig. 13.26 may serve as basis for extrapolating the wealth of theoretical and experimental results of static tests on members having these or other shapes.

13.6.2 Joints. Joints between structural members are among the most vulnerable portions of structures to repeated loading.

Test results for beam-column connections in reinforced concrete disclose that lateral restraint of concrete of the joint is essential to ensure adequate behavior, especially under repeated, alternating loads (Hanson and Conner, 1967). Unless this restraint is provided, the concrete splits under bending moments considerably smaller than one would compute ignoring this phenome-

---

[5] See the theories and test results by Royles (1966) and by Krishnasamy and Sherbourne (1966). These theories predict moment-curvature relations and low-endurance flexural fatigue of beams having a square cross section.

non. A few cycles of loading suffice then to reduce the capacity of the joint practically to zero. Moreover, large diagonal cracks develop at relatively low stresses (Corley and Hanson, 1969).

Beam-column connections of the type shown in Fig. 13.27a exhibit especially poor behavior. A well-known case quoted in the literature (McKaig, 1962) concerns a frame that collapsed under static load owing to a defective detail similar to the one in this figure despite adequate compliance with code requirements. Connections of the type in Fig. 13.27b also behave poorly under repeated load for the same reason.

Some criteria advanced for the design of ties or stirrups to prevent premature failure of connections of types in Figs. 13.27a and 13.27b (Fig. 13.27c) consider the shear or diagonal tension at the intersection as the only relevant variable (Hanson and Conner, 1967; Corley and Hanson, 1969). Since the stress distribution in the concrete differs markedly from that in a beam some distance away from a support, these criteria can only be applied without much error on the safe side, if we assume that the ultimate shear that the concrete takes exceeds considerably the one ordinarily applicable in beams (perhaps 60 percent).

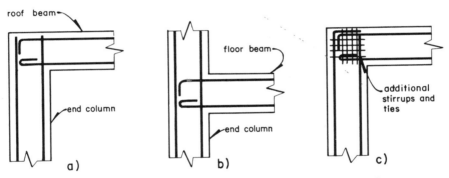

Figure 13.27. Some beam-column connections in reinforced concrete structures.

With an adjustment of this type it is conceivable that these criteria may serve not to prevent, but to limit, the growth of diagonal cracks. (Diagonal stirrups or ties would indeed prevent the appearance of such cracks, but their use is hard to justify in most jobs.) Besides, criteria based only on consideration of shearing stresses afford no means for controlling or preventing split of the concrete in planes that contain the bent bars. It is, nevertheless, relatively simple to devise conservative criteria that protect against failure of this kind.

In most beam-column connections there is a beam perpendicular to the plane defined by the column and the beam in question. Presence of the perpendicular beam ordinarily provides all the confinement that is required if the beam is properly situated relative to the connection. This conclusion was reached as a result of the tests mentioned, despite the absence of loads on the perpendicular beam. In practice, conditions will probably be even more favorable due to the negative moments acting at the faces of the support of this beam.

Connections between steel members, whether welded or riveted, usually become quickly stabilized under repeated alternating loads, so that after a few cycles the force-deformation curves hardly change from one cycle to the next until the number of cycles approaches that which causes failure (Popov, 1966). In most instances these curves closely resemble those for Masing bodies (see Section 11.3.6), which can be explained for riveted connections in view of their similarity to the corresponding conceptual model (Fig. 11.18a).

Despite the general adequacy of Masing's idealization of behavior, Popov and Pinkney (1969) point out that depending on the grade of steel, the skeleton curve obtained from incremental and cyclic loading may lie slightly above or below the skeleton curve for monotonic loading. For the diverse types of connections that they tested, the cyclic skeleton curve lies slightly above.

Most of the connections tested by these authors exhibited a skeleton moment-curvature relation that could be closely approximated by a Ramberg–Osgood curve (Eq. 11.22). Upon integrating the curvature diagram twice, this produces the same type of relationship between load and deflection. In successive cycles of unloading and reloading, the relation takes the form

$$\frac{y - y_i}{y_0} = \frac{Q - Q_i}{\beta}\left(1 + \alpha\left|\frac{Q - Q_i}{2Q_0}\right|^r\right) \tag{13.4}$$

where y denotes deflection; Q is load; α, r, y_0, and Q_0 are parameters of the skeleton curve; y_i and Q_i are the coordinates of the point where the unloading or reloading cycle begins; β is a coefficient that reflects the degree of stiffness degradation and depends on the maximum curvature attained in previous cycles, and on the number of loading cycles, due to Bauschinger and other effects. In a truly Masing type of behavior, $\beta = 1$. Popov and Pinkney find that, for the connections they tested, α is usually of the order of 0.5, r lies between 7 and 8, and β is usually close to 1. As an exception in some connection, β may decrease from slightly more than 1 to somewhat below 0.5 after more than 60 cycles of severe straining.

Not all the connections had load-deflection curves that followed the ideal shape described by Eq. 13.4. For example, a bolted connection with oversized holes and a welded connection with undersized connecting plates gave load-deflection curves of atypical shapes, with about 30 percent smaller hysteretic area than the normal curves for the same maximum load.

Despite their general stability of behavior under repeated loads, connections between structural members are especially sensitive to the quality of workmanship. Even laboratory specimens have failed under low loads due to poor workmanship (Popov and Pinkney, 1969).

The most common factor limiting the capacity of properly designed, rigid connections is the yielding, in shear, of the beam-column intersection. The average shearing stress in this portion of a connection can be computed from considerations of statics, and usually, it is sufficiently accurate to assume that overall yielding of the intersection occurs when this average shear reaches the value $1/\sqrt{3}$ times the tensile yield stress of the material, in accordance with von Mises' criterion of plasticity (Bertero, 1969). Even within the range of nearly linear behavior, shearing strains of connections may be substantial, and they should be included in frame analysis.

Most tests and analytical studies of steel connections omit consideration of the effects of gravity loads, which may be substantial (Bertero, 1969). These effects may be incorporated in calculations using approximate and relatively elementary criteria. Information is also available on the behavior of tubular joints under alternating loads (Bouwkamp, 1966).

13.6.3 Frames and Continuous Beams.
In part the behavior of frames and continuous beams under repeated loading can be predicted from that of flexural members and of joints. But we must also pay attention to redistribution of bending moments and to the phenomenon of shakedown.

Results by Bertero, McClure, and Popov (1962) of a series of tests on small, reinforced concrete frames illustrate the type of information that can be derived directly from the behavior of flexural members and joints. Anchorage of the reinforcement in the beam and columns was accomplished through the use of anchor plates because the more standard solution, consisting of bent bars, deteriorated rapidly.

It was found that the strength of these frames was not greatly affected by repeated loading. After 100 cycles at nearly 80 percent of the strength exhibited in static loading, frames still developed more than 95 percent of this strength. A frame subjected to 4 alternating cycles at 85 percent of its static capacity did not suffer any reduction in strength. These results agree with the results found for beams and columns because the frames were of such proportions that their strength was conditioned by flexure.

It is true that strength was almost insensitive to the repetition of loading. Yet, stiffness, and hence the capacity to dissipate energy through inelastic deformation, deteriorated rapidly. This observation is illustrated in Fig. 13.28.

Umemura and Aoyama (1969) have developed comprehensive criteria for the calculation of the deflections of inelastic, reinforced concrete frames from the behavior of structural members. Their detailed and realistic recommendations incorporate the shear deformations of members and joints, which are often significant.

In steel frames the lateral force-deflection relations can be predicted with accuracy, but this requires quite often that the effects of vertical loads, strain hardening, and shearing deformations of members and connections be properly taken into account (Wakabayashi, Nonaka, and Matsui, 1969).

The importance of residual stresses is brought out in results of tests performed on a steel frame of the type shown in Fig. 13.29 (Arnold, Adams, and Lu, 1966). The first load cycles gave the curves shown in Fig. 13.30. Paradoxically, in the second cycle the structure developed a much higher resistance to lateral force than on first loading. The phenomenon is due to the combined effects of residual stresses and vertical forces. The force-deformation curves shown can be predicted under the assumption that the moment-curvature relations for monotonic loading hold at all stages, but one must take due account of the effects of axial compression and of secondary moments induced by longitudinal forces. (Com-

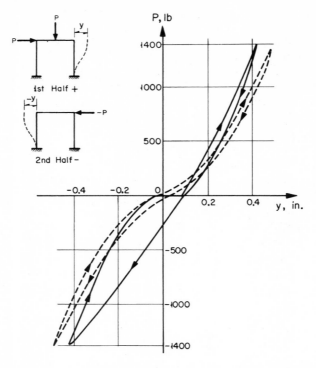

Figure 13.28. Force-deformation relations for model of reinforced concrete frame. *After Bertero, McClure, and Popov (1962).*

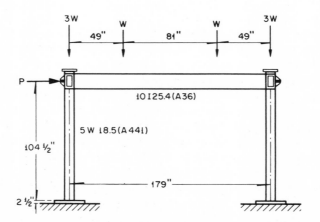

Figure 13.29. Steel frame tested under repeated loading. *After Arnold, Adams, and Lu (1966).*

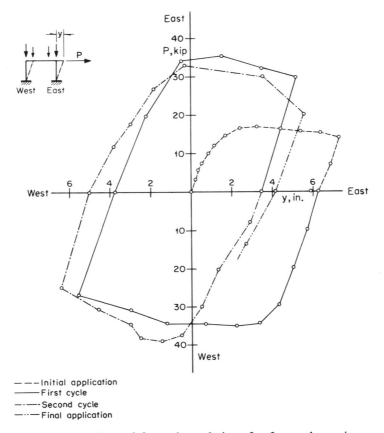

- - - Initial application
——— First cycle
—·—· Second cycle
—··— Final application

Figure 13.30. Force-deformation relations for frame shown in Figure 13.29. *After Arnold, Adams, and Lu (1966).*

parable results have been obtained by Carpenter and Lu, 1969.) After a few cycles of alternating, horizontal loading, the frame buckled in torsion.

Vertical deflection at midspan of the beam increased with each horizontal load application. It showed little recovery on unloading. This characteristic is to be kept in mind when designing frames that are expected to go through major excursions in the plastic range even if their force-deformation curves are symmetric for horizontal loading.

Under certain conditions a statically indeterminate structure may suffer inelastic strains during the first or first few cycles of loading and then, as a consequence of stress redistribution, undergo exclusively elastic strains. It is said to have *shaken down*. The phenomenon requires an increasing number of loading cycles to set in if the process is carried out with higher amplitudes of the oscillating force. At sufficiently high forces the structure never shakes down. The process can be calculated (Hodge, 1959).

Judging from tests carried out on frames and continuous beams of reinforced concrete the conventional theory of shake down errs substantially on the safe side (Gerstle and Tulin, 1966).

Methods for the analysis of frames that are based on the assumption that moment is uniquely related to curvature are unable to cope with a negative derivative of bending moment with respect to curvature. With a unique moment-curvature relation for each section independent of that for other sections, an increase in the maximum deformation (say, the deflection) of a statically determinate structure beyond the value associated with the maximum curvature would reduce the bending moments in the entire system. This would imply a general reduction in curvature, except at one section where it would increase. The situation would require a reduction in the maximum deformation; this contradicts our initial assumption. It follows that the assumption of unique dependence of moment on curvature is inconsistent with the conditions for structures in which the force-deformation curve exhibits a descending branch. Since such descending branches have been obtained in actual tests, we must use a moment-curvature relation that involves neighboring sections as well as the section under consideration or, in effect, consider that finite lengths of section are governed by the same relationship.

The paradox is similar to the one we find in stress-strain curves for mild steel bars, for example, in which striction causes the stress to drop gradually after the bar has reached a certain strain. To solve problems of this type we must deal explicitly with a dependence of stress at each section on the strains in a region near that section. A parallel, although very elementary, approach has been attempted in connection with moment-curvature relations (Rosenblueth and Díaz de Cossío, 1964).

Unfortunately, there has been very little research in this area owing to the irrelevance of the descending branch in statically determinate structures subjected to gravity loading, because these structures fail without having an opportunity to derive benefit from the descending branch. In statically indeterminate structures under gravity loading, and in all structures subjected to dynamic disturbances, the descending branch of moment-curvature and other generalized force-deformation relations signifies the disposability of reserve capacities that sometimes may be quite important. For the reasons we have mentioned, there is little information on the effects of rate of deformation on the slope of the descending branch, but the phenomenon seems to be significant.

13.6.4 Diaphragms. Some results of tests on timber diaphragms are displayed in Fig. 13.31. Such tests have led to the conclusion that a moderate number of stress reversals produce practically no deterioration. Successive cycles at a constant maximum deformation give almost unchanging force-deformation curves.

The shapes of these curves differ appreciably from those we would obtain in Masing-type structures. Equivalent hysteretic damping is of the order of 8 to 10 percent independent of amplitude. (These values were computed from the

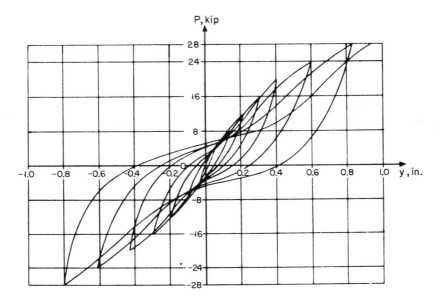

Figure 13.31. Results of repeated loading of timber diaphragms. *After Medearis (1966).*

ratio of average hysteretic energy to average strain energy under the skeleton curve.) Since wood is quite sensitive to strain rate (Hisada and Sugiyama, 1966), one would expect these curves to change materially in shape as a function of the velocity of loading.

The following remarks apply to static tests on reinforced concrete shear walls (Benjamin and Williams, 1959). These remarks are based on test results obtained from monotonic lateral loading of wall panels of various types and proportions. Some of these are shown schematically in Fig. 13.32.

The results indicate that the force-deformation relations for these elements can be reasonably well approximated as two straight lines up to the maximum force (Fig. 13.33). The slope of the first line can be computed using the standard theory of strength of materials (adding shear and flexural deformations) applied to the uncracked section of concrete. The force-deformation curve has a break at the force that causes cracking of the concrete. Empirical equations are available for predicting the second line. Roughly, the slope of the second straight line can be obtained similarly but assuming that the effective section is defined by the modular ratio transformed section, with concrete taking no vertical tension. The maximum force may be computed using the standard theory of reinforced concrete. The descending branch leaves a non-negligible reserve strength in the wall.

The same tests indicate a much greater effectiveness of vertical than of horizontal reinforcement. In fact, horizontal steel served only to limit the widths of vertical cracks and contributed little if any to resisting shear. (The tests

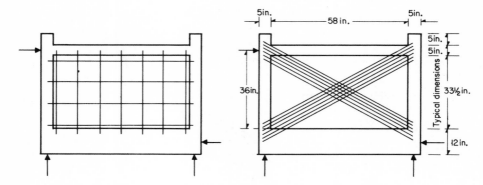

Figure 13.32. Reinforced concrete shear walls subjected to monotonic lateral loading. *After Benjamin and Williams (1959).*

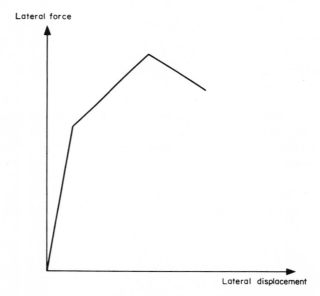

Figure 13.33. Typical force-deformation relation for reinforced concrete shear wall. *After Benjamin and Williams (1959).*

included panels whose horizontal dimension was equal to or greater than the panel height. It is debatable whether the same conclusion would apply to panels whose height greatly exceeded their length.) Inclined steel was less effective than vertical bars for resisting lateral load; when properly detailed, it was useful in decreasing the widths of cracks near the corners.

The variability of stiffness and strength of nominally identical shear walls is quite high even under laboratory control. The situation can lead to important discrepancies between any deterministic conceptual model of structural behavior and the behavior of a real building under earthquake loading, especially in what concerns torsional eccentricities. Benjamin (1969) has analyzed a simple case of this type under simplified assumptions approximately applicable up to the development of the first crack in the walls. His analysis is based on test results which disclose an extreme type 1 distribution with a coefficient of variation of about 0.2 for wall stiffnesses. The nominal shearing stress that causes wall cracking was found to have an approximately normal distribution with a coefficient of variation slightly in excess of 0.3.

Behavior of unconfined wall panels of plain masonry under repeated loads is not of particular interest. For sufficiently low shearing stresses the panels behave practically in a linear fashion under a single application of lateral force. Presumably, a few load repetitions in these walls do not introduce marked departures from the behavior of an almost linear Masing-type material. At higher stresses the walls fail in a brittle manner.

There is very little experimental information about the responses of plain masonry shear walls under the combined action of vertical and lateral forces. It is to be expected that vertical forces introduce a confinement similar to that provided by enclosing frames. From tests by Esteva (1966) on unconfined, plain masonry walls in which the vertical load has not exceeded 20 percent of the capacity of the wall, it has been found that the shearing strength equals approximately that for a wall without vertical load plus 0.38 times this load.[6]

When a partition or wall panel is enclosed by a sufficiently strong frame with which it is in close contact, the lateral force associated with any given angular deformation may be approximated as the sum of two terms: (1) the force taken by the frame as though there were no wall and (2) the force taken by the wall as though it were surrounded by a mechanism with rigid members joined by hinges (Fig. 13.34).

In a series of cycles of alternating lateral loads the force-deformation curve takes the shapes shown schematically in Fig. 13.35. Separation between wall

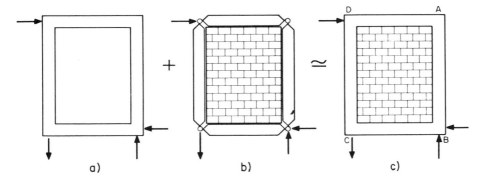

Figure 13.34. Idealization of the behavior of wall-frame systems.

---

[6] However, data by Benjamin and Williams (1958) indicate a much higher effect of vertical forces on shear strength.

418 BEHAVIOR OF MATERIALS AND STRUCTURAL COMPONENTS Chap. 13

and frame at the two corners where there is tension between these elements (*A* and *C* in Fig. 13.34c) may take place at angular distortions as low as 10^{-4} or as high as 30×10^{-4}, with an average of about 6×10^{-4}, depending primarily on the amount of shrinkage. Up to this load the force-deformation curve is practically straight. A second break in the curve, not always noticeable, occurs when a diagonal crack develops in the wall (along *BD* in Fig. 13.34c) at a shearing stress of approximately $0.75\sqrt{f'_m}$ in kg/cm², where f'_m is the direct

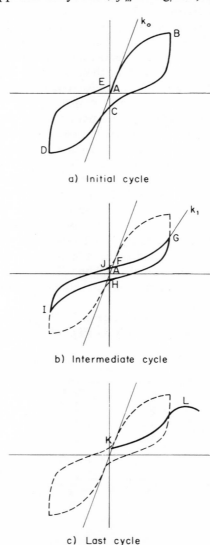

a) Initial cycle

b) Intermediate cycle

c) Last cycle

Figure 13.35. Typical load-deformation curves for frame-enclosed plain-masonry panels. *After Esteva (1966).*

compression strength of the wall. The maximum shearing force corresponds to a shearing stress of $0.9\sqrt{f'_m}$ in the wall panel. The wall-frame system may continue to deform practically without limit so long as the frame does not fail.

In successive loading cycles there is little decrease in strength, except for very slender walls or those of hollow tile. However, as shown in Fig. 13.35, even for plain masonry walls there is a drastic loss of stiffness and of energy absorption capacity.

In extrapolating results of laboratory tests to behavior under actual earthquakes we must recognize the effects of accelerations perpendicular to the walls in question. After being damaged by the action of motions contained in their own plane, unreinforced walls have little resistance to accelerations in the perpendicular direction and may easily drop, especially when they are of hollow tile (Fig. 13.36). After this the walls offer practically no resistance in their own planes. The event is particularly objectionable in facades because the falling debris may injure or kill pedestrians.

Analytical and experimental studies by Smith (1962 and 1966) on diagonally loaded, elastic rectangular panels have led him to propose estimates of the initial stiffness. There are experimental indications, however, that this criterion overestimates the initial stiffness when the frame is very flexible relative to the wall (Esteva, 1966), but the corresponding studies are available only for square panels surrounded by four equal reinforced concrete members.

Criteria are yet to be developed for determining the amount of reinforcement required at the corners of the frame to prevent diagonal cracking at these joints.

Figure 13.36. Falling out of hollow tile during the Caracas earthquake of 1967. *Photo courtesy of L. Esteva, México.*

The tests quoted have been run with frames heavily reinforced at the corners so as to prevent brittle failures of this type which were observed in preliminary tests.

Diagonal cracking at the corners of enclosing frames is not very common in practice. This may be because, except for the roof, frames are protected by the presence of important vertical loads. This contention seems to be borne out in the results of tests by Fiorato, Sozen, and Gamble (1969) on reduced-scale models of multistory frames with filter walls. Cracks developed in the reinforced concrete beams and columns, sometimes before visible cracking of the walls, but in all cases they began near, not *at*, the beam-column joints. The reason for the apparently premature cracking of the frames seems to have been a separation between frames and filler walls. According to Esteva, Rascón, and Gutiérrez (1969), cracking at the frame member intersections occurs only after diagonal cracking of the walls. Hence, if the danger of this type of wall cracking is sufficiently remote, one need not be concerned with the stress condition at the joints. Still, such failures do occur in actual buildings (Esteva and Nieto, 1967; Esteva, Rascón, and Gutiérrez, 1969). A design criterion is required in view of the brittleness associated therewith.

The presence of small gaps between the frame and the wall sharply decreases the initial stiffness of the system (Smith, 1966). There are indications that it also affects its ultimate capacity. This matter depends on factors that are difficult to control in actual construction. Hence, it introduces a source of uncertainty and of large variations in stiffness between panels of a single building.

Behavior of frame-enclosed, reinforced masonry panels is essentially like that of similar plain masonry panels. The increase in strength due to reinforcement is negligible, and the increase in ductility found in laboratory tests is not very significant (Meli and Esteva, 1968). However, well-anchored reinforcement can be expected to provide adequate protection against the fall of portions of masonry panels under actual earthquakes.

13.7 Behavior of Complete Structures

The literature abounds with reports on small amplitude, forced or free vibration tests of structures, aimed at determining their first few natural modes of vibration and corresponding degrees of damping. Many of these studies attempt to derive empirical relations between the fundamental period of vibration and the overall exterior geometry of groups of buildings. Such formulas have found their way into building codes. For example, several codes use the formulas recommended by Anderson et al. (1952)

$$T_1 = \frac{0.05H}{\sqrt{b}}$$

where T_1 is the fundamental period in seconds, H the height in feet, and b the base dimension, in the direction considered, also in feet. The expression $T_1 =$

0.1N (where N is number of stories) has also been used extensively. These and comparable formulas give an indication of the order of magnitude of the fundamental period that may be expected, but they may err systematically by a factor of 2 or more in regions of the world where soil conditions and professional practice differ significantly from those prevalent in California in the 1940's and early 1950's (Arias, Husid, and Baeza, 1963; Carmona and Herrera-Cano, 1969), and individual buildings have a fundamental period many times longer or shorter than given by the formulas (see the formal discussions of Anderson et al., 1952).

A more fruitful use of data on small amplitude vibrations of actual structures and of their response to small earthquakes consists of "reconciling" them with values computed on the basis of structural mechanical properties and thus calibrating and refining the necessarily simplified methods of calculation. This is a straightforward task that yields excellent agreement in the case of relatively simple structures, such as bare steel frames resting on firm ground, so that soil-structure interaction may be neglected (Nielsen, 1969). It becomes a formidable chore in more complex structures, such as buildings with cladding which rest on soil of appreciable deformability (Blume, 1956; Osawa et al., 1969).

Ordinarily it is found that the damping ratio associated with natural modes of vibration varies little with natural frequency. Indeed, when foundation compliance is not very pronounced, the constant-Q assumption (see Chapter 3) is found to apply with satisfactory accuracy, and the percentage of damping is found to depend almost exclusively on the structural materials and type of structure. It usually increases slightly with amplitude of vibration, especially in badly cracked, reinforced concrete structures. Representative values of percentages of damping for low amplitude vibrations are given in Table 13.1.

Exceptionally the damping coefficient has been found to vary almost in proportion to natural frequency (Nielsen, 1969), as it does in a Sezawa or Voigt body (Chapter 3). It can then be incorporated into a conceptual model of the structure as a number of dashpots in parallel with the flexible elements of the structure, assigning to the dashpot damping constant values that are proportional to the spring constants. In some cases, however, the percentage of damping decreases with increasing natural frequency (Funabashi, Kinoshita, and Aoyama, 1969), but this condition is equally exceptional.

In steel structures, practically the entire process of energy dissipation takes place hysteretically. Hence, the percentage of damping, as a function of amplitude, can be predicted accurately from alternating static tests (Rea et al., 1969).

In clad frames the problem of stiffness degradation, accompanied by a progressive loss of energy absorption capacity, plays as important a role as in reinforced concrete members.

Some tests on buildings constructed from large precast panels have been carried to sufficiently high amplitudes to cause cracking of and around the connections and have led to a revision of design criteria for this type of structure (Polyakov et al., 1969).

Few static tests with monotonically increasing deformation have been carried

TABLE 13.1. TYPICAL VALUES OF DAMPING IN NUCLEAR REACTOR FACILITIES [AFTER NEWMARK AND HALL (1969) AND NEWMARK (1969b)]

Stress level	Type and condition of structure	Percentage of critical damping
1. Low, well below proportional limit, stresses below $\frac{1}{4}$ yield point	Vital piping	0.5
	Steel, reinforced or prestressed concrete, wood; no cracking; no joint slip	0.5–1.0
2. Working stress, no more than about $\frac{1}{2}$ yield point	Vital piping	0.5–1.0
	Welded steel, prestressed concrete, well reinforced concrete (only slight cracking)	2
	Reinforced concrete with considerable cracking	3–5
	Bolted and/or riveted steel, wood structures with nailed or bolted joints	5–7
3. At or just below yield point	Vital piping	2
	Welded steel, prestressed concrete (without complete loss in prestress)	5
	Reinforced concrete and prestressed concrete	7–10
	Bolted and/or riveted steel, wood structures with bolted joints	10–15
	Wood structures with nailed joints	15–20
4. Beyond yield point, with permanent strain greater than yield point limit strain	Piping	5
	Welded steel	7–10
	Reinforced concrete and prestressed concrete	10–15
	Bolted and/or riveted steel, and wood structures	20
5. All ranges; rocking of entire structure*	On rock, $v_s > 1800$ m/sec	2–5
	On firm soil, $v_s \geq 600$ m/sec	5–7
	On soft soil, $v_s < 600$ m/sec	7–10

*Higher damping ratios for lower values of shear-wave velocity v_s.

out up to collapse of complete structures. Apparently no repeated load tests of this sort have been performed. A few notable instances of the first type deserve description.

The first concerns static tests on four, full-scale, two-story buildings of reinforced concrete having different structural solutions (Ihara and Ueda, 1965). The tests were carried up to lateral displacements of the roof of the order of 30 cm. Loading was accomplished by means of hydraulic jacks. At various stages small amplitude dynamic tests were performed.

It was found that conventional methods of analysis slightly underestimated stiffness of framed structures in the range of small deformations. These methods overestimated the stiffness of structures with shear walls. (By "conventional" we denote methods that consider member lengths center to center and disregard foundation compliance.) The same methods, coupled with standard theory of reinforced concrete, systematically underestimated the ultimate capacities of the structures.

Very high ductilities were found for all the structures; ductility factors ranged between about 15 and 40. It is well to point out that loading was essentially monotonic (with a few cycles of unloading but no reversed loading) and that the four structures had been carefully detailed and built; beams and columns were generously provided with closed ties, and only one of the structures exhibited some weak details. The shear walls cracked noticeably at much lower ductility factors than the frames, but the walls were still able to take large deformations.

Natural periods of vibration after severe cracking were about 50 to 110 percent larger than those at small deformations. Damping coefficients were of the order of 3 percent at small deformations and 4 to 9 percent after severe cracking; they were measured for small oscillations in all cases.

A second example concerns a one-fourth-scale model[7] of a four-story prestressed concrete building (Nakano, 1965). The manner of testing and the results obtained were to some extent similar to those of the first example. However, damping ratios were found to be somewhat smaller in the prestressed model. Nakano's paper contains very interesting information on the development of cracks, including some in torsion that a superficial analysis would not have suggested. Several practical design recommendations are presented in the paper.

The *Proceedings of the Fourth World Conference on Earthquake Engineering*, from Santiago, Chile in 1969, contains a few other examples of large- and full-scale tests of complete structures.

13.8 Human Reaction to Earthquakes

The reaction of human beings to earthquake motions depends on numerous factors. Paramount among these are the motion characteristics and the subjective responses of people.

Pertinent characteristics of a ground motion include its intensity and duration, its prevailing periods, the regularity along its three components of translation, and to a lesser extent, the corresponding characteristics of the rotational components of the motion. Quantitative information on the effects of these variables is apparently confined to the results of experiments performed under laboratory conditions for vertical, steady-state harmonic oscillations (Wright and Green, 1959).

The two independent variables controlled in those experiments were amplitude and frequency. Effects were evaluated according to whether the vibrations were perceptible, caused discomfort, or were hardly bearable. As we shall see in Chapter 15, at high frequencies people are essentially sensitive to acceleration; at intermediate frequencies they are sensitive to velocity; at low frequencies they are sensitive to displacement.

Other studies (Hutchinson, 1965) indicate a small correlation between the

---

[7] The scale can be said to be any number between, roughly, 1:3 and 1:4, depending on the prototype envisaged.

response curves and the natural frequencies of certain organs, such as the eyes. These correlations are weak for vertical vibrations, and they are probably not significant for horizontal oscillations. Hence, it is not unlikely that the effects of earthquake motions are essentially associated with maximum velocity.

The latter contention is strengthened by the correlation that has been noticed between maximum ground velocity and intensity, the latter expressed in subjective scales. The assumption that maximum velocity is the only significant variable ignores the influence of the duration of motion. In part this is offset by the fact that motions of exceptionally long duration normally have long prevailing periods and their nature is more regular than that of shorter-lasting quakes. Regularity breeds some anticipation of the next swaying motion and hence reduces the severity of human response.

Fear depends essentially on previous experience. Recent occurrence of a destructive ground motion increases fear, while exposure to several mild earthquakes reduces it.

Pending further research it is suggested that maximum velocity be taken as the significant variable to assess human reaction to earthquakes.

Whether human reaction should be a factor to reckon in structural design is subject to controversy. But there should be little room for doubt concerning its relevance in the design of places where large numbers of people congregate and where panic may cause further disaster.

13.9 Behavior of Soils

13.9.1 Dry Cohesionless Soils.

For purposes of the present discussion we shall recognize three types of cohesionless soils. The first group comprises soils that consist essentially of medium-sized grains of sufficient strength or those under sufficiently small stresses, so that grain breakage does not play a significant role in their behavior. The second type are those soils made up essentially of large-size grains, as is the case with rockfills, subjected to such high stresses that volume changes are significantly conditioned by grain breakage. The third type includes fine-grained materials, such as silts, that may liquify or be lubricated by air.

Behavior of the first type of dry, cohesionless soils can be described in terms of the "critical void ratio." The term was introduced by Casagrande (1936). It describes the limiting void ratio, above which a soil tends to reduce in volume when deforming in shear and below which the material exhibits a tendency to dilate. It has been shown (Whitman and Healy, 1962) that the critical void ratio is a decreasing function of the confining pressure (the first stress invariant) and an increasing function of the shearing strain. The influence of the shearing strain on the critical void ratio is such that all cohesionless soils tend to dilate for sufficiently large strains. Figure 13.37 illustrates typical relations between these variables.

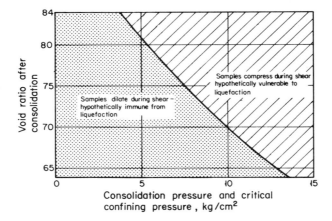

Figure 13.37. Critical confining pressure-void ratio relationship for Sacramento River sand. *After Seed and Lee (1966).*

Effects of the speed of loading are insignificant on the stress-strain relations of dry cohesionless soils within the range of speeds that is of interest in seismic problems, provided that conditions are such that the air in the voids of the soil does not develop an appreciable change in pressure. In fact, going from a time to failure of the order of minutes to one of the order of milliseconds does not raise the angle of internal friction by more than about one degree (Whitman and Healy, 1962). For the soils we are considering, pore pressures do not play a significant part.

As we shall see in connection with saturated materials, repeated alternation of loading eventually causes volume decrease in all cohesionless soils. The matter is not particularly significant for dry or partially saturated materials,[8] and we shall be content to present Barkan's (1962) theory of the effects of vibrations. This treatment has been developed for cohesionless soils under no more than a mild confining pressure.

From certain simplifying assumptions Barkan concludes that the final or equilibrium void ratio of a soil undergoing harmonic vibration and which is initially in its loosest state is given by

$$e_f = e_{\min} + (e_{\max} - e_{\min}) \exp\left(\frac{-B_e A}{g}\right) \tag{13.5}$$

where $e_{\min}$ and $e_{\max}$ are, respectively, the void ratios corresponding to the densest and loosest possible states of the material; B_e is a coefficient, of the order of 0.75 for a typical sand; A is the amplitude of the applied acceleration; g is the acceleration of gravity. Test results show good agreement with Eq. 13.5.

In this simplified theory, if the initial void ratio e_0 does not exceed e_f, no compaction occurs under vibration, and the void ratio remains unchanged.

---

[8] However, some test results under relatively high confining pressures (Seed and Lee, 1966) cast doubts on the generality of this statement.

Equation 13.5 applies only when $e_0 \geq e_f$. The theory implies, therefore, that e_f plays the role of the critical void ratio we referred to for static tests but with no dilatation occurring whatever the initial void ratio.

The time required to produce practically the maximum compaction associated with a given value of A is of the order of half a minute in dry sands. For practical purposes we may ordinarily assume that complete compaction will be accomplished by a single earthquake whose maximum acceleration slightly exceeds A.

Under high confining pressures we may expect that larger vibratory accelerations will be required to start compaction but that the final void ratios will be smaller than in the absence of confining pressure.

As a crude guide, according to Eq. 13.5, a soil of the first type having $e_{max} \leq 2e_{min}$ can be expected to be free of subsidence problems under earthquake action for which $A/g = 0.2$ if its relative density[9] is greater than 0.14, and if $A/g = 0.5$, the smallest relative density for which there would be no large subsidence, is 0.31. The corresponding numbers of blows per foot in a standard penetration test are about 10 and 30, respectively (Peck, Hanson, and Thornburn, 1953). Notice that these are minimum, not average, values in a given cohesionless formation, that they involve no factor of safety, and that the values of A/g to which they correspond are those in the formation itself, not in the underlying bedrock where they may be considerably smaller. Moreover, the presence of large confining pressures would make this treatment unsafe.

The foregoing remarks refer to compaction only. To give guides useful in the design of foundations we must also refer to the effects of vibrations on strength. Conditions with which we are concerned do not involve breakage of soil particles. Consequently we may take the angle of internal friction to be independent of normal stress. It will suffice, therefore, to describe the effects of vibration on the angle of internal friction.

Barkan (1962) has found that these effects can be expressed through an equation that is formally similar to Eq. 13.5:

$$\tan \phi = \tan \phi_\infty + (\tan \phi_s - \tan \phi_\infty) \exp \frac{-B_\phi A}{g} \qquad (13.6)$$

where ϕ is the angle of internal friction under harmonic vibrations of acceleration amplitude A, ϕ_∞ is the value of ϕ associated with $A = \infty$, ϕ_s is the angle of internal friction under static loading, and B_ϕ is a coefficient, of the order of 0.23 for medium-grained sand.

There are insufficient data to allow prediction of ϕ_∞. This angle is always small. Hence it seems wise to take $\phi_\infty = 0$ and rewrite Eq. 13.6 as

$$\tan \phi = \tan \phi_s \exp \frac{-B_\phi A}{g} \qquad (13.7)$$

Test results indicate B_ϕ is a sensitive, increasing function of grain size in the range of frequencies and grain sizes tested.

---

[9] Relative density is defined as $(e_{max} - e_0)/(e_{max} - e_{min})$.

Sec. 13.9 BEHAVIOR OF SOILS **427**

Beneath machine foundations we may take the contact pressure to be applied in part statically and in part dynamically. We may compute the corresponding bearing capacities, using for the dynamic capacity the reduced internal friction given by Eq. 13.6. Barkan suggests adoption of a criterion that can be put in the form

$$\frac{p_s}{p_s^*} + \frac{p_d}{p_d^*} = 1 \tag{13.8}$$

where p denotes contact pressure, subscripts s and d signify static and dynamic application, and the asterisk identifies capacities or design values. It is believed that this criterion is conservative.

When vibration is caused by earthquake it seems appropriate to take p_d^* as the bearing capacity for the maximum load, including its static and dynamic components. Yet, if Eq. 13.8 is conservative for machine foundations, this criterion is probably much too conservative for seismic design.

The foregoing treatment of the behavior of cohesionless soils of the first type under vibration is oversimplified. According to it the only significant variable of a disturbance is the acceleration amplitude. It is possible that the maximum velocity of the disturbance, not its maximum acceleration, is a more significant variable. However, not enough information is yet available to make a decision possible. In either case, recognition of the influence of confining pressures is inadequate.

A promising approach is that based on studies by L'Hermite and Tournon

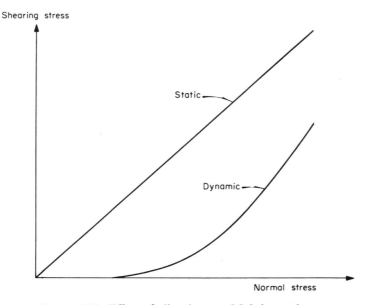

Figure 13.38. Effect of vibrations on Mohr's envelope. *After Bažant (1965).*

(1948) on fresh concrete. Bažant (1965) has attempted to apply it to saturated sands under laboratory conditions. According to this theory Mohr's envelope under vibrations is modified as shown in Fig. 13.38. The new envelope can be approximated by a second-degree parabola with the vertex and horizontal tangent at the origin, followed by a straight-line tangent to the parabola and parallel to the envelope for static conditions. The horizontal shift from the static envelope to the straight portion of the dynamic one is a function of the acceleration amplitude and of the frequency of the disturbance.

This theory is consistent with the fact that, under sufficiently high normal stresses, vibrations of given amplitude should not affect the angle of internal friction. An idea of the percentage of damping due to hysteresis in these soils can be gained from data due to I. A. Savchenko and presented by Barkan (1962).

Additional information on behavior under static testing is contained in papers by Marsal (1963b) and by Marsal et al. (1965). By performing triaxial tests on samples of unprecedented size and measuring strains with high accuracy, the authors have revised formerly established stress-strain relations. They have shown that, up to shearing stresses approaching failure, these relations are approximately of the form

$$\epsilon = K\sigma^{1-\beta} \tag{13.9}$$

provided the confining pressure is not so large as to cause appreciable grain breakage. Here ϵ denotes one of the principal components of strain, σ represents the corresponding principal stress, and K and β are constants. Approximate validity of these relations is consistent with the similarity of a mass of cohesionless soil with the model of Masing materials that we described in Section 11.3. These soils may be expected to behave in much the same way as a Masing material on unloading and reloading, even after one or more cycles have caused shear failure.

For very small stresses Eq. 13.9 cannot continue to hold exactly with a constant value of β, as it would imply an infinite velocity of wave transmission. In this range β must tend to zero.

Behavior of noncohesive soils in slopes subjected to vibration is conditioned to a large extent by grain interlocking (Bustamante, 1965). Because of this phenomenon one can build a slope steeper than the soil's angle of repose. The slope will even withstand appreciable horizontal accelerations. It will fail under static conditions, attaining its angle of repose, if it is locally disturbed. In other words, the soil is, in a way, in a state of unstable or metastable equilibrium that is disrupted when we cause a local failure, which then affects the entire slope. This condition of "unstable equilibrium" can be achieved in static conditions up to slope angles equal to the angle of internal friction.

Under transient vibrations in a shaking table, the same condition is preserved up to horizontal accelerations such that $\tan\theta + A/g$ does not exceed $\tan\phi$, where θ is the slope angle.

In these tests no appreciable effect of the dynamic disturbance on the angle of internal friction has been observed. Indeed, behavior is predicted with surprisingly high accuracy on the assumption that the dynamic angle of internal friction equals its static value. This fact contradicts other reports (Barkan, 1962) on the effects of vibrations. The situation may be due to the transient character and low frequency of the disturbance.

The second type of soil or second set of conditions we wish to consider are those in which the normal stresses are sufficiently high and grain size is sufficiently large that grain breakage occurs in an appreciable scale when applying shearing stresses. The influence of grain size is derived from the fact that, for soils having geometrically similar particles, intergranular forces are directly proportional to grain size.

When particle breakage merits consideration we must rewrite Eq. 13.9 in the form

$$\epsilon = K\sigma^{1-\beta} + F\left(\frac{\sigma - \sigma_0}{d}\right) \qquad (13.10)$$

where F is a monotonically increasing function, σ_0 is the stress at which this phenomenon begins to be important, and d is a linear measure of grain size.[10]

Finally, if air cannot escape at a sufficiently fast rate when the soil is reducing in volume due to vibrations, significant pore pressures may develop, with ensuing liquefaction of the material. The phenomenon is more likely to take place in very fine-grained soils, such as silts. The pore pressure often builds up along a surface of failure, and we speak more properly then of air lubrication. The phenomenon is familiar in the emptying of cement silos and has led to special designs tending to prevent excessive dynamic effects (Reimbert and Reimbert, 1962).

Apparently a quantitative treatment of this type of liquefaction phenomenon has not been developed, but we can urge consideration of the possibility of its occurrence in very loose silts.

13.9.2 Partially-Saturated Cohesionless Soils. As with dry soils we shall distinguish between three types of materials or three sets of conditions.

In the first group there is no significant breakage of particles and no liquefaction or fluid lubrication. Barkan (1962) treats the behavior of the material under vibration in the same manner as that of dry soils and finds that the parameters which define this behavior are functions of the water content.

For example, coefficient B_e in Eq. 13.5 decreases for small water contents from 0.75 for dry samples, to a minimum of 0.19 for the material on which he reports; next it increases to a maximum of 0.83 for a water content of 17 percent; finally it decreases to 0.59 when the material is saturated, at a water content of 22 percent. Thus, there seems to be an optimum water content of 17 percent for this sand, at which compaction under vibration is a maximum. However, the information available is scarce (see Barkan, 1962, Fig. II-17),

---

[10] Marsal (1963a) presents stress-strain curves of this sort. Notice, though, that the tests reported were monotonical. Cumulative grain breakage under alternating loads requires additional study.

and there are reasons to doubt that B_e should decrease in a cohesionless soil upon saturation.

Coefficient B_ϕ in Eqs. 13.6 and 13.7 is also found to depend on the water content. For the sand reported upon, this coefficient is apparently a minimum at a water content of about 13 percent. Data are included by Barkan on internal damping of partially saturated soils.

When grain size, material strength, and confining pressure are such that we anticipate large-scale grain breakage, the presence of moisture in quantities not approaching saturation cannot be expected to modify behavior much beyond its increasing inertia forces under vibration.

In very fine-grained, cohesionless soils the presence of moisture introduces a small cohesion that will tend to protect the material from liquefaction due to air pore pressures and from lubrication with air. At the same time the reduction in the volume occupied by air and the greater difficulty for the air to escape increase the tendency for these phenomena to take place. In large masses of silt and even of fine sand, partial saturation can be expected to raise the danger of failure in these manners.

13.9.3 Saturated Cohesionless Soils.

If pore water can flow in and out of the material at a sufficiently high rate so that appreciable pore pressures do not develop, behavior of these soils does not differ qualitatively from that of partially saturated cohesionless soils. It is merely a question of choosing the proper values of the parameters.

At the other extreme of conditions we have undrained, triaxial tests under repeated loads. To describe soil behavior under these conditions Seed and Lee (1966) and Lee and Seed (1967) distinguish between initial, partial, and total liquefaction. Up to a certain number of cycles the strains which develop in each cycle are small (less than 1 percent), but the cycle pore pressure shows a cumulative increase. After a certain number of cycles the pore pressure at zero deviator stress becomes equal to the confining stress, so that the effective stress drops to zero. The authors call this initial liquefaction. Thereafter strains increase rapidly with the number of cycles. The pore pressure equals the confining pressure over an increasing range of the deviator stress. The sand is said to be in a state of partial liquefaction. When the cycle strain reaches 20 percent the sand is said to have attained complete liquefaction. Figure 13.39 illustrates the process for sand at two different relative densities. The phenomenon is due to a loss of intergranular pressure caused by a transfer of normal stresses into excess pore water pressure.

Tests on undrained samples of cohesionless soil show that, whatever the density of the material, partial and indeed total liquefaction eventually set in after a sufficient number of load repetitions (Seed and Lee, 1966). Transition through partial liquefaction occurs only in dense materials; loose sands, for example, go directly from full shearing strength to complete liquefaction.

The phrase, "sufficient number of cycles," has entirely different meanings depending on relative density. For very loose materials it signifies no more than

Sec. 13.9 BEHAVIOR OF SOILS 431

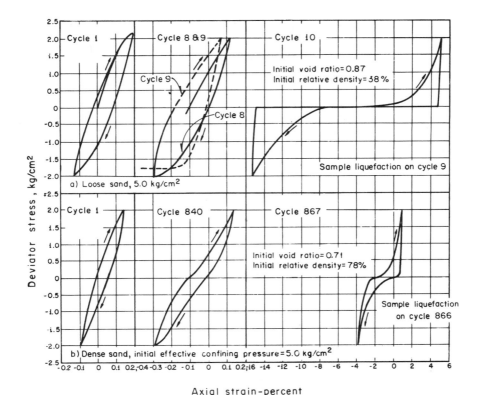

Figure 13.39. Repeated triaxial compression of undrained saturated sand. *After Seed and Lee (1966).*

about six cycles. For very dense samples it implies at least several hundred repeated applications of shear.

The influence of the confining intergranular pressure on the onset of partial or complete liquefaction seems to be qualitatively very much like its influence on volume changes of dry cohesionless materials under alternating stresses. The influence of the mean or static shear remains to be further studied, but it is probably not very significant.

Most of the information on sand liquefaction has been obtained in triaxial tests. However, as pointed out by Peacock and Seed (1968), field conditions differ significantly from those in triaxial tests. In the field there is a cyclic reorientation of the principal stress, while in triaxial tests the directions of these stresses are necessarily axial or transverse. The initial, minor principal stress in the field is usually much smaller than the major principal stress, while test samples are preconsolidated under a hydrostatic state of stress. Deformations in the field are usually associated with plain strain conditions, while test samples deform in the directions of all three principal stresses. And there may be other differences in stress history.

These differences led Peacock and Seed (1968) to conduct cyclic loading of specimens in simple shear. They found that the number of cycles required to produce liquefaction was systematically much lower than in triaxial tests. However, Finn, Pickering, and Bransby (1971) have shown that the difference may be due to the boundary conditions of the more unfavorable series of tests. Indeed, simple shear tests in which the boundary conditions are such as to cause uniform shearing strain throughout the samples give results that coincide exactly (within the range of experimental errors) with those of triaxial tests.

The effects of previous statically induced strains in sand has been studied by Finn, Bransby, and Pickering (1970). In a series of tests they compared the numbers of cycles of alternating shear required to liquefy samples that had not been subjected to strain after consolidation, with those required for samples previously liquefied and reconsolidated at the initial consolidation stresses. Although reconsolidation caused a slight gain in compaction, the numbers of cycles required for reliquefaction were systematically smaller than they were for first liquefaction. The effect was increasingly pronounced as the maximum strains in the first series of loading increased. For example, if the maximum shearing strain in the first series was 0.5 percent, the reduction in required number of cycles was about 33 percent; with a maximum previous shearing strain of 3 percent, the number of cycles for reliquefaction was about 80 percent smaller than for first liquefaction.

The precise significance of this phenomenon under field conditions is yet to be established. Its explanation may lie in a reduction of relative density.

Following A. Casagrande's suggestion, Castro (1969) refers to the phenomenon we have described as *cyclic mobility* and reserves the term *liquefaction* to the condition acquired by a soil when its practical loss of shear strength is not regained independently of strain. Under these conditions the soil flows almost as a liquid and, conceivably, the phenomenon is associated with a flow structure.

Castro has shown that the phenomena observed by Seed and Lee were essentially due to the development of gross nonhomogeneities in the samples as water migrated from the bottom to the top of the specimens causing a much looser condition in the latter portion under cyclic loading. From an extensive series of tests he concludes that liquefaction proper can only occur, even under cyclic loading, in materials having void ratios larger than the critical value corresponding to the confining pressure in question.

As we noted in connection with dry, cohesionless soils, their tendency to increase or reduce in volume under shear depends on whether their void ratio is, respectively, smaller or larger than the corresponding critical value. These tendencies produce pore water pressures when the soil is saturated and is tested in an undrained condition. These pressures are positive in soils having void ratios greater than the critical value and negative in soils having smaller void ratios. The situation leads to a decrease of intergranular pressure, and hence to earlier failure, in loose soils. Their void ratio tends to increase again only at very large shearing strains. According to Casagrande (1938) failure of these soils is accompanied by the development of a *flow structure*.

Casagrande's original determinations were based on results obtained under drained triaxial consolidation tests, "S tests" (see Fig. 13.37). Castro (1969) found that the critical void ratios in undrained or "rapid" triaxial tests, "$\bar{R}$ tests," were systematically smaller than in S tests. Some of the results he obtained with three different sands are summarized in Fig. 13.40. We see that, for the purpose of comparing the behavior of different soils, relative density is much more significant than void ratio. Also, it has been found that anisotropic consolidation leads essentially to the same results as consolidation under hydrostatic conditions (Castro, 1969).

Maslov (1957) reported results of sinusoidal vibration tests on saturated sand contained in a rectangular box. He found that the most significant parameter of the vibrations was their maximum acceleration. When this exceeded

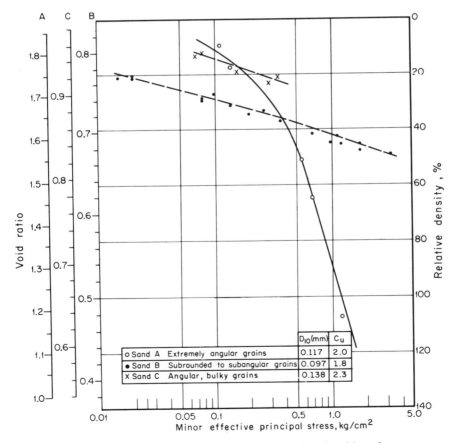

Figure 13.40. Critical void ratios and relative densities of various sands. *Data after Castro (1969).*

a critical value, the pore pressure rose sharply, and after sufficiently prolonged shaking under constant volume, the effective stresses dropped to zero. The phenomenon started at the top of the sand and progressed downward. When vibration ceased the sand consolidated as water seeped upward. The magnitude of the pore pressures that develop in tests of this type can be related to grain size characteristics of the sand (Nunnally, 1966; see also Castro, 1969).

In short, the problem of soil liquefaction (or "mobility") under cyclic loading has not been solved. Little can be said beyond the fact that it is likely to occur even during minor earthquakes in saturated noncohesive materials whose void ratio exceeds the corresponding critical value as determined in $\bar{R}$ tests (and the determination is not a trivial task). It is more likely under raft foundations than under small footings and even more likely in seams of cohesionless materials confined by relatively impervious soils. It is conceivable that water may migrate under field conditions in a manner comparable to that observed in laboratory specimens under cyclic load, and this would make every saturated cohesionless soil susceptible to liquefaction during an earthquake, but there are reasons to doubt that this would be the case, as boundary conditions and strain history differ materially from those met in the laboratory.

The three most important variables defining the behavior of saturated cohesionless soils under earthquake excitation are relative density, confining pressure, and drainage conditions. Under small confining pressures, if a soil of this nature is surrounded by material that does not liquefy, it suffices that its relative density exceed about 0.7 for it not to liquefy even when subjected to a strong earthquake, independently of grain size and drainage conditions. Perhaps this limit may have to be raised to 0.9 when drainage is especially poor and we envisage earthquakes of very high intensity and long duration—say MM intensities of 10 or more and duration of the strong phase in excess of 3 or 4 min (D'Appolonia, 1953).

For lower relative densities or high confining pressures the soil under consideration may liquefy if its grain size is so small that it excessively delays escape of pore water if the boundary conditions are such as to lengthen in excess the path that pore water must traverse in order to escape (as is the case under extensive raft foundations), or if a neighboring soil formation liquefies and consequently lets pore water out, increasing the pore pressure in the soil in question. Apparently, no quantitative criteria have been developed to permit prediction of the phenomenon under this wide range of circumstances.

The following three cases of major soil liquefaction will illustrate some of the preceding comments. The first concerns a quay wall in Puerto Montt, Chile (Fig. 13.41). The natural soil was a fine sand of medium density. The backfill was a very loose, saturated fine sand. Water could only escape upwards because of the impermeability of the retaining wall. The material liquefied completely, causing the quay wall to topple over (Rosenblueth, 1961).

The second example concerns overturning of buildings in Niigata (Fig. 13.42). Figure 13.43 compares the results of standard penetration tests before and after the earthquake, showing densification of some strata. It seems likely

Figure 13.41. Failure of a quay wall in Chile during the earthquake of 8 May 1960. *After Rosenblueth (1961)*.

Figure 13.42. Overturning of a building in Niigata during the earthquake of 1964. *After Japan National Committee on Earthquake Engineering (1965)*.

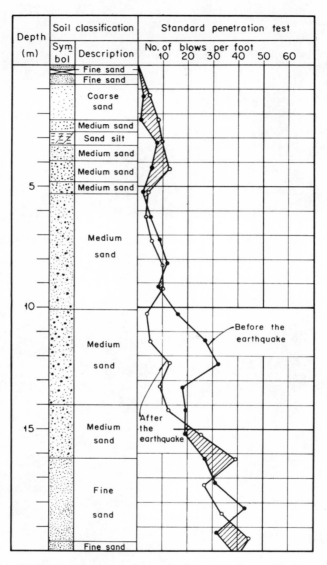

Figure 13.43. Comparison of standard penetration tests in sand under Niigata before and after the earthquake of 1964. *After Japan National Committee on Earthquake Engineering (1965).*

Sec. 13.9　　　　　　　　　　　　　　　　　　　BEHAVIOR OF SOILS　**437**

that liquefaction at some elevations where sand was especially loose caused increases in pore pressures and hence a propagation of the phenomenon at other elevations. It is also likely that the size of the foundations was such as to prevent a sufficiently rapid dissipation of pore water. It has been pointed out that, owing to this reason, the phenomenon of sand liquefaction cannot be adequately reproduced in small models (Ambraseys, 1965). Conceivably this restriction could be circumvented by using a proper time scale and, perhaps, distorted linear scales.

Finally, the spectacular slope failures in Anchorage (Shannon and Wilson, 1964) shown in Fig. 13.44 can probably be ascribed to liquefaction of seams of cohesionless soil located between practically impervious bodies of clay (Seed and Lee, 1966).

Figure 13.44. Slope failures in Anchorage. *Photo courtesy American Iron Steel Institute, New York.*

13.9.4 Saturated Cohesive Soils. In all cohesive soils reported to date, strength and stiffness increase markedly with strain rate (Figs. 13.6 and 13.7). An increase of the order of 40 percent is common for the usual strain rates of earthquakes, above the strength and stiffness of static tests.

Repetition of alternating loads decreases the strength and stiffness of all cohesive soils. Whether the increase due to the strain rate exceeds the decrease due to load repetition depends essentially on the number of repetitions, on the relative values of sustained and cycling stresses, and on the sensitivity of the soil. In very sensitive soils (sensitivity above 8), there is a net decrease under a moderate number of cycles, whereas in relatively insensitive materials, the increase prevails even after, let us say, 50 cycles (Seed, 1960).

Very sensitive clays lose so much of their strength after failure that we may speak of liquefaction. The phenomenon is associated with a reduction in effective pressure as was the case with cohesionless soils. In turn, the reduction is triggered by a collapse of the structure of the soil.

Internal damping in clays is essentially hysteretic. Hence it increases with the curvature of the stress-strain relation. For some very sensitive clays that behave almost elastically up to failure, an average value of 5.3 percent of critical damping has been reported on the basis of free torsional vibrations of small samples (Rascón, 1965). A small increase in the percentage of equivalent damping was found with increase in the amplitude of vibration. In contrast with this value, an insensitive clay loaded repeatedly well into the plastic range was found to have an equivalent damping of 31.5 percent (Taylor and Hughes, 1965). The value reported was actually 63 percent, but it was computed using a criterion that yields twice the percentage of damping obtained for the equivalent linear system whose stiffness equals the secant stiffness of the actual system at maximum deformation.

With some experience and a study of the stress-strain curve on first loading, an estimate can be made of the degree of internal damping that a saturated cohesive soil will exhibit when subjected to vibration. Laboratory test results on fissured clays subjected to forced vibrations may be very misleading unless the tests are done under the proper confining pressure. This requires the use of elaborate equipment of the type used for testing cohesionless soils. However, if only wave velocities are of interest, we may compute them from measured time of wave transmission in small unconfined specimens subjected to pulse excitation, as wave velocities are not particularly sensitive to the presence of fissures. Piezoelectric crystals are well suited as the sensing portion of instruments used in these tests.

13.9.5 Partially-Saturated Cohesive Soils.

The remarks made in connection with saturated cohesive soils apply to insensitive soils when they are partially saturated, except that the possibility of liquefaction seems remote. Data are required for a more precise description of the dynamic behavior of these materials.

13.9.6 Rocks.

The dynamic properties of rock are determined best by means of field tests. Static tests in pits and tunnels demand the application of elaborate techniques (Rocha, 1965); when properly performed they yield useful information on the strength that the material would develop under seismic loading but are not trustworthy for the determination of wave velocities. The latter are better established through seismic exploration.

Properties determined from laboratory tests deserve the same word of caution as the properties of fissured clays. Typical values of wave velocities in various kinds of rock are found in Leet (1950).

13.9.7 Field Determination of Soil Properties.

The foregoing text contains references to some laboratory procedures for the determination of the dynamic properties of soils and to static field tests from which the dynamic properties of rock may be inferred. Usually the direct determination of dynamic properties in the field has been limited to the range of practically linear behavior. Current methods use impacts produced by means of a hammer or by detonating charges of explosives and the travel times are recorded. This is done in most cases in bore holes. Duke (1969) has reviewed the literature on this subject; Kanai (quoted by Ohsaki, 1969) has published a rough correlation between the number of blows per foot in standard penetration tests and the velocity of shear waves in sands and clays (Fig. 13.45).

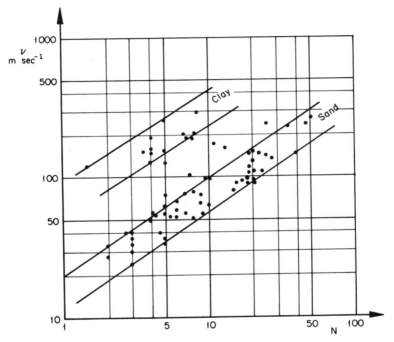

Figure 13.45. Relation between number of blows per foot in standard penetration test and velocity of shear waves. *After Kanai, quoted by Ohsaki (1969).*

PART 3

Design

Failure of Permanente Cement Bin in Alaska earthquake of March 27, 1964. (American Iron and Steel Institute.)

14

BASIC CONCEPTS IN EARTHQUAKE-RESISTANT DESIGN

14.1 Objectives of Earthquake-Resistant Design

14.1.1 Introductory Note. The aim of any purposeful activity is optimization of the outcome. When the result can be graded by a scalar quantity called *utility* and which is a monotonically increasing function of desirability, optimization is synonymous with the maximization of this quantity.[1] Accordingly, *the purpose of engineering design is to maximize the utility to be derived from the system produced.*[2]

Often it is advantageous not to deal explicitly with utility but with an *objective function* that may be any monotonic function of utility. One aims then at maximizing or minimizing this function. We may choose to measure this new scalar quantity in monetary units.

Under some conditions it is possible to formulate a criterion of design involving "zero probabilities of failure" (Turkstra, 1967). This happens when the initial cost of the system being designed is insensitive—when expressed to a certain number of significant figures—to a change, from the expected loss due to failure, to zero loss from this cause (in other words, when it is insensitive to a change from the actual probability to a zero probability of this event). This situation is often relevant in design for gravity loads. However, it is rarely met in practical problems of earthquake engineering and will not be treated here.

Translation of certain benefits or expenditures into monetary units is a simple and direct matter. Other items, such as prestige and social benefits, are pronouncedly subjective, and their grading will differ depending substantially on the person or entity for whom the work is being optimized. This aspect of the

---

[1] It is debatable that the word optimization has meaning under any other circumstances, although there is a school that deals with the optimization of processes whose outcomes are graded by vector quantities (Hausner, 1954; Thrall, 1954).

[2] When there are several possible outcomes of a decision, we take the utility associated with the decision as the expectation of the utilities corresponding to the various outcomes.

problem requires the exercise of judgment and may become a question in the realm of ethics or in that of esthetics.

Numerous procedures have been advanced for assigning quantitative values to utilities (Fishburn, 1967). Perhaps the most appropriate of these in civil engineering are the procedures developed by von Neumann and Morgenstern (1943) and the one of Churchman and Ackoff (Churchman, Ackoff, and Arnoff, 1961).

In the approach of von Neumann and Morgenstern we begin by listing the possible outcomes of our actions in order of preference. Let these outcomes be numbered $0, 1, 2, \ldots, i, \ldots, n$. Then we either prefer $i+1$ to i or are indifferent between these alternatives. Let u_i be the utility associated with the ith possible outcome. We fix u_0 and u_n arbitrarily, subject only to the condition that $u_0 = u_n$ if we are indifferent to all the possible outcomes, or $u_0 < u_n$ if we prefer n to 0. Next we find $\alpha_i (0 \leq \alpha_i \leq 1)$ such that we are indifferent between the ith possible outcome and a lottery having a probability α_i of producing outcome 0 and $1 - \alpha_i$ of producing outcome n. Then,

$$u_i = \alpha_i u_0 + (1 - \alpha_i) u_n$$

and we do this for all the possible outcomes. (It is understood that we assign no value to gambling.) Turkstra (1962) has suggested this method for application in structural design.

In the Churchman–Ackoff method we begin also by ordering the possible outcomes according to preference. Next we establish our preferences between combinations of these events. This leads to a number of inequalities between sums of utilities. (For example, if we prefer outcomes 0 and 2 combined to outcome 1 we write $u_0 + u_2 > u_1$.) Then we check the inequalities for consistency and revise them if necessary. Finally we choose tentative values for all the utilities, and using the inequalities iteratively, we arrive at the values we sought. Rodríguez-Caballero (1966) has suggested this approach for problems in earthquake-resistant design.

In civil engineering the owner, the designer, and the contractor usually have different goals. In other industries these three entities usually merge into the manufacturer. This situation lends itself to a simpler treatment, and we shall assume that it holds for the problems in which we are interested. There still remain two other points of view: that of society and that of the authority, say the building department, which sets the building code or, more generally, the norms or constraints on most of the parameters that the designer is to handle.

Hypothetically we may formulate the problem of optimization as follows. Let the following symbols denote expectations of actualized quantities in monetary units: X of the objective function, B of the positive benefits derived from the existence of the system, C the initial investment including the cost of design, and F the cost of failure and maintenance minus that of salvage. (The term *failure* is used here to include all undesirable modes of behavior, from defective appearance to collapse.) Then,

$$X = B - C - F \tag{14.1}$$

We face a problem in mathematical programming. We wish to maximize X with respect to all the design parameters, subject to pertinent constraints. If the maximum X is smaller than the net expected benefits we would derive from investing C in a different operation, the system should not be built.

Suppose that we are to actualize continuously at a constant rate of interest, c, per unit time and suppose that the time elapsed from investment until the system has been built is so short that actualization during this interval is not worth considering. Then we can write Eq. 14.1 in the form

$$X = \int_0^\infty \left(b - \sum_{i=1}^n f_i p_i\right) \exp(-ct)\, dt + B_0 - C - F_0 \tag{14.2}$$

where $b(t)$ is the expected benefit per unit time and includes such items as rent, services rendered by the system during its existence, and prestige to the owner, designer, and contractor; $f_i(t)$ is the cost of the ith mode of failure should it occur at time t; $p_i(t)\,dt$ is the probability that this mode occur between instants t and $t + dt$; B_0 are the expected benefits derived from the system during its construction, including matters of a social content when applicable; F_0 is the expected cost of failure during construction. We assume that there are n possible modes of failure outside the construction period.

Often the system can be repaired rapidly or rebuilt after failure. The benefits that are not perceived during the period of repair or reconstruction are then advantageously charged to the cost of failure. For the sake of simplicity in presentation, $F_0 - B_0$ will be charged to the capital investment or cost of construction. The task of maximizing X becomes then equivalent to that of minimizing the negative objective function

$$X' = C + F' \tag{14.3}$$

where

$$F' = \sum_{i=1}^n \sum_{k=0}^{m_i} F_{ik} \tag{14.4}$$

with

$$F_{ik} = \int_0^\infty f_{ik}(t) p_{ik}(t) \exp(-ct)\, dt$$

Here subscript k refers to the kth repair or reconstruction, f_{ik} is the cost of failure in the ith mode should it occur during the kth "reincarnation," $p_{ik}(t)\,dt$ is the probability that this event happen between instants t and $t + dt$, m_i is the number of reconstructions or repairs associated with the ith mode of failure, and F_{ik} is the present value of the expected loss due to the kth failure in the ith mode.

It is conservative to take $m_i = \infty$ for all i. Ordinarily it is also conservative to replace $f_{ik}(t)p_{ik}(t)$ with $f_{i0}(t - t_{ik})p_{i0}(t_{ik})$, where t_{ik} marks the instant of the kth failure in the ith mode; this follows from the fact that the information gained from previous failures will ordinarily lead to improved designs. Hence we may write

$$F_{ik} \leq \int_0^\infty \int_0^\infty \cdots \int_0^\infty f_{i0}(t_k) p_{i0}(t_0) p_{i0}(t_1) \cdots p_{i0}(t_k) \exp\left[-c(t_0 + t_1 + \cdots + t_k)\right]$$
$$\times dt_0\, dt_1 \cdots dt_k$$
$$\leq F_{i0} P_i^k$$

where

$$P_i = \int_0^\infty p_{i0}(t) \exp(-ct)\, dt$$

Consequently,

$$F' \leq \sum_{i=1}^n F_{i0} \sum_{k=0}^\infty P_i^k$$
$$\leq \sum_{i=1}^n \frac{F_{i0}}{1 - P_i}$$

Therefore, if we let

$$F'' = \sum_{i=1}^n F_{i0} \qquad (14.5)$$

and let F''' stand for F'' with every term divided by the corresponding p_i, we can write

$$F'' \leq F' \leq F''' \qquad (14.6)$$

since F' is at least equal to the present value of the expected loss incurred in the first set of n possible failures.

When $P_i \ll 1$ for all i, we can replace F' with F'' and according to Eq. 14.3, we can say that the design is optimum when the expected cost of construction plus the expected present value of failure is a minimum, and take for each mode of failure its first occurrence only. Notice that the p_{i0}'s are conditional probabilities and that some of the modes of failure may be mutually exclusive. In every case Eq. 14.6 affords a rapid means for deciding whether this approximation will be valid.[3]

We have spoken thus far of one objective in engineering design: optimization of the system. Sometimes the total utility or the total cost can be regarded as composed of relatively independent groups of items. We can say then that design has many objectives each consisting in the optimization of one group of items. For example, within a wide range of construction procedures the choice will affect the cost of construction but may not alter appreciably the probabilities of failure. Or in some cases, optimization may be carried out with regard to the structural capacity of a cantilever beam to carry vertical loads almost independently of the structure's capacity to resist earthquakes. This and similar circumstances entitle us to treat certain aspects of earthquake-resistant design as independent of the influence of other disturbances.

[3] This derivation of Eq. 14.6 is a generalized version of an argument due to Johnson (1953).

14.1.2 Structural Design.
In structural engineering the problem is often put advantageously in the following terms.[4] Let r denote a resistance function, a scalar quantity that increases monotonically with structural capacity, and let s denote a load function, a scalar that increases monotonically with the loads. For example, under certain conditions all loads may be proportional to s. Let these two functions be chosen so that $r \leq s$ signifies failure. Both r and s are random functions, and at least s is stochastic. Graphs are available (Johnson, 1953) which facilitate the solution of Eq. 14.3, for the design parameters on which C and r depend, under the assumption that the expectation of r is linearly related with C and assigning r and s certain classical probability distributions.

Concepts that belong in the theory of engineering reliability—such as the reliability function, the risk function, and the relation of these with the recurrence period—are quite useful in optimization of engineering systems (Von Alven, 1964; Barlow and Proschan, 1965) and more specifically, structures (Freudenthal, 1962; Freudenthal, Garrelts, and Shinozuka, 1966; Rosenblueth, 1969b). For the sake of brevity we shall make little use of them here.

14.1.3 Load Factors and Factors of Safety.
Let $\bar{r}$ and $\bar{s}$ denote certain functionals (or "characteristic" or "nominal" values) or the probability densities of the resistance and load functions in a structure. The ratio

$$L = \frac{\bar{r}}{\bar{s}} \tag{14.7}$$

is traditionally called the *load factor* of the structure. Its value depends on the criteria followed in the choice of $\bar{r}$ and $\bar{s}$. For example, if neither the structural capacity nor the loads are time functions, we may choose $\bar{r} = E(r)$ and $\bar{s} = E(s)$, or $\bar{r} = E(r) - \alpha\sigma(r)$ and $\bar{s} = E(s) + \beta\sigma(s)$, where $E(\cdot)$ denotes expectation, $\sigma(\cdot)$ denotes dispersion, and α and β are positive parameters; clearly the second choice will give a smaller load factor than the first one.[5]

When the resistance and load functions vary with time, we define L as the smallest ratio of $\bar{r}$ to $\bar{s}$ in a given time interval, usually and arbitrarily identified with the expected life span of the structure.

The common belief that $L = 1$ amounts to impending failure is valid only in such trivial cases as when r and s are deterministic variables equal to their respective functionals.

Now, in a structure subjected to a uniaxial state of stress, let r' denote the strength (yield-point stress or maximum stress in the stress-strain curve) of the material of which the structure is made, and s' the stress computed according to

---

[4] This treatment is limited to problems in which, ideally, there is only one possible load configuration, and it can be assumed that the load is applied a single time, at which instant the structure may fail, but otherwise it survives forever.

[5] The second choice forms the basis of the *Comité Européen du Béton* (CEB) building code and of other proposals (Torroja et al., 1958; Ang and Amin, 1969b). It has the advantage that, if α and β are properly chosen, the load factor required to provide a fixed, small probability of failure is relatively insensitive to the probability distributions of r and s.

the canons of strength of materials, and let $\bar{r}'$ and $\bar{s}'$ stand for some functionals of the corresponding distribution functions. The quantity

$$L' = \min \frac{\bar{r}'}{\bar{s}'} \tag{14.8}$$

where the minimum is selected from the entire structure, is ordinarily known as the *factor of safety* or the reciprocal of the *stress-reduction factor*. The same remarks apply to L' as to L concerning the case when r' and or s' are stochastic variables and to the fallacious condition of impending failure.

In simple problems it is customary to choose s as a quantity to which all loads are proportional. Then, if the criteria for selecting $\bar{r}$ and $\bar{s}$ are the same as those for $\bar{r}'$ and $\bar{s}'$, we shall find $L = L'$ in all statically determinate systems and in all structures made of materials that behave linearly. Under these conditions there is no advantage in using a load factor rather than a stress-reduction factor.

We meet the same situation in statically indeterminate structures made of elastic nonlinear materials whose stress-strain relation is a power law (Richard and Goldberg, 1965). In the more general case of statical indeterminacy combined with nonlinear behavior, the equivalence between the use of stress-reduction and load factors is not strictly correct. Differences in design resulting from one rather than the other criterion are, nevertheless, usually quite small if all nominal loads are affected by the same factor or all working stresses bear the same ratio to the corresponding nominal strengths.

There is a real advantage in the use of load factors when different load configurations can vary independently of each other. Loads due to different causes are then assigned different load factors. In some problems this is the only simple means for controlling the probability of failure. Such is the case with the overturning of a block when the block and the material on which it rests are of relatively high strength: to control the probability of overturning one multiplies the nominal lateral forces by a factor that differs from the factor applied to the gravity forces. No reasonable stress-reduction factor will allow such control. As an alternative we may use a different criterion in fixing the functionals of independent sets of forces; for example, the ratio of the nominal lateral forces to their expected maximum in a given period of time may be made greater than the corresponding ratio for vertical loads.

Conversely, stress-reduction factors are useful when there exists the possibility of different modes of structural failure—with various degrees of warning or different magnitudes of the consequences of failure—or the possibility of failure in different materials of heterogeneous structure such as one of reinforced concrete. Each material and each mode of failure—whether in diagonal tension, flexural compression, and so on—is assigned a different factor of safety. This cannot be achieved in a practical way through adjustment in load factors. We are left, then, with the advisability of using simultaneously stress-reduction and load magnification factors.

There is a third type of discrepancy between structural drawings and the com-

Sec. 14.1 OBJECTIVES OF EARTHQUAKE-RESISTANT DESIGN 449

pleted work. It concerns the structure's geometry: discrepancies in spans, cross-sectional dimensions of structural members, position of the steel in reinforced concrete, thicknesses of veneer and floor finishes, and similar concepts. The probability distributions of these differences between drawings and the prototype depend on construction practices but are almost independent of the nominal dimensions (Johnson, 1953), so they are not adequately covered by either stress reduction or load factors. A more satisfactory provision consists of assuming, in design, a set of unfavorable discrepancies between nominal and actual dimensions, to be used in conjunction with the two types of factor we have discussed. The practice of assuming reductions in effective cross section for design purposes is further justified by the possibility of abrasion and of electrolitic or bacterial attack leading to deterioration or to reduction in effective sections.

In some countries, as the United States, building codes that cover design and construction standards for structures made of specific materials or combinations of materials exert no control over the design live loads called for by municipal codes. If these loads are unrealistically low for some cities, a greater load factor is required for live than for dead loads. This hindrance is unnecessary when the same code marks design criteria, working stresses, load factors, and live loads.

Blatant mistakes, changes in type of occupancy, and drastic changes in architectural project frequently bring about conditions against which it is economically unsound to design. In other words, optimization requires that the probability of failure tends to one under conditions that depart sufficiently from the mean. Rather than attempting to modify this situation by controlling the probabilities of failure under extremely unfavorable conditions, it is advantageous to reduce the expected loss should failure occur. This is accomplished chiefly by providing the structure with high ductility. Thus we ensure adequate warning against impending failure. Warning will consist of cracking and large deflections, and it may lead to a reduction in the losses involved. By providing high ductility we incidentally produce a structure capable of large-scale redistribution of stresses, and this in itself lowers the probability of failure under unforeseen conditions of loading or oversights in design. Ordinarily, adequate ductility can be incorporated at quite a low cost.

In steel, timber, and precast concrete structures ductility requires relatively strong connections and weak members. In reinforced concrete it calls for a higher capacity in diagonal tension than in bending, in all types of structure it calls for a smaller probability of inelastic buckling than of failure in tension, and it requires that very special attention be given to details.

It should not be construed that discrepancies between the ideal and the real structure which can be ascribed to mistakes or to the other causes we have just discussed merit an essentially different treatment from the treatment we give the discrepancies due to other causes. Both types constitute random variables. The remarks concerning the advantages of providing adequate ductility over very high structural capacity apply to conditions in which the discrepancies between structural drawings and structure can be extremely high, whatever their cause.

14.1.4 Design for a Permissible Probability of Failure. A compromise between the ideal explicit optimization in structural design and the (often cumbersome) "practical" approach of load factors and related adjustments is found in the design for a probability of failure that is either implicit or is fixed on intuitive bases or by comparison with what the designer considers good practice. This probability does not necessarily coincide with the one that would produce the optimum design. Yet we have seen that under favorable circumstances the criterion of zero probability of failure leads to designs that differ little from the optimum. The criterion based on a permissible, finite probability of failure is necessarily of more general applicability.

As a further simplification, in cases when the dispersion in one type of load or in the resistance function far outweighs the dispersion in all other variables that govern design, the variable with the largest dispersion may be treated as the only random quantity, and all others are treated as deterministic variables. A simple application of this method permits establishing the dependence of the unit design live load on the loaded area (Rosenblueth, 1956).

14.1.5 Earthquake-Resistant Design. Two principal aspects of earthquake resistant design differentiate this discipline from other branches of engineering. One concerns the enormous spread or uncertainty in the disturbance. The other concerns the nature of the disturbance itself. Even if the detailed characteristics of future ground motions were known accurately, we could not be certain about the survival of given structures; at present ignorance about structural characteristics is great. These circumstances make it advisable to apply the theory of probability and optimization techniques openly in design to a more significant degree than in other engineering disciplines. The traditional deterministic disguise will do less well in earthquake engineering.

The explicit process of optimization requires assessing a number of nearly "imponderable" losses and benefits, whose value or associated utility depends on the subject for whom the optimization is done. For example, the cost of collapse to the owner usually involves that of the building, minus its insurance and salvage values, plus the expected cost of lawsuits and of indemnities. To society, it involves the cost of the building, minus its salvage value, plus the expected cost of damage to, or destruction of, its contents, and that of lives lost and injuries. And to the engineer it signifies the expected cost of lawsuits and that of his loss of prestige, plus the moral harm.

Material losses are easily estimated. The value of a human life lost may be taken as the expected contribution of an individual to the gross national product, had he lived.[6] The loss of prestige has a value that can be bounded from above as the engineer's expected profits during the rest of his lifetime had the collapse not occurred. The value of the moral harm may also be bounded from above as equal to the cost of one human life. For catastrophic earthquakes the cost of failure to the engineer is often practically nil.

---

[6] This quantity is the one which society seems willing to pay. The writers are grateful to R. L. Ackoff for calling their attention to this point.

Once the expected cost of failure has been assessed, four other factors determine the limits of earthquake intensities that mark differences in design criteria. One concerns the frequencies, or the recurrence periods, of earthquakes of various intensities; the second is the dependence for given intensities, of probability of failure, on initial costs; the third is the rate of interest, or the discount rate, at which future events are to be actualized; the fourth concerns the probabilities of failure from causes other than earthquakes. The problem is typically one of operations research or mathematical programming. Granted that a detailed quantitative solution along these lines is still too ambitious in most instances, it remains true that the engineer's intuition or educated judgment can be used to advantage in arriving at a satisfactory answer if he is conscious of the implications that his decisions carry.

There are many types of failure or damage to keep in mind in earthquake-resistant design—collapse; structural and nonstructural damage; damage to the material contents of buildings, including equipment, loss of life, and injuries to persons and property; panic; damage, by pounding, to structures other than the one being designed; indirect consequences, such as floods, fire, explosions, and malfunctioning of equipment. Among the latter the release of fissionable material in nuclear plants is especially serious. In these plants earthquake-resistant design must consider the conditions for which to trigger a "scram" to stop the nuclear reaction before the ground motion becomes so intense that there is a high probability of its causing serious damage.

Design must also contemplate the possibility of affecting the cost of repairs should damage occur. This matter is tantamount to the explicit consideration of a salvage value in other types of design when the engineer has control thereon. Its effect on the cost of failure is that of affecting one or more of the f_i's without necessarily changing the corresponding probabilities of failure.

Attention must also be given to the functions expected from structures immediately following a major earthquake. Thus, the criteria for design of hospitals and fire stations differ from those for apartment buildings.

The design of a single structure usually involves an explicit consideration of most of the types of failure we have mentioned, and it sometimes involves all of them. These phenomena may be regarded as different modes of failure. Often, however, occurrence of one of them implies occurrence of others, or it may imply that other modes of failure lose relevance. For example, structural damage may imply such large deformations that damage to nonstructural components and to equipment has to have taken place, while if the structure collapses, these kinds of damage lose significance.

In order to illustrate the quantitative treatment of these situations consider a highly idealized problem. Suppose first that we are to design, against collapse, a structure whose survival depends only on its responses to earthquakes and that these responses are proportional to the maximum ground velocity. We take the maximum ground velocity as the only random variable, while we treat structural characteristics and gravity loads as deterministic variables. We assume that we have control over only one of the structural parameters, which is a measure of its strength and which, owing to the formulation of the problem, we may take to be the maximum ground velocity that the structure withstands.

According to the first part of Eq. 7.4, if earthquakes that originate not more than a few kilometers from the station of interest do not play too decisive a role in the regional seismicity, we may assume that the maximum ground velocity is proportional to exp M, where M denotes earthquake magnitude. Suppose that earthquakes at their origins may be idealized as constituting a generalized Poisson process with mean rates proportional to $\exp(-\beta M)$. Then ground motions at the station will also constitute a generalized Poisson process, and the mean rate of occurrence of ground motion with maximum velocity in excess of, let us say, v will be of the form

$$\lambda(v) = \alpha v^{-\beta} \tag{14.9}$$

Let u denote the ground velocity at which the structure would fail; that is, failure occurs if and only if $v > u$. Suppose that the initial cost in the range of values of u that interest us can be put in the form

$$C = C_0 + ku$$

where C_0 and k are constants. If f denotes the loss involved in collapse and we use Eq. 14.3, we must minimize the objective function

$$X' = C_0 + ku + \frac{\alpha}{c} f u^{-\beta}$$

By equating dX'/du to zero and solving for u, we find the optimum value of this strength function u, which we shall denote by u_0:

$$u_0 = \left(\frac{\alpha \beta f}{ck}\right)^{1/(1+\beta)} \tag{14.10}$$

The corresponding optimal return period is found by replacing v with u_0 in Eq. 14.9 and computing the reciprocal of λ:

$$\lambda_0^{-1} = \alpha^{-1/(1+\beta)} \left(\frac{\beta f}{ck}\right)^{\beta/(1+\beta)} \tag{14.11}$$

Since β usually lies between 2 and 3 (see Chapter 7), λ_0^{-1} is not altogether insensitive to α. For example, with $\beta = 2.5$, if α is doubled, u_0 increases 22 percent and λ_0^{-1} decreases 18 percent. Notice also the dependence of u_0 and λ_0^{-1} on β, on k, on the cost of failure, and on the interest rate used for obtaining present values.

According to Eq. 7.4 we would have arrived at similar results had we dealt with a structure whose survival had been governed by maximum ground acceleration or displacement rather than velocity. We would merely have had to modify β.

Now consider the same type of structure but with nonstructural damage developing at a deformation whose spectral value is associated with a maximum ground velocity smaller than u and which we shall call u'. The corresponding loss will be denoted by f'. The loss associated with collapse will be assumed to increase from f' to $f + f'$. The initial investment will be taken as

Sec. 14.1   OBJECTIVES OF EARTHQUAKE-RESISTANT DESIGN   453

$$C = C_0 + ku + k'u'$$

where k' is a constant.

Proceeding as in the first case and assuming that u and u' may be varied independently of each other, we find that Eqs. 14.10 and 14.11 still give us the optimum value of u and the corresponding return period. The expressions for the optimum u' and for its return period are obtained by replacing f and k with f' and k', respectively.

Ordinarily f/k' is much greater than f'/k. Under these conditions we design against collapse for strong earthquakes that, on the average, occur seldom, and design against nonstructural damage for frequent, moderate earthquakes. It is not uncommon to find that the ratio $fk'/f'k$ lies between 4 and 20. Then, if $\beta = 2.5$, the ratio of the optimal design ground velocities will lie between 1.5 and 2.4, while the ratio of the corresponding return periods will be between 2.7 and 8.5.

The assumption that u and u' can be varied independently of each other is often realistic. For example, u may depend essentially on the ductility factor, and u' may depend on stiffness and on the choice of nonstructural materials and of details of connections between structural and nonstructural elements.

On the other hand, if u' should be proportional to u we would find that Eqs. 14.10 and 14.11 would require changing f into $f + \gamma^{-\beta}f'$, and k into $k + \gamma k'$, where $\gamma = u'/u$. The result is quite different from the fairly established practice that consists of checking separately for u' and u. We meet approximately the condition of proportionality when nonstructural materials and details are fixed, as is the ductility factor, and we can only change the cross sections of structural members, thus affecting simultaneously their stiffness and their strength. In general we find that independent treatment of u and u' leads to important errors when there is any appreciable interdependence of these variables.

The same idealized models may be used in a crude exploration of the benefits to be derived from the use of "defense plateaus." The conscious introduction of this type of structural solution is due to Blume (1960). Essentially it consists in the use of elements whose failure dissipates energy, but it does not in itself bring about collapse. We meet this situation when a ductile frame has relatively weak and brittle filler walls. Failure of the walls implies an economic loss, but it prevents or delays collapse. An intermediate line of defense may be secured by means of reinforced concrete fire protection in steel frame structures.

Suppose that we deal with a building in the range of initial natural periods for which we may assume that the maximum strain energy affords a satisfactory criterion of failure and that we face the dilemma of whether to tie the partitions to the frame in such a way that they will not be damaged by the deformations of the latter or to place them so that they furnish a first line of defense. In the first case we can write

$$X' = C_0 + ku + \frac{\alpha}{c}u^{-\beta}(f + f') \qquad (14.12)$$

where f is the loss due to failure of the partitions and $f + f'$ is the loss due to collapse. In the second case we would have

$$X' = C_0 + ku'' + \frac{\alpha}{c}\{[u'^{-\beta} - (u''^2 + u'^2)^{-\beta/2}]f' + (u''^2 + u'^2)^{-\beta/2}(f + f')\}$$
$$= C_0 + ku'' + \frac{\alpha}{c}[u'^{-\beta}f' + (u''^2 + u'^2)^{-\beta/2}f] \qquad (14.13)$$

where u'' is the ground velocity at which this frame would collapse if the partitions did not contribute to resist lateral load. According to Eq. 14.13, for any given u'' the loss X' is a decreasing function of u'. Hence, defense plateaus are only justified when their introduction allows us to decrease u'' by a sufficiently large amount.

It follows from Eq. 14.12 that

$$u_0 = \left[\frac{\alpha\beta(f + f')}{ck}\right]^{1/(1+\beta)} \qquad (14.14)$$

$$X' = C_0 + \left[\frac{\alpha k^\beta(f + f')}{c}(\beta + \beta^{-\beta})\right]^{1/(1+\beta)} \qquad (14.15)$$

Usually u' is fixed, and the problem consists of finding the optimum value of u''. After substituting it in Eq. 14.13 we can compare with the minimum value of X' in Eq. 14.15 and choose the better of the two designs. Suppose, however, for the sake of simplicity, that $u' = u''$. Then Eq. 14.13 gives an expression for u_0'' identical with that for u_0 in Eq. 14.14, but with $2^{-\beta/2}f$ instead of f. With the same substitution Eq. 14.15 holds for the minimum X'. We conclude that under these circumstances it is always advisable to provide the first line of defense. The advantage may be important, since f' is usually much smaller than f and, with $\beta = 2.5$, $2^{-\beta/2} = 0.42$, so that providing the first defense plateau may imply a saving of the order of 50 percent in the present value of the combined cost $ku + F$. This is effective, though, only if the frame is redesigned to take into account the existence of the first line of defense.

A more realistic idealization admits that gravity loads and structural parameters are also random variables. Some implications of this, as far as structural parameters are concerned, will be covered in subsequent articles.

When designing for gravity loads only, the usual building code format replaces a probability analysis with the use of load factors and/or stress-reduction factors. For the sake of uniformity we may use load factors also in earthquake design. Instead of stating that we design for u_0 we may stipulate a smaller design value and a load factor. The same applies to base shear coefficients and the like.

Strong winds accompanying major earthquakes are such rare events that it is not worth designing for any combination of wind and earthquake forces. Following the same line of thought we conclude that, since temporary live loads and earthquake forces are almost stochastically independent variables, it is justified to use a smaller load factor in design to resist their combined effects than for one of them alone.

A reduction in load factor is justified when designing even for the combination of permanent and earthquake forces. This reduction follows because the coefficient of variation of the sum of positive random variables is always equal to or smaller than the largest among the individual coefficients of variation, and the load factor associated with a given probability of failure is an increasing function of the coefficient of variation. Moreover, the moral and legal costs of failure are usually smaller when failure can be ascribed to an earthquake or to any other accidental phenomenon than when it occurs under operating conditions.

It is worth emphasizing that the base shear coefficients, load factors, and similar provisions in building codes constitute restraints in design. They are not directly related to optimization from the viewpoint of the owner nor from that of the designer. These provisions are intended to furnish a protection from the point of view of society, so that by complying with them there is some assurance that the design will not drift excessively from what is optimum for society. Actually, this objective is fulfilled only if code requirements are supplemented with sound engineering practice.

In most cities of the world the violation of building code provisions, if at all possible, involves such high expected losses for the designer that he will do well to respect them even if an analysis of the type we described leads to smaller base shear coefficients. There is no question about the course to follow when this analysis produces a more conservative design. Nor is there any question about the need to cover sundry aspects of design not covered by codes.

Building code provisions fulfill a second purpose: that of giving the designer a guide on various simplifications in analysis concerning aspects whose accurate calculation would be too ambitious in everyday practice. This is true of specified reductions in overturning moments, accidental torsion, detailing, and so on. As we shall see in Chapter 15, these specifications should be taken as mere restraints, since they are often based on debatable assumptions and, at best, they are only approximately valid over some ranges of the structural parameters and of ground properties.

In interpreting the base shear coefficients found in building codes we should remember that they are intended for the calculation of stresses, not deformations and that they invariably assume that the structure behaves nonlinearly. These matters will be taken up subsequently in Chapter 15.

14.2 Simple Linear Systems Having Deterministic Parameters

Studies of regional and microregional seismicity such as those described in Chapter 8 should in principle provide a set of spectra for every type of structure. Each curve would correspond to a given probability that its ordinates be exceeded at least once in a given time interval. Ideally we would be in possession of one set of curves for structures of linear behavior of each type having a particular percentage of damping, one for elastoplastic systems with a particular ductility factor, and so on. The material we presented in Chapter 9 allows these numerous sets to be reduced to one for simple structures that behave linearly; Chapter 10 describes how to deal with multidegree linear systems; Chapter 11 shows that it is unnecessary to have different sets for nonlinear structures, at least when they are nondeteriorating.

On firm ground, given the maximum base acceleration, velocity, and displacement and a measure of the duration of ground motion we can compute the expectations of the responses of a wide variety of ideal structures. The expectations practically suffice for design purposes. Indeed, we have seen that uncertainties in the earthquake parameters overshadow the deviations with respect to the expected responses for a given set of parameters. Incorporation of the effects of local geologic conditions can be accomplished in an approximate manner using the material presented in Sections 9.8 and 9.10 if we recognize the consequences of nonlinear soil behavior when these may be important.

Usually the information on seismicity is not as complete as desirable, and we must resort to crude approximations. To illustrate, suppose that we are interested in a station at which most of the important earthquakes come from groups of foci lying about 60 km from the station. Suppose that the foci belong to regions where $\beta \cong 2.5$ in Eq. 8.3. Suppose further that the group is known to be responsible for one earthquake of intensity 6 (MM scale) or greater every 3.0 yr on the average. Let us admit that this estimate incorporates all the pertinent information and allows for uncertainties in whatever correlations were used to establish it. Owing to the uncertainties the number of earthquakes with intensities greater than some given quantity will not strictly have a Poisson distribution, but we shall assume that it does for the sake of simplicity

Suppose that we are interested in the design of single-degree linear systems with 5 percent damping. We are concerned then only with the absolute probabilities of spectral ordinates. (This would not strictly be the case for multidegree or nonlinear systems, as we shall see.) From Eq. 7.2 we find that, if the station stands on firm ground, intensity 6 corresponds to a maximum ground velocity

$$v = \frac{2^6}{14} = 4.57 \text{ cm/sec}$$

According to Eq. 7.4, the corresponding maximum ground acceleration and displacement for the ground motions are

$$a = 19.0 \text{ cm/sec}^2$$

and

$$d = 8.45 \text{ cm}$$

respectively, which have been obtained by substituting v and R in the second part of Eq. 7.4 from which $M = 5.91$, and substituting M and R in the first and third parts of Eq. 7.4.

The equivalent duration ("equivalent" in the sense that a segment of white noise having this duration will furnish essentially the proper correction factors for damped systems) of these ground motions of intensity 6 is, according to Eq. 7.7, $s = 19.6$ sec.

Corresponding to these ground motions, there is an expected spectral ordinate for every given natural period of vibration and percentage of damping. For spectral ordinates we choose pseudovelocities, in the sense that their product by the natural circular frequency directly yields the maximum numerical value of the absolute acceleration. And to compute the expected spectral ordinates we decide to use values intermediate between those resulting from the rules in the first paragraph of Section 7.5 and from the assumption that the straight lines marking a, v, and d in a logarithmic plot of spectra coincide with the expected spectrum for a damping of 25 percent of critical. Admitting, besides, the validity of Fig. 9.3 for the calculation of the damping correction factor β_E for moderate and long periods, we arrive at the curve marked "return period = 3 yr" in Fig. 14.1 where spectral ordinates for very short periods of vibration have been estimated using the results of Section 9.4. This curve represents expected ordinates for earthquakes originating nearby with an intensity 6, or a magnitude 5.91, and a recurrence period of 3.0 yr.

Ordinates in Fig. 14.1 and in the next two figures carry the subscript max to indicate that they were obtained from correlations that furnish the maximum values of a, v, and d with respect to the orientation of the structure's degree of freedom.

Next we repeat the foregoing computations by taking other values for M and construct in Fig. 14.1 the curves which correspond to different recurrence periods.

Given an expected response $\bar{Q}_{max} = E(Q_{max})$, the maximum structural response has a probability $P(Q_{max} | \bar{Q}_{max})$ of not exceeding Q_{max}. Hence, on the average the maximum response will exceed Q_{max}

$$v(Q_{max}) = -\int_0^\infty F(Q_{max} | \bar{Q}_{max}) \frac{d\mu}{d\bar{Q}_{max}} d\bar{Q}_{max} \qquad (14.16)$$

times per year, where F is $1 - P$, and μ is the expected number of earthquakes per year for which $E(Q_{max})$ exceeds $\bar{Q}_{max}$. In other words, F is the probability of failure under an earthquake for which the expected response is $\bar{Q}_{max}$ and $1/v(Q_{max})$ is the return period for response Q_{max}.

In keeping with our assumptions, the number of times that the maximum response exceeds a given value during a given time interval has a Poisson distribution. Hence, the probability that a structure of strength Q_{max} fails during a time interval t is $1 - \exp[-v(Q_{max})t]$.

Application of Eq. 14.16 to the curves in Fig. 14.1 yields the full lines in Fig. 14.2, where ordinates represent spectral pseudovelocities for 5 percent damping. We have used the distribution of $Q_{max}/E(Q_{max})$ for responses to white noise, with an estimated adjustment for very rigid and for very flexible systems, so that the dispersion tends to zero as the natural period tends to zero or to infinity.

Suppose now that the same station is affected also by earthquakes originating in a second group of foci, which lie about 120 km away. Assume that this group also belongs to a region where $\beta \cong 2.5$ in Eq. 8.3. Besides, suppose that this

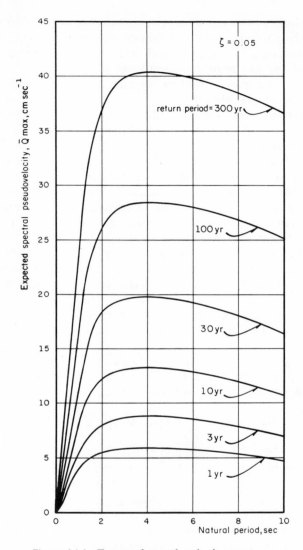

Figure 14.1. Expected pseudovelocity spectra for nearby earthquakes.

group is responsible for one earthquake of 5 *MM* intensity every 3 yr on the average.

Using the same approach as we did with earthquakes originating in the first group of foci, we arrive at a similar set of curves for those from the second group. They are shown as dashed lines in Fig. 14.2.

Now we can combine the effects of ground motions from both groups, merely by adding the values of ν (reciprocal of the return period) that correspond to given values of $Q_{\max}$. For example, according to the solid lines in Fig. 14.2, at a

Sec. 14.2 SIMPLE LINEAR SYSTEMS **459**

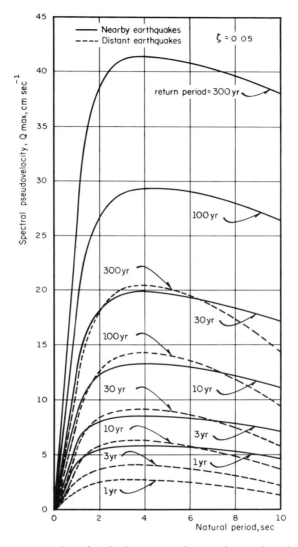

Figure 14.2. Pseudovelocity spectra for nearby earthquakes and distant earthquakes.

natural period of 7 sec, a spectral pseudovelocity of 13 cm/sec is exceeded, on the average, every 11 yr by earthquakes originating 60 km away. The dashed curves in the figure state that this same spectral pseudovelocity is exceeded every 110 yr, on the average, by earthquakes from the second group of sources. The mean yearly numbers of times that this response is exceeded are $1/11 = 0.0909$ and $1/110 = 0.0091$, respectively, which gives a total of $0.0909 + 0.0091 = 0.1$ times per year. Hence, a pseudovelocity of 13 cm/sec has a recurrence period of $1/0.1 = 10$ yr, when reckoning earthquakes from both sources. In this manner we can construct Fig. 14.3, which displays a family of design spectra for the station.

Design spectra like those in Fig. 14.3 may be used to produce an optimal design by trial and error. We may choose a value of structural reliability, which we translate into a return period.[7] We enter the figure with this period and an

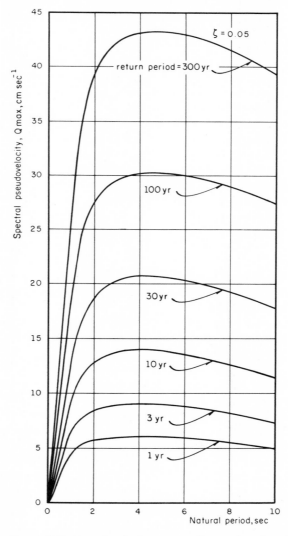

Figure 14.3. Pseudovelocity spectra for earthquakes from both sources.

---

[7] When the resistance function is deterministic and time independent and the disturbance has a Poisson distribution, the reliability function is of the form $\exp(-\nu t)$ and the return period of failures is $1/\nu$ (Barlow and Proschan, 1965).

estimated natural frequency of vibration, to obtain a tentative design pseudo-velocity. Design of the structure to withstand this response permits an estimate of the initial cost of the system, while the chosen reliability, together with a postulated criterion of actualization, yields an estimate of expected actualized cost of failure. By repeating the calculations for various tentative reliabilities, and in each case adding initial and failure costs, we can construct a curve whose ordinates give us the total expected cost as a function of reliability. From here we can choose the optimum alternative.

For example, suppose that the initial cost can be put in the form $C = \alpha_0 + \alpha_1 z$, where z is the base shear coefficient, so that the structural strength is Wz; the weight W of the structure is 40 ton and its damping ratio is 0.05, both independent of z; the structural stiffness is $\alpha_2 z$. Suppose further that the cost of failure is \$200,000, $\alpha_1 = \$50,000$, $\alpha_2 = 5$ ton/cm, and that we operate at a rate of continuous interest of 0.08 yr^{-1}. Initially let us take $z = 0.25$. Then $C = \alpha_0 + \$12,500$ and $T = 2\pi\sqrt{40/981 \times 1.25} = 1.135$ sec. Hence, from Fig. 14.3, the return period of failure would be 900 yr, so the expected cost of failure, using the relation $F' = fv/c$, would be $200,000/0.08 \times 900 = \$2,777$, and the value of the objective function X' in Eq. 14.3 would be $\alpha_0 + \$15,277$. After a few trials we find that the optimum z is 0.2, to which correspond a return period of 600 yr, $C = \alpha_0 + \$10,000$, $D = \$4,166$, and $X' = \alpha_0 + \$14,166$.

Suppose now that the stiffness is independent of z and that $T = 1.135$ sec. Using Fig. 14.3 we find, again, $X' = \alpha_0 + \$15,277$ for the tentative $z = 0.25$. However, at optimum, $z = 0.225$ and $X' = \alpha_0 + \$14,820$. Comparison with the case in which stiffness was proportional to the base shear coefficient brings out the dependence of optimal design conditions on the interrelations between the parameters that define structural behavior.

This approach may be too time consuming in practice. Whether we are interested in the design of an individual structure or in writing a building code, we may find it practical to choose a recurrence period and use the corresponding spectrum for purposes of design. The return period chosen should be a function of the importance of the structure, of the type of failure we contemplate, and of the regional seismicity of the site.

In order to be in a position to select adequate return periods, it is well to note that the probability of failure is 63 percent for $vt = 1$, 9.5 percent for $vt = 0.1$, and 1 percent for $vt = 0.01$. These values correspond to structures having deterministic parameters. In practice the design parameters would be used with either a characteristic (reduced) strength or they would be affected by a load factor to account for uncertainties in the structural parameters. The probabilities of failure would be much smaller than we would derive from the foregoing considerations. Accordingly, in highly seismic areas it would be reasonable to design unusually important structures (nuclear reactors, large arch dams) against collapse for return periods of 500 to 1000 yr; ordinary apartment or office buildings for 50 to 100 yr; and minor or provisional structures for 5 to 10 yr. Longer periods would be advisable in areas of low seismicity, and shorter return periods when designing against nonstructural damage.

The method of analysis we have sketched is applicable when there are several groups of foci that may have significant effects on the structure under consideration. It is equally applicable to the design of structures on soft ground, it being necessary only to use appropriately filtered spectra.

We have assumed throughout that the criterion of failure can be put in terms of a structural response exceeding a critical value for the first time. Methods can be developed, applicable to more elaborate and realistic criteria of failure.

14.3 Simple Linear Systems Having Random Parameters

Thus far we have treated structural parameters as deterministic quantities. Now we shall derive a general expression for the calculation of the return periods of the responses of a structure with uncertain parameters in terms of the return periods for a structure with deterministic parameters.

Let subscript c refer to the structure with deterministic parameters. Then,

$$v(Q') = -\int_0^\infty \frac{\partial v_c(Q'_c)}{\partial Q'_c} P(Q > Q' \mid Q_c = Q'_c) \, dQ'_c$$

where v is the reciprocal of the return period and Q' is a specific value of Q. This expression can also be written in the form

$$v(Q') = -\int_0^\infty \frac{\partial v_c(Q'_c)}{\partial Q'_c} P\left(\frac{Q}{Q_c} > \frac{Q'}{Q'_c}\right) dQ'_c$$

and, with $y = Q'/Q'_c$,

$$v(Q') = \int_0^\infty \frac{\partial v_c(Q'_c/y)}{\partial y} [1 - P_Y(y)] \, dy$$

where P_Y is the probability distribution function of Q/Q_c. Integrating by parts and letting $p_Y = dP_Y/dy$,

$$v(Q') = \int_0^\infty v_c\left(\frac{Q'}{y}\right) p_Y(y) \, dy$$

$$= E\left[v_c\left(Q'\frac{Q_c}{Q}\right)\right] \qquad (14.17)$$

(Esteva, 1970).

This relationship can be used to incorporate the effects of uncertainties in correlations between focal and local earthquake characteristics.

Let us first apply this result to a single-degree structure whose parameters are deterministic but whose orientation is random relative to the direction of potential earthquake foci. The correlations we used in the previous section to obtain the earthquake characteristics at the station of interest furnished us with the maximum ground accelerations, velocities, and displacements with respect both to time and direction. Consequently, such calculations lead directly to the design spectra for a structure whose survival is conditioned by its responses in the direction in which they are a maximum, which implies that our results apply directly to the design of a system whose resistance to lateral force is independent of direction. In this category fit inverted pendulums on a single, circular column, and even certain multibay, single-story buildings if we neglect coupling between vibrations in orthogonal directions and neglect rotational components of base motion. This is equivalent to having idealized the structures in question as linear systems with two identical degrees of freedom along arbitrary, horizontal, orthogonal axes.

The probability distribution function of Q/Q_c that corresponds to a uniform distribution of the angle θ between 0 and 2π, as found in Chapter 7, was used to compute $v(Q')$ for a single-degree structure. The damping ratio was taken as 0.05.

In this case Q_c is Q_{max} because we had computed v_c for Q_{max}, and Q' is $E_\theta(Q)$. The procedure was applied separately to the responses corresponding to nearby and to distant foci, as the distribution of Q/Q_c depends on the earthquake intensities, and these are different, for a given value of Q_{max}, depending on focal distance.

Application of Eq. 14.17 furnishes corrected values of v for given values of $E_\theta(Q)$. It is more convenient, however, to present the results as a family of curves of $E_\theta(Q)$ as a function of T, each curve associated with a given return period.

When plotting $\ln v_c$ as a function of $\ln Q_{max}$ for a given T, we obtain practically straight lines, which implies that $v_c(Q'_c)$ is practically proportional to Q'^{-r}_c, where r is a constant. It follows from Eq. 14.17 that, when this relation holds, if Q' and Q'_c are associated with the same return period,

$$\frac{Q'}{Q'_c} = \left[E\left(\frac{Q}{Q_c}\right)^r \right]^{1/r} \tag{14.18}$$

(Esteva, 1970).

Results obtained in this way for the example solved earlier are shown in Fig. 14.4. They are compared with the curves of Q_{max} associated with the same return periods. There is an appreciable reduction, particularly for responses associated with short return periods.

Pertinent structural parameters include mass, stiffness, damping ratio, and strength. Except for mass they are rarely known accurately. We may take into account our uncertainties about these quantities by using a treatment similar to the one we adopted for orientation. In principle, though, we would have to treat some of these variables as stochastically correlated (usually stiffness and strength are closely related) and as functions of time, at least for materials such as concrete.

For the sake of simplicity we shall assume that all the pertinent variables are independent of each other and that Q/Q_c is lognormally distributed. Suppose that the response of interest can be written in the form

$$Q = \prod_i Q_i(y_i)$$

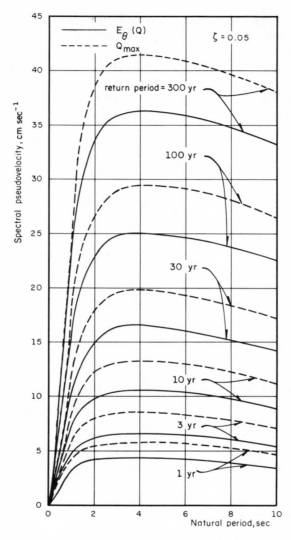

Figure 14.4.(a) Comparison of expected pseudovelocity spectra in a random direction and maximum pseudovelocity spectra, nearby earthquakes.

where the y_i's are the random structural parameters (strength, damping, etc.). Let y_{ic} denote the value of y_i previously treated as deterministic, so that

$$Q_c = \prod_i Q_i(y_{ic})$$

Then,

$$E\left(\ln \frac{Q}{Q_c}\right) = \sum_i E\left(\ln \frac{y_i}{y_{ic}}\right) \qquad (14.19)$$

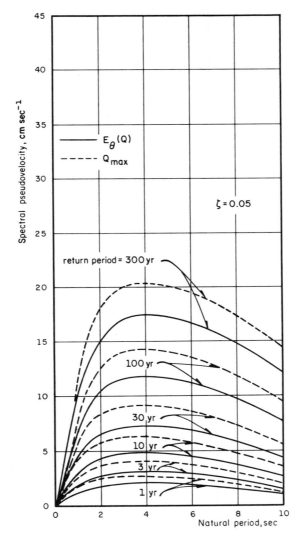

Figure 14.4.(b) Comparison of expected pseudovelocity spectra in a random direction and maximum pseudovelocity spectra, distant earthquakes.

If, for every i, Q_i and dQ_i/dy_{ic} are continuous at $y_i = y_{ic} \exp([E(\ln y_i/y_{ic})])$ and if $Q > 0$, then, save for higher-order terms,

$$\sigma^2\left(\ln \frac{Q}{Q_c}\right) = \sum_i \gamma_i^2 \sigma^2 \left(\ln \frac{y_i}{y_{ic}}\right) \qquad (14.20)$$

where $\sigma^2(\cdot)$ denotes variance and γ_i is the derivative of $\ln Q_i/Q_{ic}$ with respect to $\ln y_i/y_{ic}$ at $y_i = y_{ic}$. Using Eq. 14.18 it can be shown (Esteva, 1970) that, if

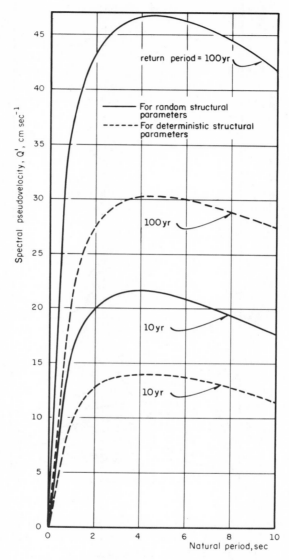

Figure 14.5. Spectra for random and for deterministic structural parameters.

$v_c(Q'_c)$ is proportional to $Q'_c{}^{-r}$,

$$\frac{Q'}{Q'_c} = \exp\left[\frac{r}{2}\sigma^2\left(\ln\frac{Q}{Q_c}\right) + E\left(\ln\frac{Q}{Q_c}\right)\right] \qquad (14.21)$$

where Q' and Q'_c are values of Q and Q_c corresponding to a given return period.

The value of γ_i for strength is 1 everywhere. For mass and stiffness it is respectively $\pm\frac{1}{2}$ times the value corresponding to natural period. For moderate damping, γ_i is approximately -0.4.

Let us apply this method to the results in Fig. 14.3. Assume that the curves in the figure correspond to modal values, $\tilde{y}_i$, of strength, mass, stiffness, and damping, so that $y_{ic} = \tilde{y}_i$, and that the dispersions of the respective logarithms of the parameters divided by their modes are 0.5, 0.1, 0.8, and 0.5. (If the dispersion of $\ln y_i/\tilde{y}_i$ is small compared with unity, it is roughly equal to the coefficient of variation of y_i.) It follows that

$$\sum_i \gamma_i^2 \sigma^2 \left(\ln \frac{y_i}{\tilde{y}_i}\right) = 0.5^2 + (0.1^2 + 0.8^2)(0.5\gamma_T)^2 + (0.5 \times 0.4)^2$$
$$= 0.29 + 0.1625\gamma_T^2$$

where γ_T is the derivative of $\ln(Q/\bar{Q})$ with respect to $\ln(T/\tilde{T})$ and T is natural period. Next we choose the recurrence periods in Fig. 14.3 and substitute in Eq. 14.21. We obtain values of Q' that lead to the spectra in Fig. 14.5 for recurrence periods of 10 yr and 100 yr. The spectra for deterministic structural parameters are shown with dashed lines.

We would have found a more important effect of uncertainty in the natural period of vibration had we dealt with a structure on a soil having pronounced prevailing periods. However, the simplifying assumptions we adopted would have introduced appreciable errors on the safe side, particularly near the peaks in the expected spectra and wherever these spectral ordinates decrease rapidly with natural period.

The ratio of the design spectral ordinates for random structural parameters to those for deterministic parameters may be regarded as a load factor or a factor of safety to account for uncertainty in the parameters. This factor is used in conjunction with an analysis based on the probability distribution modes (or, in a sense, *nominal* values) of these parameters.

In general the required load factor is a decreasing function of Q_c.

14.4 Multidegree Linear Systems

Design spectra such as those obtained in the preceding section are directly applicable, in conjunction with the approximate methods developed in Chapter 10, to the design of multidegree systems of linear behavior provided it is permissible to deal with a single mode of failure. This requires that the probabilities of failure in all other modes be much smaller, or the corresponding recurrence periods much larger, than for that one mode. When this condition is not satisfied, we must employ an approach such as that sketched in Section 11.2.

In principle it is possible to calculate all the pertinent structural responses (all generalized stresses and deformations related to design) in all the significant natural modes of vibration and combine them according to the theory of Chapter

10. However, in practice this is an overwhelming task. We can aim in this manner at computing a sufficient number of generalized forces so that all pertinent responses can be obtained from them under the assumption that the forces computed are deterministic and are applied statically. In a tier building we can compute the story shears. Assuming that these are statically applied deterministic shears, we can analyze the frames, systems of cross bracings, or shear walls, computing story torques, axial forces in the braces, bending moments, torsional moments, and shears in all the members. From the shears in horizontal members we can also find axial forces in vertical elements, be they walls or columns.

This approach introduces errors due to the simplifying assumptions on which it rests. For example, in the analysis of a building it omits consideration of the moments of forces of inertia about vertical and horizontal axes, because once the story shears have been found they are treated deterministically. Moments about a vertical axis may greatly modify the computed story torques, usually increasing them. Moments about horizontal axes tend, on the one hand, to raise stresses in girders and to raise overturning moments; on the other hand, they tend to lengthen the natural periods and hence, usually, to reduce all stresses. Considerations of this sort can be incorporated in the analysis by using for the basic generalized forces not only story shears but story torques and moments due to rotational inertia as well.

Questions mentioned in the foregoing paragraph result from treating dynamic forces as static. Other errors arise from treating random forces that act at different times as deterministic and simultaneous. For example, if we compute axial forces in columns of a building in the manner we have described, we will overestimate them systematically because the story shears that are used for the design of structural members in flexure do not act simultaneously.

Finally, design should not be based necessarily on equal probabilities of failure at all sections and in all modes, nor should it ignore the importance of the structure being designed. Variations should follow from differences in the consequences of failure and in the cost resulting from a change in the probabilities of failure. A quantitative treatment of this question is generally beyond the possibilities of a design office, at least for purposes of routine design. Yet its implicit consideration is reflected in the lateral force coefficients required by building codes, just as it is manifest in variations of load factors or of working stresses, even in design for gravity loads.

Often it is advantageous to split the problem of computing lateral forces for design into the calculation of a base shear coefficient and the calculation of the distribution of lateral forces. For many simply coupled systems this manner of carrying out the analysis allows a reasonably accurate simplified method. Suppose we are to design for a damped acceleration spectrum whose ordinates, as a function of natural period, do not exceed those corresponding to a constant pseudovelocity response spectrum. Suppose, further, that we design for the square root of the sum of squared modal responses. It can be shown that for a wide set of design spectra the base shear coefficient does not exceed the response acceleration coefficient for a single-degree system having the same percentage of damping and natural period as the given system has in the fundamental mode.[8]

A lower limit to the base shear is obtained by drawing a hyperbolic acceleration spectrum that envelopes the damped design spectrum from below and computing the base shear associated with the fundamental mode of vibration.

Frequently these bounds lie sufficiently close to each other for purposes of design, or at least of preliminary design. Consider a simply coupled system whose masses are proportional to 2 and 1 at the first and second floors, respectively. Suppose that its fundamental mode is (1, 2). Owing to the orthogonality of the natural modes, the second natural mode will be (1, −1). We can show that for a close-coupled system the squares of the natural periods are in the ratio

$$\frac{T_n^2}{T_m^2} = \frac{z_{jm} \sum_i M_i z_{in}}{z_{jn} \sum_i M_i z_{im}}$$

where T denotes natural period, m and n refer to two natural modes, z are the natural modes, M are the masses, i refers to the various masses in the system, and j is an arbitrary value of i. Using this relation we find $T_2/T_1 = \frac{1}{2}$. Assuming that the design acceleration spectrum is hyperbolic in the range of interest, we compute the ratio of the base shear coefficient to that of a single degree system whose natural period is T_1, as 0.92. The base shear is found to be 1.03 times that for the fundamental mode.

Once the base shear coefficient has been estimated or bounded, the distribution of shears along the structure can often be estimated with sufficient accuracy from the comparison with the results of dynamic analyses of similar structures. This is the basis of requirements in many building codes, in which the upper bound we have described is used for a conservative estimate of the base shear coefficient, and an approximate formula furnishes the ratio of the shear at any elevation to the base shear. Only the fundamental period of vibration need be computed or estimated then.

In very rigid systems even the estimate of the fundamental period may be omitted. Even for remotely coupled systems it is always conservative to take the base shear coefficient equal to the maximum estimate in the acceleration spectrum divided by the acceleration of gravity. If the system were infinitely rigid the lateral force coefficient would be constant throughout the system (except for variations reflecting a desired change in the probability of failure) and equal to the base shear coefficient. For practically any other system the lateral force coefficient should increase, in general terms, with elevation above ground level. Hence, it is almost invariably conservative to design relatively rigid systems for lateral force coefficients proportional to elevation above ground level and such as to give a base shear equal to the mass of the system times the maximum ordinate in the design acceleration spectrum.

[8] We can always write $S \leq K \sum_i |a_i z_i A_i/\omega_i^2|$, where S is the base shear, K is the ground-story stiffness, subscript i refers to the ith natural mode, a is a participation coefficient, z is the displacement of the first floor, A is the spectral acceleration, and ω_i is the ith natural circular frequency. It can be shown that $\sum_i |a_i z_i| = 1$. It follows that $S \leq K \max_i A_i/\omega_i^2$. Hence, the statement holds at least when $A_1/\omega_1^2 \geq A_i/\omega_i^2$ for $i \neq 1$, which is usually the case.

This criterion of design may be entirely too conservative in many cases. Yet it is satisfactory for structures whose initial cost is not especially sensitive to the seismic forces adopted in design. This is usually the case with nuclear reactors.

We have spoken of "relatively rigid" structures, and it is well that we give a natural period as reference, so we can say that such structures are those whose fundamental period does not exceed—or does not greatly exceed—the period of reference. This period is the one associated with the maximum acceleration in the design spectrum. With this criterion we ensure that the importance of the higher modes will not be excessive in the upper portions of the structure. We must introduce an exception when the design acceleration spectrum has several peaks of comparable importance; the reference period is then the lowest of those associated with such peaks.

According to this criterion we call a structure rigid and use the simplified method of analysis we have just described, when the structure lies on firm ground if its fundamental period is shorter than 0.1–0.2 sec if the most significant earthquake foci lie nearby, or when its fundamental period is shorter than 0.3–0.5 sec provided the most significant foci are distant from the station.

For structures on thick, soft mantles, the reference period may be as long as 2.5 sec (Herrera, Rosenblueth, and Rascón, 1965), which exceeds the fundamental period of structures common in urban construction. Under the circumstances, a building code may call for a purely "static" design, making the base shear coefficient independent of the estimated natural period and specifying a certain (possibly linear) variation of lateral force coefficients with elevation above ground. (Of course, nonlinear behavior should also receive due consideration in such a formulation.)

Soil-structure interaction can be incorporated in the analysis by adding degrees of freedom, as described in Section 3.15. Its neglect may introduce appreciable errors on the unsafe side when the design acceleration spectrum increases with period for natural periods equal to or longer than the fundamental of the structure.

Thus far we have introduced a single component of ground motion in dealing with multidegree systems. It is customary to design for each component of ground motion independently of the other components, as though an earthquake could be spread out in time so that when one component were acting the rest did not exist, and each did not begin until the structure had come to rest. In some cases, particularly of structures exhibiting linear behavior, this approximation introduces no more than negligible errors, although they are on the unsafe side. But even for structures of this kind, there are cases when the errors are far from negligible.

Consider first the case of only translational ground motion in a horizontal plane. Suppose that the structure resists the most intense component assumed to act along the arbitrary direction x_1. Suppose that it would also resist this component when it acted along direction x_2, at right angles with x_1. Since the structure behaves linearly, it follows that it would resist the same component along any direction. If the strength of the structure is exactly that required for it to resist the assumed single component of ground motion along each of the orthgonal directions and if the critical section or sections are the same for both directions, the strength will also be exactly the one required for the component of motion acting along an arbitrary direction. If the critical sections differ for the component of motion acting along the two orthogonal directions, the strength of the structure will be at least as large as required for any other direction of ground motion. It does not follow, however, that the structure will necessarily withstand the simultaneous action of the most intense component along one direction and even the least intense component at right angles thereto.

To illustrate this point consider a column of rectangular cross section forming part of a structure. Let S denote the maximum normal stress at the critical section of the column. When there is motion only along x_1, S at time t can be obtained from an expression of the form

$$S_1(t) = \int_0^t \ddot{x}_1(\tau)\psi(t-\tau)\,d\tau$$

where $\ddot{x}_1$ is the ground acceleration in this direction and ψ is a transfer function. Suppose that the effect of the x_2 component is given by

$$S_2(t) = \alpha \int_0^t \ddot{x}_2(\tau)\psi(t-\tau)\,d\tau$$

where α is a constant. The theory presented in Chapter 10 leads to the conclusion that, for a given probability that the design stress S^* can be exceeded at least once, under a wide set of conditions the combined effect of the two components gives

$$S^* \cong \sqrt{S_1^{*2} + S_2^{*2}}$$

which is always greater than S_1^*. Now, if the structure would barely resist the most severe of the components when this component acted along either x_1 or x_2, we must have $\alpha = 1$. And we have seen that for strong earthquakes the most and least intense components do not differ markedly in intensity from each other (Section 7.5). Hence, under conditions that we frequently meet in practice, $S^* \cong 1.4 S_1^*$. In other words, the traditional method of design for independent ground motions in two orthogonal directions often introduces errors close to 30 percent on the unsafe side in the stresses in columns of rectangular section.

There is no error of this sort in columns of circular section, and it is usually insignificant in beams. Indeed, while one component of ground motion produces important bending in a vertical plane in a beam, the perpendicular component may induce small lateral flexure, torsion, or axial force (the latter particularly when the beam constitutes the edge of a horizontal diaphragm); these effects are usually small compared with those of bending in a vertical plane.

In most designs the vertical component of ground motion is neglected. To justify this simplification it is argued that the structure must be designed to resist gravity loads with a load factor greater than the one adopted for the combination of gravity and seismic forces. The change in load factor will presumably make gravity loads govern design.

The reasoning would be correct, provided seismic design accelerations were sufficiently small, if the vertical component did not act simultaneously with other components of ground motion. The simultaneous action of the acceleration of gravity and of those of the horizontal and vertical components of ground motion should be reflected in the design of those portions of structural members which are governed by the combination of vertical and horizontal forces.

To illustrate this matter consider a section whose strength requirements were dictated 50 percent by gravity and 50 percent by the horizontal components of ground motion when neglecting the vertical component. Suppose that the effect of the vertical component of ground motion were 70 percent of the effect of the horizontal components, which is not an unreasonable figure. Recognition of this component would then require the strength to be increased from 100 percent to approximately $50(1 + \sqrt{1 + 0.7^2}) = 111$ percent.

If we were concerned with the design of, let us say, a cantilever beam within the elastic range, the effects of the horizontal components of earthquakes would probably be negligible. We would be comparing design against the acceleration of gravity by itself with design against the combination of this acceleration and the vertical component of earthquakes. It is likely that the former would often govern design because of the higher load factor that permanent loads require and because it would be taken in conjunction with a higher live load than when considering its combination with earthquake forces. Hence, the argument for neglect of the vertical component would probably be valid, but not necessarily so when the vertical component can be expected to be especially severe.

The importance of the vertical component of ground motion is usually much greater when designing with explicit recognition of inelastic behavior, as we shall see in the next section.

In Chapter 10 we described means for estimating the importance of the rotational components of ground motions and for taking them into account. When considering rotation about a vertical axis, it is well to keep in mind that inevitable inaccuracies in the calculation of rigidities and unfavorable distributions of live loads (and often also of dead loads) will usually introduce larger story-torques than the corresponding rotational component of ground motion. We shall deal with this question in some detail when dealing with the design of buildings (Section 15.6).

The assumption of linear behavior is realistic only when designing to resist earthquakes of low intensity. There is always a limiting intensity beyond which the requirement is unreasonable that there be a high probability for the structure to remain within the linear range. This intensity depends on many factors. Among them are the regional seismicity, the sensitivity of the structure's initial cost to increase in earthquake resistance, and, above all, the importance of the structure being designed. Thus, at a highly seismic site it may be reasonable to admit that an apartment building will undergo excursions in the plastic range for MM intensities as low as 6.5 or 7, while the structure for a nuclear reactor may be required to have a negligible probability of exceeding the yield point even at MM intensities of 9. But in practically every structure, extensive nonlinear behavior takes place before failure. Therefore, with few exceptions, design must reflect an explicit consideration of this type of behavior for relatively high intensities.

14.5 Nonlinear Structures

In the design stage of single-degree systems when nonlinear behavior is recognized, there are two important differences with respect to the approach described in Sections 14.2 and 14.3 for treatment of the linear range. First, there is at least one additional parameter defining structural behavior. And second, cumulative damage during successive earthquakes may play a significant role. Otherwise the approach is similar, with no more than the principles in Chapter 11 requiring incorporation.

The additional parameter or parameters referred to are those which define the force-deformation curve. For example, in an elastoplastic system, structural behavior will be governed by mass, initial stiffness, initial damping ratio, maximum force, and maximum ductility factor; the first four parameters also enter the problem of design of systems that behave linearly; the fifth one is required in the design of elastoplastic structures.

Ordinarily it is difficult to choose all the parameters so they constitute essentially independent random variables. Hence, we should often employ conditional probabilities in order to reckon uncertainty about their values.

The matter of cumulative damage during successive earthquakes merits considerably more attention than it has received. It has not been rare that a structure which withstands a major shock with visible damage collapses during a relatively mild aftershock. At present, no more can be done than to apply judgment in the light of information about behavior under repeated loading (see Chapter 13) to modify the design provisions, taking into account the probability of occurrence of several shocks of comparable intensity, the nearly certain occurrence of aftershocks, and the probable measures that may be taken to repair and strengthen the structure after every important temblor.

The degree with which a structure deteriorates in its various modes of failure under repeated loading influences design criteria not only from the viewpoint of the structure's capacity to withstand a series of earthquakes but even from that of its behavior under a single earthquake. The possibility of deterioration plays an increasingly important role as we move from designs intended to resist essentially earthquakes of type 1 (see Chapter 7) to those for earthquakes of type 2 and to designs done having type 3 in mind.

Consider, as an example, the working stresses used in connection with the design of reinforced concrete structures. At sites where only earthquakes of types 1 and 2 are anticipated it is resonable to allow, as many building codes do, a 50 percent increase in the working stresses of reinforcement and 33 percent increase in those of the concrete itself relative to values permitted under static loads alone. The difference in the increase between reinforcement and concrete proper recognizes the different ductilities to be expected as a function of the mode of failure (see Chapter 13). The allowable increase should be appreciably smaller or nil when we design against earthquakes of type 3 in connection with those modes of failure for which deterioration sets in at a rapid rate, since the number of loadings will be much greater under the action of one earthquake of this type. Such is the case with the shear or diagonal tension taken by concrete. But there is no need to modify the allowable increases in working stresses when designing against other modes of failure.

Cumulative damage is an especially serious phenomenon in inelastic systems having frankly asymmetric force-deformation curves. With elastoplastic systems whose initial natural frequency is high, a small difference (say 30 to 40 percent) in yield point in two opposite directions makes the probability of yielding in the strongest direction virtually nil. We have seen that the responses of systems that yield in only one direction are appreciably larger than those of systems having symmetrical force-deformation curves. But, more important, while the permanent deformation caused by a series of earthquakes in a structure of symmetrical force-deformation behavior grows as the sum of the squares of the increments in permanent deformation due to the individual temblors, it increases essentially as the sum (of terms of equal sign) in a system having a pronouncedly asymmetric relation. For example, if we let μ denote the ductility factor and assume that the expected permanent deformation due to a single earthquake is $(\mu - 1)y_y$, where $y_y = $ yield-point deformation, the assumption is valid asymptotically for exceptionally large values of μ; it is conservative for the expected value. A series of n earthquakes of equal duration and intensity would cause an expected permanent deformation of $[(\mu - 1)\sqrt{n} + 1]y_y$ in a structure having a symmetric force-deformation curve and of $[(\mu' - 1)n + 1]y_y$ in one having a strongly asymmetric curve, with μ' denoting the corresponding ductility factor ($\mu' > \mu$ if $\mu > 1$). If, for example, $\mu = 2$, $\mu' = 4$ (see Chapter 11), and $n = 9$, the ratio of the two expected maximum deformations would be $19:4 = 4.75:1$.

This phenomenon manifests itself in the behavior of slopes (for example, in earth and rockfill dams) and in that of cantilevers and of simply supported beams when the yield moment can be expected to be reached, on the average, every few years. Restoration may be relatively inexpensive (for example, the crest of a dam may be raised to its original elevation through additional filling after every major earthquake). If so, design need not be influenced seriously by consideration of the cumulative nature of yielding. Otherwise, as with a structural member in a costly building, for example, a specially conservative practice will be in order so as to account for such behavior.

As an alternate solution in structural members we can produce practically symmetric curves of vertical acceleration vs. deformation by appropriately adjusting the positive and negative yield moments.

A few examples of explicit optimization are found in the literature on earthquake-resistant design. An interesting application to the revision of the earthquake resistance of a gravity dam built in masonry is due to Esteva, Elorduy, and Sandoval (1969). Their treatment is similar to the one presented in the foregoing paragraphs. The strength of the dam was assumed to be governed by cohesion and internal friction until the beginning of breakage and sliding over a surface of failure, after which only friction remained active. Sixty-six alternatives were studied, including the possibility of leaving the dam as it stood, increasing its section by different amounts up and downstream, and increasing its cohesion by grouting. For each alternative the sum of initial investment and expected present value of losses due to failure was computed, and the alternative giving the lowest sum was recommended as the optimum course of action. This alternative raises the expected lifespan of the dam from its original 400 yr to about 1000 yr.

For multidegree-of-freedom systems whose behavior beyond the linear range is recognized in design, the foregoing remarks apply, and there is an additional difference with respect to the corresponding linear-behavior systems. Some criteria for relating the responses of nonlinear to "equivalent" linear systems hold when there is a single degree of freedom. The same criteria tend to underestimate the responses of nonlinear multidegree systems. This is especially noticeable when the structure is relatively weak in one or in a few degrees of freedom, so that only in these instances does it dissipate energy through inelastic behavior.

We have discussed the matter for closely coupled systems in which only one of the spring elements yields and thus functions as a fuse, preventing the appearance of sufficiently large stresses in the other spring elements to cause them to yield (Section 11.5). As a second example, consider a structure the distribution of whose stiffnesses is such that it undergoes no torsion in the range of linear behavior. If one side of the structure should yield at a substantially lower deformation than the other, the ensuing rotation about a vertical axis would induce torsional moments that would in turn produce additional torsional vibrations. Even in a structure whose elements have yield levels such that, in theory, under translatory excitation there would be no torsional oscillations, there may appear, accidentally, a small torque that would be magnified through this process.

The phenomenon is especially important in structures of the type of X-braced towers. The literature abounds in descriptions of failures of elevated water tanks, attributable only to torsional failure, which nominally could not have been brought about.

Failure of Mansion Charaima near Caracas in earthquake of July 29, 1967. (American Iron and Steel Institute.)

15

EARTHQUAKE-RESISTANT DESIGN OF BUILDINGS

15.1 Design Spectrum

Modal analysis requires that the design spectrum be specified. Many building codes stipulate either a design acceleration spectrum or a base shear coefficient as a function of natural period. These coefficients are essentially ordinates of acceleration spectra divided by the acceleration of gravity; the relationship holds exactly in single-degree systems.

The intention in building codes is that spectral ordinates be used for the computation of inertia forces which serve to obtain stresses against which one must provide, under the assumption of linear behavior. Accordingly, design spectra imply certain amounts of damping and certain amounts of inelastic action or, with reference only to systems idealized as elastoplastic, certain ductility factors.

The amounts of damping are not stated explicitly, and present codes do not contain provisions for recognizing variations from one material or one structural solution to another. Design spectral ordinates are independent of these variables. We could argue that there is often a lesser dispersion of strength in the less highly damped materials and structures (say in steel frames as compared with reinforced concrete frames or in reinforced concrete frames as compared with masonry shear walls). The two effects may therefore cancel out each other. Although the argument is qualitatively true in many cases, it is not correct with absolute generality nor can we expect it to apply in a quantitative sense.

The use of special devices intended to increase damping is not given attention in present building codes. Some codes allow changes in design spectra as a function of anticipated ductility factors and the possibility of having more than one line of defense, making both concepts functions of the structural solution chosen. Thus, the Uniform Building Code of the United States (*Earthquake Resistant Regulations, A World List*, 1966a) uses a coefficient, in the design accelerations, of 0.67 for "buildings with a ductile moment resisting space frame" with respect to ordinary moment resisting frames. The "space frames" must be capable of resisting the lateral forces without the aid of diaphragms, walls, or braces. The same code allows a factor of 0.80 for buildings with a dual bracing system in which shear walls can resist the entire lateral forces, and frames 25 percent thereof. It specifies a magnification factor of 1.33 in the design accelerations for box-type structures (which resist lateral forces solely by means of shear walls or cross braces) relative to ordinary moment-resisting frames. For inverted pendulums and other special structures, there is a magnification factor of 2.00, and for elevated tanks this factor is 3.00.

The reduction allowed for space frames reflects their high ductility. It also recognizes the reduction in probability of collapse for structures having several lines of defense. The same is true of buildings with a dual bracing system. In these buildings failure of, let us say, structurally effective cross braces, walls, and partitions does not leave the frames incapable of resisting the last portion of a strong ground motion nor its aftershocks or additional earthquakes. The increase for box-type structures is intended to account for their usually smaller ductility. The same criterion applies to inverted pendulums.

Other building codes have comparable variations in design spectra (*Earthquake Resistant Regulations, A World List*, 1966b, 1966c, and 1966d).

The existence of "defense plateaus" (see Section 14.1) is tantamount to a measure of ductility in the overall behavior of a structure. The analogy is illustrated in the very schematic Fig. 15.1.

For any finite, constant ductility factor, as the initial stiffness increases toward infinite values the acceleration that the system must withstand to survive tends to the maximum ground acceleration. Apparently it should follow that, whatever the reason for a reduction in design spectral accelerations—whether high ductility or the availability of defense plateaus, there should be no reduction at very short natural periods.

The argument is not altogether valid, however. For structures on firm ground all codes that take ductility into account through a reduction or magnification

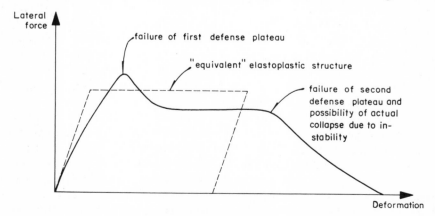

Figure 15.1. Equivalence of ductile behavior and defense plateaus.

factor use a design acceleration independent of natural period in this range of very short natural periods. Actual acceleration spectra have ordinates that increase with natural period in this range. Beyond the peak acceleration the design values should indeed vary, in a rough way, inversely with the ductility factor. Therefore the provision is tenable for structures on firm ground.

On soft soil a similar argument applies. Although the design acceleration is sometimes made to increase with natural period for short values thereof, this increase is far slower than that in actual acceleration spectra.

There are reasons for a conscious overestimate of spectral accelerations in the range of very short natural periods. First, this provision protects against a serious underestimate of natural periods; such errors are particularly likely to be introduced in the analysis of rigid structures because it is common to overlook relatively important soil-structure interactions, play at joints, shrinkage cracks, and other phenomena that lengthen natural periods of vibration. Second, if the shape of actual acceleration spectra in this range were taken for design, one would produce structures that would "fail unsafe," in the sense that incipient failures that are associated with a lengthening of natural periods[1] would leave the structures in a less favorable condition than they had at the outset of the first major shock to affect them. By making the design acceleration constant in the range between zero natural period and the period associated with the peak spectral aceleration, one ensures that lengthening of the periods of vibration will not produce a situation worse than those foreseen in design. Lengthening beyond the period associated with the peak leads to structures that "fail safe." Third, very rigid structures are usually such that overdesign in earthquake resistance does not involve an excessive initial cost.

Adoption of a constant design acceleration in the range of short natural periods is consequently justified for structures on firm ground; for them the peak spectral acceleration is associated with natural periods in the range of 0.05 to about 0.5 sec. For structures on such soft and thick mantles that the peak occurs at a period of, let us say, 2.5 sec, it would be too costly to adopt a constant design acceleration in the entire range below this period. By specifying a slow increase in the design acceleration with the natural period in this range, we supply the same type of protection as the provision we described for structures on firm ground.

Some building codes make design accelerations a function of the importance of the building under consideration. The magnitude of this differentiation is open to question, as it has invariably been fixed on intuition alone. But the principle is sound because the probability that serious failure occurs during any finite time interval in any given site is finite; the conditional probability of this event, given that a major earthquake occurs, is not negligible.

Figure 15.2 compares simplified spectra for the NS component of thé El

---

[1] The literature is filled with examples of buildings whose natural periods have lengthened appreciably as a consequence of damage (e.g., Ihara and Veda, 1960; del Valle and Prince, 1965). Even minute cracking, undetectable to the naked eye, has a noticeable effect, as witnessed by several buildings in California (Carder, 1936).

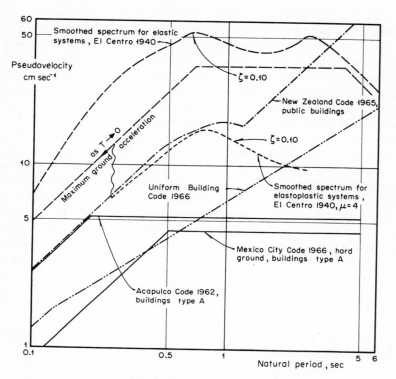

Figure 15.2. Comparison of El Centro spectrum NS component with code spectra for structures on hard ground.

Centro 1940 earthquake with the design spectra specified by various codes for buildings on firm ground. The code accelerations have been multiplied by the approximate overall load factors which they specify or imply.

Relative to this particular earthquake, most codes err on the unsafe side in the range of short natural periods. In areas where earthquakes having at least this intensity can be expected to occur relatively often, the situation should be corrected.

All the design spectra err on the safe side for very long natural periods. The error involved in the use of a frequency-independent pseudovelocity spectrum for these periods is ordinarily not objectionable because few urban structures have such a long fundamental period of vibration as to make the matter significant. Besides, the cost of taking such small accelerations for very long periods is already small for buildings, so that a reduction would be hard to justify. The same remarks do not apply in connection with some long-period structures that are not buildings. They do not apply either, even in building design, to the cutoff or minimum acceleration that some codes specify nor to the monotonic increase in pseudovelocity that the Uniform Building Code requires.

Use of a constant lower bound of design acceleration has been defended on the basis that it does not involve excessive additional cost for long-period structures and that it protects structures against crass mistakes in dynamic analysis. Since constant pseudovelocity already does this to a very major extent, the arguments are debatable.

There is another justification for the increase in conservativeness with the natural period that characterizes the Uniform Building Code. Systems having a high number of degrees of freedom are likely to be overdesigned in many of these degrees and/or underdesigned in the rest. The situation gives rise to unusually high demands on energy absorption in the degrees that are not overdesigned, as we described earlier (Section 11.5). Since the fundamental period increases, in very general terms, with the number of degrees of freedom, a simple criterion leading to uniform safety makes the ratio of design acceleration to the one actually required an increasing function of period. Whether the increase with the two-thirds power of period, as in the Uniform Building Code, is adequate, is another matter. And surely it would be desirable to introduce recognition for the phenomenon in question in a more direct manner, bypassing the fundamental period of vibration.

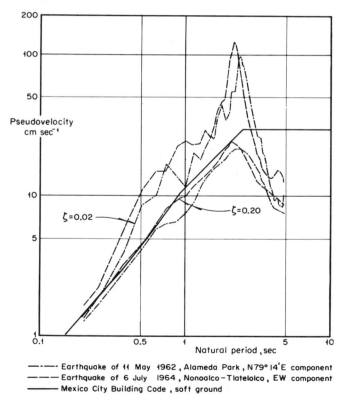

Figure 15.3. Comparison of spectra for actual earthquakes and code requirements on soft ground.

The preceding remarks suggest that additional protection of tall buildings against excessive demands on plasticity are unwarranted. It should suffice to use the design spectrum that makes the pseudovelocity vary with the two-thirds power of the natural period, and neither forbid structures whose height exceeds some figure nor change the design criteria therefor.

A second comparison of spectra is displayed in Fig. 15.3 for buildings on one soft formation. Most of the remarks made in connection with hard ground carry over to this condition, with a mere change in the scale of periods. The flat portion of the acceleration design spectrum between periods of 1.0 and 2.5 sec reflects the presence of several important prevailing periods of the ground that may change from site to site in the city in question (see Esteva, Rascón, and Gutiérrez, 1969).

15.2 Base Shear Coefficient

In a manner more or less directly related to the design acceleration spectrum, most building codes specify the base shear coefficient as a function of the fundamental period of vibration of a building. It follows from the remarks in Section 14.4 that we overestimate the base shear in multistory buildings slightly when we multiply the mass of the building by a coefficient equal to the ordinate of the design acceleration spectrum that corresponds to the fundamental period of vibration if, for all periods shorter than this, the design spectrum does not lie above the hyperbola

$$A = \frac{A_1 T_1}{T}$$

where A is the design acceleration associated with the natural period T, and subscript 1 identifies the fundamental mode of vibration. This is the case for all building code design spectra published to date.

A comparison by Bustamante (1961a) between the results of modal analysis with the use of a base shear coefficient substantiates the conclusion just stated.

15.3 On the Calculation of Natural Modes

It would seem from the foregoing comments that dynamic analysis is not justified for the calculation of base shears. Apparently it should be sufficient to estimate the fundamental period. However, the fundamental period of vibration of a building cannot be estimated with satisfactory accuracy from exterior geometry alone but practically demands a dynamic analysis (see Section 13.7). Commonly used formulas introduce inadmissible errors in base shears.

Relative displacements between consecutive floors must be computed in many practical cases. The purpose is to verify the design of connections between struc-

Sec. 15.4 SHEAR DISTRIBUTION 483

tural and nonstructural elements as well as the gap that one must provide between adjacent structures. Calculation of relative displacements under a set of lateral forces involves much of the work that goes into the calculation of a building's fundamental mode of vibration. Hence there is little excuse for omitting dynamic analysis in these cases, except for preliminary design. Moreover, computer calculation of higher modes of vibration involves little additional labor above that required for the fundamental mode.

If natural modes are to be calculated, it is worth taking into account those concepts that make the dynamic behavior of a building significantly different from that of a simply coupled system on a fixed base, whenever the following concepts are important: soil-structure interaction, axial deformations of structural members, rotational inertia of floor systems, deformations of supposedly rigid diaphragms in their own planes, and so on.

15.4 Shear Distribution

Within the range of linear behavior we can always compute story shears, as well as all other generalized stresses and deformations, using the methods presented in Chapter 10. In practically all multistory buildings the significant natural frequencies of vibration are neither so high nor so close together that there be serious objections to the use of the root sum of squared responses in the natural modes to compute the story shears. Results obtained through this simplified method are closely in accord with those of more refined methods and with the shears computed as responses to several actual earthquakes (Jennings and Newmark, 1960; Clough, 1962). These results pertain to buildings idealized as close-coupled systems.

The shear distributions in a uniform shear beam may serve as guides for estimating the distribution in buildings whose behavior may be idealized as that of a close-coupled system, and for suggesting approximate methods of analysis. The shear beam is assigned uniform stiffness and mass per unit length through its height. Figure 15.4a shows the shear distributions in a uniform shear beam with fixed base for the four design spectra in Fig. 15.4b. The distributions were obtained as the root sum of squared modal responses. They are compared, in the figure, with the shear distributions obtained from two different static methods of analysis: one with constant lateral force coefficient and one with a linear variation of this coefficient with height above ground. The curves shown have been adjusted to scale so that they all give the same base shear.

The shear distribution for buildings on soft ground subjected to distant earthquakes (4 in Fig. 15.4a) differs markedly from all the rest. The difference is due mostly to the outstanding influence of the second natural mode of vibration in this case (see Fig. 15.4b).

Static methods of analysis attempt to furnish a distribution of seismic shears and sometimes of other generalized forces, approximating the distributions

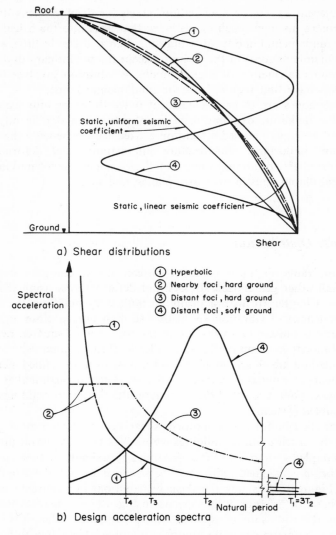

Figure 15.4. Comparison of shear distributions in uniform shear beam.

obtained from accurate dynamic analysis. To this end these methods assume the action of statically applied horizontal accelerations, or *lateral force coefficients*, when the accelerations are expressed in terms of the acceleration of gravity, and in many instances make certain adjustments.

The oldest of these methods was applied early in the twentieth century by T. Naito (Naito, 1960). It gained popularity because most of the buildings designed by him withstood the 1923 Tokyo earthquake satisfactorily. The method uses a uniform, lateral force coefficient. According to Fig. 15.4a, this criterion systematically and seriously underestimates the shears in the upper stories when the base shear has been correctly assessed. We would reach the same conclusion in buildings having almost any mass and stiffness distributions within practical limits. Initial success of the method must be ascribed essentially to two factors: use of a high base shear coefficient together with low working stresses, and neglect of the contribution of "nonstructural" elements (mostly walls and partitions), which is important particularly in the upper stories.

Adoption of functional architecture in the 1930's and 1940's together with the increased popularity of architectural solutions in which partitions were light, weak, and movable made it evident near the west coast of the United States that a uniform, lateral force coefficient would necessarily leave the upper stories unprotected, or it would lead to excessive overdesign of the lower stories, or both. Experience and judgment coupled with results of simplified analyses confirmed the inadequacy of the old static method. In 1943 the developers of the Los Angeles Building Code, following the practice of some advanced structural engineers, introduced a formula for calculation of seismic shears that amounted to a raising of the lateral force coefficient as a function of floor elevation above ground.[2] The formula was adopted by the Uniform Building Code in 1946.

When results of dynamic analyses of buildings became available, in large measure as a consequence of the efforts of Biot (1943) and of Alford, Housner, and Martel (1951), it was possible to develop a simple static method of wider applicability. The method essentially advocated the use of lateral force coefficients corresponding to an acceleration varying linearly from zero at the base to a maximum at the roof of the building (Anderson et al., 1952). This criterion was incorporated in the San Francisco Building Code in 1956 and replaced the Los Angeles criterion in the Uniform Building Code in 1964.

The comparison in Fig. 15.4a shows that use of a lateral force coefficient proportional to elevation produces shears better in accord with the results of dynamic analysis. This is especially true when the arithmetic plot of the design acceleration spectrum is hyperbolic (corresponding to equipartition of energy as a function of frequency), and when we adopt the two spectra in Fig. 15.4b that are representative of conditions on firm ground. Agreement is good when the most important earthquake foci lie relatively far from the site of interest, so that the design acceleration spectrum is constant over the range of the natural periods of the structure. Even so the static method still underestimates shears in the upper stories. This is particularly true when the fundamental period of the structure is long and the most significant earthquake foci lie close to the site.

This result has led some structural engineers to suggest that building codes specify methods of analysis leading to greater shears in the upper stories than those produced by the linear variation of lateral force coefficients. This tendency

---

[2] On the development of American' earthquake-resistant code provisions see Binder and Wheeler (1960).

is reinforced by the fact that in most tall buildings the mass varies little from one floor to the next, beginning at some moderate elevation above ground, while story stiffness decreases rapidly with elevation. The variation leads to greater shears in the upper stories for a given base shear.

The natural modes of a shear beam whose stiffness and mass per unit length vary in an arbitrary manner but are proportional to each other along the entire height of the structure are the same as in a uniform shear beam (see Section 3.4). It follows that the lateral force coefficients for a shear beam will be proportional to those for a uniform shear beam if stiffness and mass per unit length vary as a power of the distance from the top. If stiffness decreases faster with elevation than does the mass per unit length, there will be need for a more pronounced increase of lateral force coefficients with elevation than in the case of a uniform shear beam.

One simple means for covering this need for higher shears near the top of buildings idealized as closely coupled systems and lying on firm ground consists in assuming that a fraction of the base shear is caused by a concentrated force applied at the top of the building while the rest is distributed in accordance with a triangular variation of horizontal accelerations. This approach is used in the Uniform Building Code. Bustamante (1965b) has shown that, if we neglect axial deformations of columns (that is, overall bending of the building) the assumption that the top concentrated force is 5 percent of the base shear takes care of almost all situations likely to be encountered in practice and apparently does not penalize the uppermost stories in excess.

Favoring the increase in lateral force coefficients for the upper stories are the phenomena of axial deformations of columns and flexural deformations of shear walls. As we shall see in Section 16.1, upstanding flexural beams develop much higher shears at moderate and high elevations above ground than shear beams having the same base shear. The deformations in question make the behavior of buildings intermediate between those of closely coupled systems and of flexural systems (see Blume, 1967). Hence, we recognize the need for the increase in shears at high elevations.

Assuming that the effects of these deformations are almost universally related to the building height-to-base ratio, some building codes specify that the concentrated force assumed to act at the top of the building be made a function of this ratio alone. This criterion is oversimplified, as the height-to-base ratio is not the only significant variable involved. When there is doubt about the importance of overall flexural deformations, it seems wise first to analyze the building neglecting this effect and then employ more refined methods when the preliminary analysis shows that they are justified.

Another consideration favoring a faster-than-linear increase of lateral force coefficient with elevation operates when the design acceleration spectrum decreases with period at a faster-than-hyperbolic rate within part of the range of the structure's natural periods. This happens when the fundamental period of the building is relatively long and the most significant earthquakes are likely to be shallow, of moderate magnitude, and originate nearby.

Despite the remarks above, there are reasons against using a very pronounced increase in the lateral force coefficients for the upper floors. It may be justified to stay below what a conventional dynamic analysis indicates. For one thing, base rotation will tend to make the shear distribution approach the shear distribution associated with a linear variation of the lateral force coefficient. For another, the consequences of underdesign in shear in the first few stories are more serious than those of a similar underdesign near the top of the structure; this is true whether we contemplate the possibility of total vs. partial collapse or practically any other type of damage.

As indicated by the comparison in Fig. 15.4, the shear distributions in buildings that rest on compressible ground are not amenable, in general, to the type of simplified treatment that applies to buildings on firm soil. Serious errors may be introduced when attempting to proceed in like manner, especially if a strongly prevailing frequency of the ground falls near one of the first few natural frequencies of the building. Still, if the fundamental frequency of the ground is certain to exceed the first harmonic by a large margin and there are no very pronounced peaks in the design spectrum at higher frequencies, the criteria expounded for buildings on hard ground will apply without excessive error and, sometimes, with even better accuracy than on firm ground, because of the relative preponderance of the fundamental mode. Under more general conditions dynamic analysis should not be waived.

Particularly on soft ground, soil-structure interaction may be important. When the building is analyzed dynamically, explicit recognition of this phenomenon is, in principle, a simple matter. It suffices to introduce the additional degrees of freedom that we discussed in Section 3.15. We assume then that the base of the building does not rest directly on a rigid formation but that it does so through a set of springs and dashpots and that there is a rigid mass of soil of a certain shape attached to the base.

According to Section 3.15, the characteristics of these springs, dashpots, and masses can be computed with fair accuracy under the assumption that the soil behaves linearly, provided we deal with a shallow foundation and provided the soil can be idealized as a homogeneous halfspace. In the case of deep compensated foundations, of foundations on piles or piers, or of buildings on well-stratified soil, our estimates of the structural parameters will necessarily be cruder.

Once we have decided on these parameters, we may either use an approximate modal analysis of the building or perform Monte Carlo analyses using actual and simulated earthquake records and recognizing, if we wish, nonlinear behavior of the structure. (The modal analysis would not be strictly correct, since the dashpot constant that partly replaces the ground will ordinarily be too large for the system to have classical modes, but we would not expect excessively large errors from this source.)

These types of analysis present no basic difficulties. However, the amount of computational work may be appreciably increased with respect to that required for a building on rigid ground. This is because, in many cases in which we neglect soil-structure interaction, we are content to take into account the degrees of freedom associated with horizontal translation of the floors in one direction only (e.g., one degree of freedom per story) when we do not consider torsion. When we include rocking motion of the base, the rotational inertia of the floors can hardly be neglected, except in unusually slender buildings. Hence, ordinarily we must more than double the number of degrees of freedom.

Unless we perform a dynamic analysis that explicitly incorporates soil-structure interaction, we cannot predict accurately the effect of this interaction. We can be sure, nevertheless, that rocking will lengthen the fundamental period of vibration and usually increase the percentage of damping in the fundamental mode without greatly affecting the higher modes. Rocking will also tend to straighten the shape of the fundamental mode, generally making the assumed triangular distribution of horizontal accelerations more accurate for buildings on soft ground than on hard ground. At the same time rocking tends to magnify the effects of gravity loads (the "P–Δ effect").

15.5 Appendages

Consider the structure shown schematically in Fig. 15.5a. A naive application of the second static method described in the preceding paragraphs would give the lateral force coefficients depicted in Fig. 15.5b. However, dynamic analyses of systems of this type usually indicate larger shears at the base of appendages on buildings than we would find if the appendages stood directly on the ground. Owing to this reason, some building codes require that appendages be designed for at least twice the local lateral force coefficient corresponding to static analysis of the building but not less than the coefficient that one would use if the appendage stood directly on the ground. This would require lateral force coefficients comparable to those shown in Fig. 15.5c, although this criterion is often not even approximately correct. Moreover, should the common platform undergo torsional oscillations and should the appendage stand eccentrically on the platform, we would find that the amplitudes of the base motion of the appendage differ markedly from what they are at the center of the platform.

The most severe conditions for an appendage arise when its fundamental period of vibration lies close to that of the rest of the structure. Magnification factors of as much as 8.0 in the appendage's base shear, relative to the result of conventional statical analysis, have been reported by Bustamante and Rapoport (1961) for realistic situations of this sort, assuming a design acceleration spectrum that is typical for structures on firm ground. Even higher magnifications can be expected for structures of these characteristics when they rest on soft mantles whose prevailing period approximately coincides with the fundamental of the structure. The same observations apply, albeit to a lesser degree, when near coincidence of natural periods—of the appendages, the rest of the structure, and the ground—refers to the corresponding harmonics.

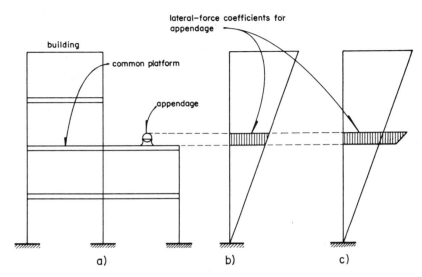

Figure 15.5. Lateral-force coefficients for appendages:
  a) Structure
  b) Naive solution
  c) More realistic lateral-force coefficients.

These very high magnification factors were computed assuming that design responses were equal to the root of the sum of the modal values. However, as shown by Penzien (1969), this criterion greatly overestimates the appendage responses when the mass of the appendage is small compared with that of the building (say, less than 1/100 of the building), and there is near coincidence of the fundamental periods of both subsystems. Nevertheless, the results of these more accurate analyses still show a pronounced magnification of appendage responses under these conditions. It is precisely the closeness of natural frequencies that causes the root of the sum of squared modal responses to be seriously in error; in these cases considerably more accurate results may be expected from the second approximate method of analysis expounded in Chapter 10 and which leads to Eq. 10.10.

In the same paper Penzien describes an approximate method of analysis applicable to buildings having offsets or appendages as well as to structures that respond in a combination of torsion and translation. The method is relatively accurate. It makes use of two-degree-of-freedom response spectra of the motion for which the system is being analyzed. Such spectra are available for the NS component of the 1940 El Centro earthquake.

The unfavorable circumstances that may lead to abnormally high stresses in appendages are only partially covered by designing in accordance with the static method of analysis and assuming that a concentrated horizontal force of 0.05 to 0.1 times the base shear acts at the top of the structure. The most critical situations can be identified only through dynamic analysis. Due margin should be left for uncertainties in structural parameters, especially in stiffnesses. No reassurance should be derived from the possibility of nonlinear behavior of the appendage, for such behavior may actually tend to increase its maximum deformation.

The situation can be so serious that, apparently, design of all buildings having a relatively flexible top should be required to rest on accurate dynamic analysis. This is not necessarily the case, unless we restrict our attention to those structures in which failure of the top portion may be exceptionally objectionable. Often the consequences of failure of the top portion are not very costly, so that this consideration alone does not justify the requirement of nearly universal dynamic analysis, particularly since the probability is small that extremely high response of the uppermost portion of a structure be met for an arbitrarily selected building.

15.6 Story Torques

Dynamic analysis will yield the torsional moments acting at every story of a tier building with less accuracy than it does the story shears. For one thing, the rotational components of ground motion introduce a proportionally higher fraction of the disturbance from the viewpoint of story torques than they do from the viewpoint of shears; in fact, the entire disturbance that causes torques can be of this type. And at present we can only infer characteristics of such components from records of the translational components (see Section 7.7). For another, a small error or inaccuracy in the calculation of relative rigidities or an unforeseen load distribution can introduce increases of several hundred percent in story torques when computations indicate that the resultant of horizontal forces passes near the center of torsion in every story. The situation is particularly serious in nominally symmetrical buildings, in which elementary analysis does not disclose the slightest torque, while actually, with probability one, there will be such generalized forces during an earthquake. We have also seen how nonlinear behavior can introduce torques not accounted for in conventional analysis.

The present state of knowledge precludes an accurate estimate of these "accidental," additional torsions. Hence it is not justified to include explicit recognition of the nominal eccentricities in a dynamic analysis when they are relatively small, since such recognition increases the number of degrees of freedom by a factor of 2 to 3. In order for the nominal eccentricities to be negligible in dynamic analysis, they must be small compared with the accidental eccentricities.

Actual eccentricities in buildings usually differ substantially from the computed values. The sources of these differences may be classified in two groups. The first gives rise to "accidental" torsions that affect even nominally symmetric buildings and that are not ordinarily taken into account in dynamic analysis.

The second group reflects differences between the results of static and dynamic methods of analysis; it is often referred to as dynamic magnification (or dynamic abatement) of eccentricity.

The main causes of accidental eccentricity include the rotational component of ground motion about a vertical axis, the differences between assumed and actual stiffnesses and masses, and asymmetrical patterns of nonlinear force-deformation relations. Other sources, such as asymmetry of damping constants and the deformation in a direction perpendicular to the one being analyzed, usually play an insignificant role.

Let us consider first the matter of ground rotation about a vertical axis. The following treatment is based on Section 7.7 and on Newmark (1969a). We concluded in Eqs. 7.13 and 7.14 that

$$\max_t |\ddot{\phi}| = \frac{a_1}{v_s} \tag{15.1}$$

and

$$\max_t |\dddot{\phi}| = \frac{h_1}{v_s} \tag{15.2}$$

where t is time, ϕ denotes base rotation, a_1 and h_1 are, respectively, the maximum numerical values of $\ddot{x}_1$ and $d^3 x_1/dt^3$, x_1 is the ground displacement in a horizontal direction, and v_s is the velocity of shear waves. Similarly,

$$\max_t |\dot{\phi}| = \frac{v_1}{v_s} \tag{15.3}$$

where $v_1 = \max_t |\dot{x}_1|$.

Let b denote the long dimension of the building plan. Then $\max_t |\phi| b/2$ may be regarded as an added displacement, due to torsion, of the base of the building. According to Eq. 15.3 and omitting subscript 1, we can write

$$\max_t |\dot{\phi}| \frac{b}{2} = \frac{vb}{2v_s}$$

$$= \frac{vt_b}{2} \tag{15.4}$$

where t_b denotes the transit time of the wave motion to pass over the long dimension of the building. Similarly, from Eqs. 15.1 and 15.2,

$$\max_t |\ddot{\phi}| \frac{b}{2} = \frac{at_b}{2} \tag{15.5}$$

$$\max_t |\dddot{\phi}| \frac{b}{2} = \frac{ht_b}{2} \tag{15.6}$$

On the basis of Section 7.4 we conclude that, for about 7 percent of critical damping at moderate focal distances, expected spectral ordinates for a ground motion that is pure translation can be found from the approximate relations $D/d \cong 1.2$, $V/v \cong 1.6$, and $A/a \cong 2.0$, where we have omitted the "expectation" symbol. We may expect slightly higher amplification factors in response to the rotational components of ground motion because of the greater number of oscillations. If we adopt 1.33, 1.67, and 2.40 for these amplifications, Eqs. 15.4–15.6 lead, respectively, to the following expected spectral values of added displacement and velocity relative to the ground and added absolute acceleration, due to torsion:

$$D_\phi = 0.67vt_b$$
$$V_\phi = 0.83at_b$$

and

$$A_\phi = 1.20ht_b$$

(15.7)

As we saw in Chapter 7, for most earthquakes of engineering significance, ad/v^2 lies in the range of 5 to 15. It can be shown that in all cases hv/a^2 must exceed 1, and it seems reasonable that it should be of the same order of magnitude as ad/v^2. If we adopt the assumption that hv/a^2 also lies in the same range, or even that it is equal to ad/v^2 for every given earthquake, we have a means for estimating h from a and v and hence for finding estimates of D_ϕ, V_ϕ, and A_ϕ using Eqs. 15.7.

For the NS component of the El Centro 1940 earthquake, for which $d \cong 10$ in., $v \cong 15$ in./sec, and $a \cong 120$ in./sec², application of this method with the assumption that $hv/a^2 = ad/v^2$ gives $D_\phi = 10t_b$, $V_\phi = 100t_b$, and $A_\phi = 6000t_b$. If we deal with a building whose long dimension is 100 ft and rests on a ground having $v_s = 1000$ ft/sec, so that $t_b = 0.1$ sec, we find $D_\phi = 1.0$ in., $V_\phi = 10.0$ in./sec, and $A_\phi = 600$ in./sec² $= 1.55$ g. These values are by no means negligible, although as we shall see, the value we should take for v_s is open to question.

The spectra for $t_b = 0.1$ and 0.05 sec are shown in a logarithmic plot in Fig. 15.6 for comparison.

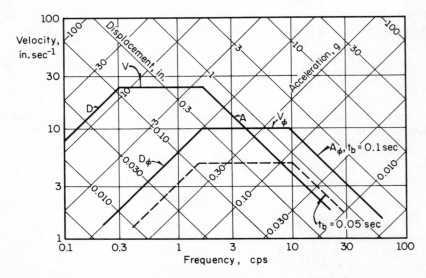

Figure 15.6. Response spectra for linear motion and additional torsional motion. *After Newmark (1969a).*

In order to compare the response of symmetrical buildings to the rotational excitation with those due to translational components we must compare the natural frequencies of vibration. This was done in a straightforward fashion by Newmark (1969a), arriving at the sample results in Table 15.1 for the single-story buildings shown schematically in Fig. 15.7. In all cases the mass is assumed to be uniformly distributed in plan. In the table, ω_T and ω_x are, respectively, the natural circular frequencies of vibration in torsion and in translation, parallel to the x axis, while in the figure k_x and k_y denote stiffnesses in the x and y directions.

TABLE 15.1 PARAMETERS FOR STRUCTURES STUDIED
[AFTER NEWMARK (1969a)]

Condition			Structure			
b'/b	$\sum k_y / \sum k_x$	Quantity	Uniform (Fig. 15.7a)	Perimeter (Fig. 15.7b)	9-Column (Fig. 15.7c)	4-Column (Fig. 15.7d)
1.0	1.0	ω_T/ω_x	1	1.732	1.414	1.732
		bD_ϕ/e_yD	3	1	1.5	1
0.5	1.0	ω_T/ω_x	1	1.897	1.414	1.732
		bD_ϕ/e_yD	4.8	1.333	2.4	1.6
0.0	All	ω_T/ω_x	1	1.732	1.414	1.732
		bD_ϕ/e_yD	6	2	3	2

With the ratio ω_T/ω_x known for a particular structure, we can enter the response spectra in Fig. 15.6 and for a particular value of t_b determine the relative value of the added response D_ϕ to the translational spectral displacement D. Values of D_ϕ and D are always to be read from the displacements scales in the figure.

For a particular structure, when D_ϕ/D is known we can determine the eccentricity e_y of the inertia force F_x in the following way. We can write

$$D = \frac{F_x}{\sum k_x} \qquad (15.8)$$

Let $F_x e_y$ be the torsional moment associated with a displacement D_ϕ at one end of the building in the long dimension and $-D_\phi$ at the other:

$$2D_\phi \sum (k_x y^2 + k_y x^2) = F_x e_y b \qquad (15.9)$$

Then, from Eqs. 15.8 and 15.9 we derive

$$\frac{D_\phi}{D} = \frac{e_y b \sum k_x}{2 \sum (k_x y^2 + k_y x^2)} \qquad (15.10)$$

For the special case of Fig. 15.7a, Eq. 15.10 gives

$$\frac{D_\phi}{D} = \frac{6 e_y b}{1 + b'^2 k_y/b^2 k_x} \qquad (15.11)$$

When b' or $k_y = 0$, $D_\phi/D = 6e_y/b$; when $b = b'$ and $k_y = k_x$, $D_\phi/D = 3e_y/b$.

494 EARTHQUAKE-RESISTANT DESIGN OF BUILDINGS Chap. 15

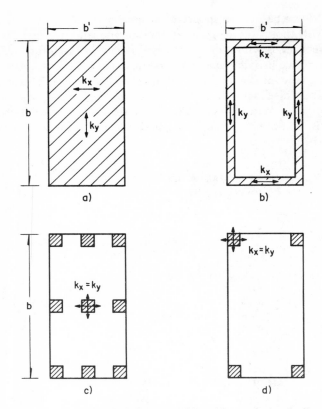

Figure 15.7. Types of buildings considered in analysis of effects of rotational component of ground motion:
 a) Uniform resistance
 b) Resistance on perimeter only
 c) Nine-column building
 d) Four-column building
 After Newmark (1969a).

Values of $bD_\phi/e_v D$ for several special cases are given in Table 15.1. Typical results for $t_b = 0.1$ sec are given in Table 15.2, where f_x and f_T are the natural frequencies of vibration in the direction of b' and in torsion, respectively, both in hertz.

From these results approximately linear relationships can be established between the ratio e_y/b and $f_T t_b$ for each type of building, aspect ratio, and range of natural frequencies considered. In general an accidental eccentricity of 5 percent of the longer plan dimension, required as minimum by the Uniform Building Code, seems reasonable for framed buildings having fundamental periods in excess of about 0.6 sec or shear-wall buildings with fundamental periods greater than about 1.0 sec. Accidental eccentricities of about $0.1b$ would be reasonable for shorter fundamental periods, perhaps increasing to

TABLE 15.2. RESULTS OF CALCULATIONS FOR TORSIONAL ECCENTRICITY; ALL VALUES ARE FOR $t_b = 0.1$ sec [AFTER NEWMARK (1969a)]

Structure		Uniform (Fig. 15.7a)			Perimeter (Fig. 15.7b)			9-Column (Fig. 15.7c)		4-Column (Fig. 15.7d)	
b'/b		1.0	0.5	0	1.0	0.5	0	1.0	0	1.0	0
f_x(Hz)	Item										
0.318	f_T	0.318	0.318	0.318	0.550	0.600	0.550	0.450	0.450	0.550	0.550
	D_ϕ/D	0.083	0.083	0.083	0.083	0.083	0.083	0.083	0.083	0.083	0.083
	e_y/b	0.028	0.017	0.014	0.083	0.062	0.041	0.056	0.028	0.083	0.042
0.5	f_T	0.5	0.5	0.5	0.87	0.95	0.87	0.71	0.71	0.87	0.87
	D_ϕ/D	0.131	0.131	0.131	0.131	0.131	0.131	0.131	0.131	0.131	0.131
	e_y/b	0.044	0.027	0.022	0.131	0.098	0.065	0.088	0.044	0.131	0.065
1.0	f_T	1.00	1.00	1.00	1.73	1.90	1.73	1.41	1.41	1.73	1.73
	D_ϕ/D	0.262	0.262	0.262	0.240	0.219	0.240	0.262	0.262	0.240	0.240
	e_y/b	0.087	0.055	0.043	0.240	0.180	0.120	0.174	0.087	0.240	0.120
1.59	f_T	1.59	1.59	1.59	2.75	3.0	2.75	2.25	2.25	2.75	2.75
	D_ϕ/D	0.416	0.416	0.416	0.241	0.220	0.241	0.294	0.294	0.241	0.241
	e_y/b	0.139	0.087	0.069	0.24	0.18	0.12	0.196	0.098	0.241	0.120
3	f_T	3.0	3.0	3.0	5.2	5.7	5.2	4.2	4.2	5.2	5.2
	D_ϕ/D	0.79	0.79	0.79	0.46	0.42	0.46	0.56	0.56	0.46	0.46
	e_y/b	0.26	0.16	0.13	0.46	0.35	0.23	0.38	0.19	0.46	0.23
5	f_T	5.0	5.0	5.0	8.7	9.5	8.7	7.1	7.1	8.7	8.7
	D_ϕ/D	1.31	1.31	1.31	0.75	0.69	0.75	0.92	0.92	0.75	0.75
	e_y/b	0.44	0.27	0.22	0.75	0.56	0.37	0.62	0.31	0.75	0.37

$0.15b$ at a fundamental period of 0.2 sec. These conclusions were derived from the analysis of single-story structures. They carry without essential modification into multistory buildings.

The conclusions would probably be too conservative if the rotational component of ground motion about a vertical axis were the only cause for accidental torsion. Indeed, as we saw in Section 7.7, the magnitude of this rotational component is about 29 percent smaller than is estimated by the present approach because ground motions are not equally intense in two orthogonal directions. Moreover, the assumption that v_s is of the order of 1000 ft/sec would be correct for many soils found under important cities, but because of the usual path of refracted waves, shear velocities of about 10 times this value are probably adequate for the S-wave phase of most accelerograms, since the waves to consider are those traveling through base rock and being refracted in an almost vertical direction. The smaller velocities are reasonable for the surface wave phase, which contributes a relatively small portion of the total disturbance.

However, there are other important contributions to accidental torsion. The salient cause among these is the randomness of stiffness under lateral loads (see Section 13.6.4 in connection with the random nature of this variable in masonry shear walls). Unfavorable distribution of mass may also tend to increase the accidental torsion, especially in warehouses. In all, then, the foregoing recommendations on the order of magnitude of the accidental torsion to assume in design are probably correct, except that they overestimate the importance of the fundamental period of vibration. There is little justification for assuming that this torsion should be taken as minimum, as some codes specify, rather than as a quantity to be added to the torques found from standard dynamic analysis.

Static analysis often gives substantial differences with respect to the dynamic, even for ideal, linearly elastic structures subjected only to translation of their bases. However, the results of modal analysis are also grossly in error when design responses are taken as the square root of the sum of squared modal responses, especially when pairs of natural periods lie close to each other (Section 10.2).

In order to study the effects of shear and torque separately, single-story buildings like the one shown schematically in Fig. 15.8 have been analyzed under the action of earthquakes having flat and hyperbolic acceleration spectra by Elorduy and Rosenblueth (1968); see also Rosenblueth and Elorduy (1969a). Results in Figs. 15.9 and 15.10 refer to the shear force and the magnification factor for eccentricity, relative to its "static" value e_s (see Fig. 15.8), including a comparison with taking the total response equal to the sum of squared modal responses for different values of the parameters.

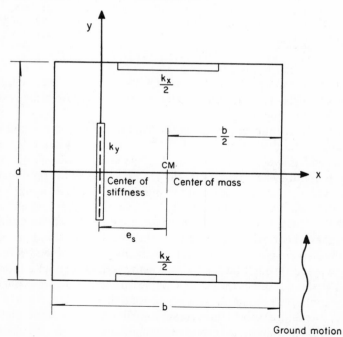

Figure 15.8. Plan of single-story buildings considered. *After Rosenblueth and Elorduy (1969a).*

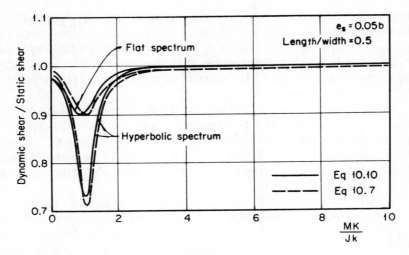

Figure 15.9. Relation between the dynamic and static shears.
After Rosenblueth and Elorduy (1969a).

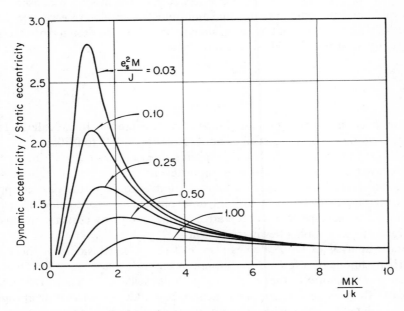

Figure 15.10. Magnification factor for eccentricity.
After Rosenblueth and Elorduy (1969a).

This study shows that shear forces computed by the latter criterion are practically equal to those obtained from the more accurate Eq. 10.10 and that, for these spectra, it is always conservative to ignore torsion when computing base shear. However, omitting the last term in Eq. 10.10 seriously overestimates torques in ranges where natural frequencies are nearly equal to each other, but in these same ranges even the complete Eq. 10.10 yields appreciably higher torques than may be obtained from the product of e_s and shear. Results for buildings of arbitrary shape indicate that under some conditions much higher magnification factors may obtain for the design shear in certain walls or frames, relative to the results of static analysis.

One way of overcoming the deficiencies of the static method consists in specifying that the design eccentricity be taken equal to e_s or $1.5e_s$, whichever is more severe for the structural member being designed, and that the static method be applied only when $K/k \geq 36J/M - 25e_s^2 M/J$, where $K =$ torsional stiffness about the center of mass, $J =$ polar moment of inertia about the same center, $k =$ translational stiffness, and $M =$ mass.

This treatment is confined to buildings having natural periods that are neither too short nor too long. When the fundamental period is extremely short, all maximum accelerations approach that of the ground, so that the static method of analysis yields satisfactory results; in other words, design responses approach the algebraic sum of the modal responses. When the periods of the natural modes that contribute significantly to the overall responses are very long, the design responses approach the numerical sum of the modal responses.

Combining the dynamic magnification and the accidental eccentricities, we conclude that, within the limitation stated about K/k, the design story eccentricity is the most unfavorable of the quantities

$$e_d = 1.5e_s + \alpha b$$

and

$$e_d = e_s - \alpha b$$

where b is the building's plan dimension measured perpendicularly to the ground motion, and α is a coefficient of the order of 0.05 to 0.10.

The use of two-degree-of-freedom spectra, due to Penzien (1969) and to which we referred in Section 15.5, is also applicable to the analysis of buildings subjected to torsion under translational excitation. And, as in the case of appendages, yielding is likely to increase the unfavorable effects of torsion. Hence it is advisable to be especially conservative in the design of corner columns and peripheral shear walls (Newmark, 1969a).

Static analysis of tall buildings may lead to the conclusion that there is eccentricity in one or a few stories but not in the rest. There are indications that, under these conditions, the entire building will oscillate in torsion (Bustamante, 1961b). To provide for this phenomenon, one code, which is now abrogated, has stipulated that the eccentricity at all stories be taken at least equal to half of the maximum value of e_s throughout the building, or the story torque be taken at least equal to half the maximum statically computed torque in any one story, whichever of the two lower bounds gave the smallest torsion (Rosenblueth, 1960). Although this rule took care of the phenomenon to some extent, it is probably too crude in many cases. If static methods of analysis are to continue in use, these matters merit additional study.

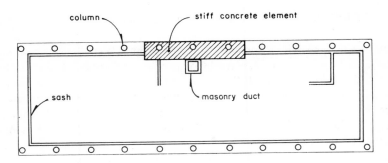

Figure 15.11. Extreme case of torsion in a building.
After Esteva and Nieto (1967).

Figure 15.12. Torsional failure of column in building shown in
Figure 15.11. *After Esteva and Nieto (1967).*

Story torsion is usually associated with comparatively small relative rotations. In most cases it is essentially resisted by pairs of shear forces in flexural or braced frames and shear walls, and the contribution of the torsion of individual columns is ordinarily negligible. However, when one or a few columns are much shorter than the rest in a story, the shorter columns may be subjected to important torsional moments. Esteva and Nieto (1967) and Esteva, Rascón, and Gutiérrez (1969) describe the case of one reinforced concrete building in which the existence of an asymmetrical concrete block was the main cause of a pronounced eccentricity in plan (Fig. 15.11), and this was responsible for the torsional failure of one column (Fig. 15.12).

15.7 Overturning Moments

Overturning moments computed directly from dynamic analysis should be taken generally at face value for purposes of design.

Important reductions have been advocated, reasoning that the actual overturning of buildings during earthquakes is a rare event and that results of rational analysis should be cast aside. Such complete overturns as those observed in the Niigata earthquake of 1964 (Fig. 13.42) have been rare indeed and are rightfully disregarded when speaking of the valuation of overturning moments for the design of structural members, since those phenomena point to soil liquefaction rather than to alarmingly high moments. However, even the trained eye can hardly discern the role played by axial forces in bringing about column failures such as those illustrated in Fig. 15.13, in which flexure and shear complicate the evidence. If a column is subjected to an important axial tension its capacity to resist shear or flexure, or even torsion, is drastically lowered with respect to a similar column under moderate axial compression; if it fails under the combined action of a sizeable longitudinal compression, shear, and flexure, its aspect will not differ much from that of a column under a small vertical force, failing in flexure and shear.

It follows that the consequences of structural failure directly caused by overturning moment are at least as serious as those ensuing from failures caused by story shears and other generalized forces. Hence, the load factors to be used in design to resist overturning moment should not be substantially lower than those adopted to provide against a high probability of failure under these other generalized forces. The only justifiable reduction rests on the higher cost sometimes associated with providing resistance in the columns to overturning moment, and usually the reverse is true.

A different situation operates in provisions to prevent separation of a building foundation from the ground. The consequences of this phenomenon do not surpass damage to surrounding sidewalks and to connections between the building and public utilities, provided the foundation is sufficiently strong and rigid to prevent excessive redistribution of forces in the superstructure. Even

Figure 15.13. Failure of a reinforced concrete column. *After Esteva, Díaz de Cossío, and Elorduy (1968).*

if the latter condition is not fulfilled, the damage that a moderate redistribution of forces causes in the superstructure may not amount to more than mild cracking of some beams. Yet the cost may be prohibitive if we try to make partial, temporary separation between the foundation and the ground virtually impossible. An apparently unconservative practice is thus sometimes justifiable when providing against this event.

When using the static method of analysis there is no reason—other than economical—for reducing the overturning moment computed from the design shears in single-story buildings. Considerations of equilibrium show immediately that in these structures the overturning moment at the base must equal the product of base shear and story height. In multistory buildings it is undoubtedly erroneous, on the conservative side, to take this moment equal to the integral of the design shears.

According to the remarks made in the previous paragraph, the consequences of this overestimate may be serious. The main cause of the overestimate is evident: the shear envelope is the locus of maximum shears that do not occur simultaneously. In fact, in all natural modes except the fundamental mode, some story shears are positive while others are negative.

There is a second cause: the design shear envelope is necessarily a simplified approximation to the ideal envelope. If the design envelope is "correct" at one story and is not unconservative anywhere, it is almost certainly conservative everywhere except in that one story. Hence, its integral exceeds the area under the ideal shear envelope. If the design shear envelope is unconservative in one or more stories, yielding of that portion of the structure will tend to prevent the development, elsewhere, of story shears in excess of the design values. It will usually lead to appreciably smaller shears in the stories that do not yield. Again, the area under the design shear envelope will exceed the one under the ideal envelope. (The same argument applies, although in a lesser degree, to design based on dynamic analysis if overturning moments are computed from the design shears rather than directly from a combination of modal moments, since the shears produced by any earthquake will certainly differ from the results of analysis.)

The fact that the integral of the design story shears should exceed the actual demands on resistance to overturning moment does not justify the exaggerated reductions that some codes formerly allowed. Let us discuss only two reduction criteria that enjoy popularity at present. On the basis of the former Los Angeles code,[3] some codes (*Earthquake Resistant Regulations, A World List,* 1966b, 1966c, and 1966d) call for an overturning moment equal to the product of the design shear at the elevation considered and the distance from that level to the center of mass of the portion of the building that lies above the level in question (Fig. 15.14). The same codes require that story shears be computed from an assumed diagram of horizontal accelerations that increases monotonically with

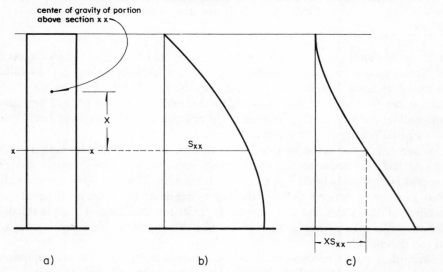

Figure 15.14. Criterion for calculation of design overturning moment:
    a) Structure and section considered
    b) Shear distribution
    c) Overturning moments.

[3] That code also allowed a 20 percent reduction with respect to the results of the criterion described in this paragraph.

Sec. 15.7 OVERTURNING MOMENTS **503**

height above ground level. Therefore, this criterion always involves a reduction relative to the integral of the design shears, except fot the uppermost story, where, properly, there is no reduction.

Let us apply the criterion to a building having a uniform mass per unit height. Let the diagram of horizontal accelerations be linear, passing through zero at the base (Fig. 15.15a). We obtain the shear and overturning moment diagrams in Figs. 15.15b and 15.15c. For comparison we have drawn in Fig. 15.15c the diagram of overturning moments resulting from integration of the shear envelope of Fig. 15.15b. There is a reduction of 25 percent in the moment at the base, and of 10 percent at midheight.

If the building were idealized as a uniform shear beam, if it were analyzed dynamically taking design responses equal to the square root of the sum of squared modal responses, and if the arithmetic plot of the design acceleration spectrum were hyperbolic, the shear diagram would be parabolic with the vertex at the top of the building. The area under the shear diagram would equal $\frac{2}{3}$ of the building height times the base shear. On the other hand the overturning moment at the base of the building is $\sqrt{\frac{1}{3}} = 0.577$ times the same product (see Section 10.4).[4] We would thus find a reduction of 13.4 percent at the base.

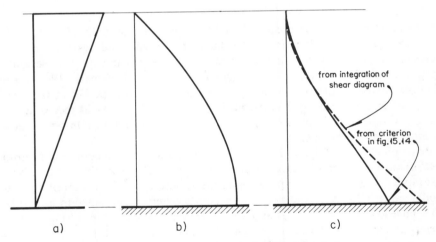

Figure 15.15. Comparison of overturning moment diagrams:
     a) Lateral-force coefficients
     b) Shear distribution
     c) Overturning moments.

[4] This coefficient can be derived immediately from the relation

$$Q = V\sqrt{\frac{2}{T}\int_0^T \psi_q^2\, dt}$$

where ψ_q is the transfer function of Q, if ψ has the period T (see Eq. 10.5). For the overturning moment at the base, ψ is triangular with maximum ordinate equal to the base shear times the building height, while the transfer function for the base shear is rectangular (see Section 10.1, particularly Fig. 10.2). Hence the factor $\sqrt{\frac{1}{3}}$.

(Curiously, this method of analysis gives the same reduction factor throughout the building.)

The fact that the design shear envelope has its vertex at the bottom rather than at the top of the building is consistent with a flatter design acceleration spectrum. This leads to an even smaller reduction on the basis of the modal analysis used. The reduction at midheight would be negligible in the light of a dynamic analysis using such an acceleration spectrum.

The discrepancy found for the uniform shear-beam should not be disquieting. Analysis of a large number of structures points out that the criterion expounded is conservative for most cases to be met in practice (Bustamante, 1965b). When it errs on the unsafe side, the error is not important for buildings that deform essentially in shear, except in the uppermost stories, where it can be taken care of by assuming that a concentrated force acts at the top, as described previously.

The second criterion of interest was advanced by Rinne (1960) and has been adopted in many a building code. According to it, the overturning moment computed at the base of the building as the integral of the shear envelope should be affected by a factor J, defined by the expression

$$J = \frac{0.5}{T_1^{2/3}} \tag{15.12}$$

but J should not be taken greater than 1.0 or smaller than 0.33. Here T_1 is the structure's fundamental period in seconds. In typical buildings this criterion yields adequate correction factors (Bustamante, 1965b) but may err seriously on the unsafe side in very flexible structures (see also Newmark, 1965a). For example, a single-story mill building having a fundamental period of the order of 0.7 sec is not inconceivable. According to Eq. 15.12 it would receive a correction of 0.63; this implies an error of 37 percent on the unsafe side relative to equilibrium requirements.

Rosenblueth, Elorduy, and Mendoza (1967) and Rosenblueth and Elorduy (1969a) have analyzed buildings idealized as uniform shear beams, using both the assumption that design responses equal the square root of the sum of squared modal values (Eq. 10.7) and the more accurate criterion contained in Eq. 10.10. They have assumed a flat and a hyperbolic acceleration spectrum. The hyperbolic spectrum has a cutoff that makes $A = $ constant for $T \leq 0.1T_1$, where $T_1 = $ fundamental period of vibration. It has been assumed that the damping coefficient is 5 percent in all the natural modes, that the equivalent segment of white noise has a duration $s = 20T_1$, and that the spectra already include effects of damping. Results appear in Figs. 15.16 and 15.17 which include a comparison with the shears and moments computed from statically applied accelerations proportional to elevation above ground. The base shears have all been computed for the same spectral acceleration, A_1, associated with the fundamental period.

These results lead to the following conclusion. The base shear is from 0.877 (for a hyperbolic acceleration spectrum) to 0.816 (for a flat acceleration spectrum) times $(A_1/g)W$, where g is the acceleration of gravity and W is the weight of the building.

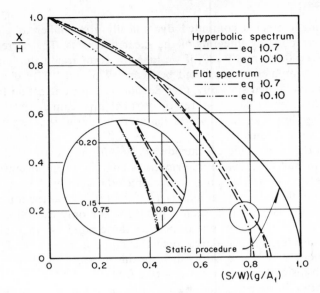

Figure 15.16. Distribution of shear forces. *After Rosenblueth and Elorduy (1969a)*.

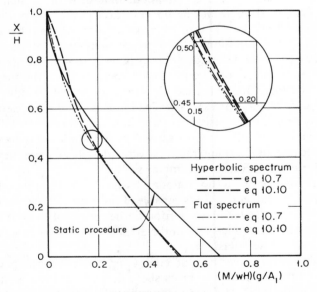

Figure 15.17. Distribution of overturning moments. *After Rosenblueth and Elorduy (1969a)*.

The static method is satisfactory for computing the shear distribution in buildings idealized as uniform shear beams, except near the top of very tall buildings subjected to nearby earthquakes. (The same conclusion does not apply to nonuniform buildings in general.)

Overturning moments computed dynamically are smaller at all elevations than the corresponding integrals of the shear diagrams. The correction factor ranges between 0.886 and 0.989, if we take for comparison the integral of the dynamically computed shears,[5] and between 0.773 and 0.779, if we take that of the shears computed statically. (See Section 12.2 for correction factors to be applied to hydrodynamic pressures on vertical upstream faces to obtain the corresponding shears, as the pressures against dams are in this respect analogous to shears in shear beams.)

Errors on the unsafe side are introduced by both of the approximate criteria expounded in buildings whose deformations contain an important component of overall flexure. This situation is often corrected in part by the assumption that 5 to 10 percent of the base shear acts as a horizontal force at the top of the structure. But when codes contain such a provision for only exceptionally slender buildings, they leave without due protection those buildings that, although not particularly slender, still have an appreciable component of overall flexure due to their structural characteristics, such as the presence of slender shearwalls coupled with insignificant moment-resisting frames.

Again, both criteria bring about errors on the unsafe side in cases when the acceleration spectrum has a pronounced peak in the neighborhood of the building's fundamental period of vibration. Indeed, if the fundamental mode alone were excited, there would be no reduction in the overturning moment.

Rinne's method for reduction of overturning moments is often used in conjunction with a linear variation of the overturning moment from zero at the roof of the building to the value computed at its base. Normally the actual overturning moment envelope is concave and has a small slope at the top. The linear variation would therefore be conservative throughout the structure if it passed through the correct value at the base. The latter is often underestimated when using this criterion, but the design values computed in this manner for the upper stories are still conservative in most cases. The economic consequences are not significant because overturning rarely governs design at these elevations, and even when it does, it affects it in a minor scale.

This procedure provides no rational basis for distributing the moment among the various frames and walls capable of resisting it. The code requirement that overturning be allotted in proportion to the shears taken by these elements or subsystems may introduce substantial errors. These same conclusions apply to the first criterion described.

A preferable method consists of computing the overturning moments in every wall and frame directly from its shear diagram and modifying the computed values in the ratio of the (usually reduced) overall overturning moment to the computed overall value at the elevation in question.

As a practical recommendation it is suggested that the first criterion be taken for design to resist overturning moment, distributed in proportion to the statically computed values. However there should be no reduction if overall flexure is

---

[5] This factor cannot be smaller than the value we would obtain if the overturning moment were due only to the fundamental mode, which would give 0.775 for a hyperbolic acceleration spectrum.

important. If a preliminary analysis of this sort indicates that there will be substantial savings in capital investment by designing to resist a reduced overturning moment, a reduction of 5 to 10 percent may be justified in base moment even with respect to the results of dynamic analysis.

Fenves and Newmark (1969) have analyzed the shear and overturning moment distributions in four types of shear beam. In one type the mass and stiffness are distributed uniformly throughout its height, while in the other three types either the mass or the stiffness or both decrease with elevation above ground. Three design spectra were used in the analysis. They are identified as acceleration, velocity, and displacement branches and correspond, respectively, to spectral accelerations independent of natural period and are inversely proportional to T and to T^2. Responses were assumed to be proportional to the square root of the sum of squared modal values. The beam mass was taken as concentrated at 20 equally spaced points; hence, the structures could be regarded as idealizations of 20-story buildings.

In all cases the correction factor for moment was found to be practically equal to 1 for acceleration branch spectra; it varied between approximately 0.85 at the base and 1.00 at the top for velocity branch spectra and between 0.25 and 0.35 at the base to 1.00 at the top for displacement branch spectra.

In practice the displacement branch spectra are associated with extremely long natural periods. For most buildings of interest, founded on firm ground, the first few natural periods fall in the velocity branch and the rest on the acceleration branch. Accordingly, as long as a building can be idealized as a shear beam, the conclusions that we presented earlier on overturning moment are confirmed in this study.

The same authors studied a group of flexural beams, to which we shall refer in Section 16.2, as well as a 10-story, slender, frame building. Axial deformations of the building columns made its behavior intermediate between those of shear and flexural beams. The correction factors are also intermediate between those for the two types of beams, but they still indicate that the code values of J are usually too low when the design spectrum is idealized as either an acceleration or a velocity branch.

15.8 Drift Limitations[6]

Several reasons have been advanced for limiting drift during earthquakes. The first purpose is to limit damage to nonstructural elements such as partitions, plastering, veneer, window panes, and installations. Given a type of connection between structural and nonstructural elements, given a set of gaps between them, and given a connection design for piping and other installations, there is a

---

[6] We understand by *drift* the relative displacement between consecutive floors divided by the corresponding story height.

508 EARTHQUAKE-RESISTANT DESIGN OF BUILDINGS Chap. 15

drift beyond which the probability of damage is very high. Revision of a tentative design under the action of earthquakes of various intensities may indicate that, despite the high likelihood of frequent nonstructural damage, the design is adequate or else that it requires revision. The decision should be based on economic considerations, giving proper weight to the probability of injury and loss of life from falling debris.

If it is decided to redesign, one may choose between making the structure more rigid—usually at the added expense of increased design accelerations—or changing the details of its connection to nonstructural elements or even changing the specifications for the latter elements. An apparently trivial change, such as deciding in favor of light-colored rather than dark paint on plaster, may have an influence in reducing the frequency with which repairs are needed.

Common means for achieving "floating" partitions are illustrated in Fig. 15.18. Usually when this condition is sought it is simple and economical to place the partitions in planes that do not contain columns. In this manner only the top and bottom of every partition need a special treatment to allow play in either of them between the partition and the structure. Wherever a gap between the partition and the structure is to be visible, there is often a need for an element to hide it or to fill it and prevent unsightliness and dust gathering.

Especially in buildings that are repaired and strengthened after suffering earthquake damage there is sometimes advantage in using a peripheral metal band of the type shown schematically in Fig. 15.19. If the shearing and normal forces required to make the band yield are chosen properly, it is possible to limit the lateral force that the structure will transmit to the partition and at the same time take advantage of the capacity of the partition to resist such forces and make use of the energy absorbing capacity of the band.

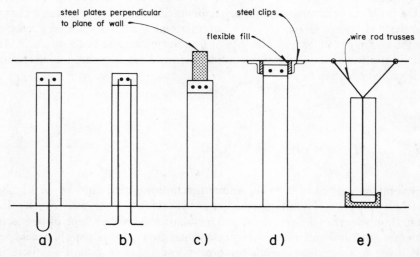

Figure 15.18. Common solutions for isolation of partitions.
After Rosenblueth and Esteva (1962).

Sec. 15.8 DRIFT LIMITATIONS 509

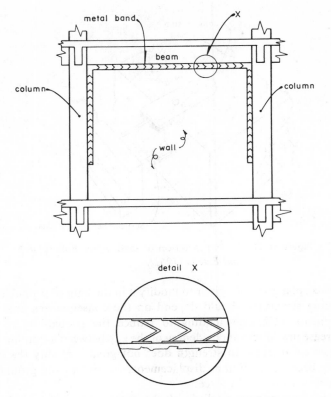

Figure 15.19. Metal band to protect partitions.
After Guerrero y Torres (1965).

Protection of window panes from the lateral distortions of the structure has sometimes been achieved by mounting the window frames on springs that hold them against the structural frame. A detail such as the one shown in Fig. 15.20 has also been used for this purpose. More often, mastics that retain their plasticity to allow play of the panes in the window frames have been used. In every case there is need to design against forces perpendicular to the partition or window whether these forces can be expected from earthquakes or from wind.

The second reason given for limiting drift is that this in turn limits the effects of eccentric gravity loads that magnify those of the lateral forces. However, rather than limiting drift for this reason, it is preferable to consider the effects in question and to design for them, as explained below.

Third, it is sometimes argued that drift limitations tend to limit pounding between adjoining structures. Clearly it is not drift but total sway that constitutes the pertinent response from this point of view.

Consequences of pounding may be minor, when the collision takes place between slabs at areas not covered by expensive veneer, or catastrophic, when a protruding slab may strike one or more columns and fracture them. The

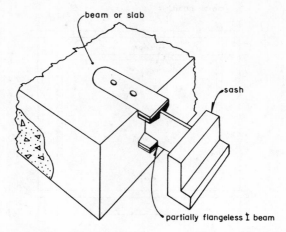

Figure 15.20. Partial isolation of sash. *After Rosenblueth and Esteva (1962).*

decision to accept a given design or to modify it in the light of probable pounding with adjoining structures should depend on the consequences anticipated for this phenomenon. Changes tending to reduce the probability of pounding include increase in stiffness and widening of the gap between adjoining structures. In most cases an increase in strength does not greatly modify the probability of pounding because maximum displacements relative to the ground are relatively insensitive to the yield force.

Alternately one may tend to diminish the consequences of pounding without appreciably changing the probabilities of its occurrence. Figure 15.21 illustrates a solution of this type.

When adjoining structures have nearly the same natural modes and periods of vibration, it may be advisable to tie them together in a variation of the solution shown in Fig. 15.21, so as to force them to vibrate in phase. This is seldom the case, and vibration of adjoining structures out of phase has particularly objectionable consequences unless adequate measures are adopted to prevent excessive pounding.

Pounding between buildings that belong to different owners has special legal implications. From the viewpoint of the individual designer, often the most convenient criterion consists in leaving horizontal distances between the building in question and the property lines sufficiently wide so that pounding is unlikely except under such violent ground motions that it will be ascribed to an act of God.

It is well to emphasize that the gap should be measured to the property line in order to take into account the possibility of demolishing whatever structures exist in the adjoining properties and whatever construction of new buildings that could just comply with code requirements on separation from the property lines. In deciding the width of gaps, the designer should also consider the

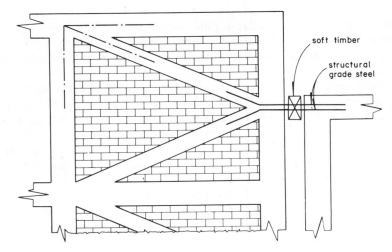

Figure 15.21. Protection against pounding. *After Rosenblueth and Esteva (1962).*

possibility of tilting due to differential settlement of the structure in question as well as that of the adjoining buildings.

Finally, it is adduced that drift limitations tend to reduce discomfort, alarm, and panic among the occupants. It is not drift but the total motion of inhabited floors that plays a significant role in this problem. The discussion in Section 13.8 suggests taking the maximum absolute velocity of floor motion as the response to control with an aim to limit psychological effects.

Maximum velocity relative to the ground is a quantity more readily estimated from pseudovelocity spectra. It has an advantage over maximum absolute velocity in that it reflects the fact that the limitation should be less strict when shaking of the ground itself is more severe.

In calculating drifts and sways it is well to keep in mind that the stresses obtained by applying the coefficients and design acceleration spectral ordinates found in building codes imply the acceptance of relatively major inelastic behavior. According to what we saw in Chapter 11, within the ranges of greatest interest for the design of buildings, the maximum deformations are of the same order as those in "equivalent" linear systems, while the accelerations are approximately equal to $1/\mu$ times those for these systems, where μ is the structure's ductility factor. Hence, the deformations in the actual structure are approximately μ times the ones we obtain by applying conventional elastic analysis with the stresses we use for design.

Roughly, the ductility factors implied in building codes are about 4 to 6 for moment-resisting frames, and inversely proportional to the base shear coefficients for other types of structures. It does not follow, however, that the conventionally computed drifts and sways should be multiplied by 4 to 6 for moment-resisting frames or by the corresponding ductility factors for other types of structures. Coefficients in codes lead to structural designs in which the intention is to limit the probability of collapse while, in computing drifts and sway, we are ordinarily concerned with much less severe types of damage. A substantial increase is nevertheless justified with respect to the results of conventional analysis.

The total drift in any given story is the sum of shearing deformation of that story, axial deformations of the floor systems, overall flexure of the building (axial deformations of columns), and foundation rotation. Only the first two types of deformation are effective from the viewpoint of damage to nonstructural elements, and axial deformations of the floor systems may usually be neglected.

Once the design drift has been computed, the amplitude of gaps between the structure and partitions or similar nonstructural elements can be established by adding the estimated local deformations of structural members and the construction tolerances (see Fig. 15.22). The required gap between window panes and their frames can be determined from the following expression, developed by Bouwkamp and Meehan (1960),

$$\Delta y - \phi H = 2c\left(1 + \frac{H}{B}\right) \tag{15.13}$$

where

Δy = displacement of head of sash with respect to sill
ϕ = rotation of sash frame
H = vertical sash dimension
B = horizontal sash dimension
c = clearance between glass and sash

The expression is derived from geometric considerations and does not provide for rigidity of the mastic nor for inaccuracies in the manufacture of the window frame nor in the cutting of glass.

When the mastic is not especially chosen so as to preserve its plasticity, Bouwkamp and Meehan replace the foregoing expression with

$$\Delta y - \phi H = 2c\left(1 + \frac{H}{B}\right)\left(0.36 - \frac{0.115B}{H}\right) \tag{15.14}$$

The calculation of sway would yield slightly conservative results if it were obtained as the sum of the computed (and properly magnified) relative displacements between consecutive floors. Either a reduction similar to that used for overturning moments or a recalculation using dynamic analysis is in order when we desire more accurate results.

Because the use of the maximum velocity relative to the ground affords a very rough and indirect guide for the present purpose, there is little point in adopting refined methods for its estimate. Within the linear range of behavior we can estimate the maximum relative velocity in each natural mode from the pseudo-velocity spectrum. The modal components can be combined as other responses are combined.

Let T_{av} denote the natural period at which the lines marking maximum ground acceleration and velocity intersect in a spectrum, T_{vd} the natural period at which

Sec. 15.8 DRIFT LIMITATIONS 513

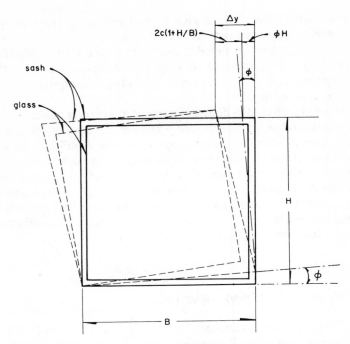

Figure 15.22. Gap between glass and sash. *After Bouwkamp and Meehan (1960).*

the lines marking ground velocity and displacement intersect, and T the fundamental period of a building. On firm soil, and speaking in very rough and broad terms, if $T < T_{av}$, an increase in stiffness causes a reduction in maximum floor velocities. (However, if T is extremely short, human reactions are probably more sensitive to acceleration than to velocity, so that the reduction in maximum velocity will not greatly affect the importance of psychological effects.) The reverse is true if $T > T_{vd}$. When $T_{av} \leq T \leq T_{vd}$, the maximum floor velocity is relatively insensitive to initial stiffness for a given ductility factor; curiously, it tends to decrease with an increase in ductility factor if the initial stiffness is not modified because the spectral acceleration tends to decrease while the maximum displacement relative to the ground is not very sensitive to the change in ductility factor. Consequently, in the range of greatest practical interest, psychological effects change little with modifications in structural characteristics and tend to increase with those modifications that improve behavior on other counts. It follows that it is usually unwise to design with the view of limiting the psychological consequences of earthquakes.

Two obvious exceptions to this approach concern the favorable effects of isolating the structure from the ground (see Section 15.12) and those of increased damping in the range of short and moderate fundamental periods of vibration. Changes in the percentage of internal damping are attained by modifying the choice of structural materials. Use of special devices for this same purpose—

such as the ductile steel band between partitions and structure which we mentioned earlier—deserves serious consideration.

In practice, drift limitations aimed at controlling psychological effects are often dictated by design to resist wind pressure rather than to resist earthquakes. Under the action of wind force, increase in rigidity is unequivocally a remedy for limiting excessive sway.

Thus far we have referred to design of structures on firm ground. For buildings that rest on soft soil having pronounced prevailing periods, the choice of structural rigidity may be decisive. For example, consider a building whose fundamental period of vibration coincides with the only significantly prevailing ground period. Clearly, an important increase in the stiffness of the building will reduce the amplitude of its oscillations and reduce, accordingly, the maximum floor velocities. A reduction in stiffness would apparently have the same effect, but it would entail the possible danger of making a higher natural period of vibration coincide with the prevailing ground period.

15.9 Analysis of Common Structures

The choice of the method of analysis depends on the desired accuracy, the regularity of the frame, the computational facilities avaiable, and personal preference.

Static analysis of individual elastic, purely flexural frames subjected to lateral forces has received much attention. Methods of static analysis have been developed which encompass the whole gamut of conceivable degrees of accuracy. Such crude means for estimating the distribution of forces, as the portal and cantilever methods, are rarely useful. Not only do they often entail inadmissible errors but they are of no use for estimating deformations. Although Bowman's method (Sutherland and Bowman, 1958) is more accurate it still has the latter limitation.

For not very accurate analyses one may use Wilbur's formulas (Wilbur, 1934), the factor method (Norris and Wilbur, 1960), or other procedures (Rosenblueth and Esteva, 1962) that rest on assumed relations between the rotations of certain groups of joints. When the beams are not very flexible relative to the columns and the frames are reasonably regular, these methods give errors smaller than about 10 percent in the story drifts. The errors are considerably greater in the bending moments that act on individual structural members, but they can be reduced by using, let us say, moment distribution to make these errors of the same order as those in drifts.

When the beams are very flexible in comparison with the columns, much greater errors can be expected near the base and near the roof of the building. This situation is considerably improved by replacing the methods we have mentioned with the use of tables (Blume, Newmark, and Corning, 1961) that permit estimating the positions of inflexion points in the columns.

Variations of the "principle of multiples" used in conjunction with the cantilever distribution method (Lightfoot, 1956) lead to considerably greater accuracy.

Exact results ("exact" for the idealized properties of the frame under analysis) can be attained by using methods of systematic successive corrections (such as adaptations of moment distribution), of iteration (such as the Parcel–Maney or Kani methods), or of relaxation.[7]

Examination of some damaged buildings indicates that secondary beams participate appreciably in resisting lateral forces, perhaps to a greater extent than elementary analysis would indicate. Trustworthy methods are required for calculating the contribution of these structural members to rigidity and to strength. The same is true of the contribution of slabs. Corresponding bounds can be computed easily, but often they are not as close to each other as one would like them to be.

An empirical formula has been derived by Khan and Sbarounis (1964) from tests on elastic models of flat plates to determine their effective width from the viewpoint of stiffness under lateral forces. There is still a need for methods to calculate accurately the contribution of joist floors and other commonly used structural systems. Staircase ramps afford one of the most critical examples of a structural solution met in everyday practice for which there is still no satisfactory method of analysis.

Shearing deformations of structural members are easily taken into account in an approximate way by modifying the distribution constants. As we have seen, the shearing deformations of beam-column intersections may be important, but there is no difficulty in recognizing them in analysis.

Axial deformations, especially those of columns, may be quite significant. Their inclusion in analysis using iterative methods presents serious difficulties of convergence. When this happens one must resort to artifices or use relaxation techniques or certain forms of matrix analysis.

Digital computer analysis of even highly redundant frames has been developed to the point that it can accommodate all relevant refinements (Clough, King, and Wilson, 1964; Tezcan, 1966; Muto, 1969). It knows practically no limitation, except those resulting from the lack of knowledge of the effects of slabs, secondary beams, ramps, and other structural members that do not form part of a frame but contribute to its stiffness.

General methods are also available for "skeleton" structures (those formed entirely of bars and arches)[8] that can analyze moment-resistant frames, as well as those with any type of bracing, including all pertinent types of deformation. For purposes of desk-computer or slide-rule analysis of frames with A or X braces, formulas have been published which permit taking into account buckling of the compression braces and cracking of the tension members (Rosenblueth and Esteva, 1962).

Analysis of structures having shear walls presents problems of its own. Digital

---

[7] For a description of some of these methods see, for example, Rosenblueth and Esteva (1962).

[8] "STRESS" is ideally suited for the purpose; see Fenves (1964).

computer programs for skeleton structures can be made to fit them by replacing the shear walls with systems of bars. In desk-computer analysis the main difficulty lies in the fact that, unless the shear walls are very stocky, they interact to an appreciable degree among themselves and with moment resisting frames. Neglect of this interaction in slender shear walls causes an overestimate by several fold of the wall deflections.

These structures are conveniently analyzed using an iterative procedure. We begin by assuming that at each floor, part of the lateral force is taken by the shear walls and the rest is taken by the frames. Next we force the connecting beams and the frames, in our calculation, to conform thereto. We find a set of generalized forces that tend to reduce the wall deflections. We compute a new wall configuration, adjust the connecting beams and the frames thereto, and so on.

When the walls are not too flexible relative to the beams and frames, the procedure converges whatever initial assumption we make of the distribution of forces between walls and frames. With somewhat more flexible walls it converges only for certain initially assumed distribution of forces. Very flexible walls are best treated as columns to obviate questions of convergence.

All the foregoing methods of static analysis give rise to their counterparts in dynamic analysis in the linear range. The material presented in Chapter 4 supplements the resulting methods. Static analysis of nonlinear systems can be reduced to an incremental procedure by replacing actual force-deformation curves with sets of straight lines. Accordingly we shall omit discussion of this question.

15.10 Effects of Gravity Forces

By generalizing the methods presented in Chapter 4 we can incorporate gravity forces into dynamic analysis. A simple, approximate method, due to Rosenblueth (1965), is adequate in many cases.

Suppose that the drift ψ is limited either by code requirements, by established practice, or by considerations on damage of nonstructural elements, or that a conservative estimate of it has been made from comparison with previous designs. Then we can write

$$M = (S + \psi W)H \tag{15.15}$$

where M is the story moment (sum of both end moments in all the columns in the story), S the story shear, W the sum of all vertical forces down to the story under consideration, and H the story height. Whatever method of analysis was deemed adequate to compute the effects of the story moment HS, when neglecting the effects of gravity loads, it can be applied to find the effects of M and thus establish an upper bound to the combined effects of lateral and vertical forces.

If we take the symbols entering Eq. 15.15 to correspond to the structural elements contained in a single vertical plane, we can assume different values of ψ at various points in a given story and thus account for story torsion and deformations of the floor systems in conjunction with the vertical forces.

After analyzing the structure under the action of the story moments M, we can compute an improved value of ψ for every story and every frame in the structure. Thus we can verify that allowable values are not exceeded, that our preliminary estimates of ψ did not err on the unsafe side, and that they were not excessively conservative. When justified, we can use the improved story drifts in an iterative process in which we successively improve the estimated values.

A comparison of S with ψW tells us either at the outset of our computations or at the end of the first cycle whether this conservative approach is not entirely too conservative. The fact that this is rarely the case is brought out by substituting typical values into Eq. 15.15. Thus, S/W is ordinarily in the range of 0.05 to 0.20, while ψ rarely exceeds 0.01 (after multiplying the conventionally computed values by a reasonable ductility factor). Hence, $\psi W/S$ is ordinarily of the order of 5 percent and rarely more than 20 percent, so that the maximum percentage error in M is considerably smaller than these values. The error is usually quite tolerable. This obviates the need for the iterative process or other more refined methods of analysis.

This approach is not applicable when design is governed by instability, which may happen in some very flexible structures. For single-story buildings the treatment presented in Section 11.3.9 affords an approximate method of analysis.

15.11 Foundation Design

The design of foundations to resist the combined action of gravity and overturning forces is generally straightforward, although it often involves special problems. For example, it may require the consideration of ballast to prevent lifting of one side relative to the ground, or the design of piles in tension for the same purpose (or of piles in compression), or of a foundation that spreads beyond the area of the building to limit contact pressures.

The criteria do not differ essentially from those which one may deem appropriate under the action of gravity loads alone, except on three major matters: the speed of loading, its repeated nature, and the choice of load factors—possibly different for lateral than for vertical forces—to control the probability of tilting.

Design against differential horizontal displacements does present serious difficulties of analysis. No more than the application of crude rules is feasible at present. On noncohesive soil the tendency for isolated spread footings to move horizontally with respect to each other requires the provision of structural members in the foundation, capable of taking tension as well as compression, in order to limit damage to the superstructure. Some building codes specify the design axial force in these members as 10 percent of the greater of the vertical forces acting on each of the two footings connected thereby. Since it is the lighter of the two footings that should be forced to move essentially as the heavier one, and not vice versa, the axial force should probably be, in any case, a fraction of the smaller load.

On cohesive ground the danger of the phenomenon we have referred to is unlikely. Often no tie girders are required in the foundation. In some conditions, though, there is danger that cracks may open in the soil. To minimize the probability of crack formation under the building, it is advisable in these cases to use tie girders or other means of tension reinforcement. The amount of reinforcement required is essentially independent of the vertical loads transmitted to the ground even when the soil has an important angle of internal friction. This paradoxical situation results from the function of the reinforcement. The function is not to maintain an integral foundation while the subjacent ground is displaced laterally by cracking but to change the direction of cracks in the surrounding ground so that they circumvent the structure. For the same reason the amount of reinforcement required may be a small fraction of the tension that the soil would resist before cracking. It is also an increasing function of the variability of this strength in the direction of the reinforcement. The authors are unaware of satisfactory criteria for deciding on the amount of reinforcement that should be provided in order to divert the ground's cracks.

15.12 The Choice of a Structural Solution

The optimum structural solution is dictated by economic and architectural considerations and depends markedly on the seismicity of the site.

Some materials behave in a distinctly more favorable way than others under the action of repeated alternating loads. Yet by designing the latter materials according to more conservative criteria it is possible, sometimes, to arrive at a solution involving lower capital investment without a comparable increase in the expected actualized cost of failure. On the other hand, architectural advantages may favor choice of the first material. Commercially oriented arguments favoring one material over another, even when supported by test results, should not always be taken at face value.

A structural solution that is optimum when designing without regard for earthquakes does not necessarily remain even acceptable when one designs to withstand intensive ground motions. For example, architectural demands may lead the structural engineer to favor shallow floor systems, perhaps flat plates, when he designs a tall building for modest earthquakes or decides to ignore these phenomena. If he designs to resist strong ground motions, he will try to convince the architect that concessions are in order; otherwise he will produce a very flexible structure requiring wide gaps with nonstructural elements and hence a special treatment. And if he does not provide the gaps, frequent cracking of walls, as in Fig. 15.23, is sure to occur, and wide separations will be required with respect to property lines. The columns will have huge sections in the first several stories, since their lowest points of contraflexure may be two or three stories above ground level and the effects of vertical loads will greatly magnify those of lateral forces, and special provisions will also be needed for windows, piping fixtures, and so on.

A less obvious dependence of the optimum solution on the design intensities concerns the matter of concentration of rigidity. Take a moment-resisting frame as shown in Fig. 15.24a. If we do not design to resist earthquakes, there will rarely be a reason for supplying it with cross bracing, other than to resist wind pressures and to reduce the probability of overall buckling.

If we design for mild shocks, an arrangement of braces as in Fig. 15.24b may be adequate and architectural requirements may make it difficult to choose a different arrangement. It is true that overturning moment will induce axial compression in one of the columns at the sides of the panels that contain the braces, and it will induce tension in the other column. But the compression will be sufficiently small compared with the compression induced by gravity loads alone that its presence may not even alter the column's design, and the tension will surely not surpass the gravity-load compression.

If we design for higher earthquake intensities these considerations will not apply, and there will be a distinct advantage to distributing the cross braces

Figure 15.23. Typical cracks in partitions. *After Esteva, Díaz de Cossío, and Elorduy (1968)*.

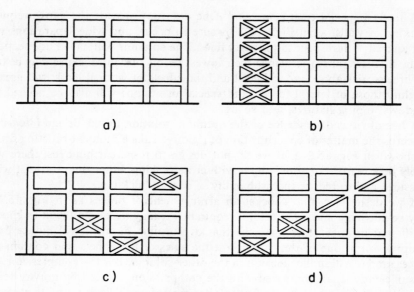

Figure 15.24. Different structural solutions for frame with and without diagonal braces:
(a) For no earthquakes or very strong earthquakes
(b) For mild earthquakes
(c) and (d) For moderate and strong earthquakes.

as in Fig. 15.24c, or as in d, so as to make a large number of columns participate in resisting the overturning moments, simultaneously lengthening the effective moment arm. Both measures will alleviate conditions in the foundation and also increase the structure's rigidity, perhaps making it unnecessary to take special provisions to protect nonstructural elements.

Design for even stronger earthquakes may lead us to dispense with braces altogether, reverting to Fig. 15.24a. The choice might be due to architectural limitations that prevent us from adopting a solution such as those in Fig. 15.24c and d, forcing us to a solution similar to the now very objectionable one in Fig. 15.24b. It might also follow from the importance of ductility under the new conditions of design and the awkwardness of large-section braces and heavy details.

As a second example, consider the building represented in the plan view in Fig. 15.25a. If the ratio of height to least dimension of the base is small or the design intensity is low, this arrangement of shear walls may be good. We must, in this case, take into account the deformations of the floor systems in their own planes and design these systems and the central transverse frames accordingly. For slender buildings designed to resist strong earthquakes the arrangement will no longer be desirable. Concentration of overturning moments in the two shear walls will cause considerable difficulty in the design of the foundation —to prevent uplift and excessive contact pressures—and of the corner columns—

to take important vertical tension and compression. A preferable solution is illustrated in Fig. 15.25b, in which shear walls have been distributed along the entire plan. Should there be architectural objections to this alternative, or should the building be very slender and we wished to design for exceedingly high intensities, it is likely that the total omission of shear walls would prove advantageous.

The desirability to limit the vertical forces induced by overturning moments, coupled with architectural restrictions on shear walls and cross bracing, will often suggest an arrangement of either these walls or these braces distributed in different bays and different vertical planes from one story to the next. The advantages of increased lateral stiffness and of reduced vertical forces more than compensate for the additional stresses that appear in the floor systems in their capacity as horizontal diaphragms as well as in their work as parts of the wall or bracing systems.

Figure 15.26 illustrates another situation in which the earthquake intensities that one wishes to resist influence the choice of a structural solution. Suppose that architectural limitations permitted the distribution of columns shown in Fig. 15.26a. If we do not design for sizeable lateral forces or if the building is relatively short, the most economical arrangement for a reinforced concrete frame is probably as schematized in this figure. Here we take advantage of the intermediate column and make the beam of uniform depth throughout. For intermediate building heights and earthquake intensities, whether we use a steel or a concrete frame, it is preferable to adopt a variable beam depth (Fig. 15.26b). This obviates high concentrations of bending moments and shears in the shorter spans and in the contiguous members. And in the case of tall buildings designed for high seismic intensities, it is even better to omit altogether the column that causes the disparity in spans (Fig. 15.26c).

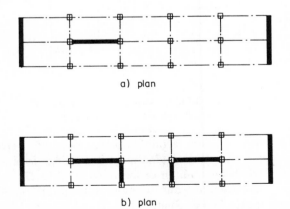

Figure 15.25. Two different arrangements of shear walls:
    (a) For mild earthquakes and short buildings
    (b) For strong earthquakes or moderately tall buildings.

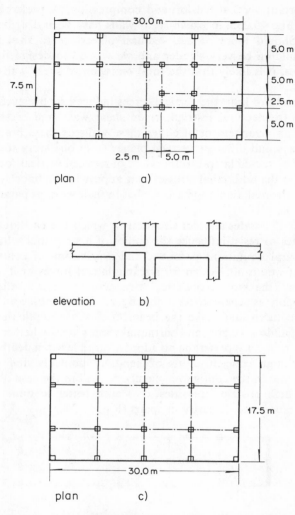

Figure 15.26. Different structural solutions for building with uneven spans.

Contemporary trends in architecture employ a number of structural details of design for which satisfactory criteria are still lacking. Typical among them is the use of eccentric beam-column connections (Fig. 15.27), whether in concrete or in steel. There are numerous examples of local failures of these connections during earthquakes (Rosenblueth, Marsal, and Hiriart, 1958), which show that ignoring the eccentricity produces seriously objectionable designs.

Some structural solutions in reinforced concrete seem ideally suited for certain buildings, except that they call for exceedingly high ductility factors in a few structural members. Sometimes this happens with the combination of shear

Sec. 15.12 THE CHOICE OF A STRUCTURAL SOLUTION 523

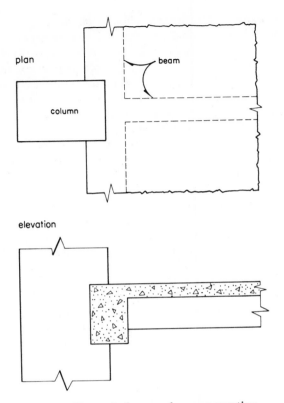

Figure 15.27. Eccentric beam-column connection.

walls and frames that may be entirely satisfactory except for the ductility demand at the beams that are directly connected to the walls in their own plane. The situation may be taken care of by designing these beams in reinforced concrete with sufficient confinement through the use of closely spaced lateral reinforcement and very careful detailing, or by replacing these beams with steel members. If confined concrete is chosen, it may be justified to hide the beams so that spalling at regions of large strains will not be visible following a strong earthquake. If steel is preferred and earthquakes of long duration are expected, the regions near the supports may require special stiffeners to prevent likelihood of the type of failure under repeated loads that we described in Section 13.6.2.

Another example is found in staircase ramps. Very often there is an advantage in replacing them with steel members that carry precast steps.

In many building codes we still find a relic whose origin and meaning are difficult to understand—the demand that the structure be designed so that it "move as a unit." Presumably this requires design of horizontal diaphragms satisfying compatibility. Codes frequently specify also that buildings of irregular plants (say, $E, I, L,$ and U shapes) be provided with construction joints—essentially expansion joints—so as to divide them into rectangular units.

Obviously, horizontal diaphragms should be designed to withstand the forces that, according to rational analysis, will act on them, consistently with those acting in the vertical resisting elements. And if this is done, it will often be found uneconomical to divide the plan into regularly shaped units. The decision should proceed from a comparative study of alternate solutions. Again the optimum will be found to depend on the intensity of earthquakes that the building is expected to resist without serious damage.

For example, the narrow band between axes A and B and between 1 and 2 in Fig. 15.28 will be called upon to resist small stresses (axial, shearing, and bending) if this structure is designed for small lateral forces. Horizontal bending may become quite high if we design for moderate intensities, making it desirable to add a beam at every floor along axis 4 between A and B. For higher design intensities we shall probably find it convenient to do without the added beam, introduce a wide expansion joint between A and B, as shown by the dashed lines in the figure, and cantilever the floor from both A and B toward this joint.

Figure 15.29 represents a solution commonly used for one- and two-story school buildings in tropical and semitropical countries. When the possibility of strong earthquakes is remote, this arrangement may be desirable because of its functional assets. However, when the structure is called upon to resist strong ground motions in the longitudinal direction, the solution has severe drawbacks. The least of these lies in the torsion induced by the difference in stiffnesses between axes A and B, because transverse walls and partitions usually have ample strength and rigidity to resist it.

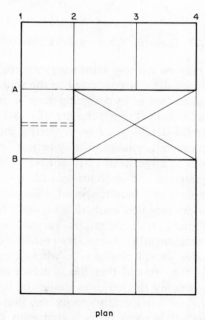

Figure 15.28. Strangled slab.

Sec. 15.12 THE CHOICE OF A STRUCTURAL SOLUTION **525**

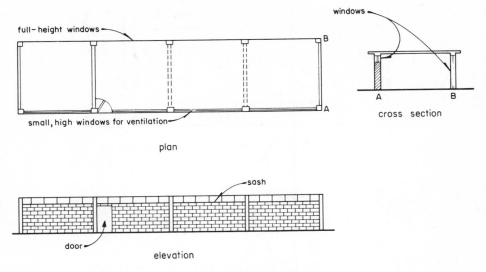

Figure 15.29. Typical school building in tropical country.

The chief disadvantages of this type of structure are: (1) the stiffness of axis B is of a higher order of magnitude than that of axis A and hence is subjected practically to the entire longitudinal force; (2) unless the columns in axis B are designed to resist extremely high shears, their strength in shear will be much smaller than that in flexure, so that they are likely to fail in shear, which involves an undesirably low ductility; (3) the increased stiffness in the longitudinal direction, relative to a solution in which the curtain walls would not participate in resisting horizontal motion, subjects structures of this type, resting on firm ground, to considerably greater spectral accelerations. Failure of columns of axis B in diagonal tension is practically inevitable during a strong earthquake. The literature abounds in examples of this sort (Rosenblueth and Prince, 1965). This is tragic because schools normally belong to this type of structure, and the cost of preventing this sort of failure, by making the curtain walls independent of the frames for horizontal motion, is quite low.

A similar situation arises in hot climates when the central longitudinal partitions are interrupted near the top slab to allow cross ventilation (Fig. 15.30). Countless cases of failures of the row of columns having interrupted partitions in structures of this type are found in the literature (Esteva and Nieto, 1967; Esteva, Rascón, and Gutiérrez, 1969).

These situations usually result from wishful thinking, which leads to assume that structural elements—in these examples the curtain walls and central partitions—regarded as nonstructural during the design stage, will not partake of structural action during an earthquake. The same fallacy is responsible for the appearance of large diagonal cracks at the corners of some frames that enclose unreinforced masonry walls (Fig. 15.31).

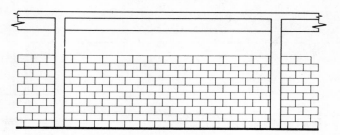

Figure 15.30. Interruption of partition for ventilation in hot climates.

Figure 15.31. Cracking of reinforced concrete frame due to its interaction with infilling wall panel. *After Esteva and Nieto (1967).*

The number of examples could extend indefinitely to show that dogmatic postures are untenable in the election of optimum structural solutions. Such matters as regional seismicity, local soil conditions, local economic situation, and architectural requirements determine the choice of solutions that are apparently objectionable when judged in the frame of a different set of circumstances.

Unprejudiced experience is valuable in lending orders of magnitude to such terms as "tall," "slender," and "intense," which we have used with vagueness in the foregoing paragraphs, as these orders change from site to site. Such experience is also of use in pointing out particularly vulnerable spots in some structural solutions that would otherwise be judged optimistically. This applies to connections between beams and shear walls, for example. Based on experience

Sec. 15.12   THE CHOICE OF A STRUCTURAL SOLUTION   **527**

with the construction difficulties that X-bracing presents at joints rather than with regard to the increase in ductility (at the expense of rigidity) it seems that an engineer will lean toward the use of A-braces instead (Fig. 15.32b vs. a).

There has been sustained interest in solutions that may drastically reduce earthquake stresses throughout the structure (see Matsushita and Izumi, 1969). The first analytical attempts in this direction advocated what became known as the "flexible first story" (Green, 1935). It was contended that a sufficiently flexible first story would so lengthen a building's natural periods of vibration that it would reduce the base shear, and hence all stresses in the superstructure, to significantly lower values than is possible with more conventional structural solutions. Important savings would ensue. In order to prevent excessive sway under wind and mild earthquakes, a "fuse" would be provided, consisting for example of brittle and weak, hollow tile partitions. These would fail under a strong shock.

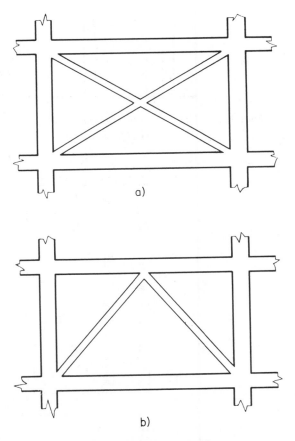

Figure 15.32. Two types of braces:
    (a) X-braces
    (b) A-braces.

Biot (1943) showed that the first story would have to be impracticably flexible to achieve a significant economy in higher stories. Typically, in a 20-story building a 10-fold increase in the flexibility of the first story will reduce stresses at all higher elevations by no more than about 30 percent. The solution is even less effective than might seem at first, since the large deflections of the first

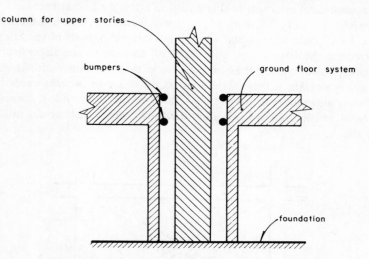

Figure 15.33. Use of hollow basement-columns.
After Matsushita and Izumi (1965).

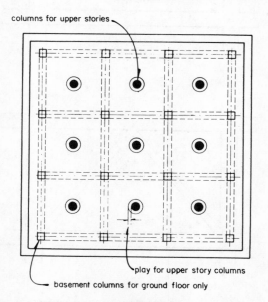

Figure 15.34. Use of double system of columns.
After Matsushita and Izumi (1965).

story bring about large story moments due to the action of gravity forces.

A more practical version of the same solution was proposed at a later date (Matsushita and Izumi, 1965). In it the basement columns are hollow and quite rigid; they enclose very flexible columns that carry the whole superstructure. In this way it is simple to limit the ground-slab displacements, giving rise to bilinear behavior of the system (Fig. 15.33). A variation of this solution is shown in Fig. 15.34. Despite the improvement there is still the matter of increased story moments in the flexible columns.

Other proposals similarly oriented employ the use of soft pads under the basement or ground-story columns [(Fig. 15.35) Joshi (1960)], the use of rollers [(Fig. 15.36) González-Flores (1964)], and the adoption of suspended supports [(Fig. 15.37) Garza-Tamez (1968)]. Some of these solutions are patented. The last two mentioned have been proposed in conjunction with dashpots tending to reduce deflections. To the authors' knowledge no major applications of any of these alternatives have been attempted.

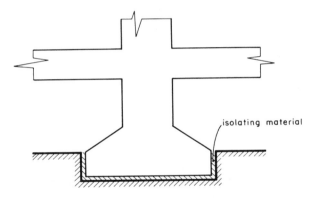

Figure 15.35. Use of rubber pads for partial isolation from earthquake motions. *After Joshi (1960).*

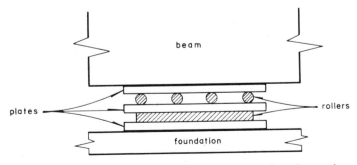

Figure 15.36. Use of rollers for isolation from earthquake motions. *After González-Flores (1964).*

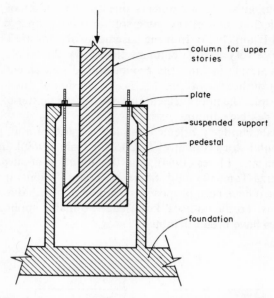

Figure 15.37. Suspended supports. *After Garza-Tamez (1968)*.

15.13 Structural Synthesis

A very promising approach to earthquake-resistant design of buildings lies in structural synthesis ("direct design"). At least two important steps have been made in this direction. One consists of fixing allowable story drifts in a multi-story building and using a computer program that, iteratively, selects story shear stiffnesses so that the envelope of the drifts produced by a given family of earthquake records is no more and no less than the allowable values (Matsushita and Izumi, 1965). Thus far the method has only been applied in the range of linear behavior.

The second contribution is applicable to single-story buildings of reinforced concrete. Through the use of graphs, it directly furnishes the required column sections and reinforcement given the design spectrum, the mass of the building, its height, the allowable drift, the allowable ductility factor, the concrete strength, and the yield-point stress of the reinforcement (Borges, 1965).

Collapse of bridge during Alaska earthquake (1964).

16

OTHER TOPICS IN EARTHQUAKE-RESISTANT DESIGN

16.1 Inverted Pendulums

16.1.1 General Considerations. Under the present heading we include structures of the types depicted in Fig. 16.1. They have three features in common: (1) their mass is mostly concentrated at the top, (2) the formation of a single plastic hinge in a column suffices to bring about collapse, and (3) the influence of gravity forces is decisive in lowering the capacity to withstand lateral forces. When these structures rest on soft ground, the condition of soil-structure dynamic interaction (Rascón, 1965) plays a dominant role unless the foundation has exceptional rigidity and strength and spreads out far beyond each column base or ties several such bases together.

As shown by Rascón (1965) the bending moments at the tops of the columns of structures in Fig. 16.1b and 16.1c deserve special attention. Earthquake-induced moments of this kind govern, in large measure, the design both of the column and the roof slab or shell. In both types of structure, practically unavoidable differences in thickness of the roof system from one side to the other introduce asymmetry in the mass distribution and, hence, additional bending moments at the column top that would not be present if the structure were strictly symmetrical. These moments are often more important than those due to eccentric live load because the latter may be absent at the time of a major earthquake.

Accidental torsion is particularly significant in structures of the type shown schematically in Fig. 16.1b. The most obvious cause of these horizontal moments about a vertical axis at the column tops, as that of dead and live load bending moments at the column top in nominally symmetric inverted pendulums, is found in the unavoidable asymmetry of mass distribution. There is also the rotational component of ground motion about a vertical axis.

Structures represented in Fig. 16.1b have two additional special problems—

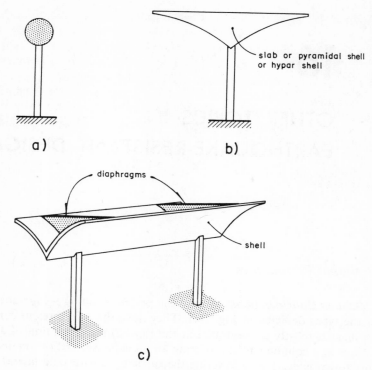

Figure 16.1. Three types of inverted pendulum:
(a) Simple inverted pendulum
(b) Umbrella roof
(c) Butterfly shell.

horizontal bending moments and torsion about a horizontal axis in the roof system—both of them due to horizontal inertia forces.

For all these reasons, particularly due to dependence of the structures' stability on the strength of a single plastic hinge, inverted pendulums have been quite vulnerable to earthquakes.[1] Several building codes are especially conservative in connection with these structures in comparison with the requirements for design of buildings.

16.1.2 Simple Inverted Pendulum. Consider the system in Fig. 16.1a. Let us idealize it as consisting of a rigid mass concentrated at a point, supported by a prismatic, elastoplastic, massless column fixed at its base. We shall consider excitation through unidirectional translation of the ground. We shall ignore the change in elevation of the roof as the system oscillates.

---

[1] See Rosenblueth, Marsal, and Hiriart (1958). Notice, though, that the case described there was aggravated by near coincidence of the structure's fundamental period and that of the ground.

Elastoplastic behavior implies that the base moment-deformation curve in the absence of vertical loads would be as the upper line in Fig. 16.2. Under the circumstances we could analyze the system as any other elastoplastic one. The action of the gravity force introduces an additional bending moment at the column base of unfavorable sign, equal to the weight of the point mass times its horizontal displacement. We can take it into account by subtracting, from the ordinates of the base moment–deformation curve, those of a straight line through the origin, with slope equal to the weight in question, as indicated in the figure.

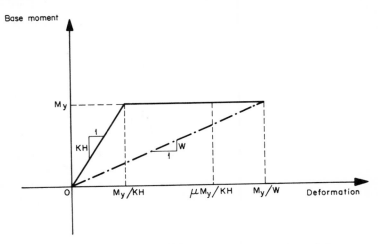

Figure 16.2. Idealized behavior of simple inverted pendulum.

Unless there is reason for earlier failure, the structure will reach a state of unstable equilibrium when the two lines cross each other. This occurrence does not necessarily signify collapse; since the immediately succeeding ground motion may bring the system back to a stable configuration, but the probability of this return is small. Consequently we may take the intersection as tantamount to failure.

In Section 11.3.9 we analyzed structures of this type under two additional simplifying assumptions: that their strain energy was the same as if their initial stiffness were infinitely large, and that the disturbance was a segment of a stationary stochastic process. In the following sections we shall describe an approximate treatment that dispenses with these two simplifications and is consistent with the rest of Section 11.3. However, the approach of Section 11.3.9 has received ample confirmation in simulation studies which the present treatment has not.

Let us assume that the system is a truly elastoplastic one with an initial stiffness equal to the difference in initial slopes between the two force-deformation lines and a ductility factor such as to give a strain energy equal to the area between both lines up to the point of their intersection. It follows from inspection of Fig. 16.2 that the deformation associated with a bending moment equal to the yield moment, acting at the base, due to the gravity force, is M_y/W, where M_y denotes the yield moment and W is the weight. The deformation at which yielding begins is M_y/HK, where H denotes height and K is the initial stiffness. Therefore the ductility factor of the "equivalent" elastoplastic system is

$$\mu' = \frac{1}{2}\left(1 + \frac{HK}{W}\right)$$

This is actually the highest possible value of μ'. If the original structure without gravity effects had a ductility factor $\mu < HK/W$, μ' will be less than the foregoing value and smaller than μ:

$$\mu' = \frac{2(HK/W)\mu - \mu^2 - 1}{2(HK/W - 1)}$$

The yield moment of this system is

$$M'_y = M_y\left(1 - \frac{W}{HK}\right)$$

The error involved in neglecting the effect of the gravity force may be quite serious. Suppose that $\mu = 6$ and $HK = 6W$. Neglecting the effects of the gravity force might lead one to believe that the structure could be designed for a spectral acceleration equal to $\frac{1}{6}$ that of a simple linear system having a stiffness K. We would find the required yield moment given by

$$M_y = \frac{AW}{6gH}$$

where A is this spectral acceleration and g is the acceleration of gravity. The ductility factor of the equivalent elastoplastic system is only 3.5. Its initial stiffness is $K - W/H$. Hence, its initial period is

$$\left(1 - \frac{W}{HK}\right)^{-1/2} = 1.095$$

times the initial period that the system would have in the absence of the gravity force. If the pseudovelocity design spectrum is period independent in the range of interest, the design acceleration will be $A/1.095$. Consequently, the elastoplastic system would have to be designed for a yield moment

$$M'_y = \frac{AW}{1.095 \times 3.5gH}$$

whence,

$$M_y = \frac{AW}{1.095 \times 3.5gH} \frac{1}{1 - 1/6}$$

which is 1.88 times the yield moment we had found when neglecting the acceleration of gravity.

16.2 Towers, Stacks, and Stack-Like Structures

The structures we consider in this section differ from inverted pendulums in that much of their mass is distributed along their height. They differ from buildings in that overall flexural deformations prevail.

Some problems typical of these structures are brought out in the analysis of a simple example. Consider a uniform, purely flexural beam, fixed at its base. Neglect shearing deformations and the effects of rotary inertia, of damping, and of gravity forces. Using the results of Section 3.6 we can show that the base shear in the nth natural mode is given by

$$S_n = B_n A_n T_n$$

where B_n is a coefficient that tends asymptotically to a constant, finite value as n tends to infinity (it is essentially constant beyond the third natural mode), A_n is the spectral acceleration associated with the nth natural mode, and T_n is the nth natural period of vibration. For large enough n, T_n varies almost exactly as $(n - \frac{1}{2})^{-1}$.

It follows that, if the spectral acceleration remains finite as the natural period tends to zero and if we combine the natural modes by taking the sum of the absolute values of the corresponding responses, the base shear will tend to infinity.

We shall also find that it tends to infinity according to any of the criteria for combining natural modes that we presented in Chapter 10 if A_n varies inversely as T_n. However, if A_n remains bounded as T_n tends to zero, the base shear will also be finite whether we use for design the square root of the sum of squared responses or any of the more refined criteria advanced in Chapter 10.

Base shears in real structures are finite for several reasons: (1) acceleration spectra do not have ordinates that tend to infinity with natural frequency, (2) real structures are always damped, (3) real structures are not strictly fixed, and (4) shearing deformations and rotary inertia tend to reduce the contribution of higher modes. (Gravity forces will significantly affect vibrations in only the fundamental or the first few natural modes.) Still, the result obtained for the ideal system analyzed points out the importance of taking account of very high harmonics as well as of incorporating the differences between the ideal system and actual structures, particularly when the fundamental period is so long that the design acceleration spectrum is hyperbolic over several natural periods of vibration.

The approach found in some codes (Rinne, 1960), such as the Uniform Building Code, to analyze chimney stacks is not defensible on rational grounds. About half of the correction coefficient specified therein for overturning or bending moment comes from the difference between design accelerations for base shear marked by the code and an assumed, much less conservative response spectrum. As a result the criterion of design for moment is far less conservative than that for shear. Moreover, the analysis is based partly on stresses obtained from a divergent series of which only the first three terms are retained.

A second method of analysis, due to Housner (1956), is free from the objection that concerns the dual acceleration spectrum but still uses the first three terms of the divergent series. It includes the use of coefficients adjusted to give design stresses consistent with available experience.

The following summary is taken from an application of the methods of analysis in Section 10.2 by Montes and Rosenblueth (1968) and by Rosenblueth and Elorduy (1969a). Gravity effects, shear deformations, and rotational inertia were neglected. These matters are often important and should be incorporated in future studies.

The transfer functions are not periodic. Yet there is little difference in the shears and moments computed from Eq. 10.7 (square root of the sum of squared modal responses) and from Eq. 10.10 (the more refined treatment). The error is less than 1.5 percent in the base shear and less than 0.4 percent in the base moment.

Two design acceleration spectra were considered: hyperbolic with a cutoff at a natural period equal to the $\frac{1}{10}$ of T_1, the structure's fundamental period, and flat. All modes were assumed to have 5 percent damping. Earthquakes were idealized as stationary Gaussian processes of duration $s = 4.78T_1$.

Computed shears and moments are shown in Figs. 16.3 and 16.4. Simple expressions can be used to approximate them (Montes and Rosenblueth, 1968).

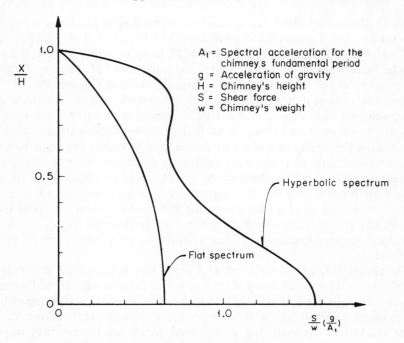

Figure 16.3. Distribution of shear forces.

The code formula for correction of the integral of the shear envelope to obtain the base overturning moment ($J = 0.6 T_1^{-1/2}$ but ≥ 0.4) often errs seriously on the unsafe side (the computed J is 0.587 and 0.972 with hyperbolic and flat acceleration spectra, respectively).

For spectra of arbitrary shape we need not take design shears greater than those for either a hyperbolic or a flat spectrum that constitute upper bounds to the design spectrum in the range of periods $T \leq T_1$. This consideration leads to a simple method for specifying a conservative shear envelope for an arbitrary spectrum (Fig. 16.5). The same criterion may be used for the design moments.

Analysis of several tapered chimneys (Rumman, 1967) subjected to records of actual earthquakes confirms that the root of the sum of squared modal responses gives a satisfactory approximation to the total response. The base shear coefficients are given with sufficient accuracy by the same expressions as for cylindrical chimneys. The moment correction coefficients for tapered chimneys also exceed the code J's by an appreciable margin (Montes and Rosenblueth, 1968).

Fenves and Newmark (1969) have made a study of shears and moments in flexural beams along the same lines as the one we described for shear beams. They find, in general, smaller correction factors than the ones we presented above. However, the two sets of results are not directly comparable, mostly because Fenves and Newmark discretize the beams as 20-story structures,

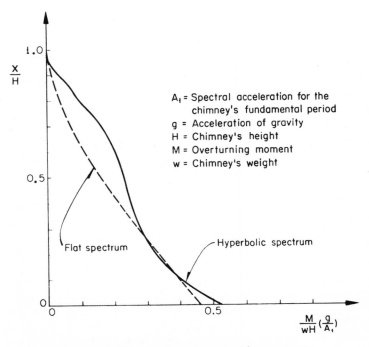

Figure 16.4. Distribution of overturning moments.

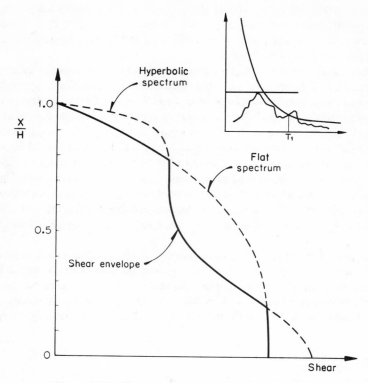

Figure 16.5. Shear envelope for arbitrary spectrum.

while Montes and Rosenblueth treat them as having distributed mass, and in flexural beams the importance of the higher modes is great (Elorduy, 1969).

In very slender chimneys, gravity effects may be important. They can be incorporated in an approximate manner through a procedure of successive approximations, taking due account of the fact that base rotation and nonlinear material behavior will increase deflections beyond the values given by elementary linear analysis.

The effect of nonlinear behavior in these structures is not very important. Large ductility factors, say in excess of 2.0, are unlikely in chimney stacks. Even moderate ductility factors, if associated with inelastic action in the stack itself, involve the appearance of visible cracks in masonry and reinforced concrete chimneys. In thin steel stacks they involve objectionable local buckling (Blume, 1963). Since ductility is not to be taken much advantage of, extremely high design accelerations must be adopted.[2] Energy absorption through yielding of the anchor bolts offers a means for utilizing ductility without impairing the

---

[2] Blume (1963) presents an interesting analysis of damage undergone by several Chilean chimney stacks that had been designed apparently in accordance with very conservative criteria.

appearance of the stack itself (Rosenblueth, 1961) but deserve special detailing; otherwise replacement of elongated bolts may call for major repair operations in the foundation.

As is the case with inverted pendulums, stack-like structures are especially vulnerable to earthquakes; failure of one section brings about collapse.

The foregoing remark is apparently substantiated by the failure of an appallingly high percentage of chimneys during earthquakes in New Zealand (Bennett, 1965) and in Chile (Steinbrugge and Flores, 1963). But the value of these experiences should not be overrated, as they are mostly associated with structures that were either not designed to resist earthquakes or were seriously underdesigned. Towers whose structure constitutes frames behave much like buildings and can be analyzed accordingly.

Cross-braced towers are particularly vulnerable to earthquakes. These structures present problems of accidental torsion due mostly to asymmetrical yielding of the braces. Such matters have been discussed previously in this work. Others pertain to structures that support storage tanks and will become apparent in Section 16.6.

Attempts have been made at decreasing the amplitude of responses of towers having cross braces by providing them with damping devices (Ruge, 1938). Results have not been encouraging because towers lie usually in remote locations where it is not realistic to assume adequate upkeep of the devices.

16.3 Bridges

Within their diversity of dimensions, materials, and structural solutions, bridges have certain features in common from the viewpoint of their response to earthquakes. Their salient characteristic lies in the fact that their supports tend to undergo differential motions during earthquakes. This is partly due to distance between supports and partly to the differences in geologic and topographic features at the supports or surrounding them. Therefore, even for short spans, there is a tendency for the abutments to move differentially.

In Section 2.4 we described and illustrated a method for analyzing structures whose supports undergo different motions. In the absence of trustworthy bases for the formulation of relations between the motions of various supports, one may analyze the structure for a family of simulated earthquakes, assuming that the ground motion consists of a wave that travels in a direction θ relative to the bridge axis. Assign θ various values between 0 and $\pi/2$ and design for the envelope of the responses computed in this manner. Each phase of the ground motion may be assumed to travel at the corresponding wave velocity.

Depending on the type of bridge, various aspects of behavior will deserve special attention. For example, bridges having simple supports are prone to fail as shown in Fig. 16.6, through loss of these supports, unless the corresponding details are carefully designed against such an event. Relative displacements

Figure 16.6. Failure of bridge decks through loss of supports. *After Japan National Committee on Earthquake Engineering (1965).*

between independent piers are likely to attain several inches, or even feet. (The statement is by no means exaggerated. Recall that the maximum ground displacement at El Centro 1940, in the NS direction, obtained by twice integrating the accelerogram was computed as 8.3 in., but it may have been 12 or 15 in. The maximum displacement of a lightly damped, long-period system relative to the ground may easily be 1.5 times the maximum ground displacement. Its maximum absolute displacement is greater.) The tilting of piers may tend to increase the seismic disturbance on the deck and the relative displacement between deck supports. A more significant cause of relative displacements is the occurrence of slips at geologic faults. This may prove to be a phenomenon which occurs frequently enough to consider in bridge design as more bridges are built in seismically active areas, since it is not uncommon for these structures to span over major faults.

During a single major earthquake the slip across a fault may be of the order of several feet. However, slips occur across some active faults even in the absence of earthquakes. A method is available for estimating the rate of slip on the basis of *seismic moment* (Brune, 1968).

Average rates of slip, either observed, calculated, or inferred, at various locations are given in Table 16.1. These values are areal averages; locally, the rates of slip may be somewhat larger. At some sites the direction of slip is inclined relative to the horizontal. And it is worth keeping in mind that many earthquakes are not associated with surface faults.

TABLE 16.1 AVERAGE RATES OF SLIP ACROSS GEOLOGIC FAULTS*
[AFTER BRUNE (1968)]

Zone	Time interval	Average rate of slip (cm/yr)
Southern California	1934–1963	
Imperial Valley		3.2
San Andreas fault		0.03
Kern County		17
Total area		5.8
San Jacinto fault	1912–1963	1.5
Southern California area	1912–1963	3.7
San Andreas fault	1800–1967	6.6
New Zealand	1914–1948	7.2
Turkey (Anatolian fault)	1939–1967	11
Oceanic transcurrent or transform faults		
Romanche (2°N, 30°W to 2°S, 14°W)	1920–1952	3
South Pacific (50°S, 137°W to 60°S, 120°W)	1925–1952	9
San Mayen (70.5°N, 15°W to 72.5°N, 0°W)	1955–1964	2
Island arcs		
Tonga 0 to 60 km depth	1920–1954	5.2
Tonga 100 to 700 km depth	1920–1954	0.23
Japan	1905–1955	15.7
Aleutians	1905–1967	3.8

*not including creep

Even when a bridge does not span a potentially active fault it is impractical to provide such long and wide supports that the simply supported spans do not slide out of the piers during a strong quake. Consequently, it is important to limit the relative motions at the tops of the piers, by providing butts or stop plates that permit sliding of the bridge supports only to take care of length variations due to temperature changes, and to shrinkage when applicable. This may not always be possible.

Some bridges have collapsed during earthquakes; others have collapsed as a consequence of failures of pier foundations.[3] The phenomenon is mostly associated with scour when it occurs in the absence of earthquakes; it is mostly associated with soil liquefaction when caused by temblors. Information in Section 13.9 permits the engineer to decide on design criteria to make the latter phenomenon sufficiently improbable. When investigating the stability of piers, it is important to consider the dynamic effects of the surrounding water (see Sections 6.4 and 12.3).

Suspension bridges present problems of their own. Some problems are due to such bridges having a very long fundamental period of vibration.[4] As a consequence, damping, which is small for these structures, is practically ineffective

---
[3] Kodera (1965) cites an example of this sort of failure brought about by an earthquake.
[4] The fundamental period of Golden Gate Bridge, for example, is 19 sec (Carder, 1947).

in reducing their responses, so their deflections are large compared to those of other civil-engineering structures. It would be undesirable to allow suspension bridges to undergo important permanent deformations with recurrence periods of a few decades. Consequently these bridges should be designed to withstand moderate shocks without exceeding a ductility factor of 1.

Another consequence of the length of their fundamental period is that experience gained from behavior of other structures cannot be directly extrapolated to the design of these bridges. Since earthquake spectra are not very trustworthy in this range of periods and since the total number of suspension bridges that have undergone major earthquakes is small, they should be designed following conservative criteria.

The second cause for special problems in suspension bridges is found in their appreciable departure from linear behavior within the elastic range. The situation stems from changes in geometry during vibration. Either nonlinear elasticity must be brought explicitly into the analysis of these structures or conscious errors on the safe side must be introduced to make up for neglect of this property.

Finally, suspension bridges are difficult to analyze dynamically because they possess a large number of significant natural modes of vibration, some of which are excited only by vertical ground motion, some only by out-of-phase vibration of the supports, and some are not immediately obvious to a mind trained only in the analysis of more conventional structures.[5]

16.4 Retaining Structures

After the end of an earthquake motion we may find a retaining wall in the same condition it had before the motion began, or we may find that it has moved, as illustrated in Fig. 16.7, or it may have turned over entirely. In analyzing these possibilities we must use the soil properties as affected by vibration. Section 13.9 contains information that is useful for estimating these properties.

The information is particularly incomplete and sketchy for saturated soils. The question is then complicated by the interaction of pore and effective pressures and by problems of dynamic seepage and drainage. Some efforts have been made to measure total and pore water pressures in backfills of retaining walls (Matuo and 0-Hara, 1965), but the results are not trustworthy because of difficulties in techniques of measurement and in translation from models to prototypes. Nevertheless, whatever information is available will permit a rough estimate of the relevant soil properties.

If the maximum ground acceleration is insufficient to cause inelastic strains in the soil, we shall find the retaining wall unaffected by the earthquake in question and no additional analysis will be required. If the maximum ground

---

[5] Konishi and Yamada (1965) describe the analysis of a long suspension bridge under the assumption that it behaves linearly. Its study may serve as a guide for the analysis of other structures of this type.

Sec. 16.4 RETAINING STRUCTURES **545**

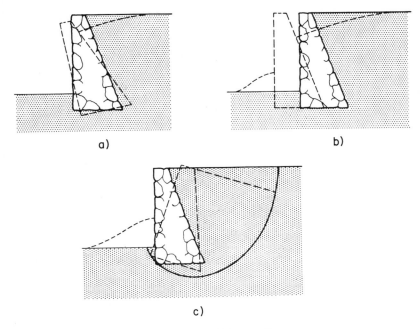

Figure 16.7. Movement of retaining walls:
    (a) Tilting over
    (b) Sliding
    (c) Partial failure along deep surface of sliding.

acceleration exceeds this value, there will be yielding of the system according to one of the mechanisms depicted in Fig. 16.7. The probability that the system yield upward is negligible. Hence we may assume that it can only yield downward.

As we shall see, even in large earth dams, elastic deformations can often be neglected in comparison with deformations in the plastic range. Because of the relatively small dimensions of most retaining structures, the simplification which consists of neglect of elastic deformations is usually justified in the analysis of these structures. Hence, we may idealize the soil-wall system as rigid-plastic, that is, assign it a force-deformation relation as shown by the continuous line in Fig. 16.8.

Gravity effects and the sustained effects of vibration will ordinarily lower the soil strength with increasing strain, at least in the case of failure by tilting, as indicated by the dashed line in the figure. But explicit consideration of the negative slope introduces an additional variable and leads us into an idealization for which no criteria are available. It seems advisable, therefore, to preserve the idealization of the force-deformation relationship as rigid-plastic.

For the case of strain softening, following an approach similar to the one we adopted for inverted pendulums, we may speak of an equivalent rigid-plastic system yielding in one direction only. Its yield force is the same as that of the real system, and its maximum allowable deformation is half of the one associated

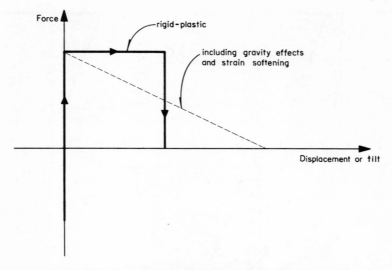

Figure 16.8. Force-deformation relations for retaining wall.

with zero lateral force in the latter system. For the equivalent rigid-plastic system we may use the results in Section 11.4 in order to compute, in an approximate manner, the maximum deformation or tilt and compare it with the allowable value to determine whether the wall is stable under a given earthquake.

If failure can occur by a mechanism different from that of tilting, the slope of the second branch in the force-deformation relationship may be negligible, whether it is positive or negative. In such cases behavior of the equivalent rigid-plastic system may practically coincide with that of the actual soil-wall system. In all retaining walls it is important to remember that the effects of successive earthquakes are additive.

The foregoing treatment has assumed that the retaining wall can yield without suddenly losing its capacity. If the wall behavior is brittle, failure will occur as soon as its capacity is exceeded. Analysis may then be confined to the assumption of rigid behavior and the effects of inertia forces assuming a reduced soil strength. The possibility of soil liquefaction merits a similar treatment.

Conditions near the top of a silo are similar to those in the neighborhood of a plane retaining wall, except that the structure is not capable of extensive yielding while the force-deformation relation of the fill material may be practically symmetrical. At lower elevations grain pressures may be computed under the assumption of a reduced strength or angle or internal friction, otherwise neglecting the strains induced by earthquake in the stored material.

16.5 Tunnels and Pipes

The design of tunnels and pipes to resist earthquakes can be divided into four groups of problems: (1) transverse stresses, for the calculation of which the structure may be treated, approximately, as a closed ring; (2) longitudinal bending and shear in the tunnel regarded as a beam; (3) axial tension and compression; (4) matters concerning connections of the tunnel with other tunnels, with inlets, and with other appurtenances.

Transverse stresses due to earthquake will rarely govern design. Perhaps an exception is found in tunnels that are very rigid in the longitudinal direction relative to the surrounding ground. If lateral contact stresses can be estimated, a static analysis of the tunnel regarded as a ring will probably suffice for design purposes.

In Section 10.6 we showed how the curvatures of a very flexible pipe could be estimated under the assumption that the structure offers no restriction to soil motion. On that basis, the maximum curvature produced by a shear wave traveling in the direction of the pipe axis can be written as

$$\max_t \left| \frac{d^2 x_2}{dX_1^2} \right| = \frac{a_2}{v_s^2} \tag{16.1}$$

where X_1 is a coordinate along the pipe axis, x_2 is the ground displacement perpendicular to X_1, a_2 is $\max_t |\ddot{x}_2|$, and v_s is the shear-wave velocity in the direction of X_1.

Sakurai and Takahashi (1969) report field observations that confirm the conclusions of this treatment.

The matter of dynamic soil-tunnel interaction with relatively rigid tunnels has apparently not received attention. Under the assumption that the soil and the pipe behave linearly and elastically an approximate solution is possible if we replace the soil with a discrete system[6] and assume, for example, that in the absence of the pipe the soil would vibrate horizontally describing a sine function of a coordinate parallel to the pipe axis. The solution requires the use of a large-capacity computer.

A rough approximation to the answer can be obtained treating the problem statically and replacing the soil with independent spring elements that give rise to a subgrade reaction treatment. The coefficients of subgrade reaction can be estimated from studies concerning piles (Penzien, Scheffey, and Parmelee, 1964).

A similar study can be performed for longitudinal straining of the tunnel lining, replacing S with P waves, and contact pressures with contact shearing stresses. Often it is also worth investigating the effects of waves traveling at an angle relative to the tunnel axis; these introduce combined bending and axial forces.

In one instance at least, a model study has been carried out to investigate the effects of an irregular sloping soil-rock interface a short distance below the structure (Bustamante, 1965d). It served also to determine, in an approximate manner, the effects of nonlinear behavior of the soil as well as the amount of play required at connections between the structure and its inlets and outlets.

In general terms the problem of design of pipes on rigid supports placed at

---

[6] Ang and Rainer (1964) have developed a computer program available for analyzing the dynamic behavior of a continuous mass idealized in this manner and subjected to an arbitrary disturbance.

frequent intervals is similar to that of tunnels, with obvious modifications. If the supports are relatively wide apart, we may resort to the method of analysis presented in Sections 2.4 and 10.5 for systems whose supports undergo differential vibrations. The ground motion may be assumed to constitute a traveling wave, much as we assumed for very flexible tunnels.

When a liquid flows through a pipe or a tunnel, its interaction with the vibrating structure introduces a feature that has apparently not been investigated. The nature of this interaction can be inferred from a study of wind-induced vibrations of pipes through which flows a fluid (Housner, 1952b).

Sliding along a geologic fault that is traversed by a pipe or a tunnel introduces stresses that may be many times more important than the ones we have discussed. To the authors' knowledge, no criteria have evolved for design under these conditions.

Where soil strains are so large as to cause appreciable permanent deformations, severe damage to utilities can be expected unless they are specially designed to withstand them. The following instances illustrate the damage experienced by piping.

According to Bromson (1959, p. 31), "... all but one of San Francisco's water arteries were broken by the quake..." of 1906, and (p. 37) there was no water to fight the fires because "... the mains were broken." These remarks emphasize the importance of a conservative design and careful construction of those utilities whose service is critical following a major earthquake.

Eckel (1967) reports on the extensive breakage of petroleum and natural gas facilities in Alaska during the 1964 earthquake. Causes of failure included axial tension, compression, and flexure. In repairing the pipes it was necessary to take into account the constraints imposed by the ground, which was frozen to a depth of 6 ft. This imposed residual stresses due to the earthquake that would later combine with additional stresses due to thawing and to subsequent freezing. The residual stresses were only relieved at ground breaks. Hence, on the basis of maps showing where repairs had been made, it was possible to identify locations where residual stresses were probably high. They were then relieved by cutting the pipes at these locations. Some pipes were under such tension that the cuts opened as much as 2 in.; elsewhere compressive stresses shortened the pipes. The presence of ground cracks also indicated sections where residual stresses were high, and so, pipes were also cut at those locations. The pipe displacements at the cuts served as indications of the adequacy of these measures.

16.6 Tanks and Hydraulic Structures

For tanks supported directly on the ground, manners of failure (in the broadest sense of the word) comprise structural damage and excessive sloshing. We have covered the theories that permit estimating hydrodynamic pressures

(Sections 6.2 and 6.3) and the amplitude of sloshing (Section 6.3) during earthquakes.

Wide uncertainty typifies the corresponding criteria of failure. For example, it may be of only minor consequence that a reinforced concrete tank should crack during an earthquake and let liquid escape. The structure can be repaired later, and the foundation will not have suffered damage if the soil is sufficiently cohesive or if it is adequately protected. The same applies to sloshing. Yet, a different set of circumstances may make these events most objectionable.

The main feature characterizing seismic behavior of elevated water tanks concerns the fact that the liquid dissipates an insignificant amount of energy during vibration. Considered by itself, water in an ordinary tank behaves with a fraction of 1 percent of critical damping. Unless baffle plates are provided, the structure must be designed to dissipate practically the entire energy fed by the earthquake. Moreover, most structures that support elevated tanks belong to the inverted pendulum type. Others are cross-braced towers in which extensive yielding of the braces has a high probability of being asymmetrical and hence of causing accidental torsion with serious consequences. We conclude that it is advisable to design these structures for spectra associated with small degrees of damping (not more than about 2 percent) and small ductility factors (perhaps 1.5 or less). It is always advisable to tend to avoid the coincidence of the first two natural periods in any direction.

Earthquake-resistant design of hydraulic structures presents many special and interesting problems that deserve attention:

1. In canals there is the possibility of base failures and of soil liquefaction (see Section 13.9). The latter may cause cracking of the lining and consequently piping. Diversion tunnels are vulnerable especially to differential movements of fractured rock.
2. In submerged structures, such as intake towers, there are the problems of analysis discussed in Sections 6.4 and 12.3.
3. In all concrete structures one must be concerned with the possibility of cracking. If the consequences thereof may be serious, it is well to design against major cracking permitting only a small amount of nonlinear action.
4. In the design of practically all dams the difficulties of analysis are so great, the consequences of failure so serious, and the cost of appreciably changing the probabilities of failure are of such high magnitude that refined model and analytical studies are almost always in order.
5. The possibility of slope failures in the reservoir, creating fresh-water tsunamis, is inherent in many dams. The consequences are usually catastrophic. Yet the criteria for predicting the event, for calculating its characteristics, and for modifying the probabilities of its occurrence are still not developed satisfactorily (see Sections 6.2 and 6.5).

The stability of buttress dams is especially sensitive to the combination of the horizontal components of ground motion. With reference to Fig. 16.9 we notice

550 OTHER TOPICS IN EARTHQUAKE-RESISTANT DESIGN Chap. 16

that accelerations perpendicular to the river axis will endanger the stability of the buttresses to a degree that depends on the compression to which they are subjected as a consequence of gravity, hydrostatic, and hydrodynamic forces, the latter due essentially to ground motion in the direction of the flow of the river. Since the phenomenon is strongly nonlinear, its calculation will hardly yield to modal analysis, except by making important concessions to expediency. Monte Carlo analysis in an analog computer lends itself well to investigation of this problem.

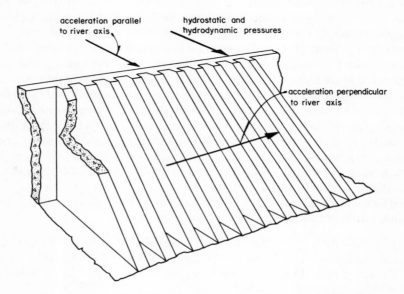

Figure 16.9. Buttress dam subjected to two components of ground motion.

In the analysis of arch dams, we are confronted with the complications implicit in calculating hydrodynamic pressures under conditions that must be idealized as three dimensional (see Sections 6.2 and 12.2). Ground motion transverse to the axis of flow, the matter of differential motions between abutments, and the problem of collapse toward the upstream face are also questions that require special treatment.

In both buttress dams and arch dams the presence of contraction joints causes dynamic behavior to differ significantly from that of the corresponding monolithic structures (Raphael, 1960).

Earth and rockfill dams present many aspects of their own (Sherard et al., 1963; Ambraseys, 1960 and 1962). The salient questions are

1. The possibility of soil liquefaction or other forms of foundation failure.
2. The danger of brittle fracture or cracking of thixotropic materials, followed by large-scale piping. The core of hydraulic-fill dams is most prone

to suffer this kind of failure. This type of construction has been practically abandoned for this very reason, but not all dams of modern types are exempt from this danger. Consciousness of its possible occurrence leads to careful selection of the core material, sometimes at a high premium.

3. The fact that deformations within the linear range of material behavior are almost negligible compared to those produced by incipient slides and by volume changes due to rock breakage at points of grain contacts. The latter phenomenon is most relevant to rockfill dams.[7] Hence, applicable methods of analysis depart from those which have become accepted for other structures. Practical methods adequate for predicting volume changes due to rock breakage are yet to be developed.[8] The transfer of effective stresses into pore water pressure caused by these volume changes and the consequent lowering of shear strength are also practically unexplored matters but of great importance to the stability of rockfill dams.[9] This is actually true of hydrodynamic seismic pore pressures in general and of their effects on the stability of all earth, gravel, and rockfill dams.[10]

4. The criteria of failure. Suppose we take measures to reach negligible probabilities of failure through brittle fracture of the core, soil liquefaction, and piping, and reserve for separate treatment the matter of fresh-water tsunamis. Practically the only type of failure left that merits consideration is the loss of freeboard through lowering of the crest due to a series of small slides and to a reduction in volume of coarse-grained materials. The loss of freeboard may be of relatively little consequence if the crest is not overflowed during the earthquake in question, since there may be an opportunity to implement repairs and restore the original crest level before another strong earthquake strikes. If there is overflow, the consequences may or may not be serious, depending on how well protected the down-

---

[7] The overwhelming importance of nonlinear deformations relative to those in the linear range is brought out by field and by laboratory observations. For example, a well-instrumented dam in Japan has been subjected to several earthquakes that have not caused visible, permanent distortion (Okamoto, Citamura, and Kato, 1968). Yet the ratios of maximum crest acceleration to maximum ground acceleration reported for this structure are found to be a frankly decreasing function of the latter acceleration. The response of scaled models of rockfill dams has been predicted reasonably well by idealizing them as rigid-plastic systems (Bustamante, 1965a). (There was no appreciable crushing at grain contacts in these models. Had this phenomenon been reproduced, irrecoverable deformations would have occurred at lower accelerations.)

[8] Grigorian (1960) has proposed a very general and promising treatment of the relation between stress history and volume change. However, there are some practical difficulties associated with its application.

[9] Davis (1960) has published a short discussion of the implications of hydrodynamic pore pressures, induced by volume changes, on the stability of rockfill dams, with mention of model-prototype relations.

[10] Patel and Bokil (1962) and Patel and Arora (1965) have attempted to incorporate this factor in stability analyses of earth dams. However, the present authors find some of the assumptions made in those publications questionable.

stream face of the dam is against scour and on the possibility of flooding of downstream villages.

5. The three-dimensional nature of the problem which characterizes many earth and rockfill dams. This introduces complications beyond the capabilities of present methods of stability analyses with recognition of nonlinear behavior.[11] It also taxes the ordinary facilities for model testing.

6. The large dimensions of the interface between a major dam and its foundation. We need not consider differential motion of the abutments here; even the lengths of lines in the interface, parallel to the river axis, are so large that seismic waves that may significantly affect the dam's behavior reach the ends of these lines at significantly different times. Again difficulties in analytical and model-testing studies are yet to be surmounted to incorporate this feature.

7. The mass of large earth and rockfill dams is such as significantly to affect the motion of their foundations (Chopra and Perumalswami, 1969).

8. The large expenditure required in order significantly to reduce the probability of failure. As a consequence, apparently small factors of safety are truly justified for these structures, whose failure entails losses of impressive proportions and about whose seismic behavior there is enormous uncertainty. The situation calls for extensive analytical and model studies, coupled with exceptionally careful construction practices and a strong research effort, of which recording of prototype responses to earthquakes must play a major part.

16.7 Strengthening Damaged Structures

Few problems in structural engineering offer a challenge comparable to the one that faces the engineer who must decide whether to patch up a structure damaged by earthquake, to strengthen it and to what extent and in which manner, to demolish it in part, or to condemn it. First, he must estimate the

---

[11] Finite-element techniques have been successfully applied by Clough and Chopra (1966) to the linear analysis of two-dimensional dams (see also Section 4.2) and can be extended to nonlinear behavior. The same can be said of simpler approaches that idealize the dam as rigid-plastic (Bustamante, 1965a) or that introduce some refinements relative to this assumption (Seed and Goodman, 1964). Dams idealized as homogeneous, linearly elastic, three-dimensional bodies bounded at the abutments by rigid vertical planes and resting on a rigid horizontal plane have also been analyzed successfully (Ambraseys, 1960). These analyses are based on simplifying assumptions that introduce appreciable errors, as may be concluded from a comparison with the finite-element analysis of two-dimensional dams (Clough and Chopra, 1966). Still, they are useful in illustrating certain features of three-dimensional behavior. It is found that the influence of the abutments is decisive on the natural modes and periods even when the distance between abutments is as large as five times the height of the dam.

Sec. 16.7 STRENGTHENING DAMAGED STRUCTURES 553

capacity of the structure to withstand gravity, wind, and earthquake forces in its present conditions as well as after the adoption of alternative strengthening solutions. Then, he must weigh the social and economic consequences of the various alternatives. And usually he must reach a decision in a very short time.

An engineer who is confident when predicting the capacity of a future structure on the basis of its drawings and calculations will, paradoxically, experience great uncertainty when confronted with an existing structure, even if he has the drawings before him. He will notice on inspection that cross-sectional dimensions are not in strict accord with those shown on the drawings and that connections are not welded exactly as specified. If he has a chance to verify it, he will find that steel yield-point stresses differ considerably from nominal values; that the same is true of concrete strengths and moduli of elasticity; that the reinforcement is not where one would expect it to be; that even dead loads are markedly different from those assumed in the original design; that live loads bear little resemblance to those specified by the local building code.

Estimating structural capacity will be a difficult task, especially if the structure presents cracks, buckled flanges, or other signs of damage. Things are even more difficult in the absence of structural drawings, and this is not an uncommon situation.

Load tests supply additional information for estimating structural capacity. One must not place much weight on results of these tests, though. For one thing, lateral load tests are nearly always out of the question, and we must be content with tests using vertical loads, so the information on earthquake resistance is most indirect. For another, a severely cracked or otherwise damaged structural member will often withstand practically the same statically applied load as when undamaged, but its energy-absorbing capacity will be appreciably lower. Figure 16.10 illustrates this situation by comparing two schematic load-deformation curves, the shapes of which have been derived from information in Chapter 13 and in some of the references of that chapter. Over a large range of structural properties and earthquake characteristics, energy absorption is more significant for earthquake resistance than for load carrying capacity; accordingly, the damaged member is usually less capable of contributing to the capacity to resist earthquakes. The matter is not brought out in the usual interpretation of load tests. The cost and time involved in performing a large number of load tests further limits their applicability.

Sometimes the crude analyses that can be performed, in the short time usually available before action must be taken and in the light of meager information, will disclose major weaknesses of the structure, consistent at least qualitatively with the damage observed. For example, one may find a pronounced torsion or faulty connection details. Remedial measures will then be relatively obvious. When this is not the case, the best information at hand is the extent of damage undergone by the structure, together with some characterization of the earthquake motion.

Under the circumstances we can employ an almost qualitative type of reasoning. To illustrate it, consider a structure idealized as an elastoplastic single-

554 OTHER TOPICS IN EARTHQUAKE-RESISTANT DESIGN Chap. 16

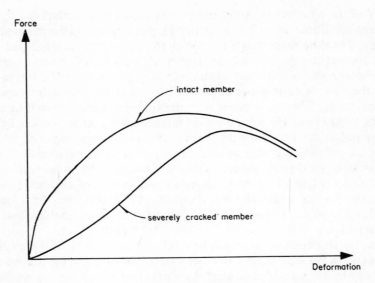

Figure 16.10. Effects of cracking on the force-deformation relation.

degree system. From inspection of the structure, a comparison with the aspect of specimens tested in the laboratory under repeated loading,[12] and a study of the drawings, if available, we can estimate its original natural period and the maximum ductility factor that we can assign to it, taking into account the expected duration of prevailng earthquakes at the site in question. Call this ductility factor μ. From inspection we conclude that the structure developed a ductility factor $\mu' \leq \mu$ and that cumulative damage has reduced the maximum ductility factor for subsequent earthquakes to $\mu'' \leq \mu$. (In many cases, $\mu'' = 1 + \mu - \mu'$.)

Suppose now that the acceleration spectrum associated with the earthquake that caused the damage has a recurrence period θ_1. Suppose that it has the corresponding shape in Fig. 16.11 for a ductility factor of 1. The original condition of the structure would be represented by point A in the figure, whose period is the estimated original natural period of the structure and whose ordinate is $1/\mu'$ times the corresponding ordinate in the curve labeled θ_1. The condition of the damaged structure will be represented by point B. The structure would altogether fail if subjected to an earthquake whose ordinate, for the same period as B, is μ'' times that of point B. The corresponding spectrum would have a recurrence period of, let us say, θ_2 and is identified in the figure accordingly. The actual recurrence period for collapse of the structure can now be estimated as somewhat shorter than θ_2 because of the possibility of additional cumulative damage.

[12] For example, from photographs of reinforced concrete members published by Blume, Newmark, and Corning (1961); of steel members, by Bertero and Popov (1965); of masonry walls by Esteva (1966); and from similar information in other papers.

Sec. 16.7 STRENGTHENING DAMAGED STRUCTURES 555

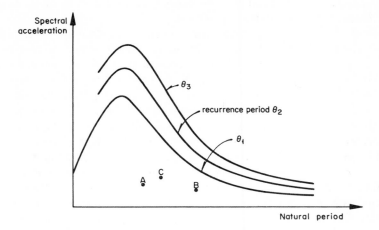

Figure 16.11. Acceleration spectra causing damage and collapse.

With this information we can estimate the social and economic consequences of leaving the structure in its present state, bringing it back to its original condition, strengthening it to a condition represented by, for example, point C in the figure, which could be associated with a larger ductility factor and the corresponding spectrum for a recurrence period θ_3, or condemning the structure. The same type of reasoning can be extended to multidegree systems and to other types of behavior.

If the only information available consists of the general type of structure, the extent of damage, and the regional seismicity, strengthening should preserve the structural solution and its relative rigidities and relative strengths as much as possible; otherwise unknown modifications are introduced which may worsen rather than meliorate the situation.

Whenever we have more information than this, we meet in practice an enormous variety of conditions. Some of the most representative among them will be illustrated with reference to buildings.[13]

The first example concerns a 16-story hotel having a reinforced concrete, moment-resistant frame and unreinforced masonry filler walls. The building is founded on firm rock at a relatively shallow depth. There are active nearby foci, so that most of the important earthquakes to be expected have prevailing periods, as disclosed in acceleration spectra, that are much shorter than the building's fundamental period of vibration.

Under the action of several earthquakes, the strongest of which had an intensity 6 in the MM scale, the building underwent no visible signs of damage. Then an earthquake of intensity 8 produced damage in the elevator machine rooms; diagonal cracking of some and peripheral cracking of all partitions in the longitudinal direction of the building in the 11th, 12th, and 13th stories; and

[13] Based on information contained in the authors' files.

inclined cracking in the beams of the floors that limit a story of small height, which houses the air-conditioning system.

Analysis based on the structural drawings and on information derived from inspection indicated that (1) damage in the structures of the elevator machine rooms could be ascribed to near coincidence of their period of vibration with the building's second natural period, (2) damage to the partitions was mostly associated with vibration in the building's second and third natural modes, (3) the center of torsion in stories 11–13 almost coincided with that of the longitudinal partitions in these stories, and (4) the columns in all stories had considerably greater capacity to resist lateral forces than the beams.

In the light of this information it was decided to (1) reinforce the structure of the machine house by locally repairing damage and adding cross braces in both directions, designed for a horizontal acceleration equal to four times the building's base shear coefficient, (2) repair the damaged partitions and reinforce them with welded wire fabric and a layer of cement mortar, and (3) repair and reinforce the damaged beams by adding vertical diaphragms of reinforced concrete.

In the future the partitions that were strengthened will not develop high deformations beyond their range of linear behavior, but partitions in other stories will surely be damaged. Extensive damage is not expected with ground motions of intensity smaller than that associated with a recurrence period of the order of 25 yr. It is anticipated that the partitions in stories 10 and 14 will crack under earthquakes of intensities whose recurrence periods lie between 25 and 55 yr. Considerably more extensive damage is expected for higher intensities, but the probability of collapse during the next century is insignificant. It is also anticipated that the beams in the floor immediately above those which were strengthened will be damaged during a strong earthquake.

The case described above illustrates a situation in which the engineer feels that he is playing a game of chess. Nature is his opponent. Every earthquake and every set of repairs constitute moves in the game; the engineer must guard against checkmate.[14] Indeed, the decision not to strengthen undamaged partitions is tantamount to a gambit, for this invites severe damage concentrated in the partitions of two stories rather than minor cracking throughout a greater number of stories during future earthquakes. The situation provoked in this manner is desirable especially because we are concerned with a hotel that lends itself better to major repairs and strengthening in two stories rather than minor work of this type in many stories.

Now consider a type of damage that is frequently met in reinforced concrete beams that follows from cutting tensile reinforcement in tension zones. Near the points of cutoff, "diagonal tension cracks" are nearly transverse and develop at an appreciably lower stress than do inclined cracks at other sections along the beam (Ferguson and Matlob, 1959). When the transverse cracks are found following an earthquake and their direct cause is identified, the common

---

[14] The simile was drawn by R. W. Binder, a structural engineer, in Los Angeles, Calif., apropos of this very building.

solutions used for raising the structural capacity consist of increasing the section with reinforced concrete (Fig. 16.12 a or b when there are no limitations on beam depths; Fig. 16.12c where there is such a limitation) or in applying exterior longitudinal prestress (Fig. 16.13).

The second solution is especially interesting. At the same time that longitudinal prestress increases the capacity to resist shear and flexure, and forces new diagonal tension cracks to be inclined, it reduces the beam's ductility. A

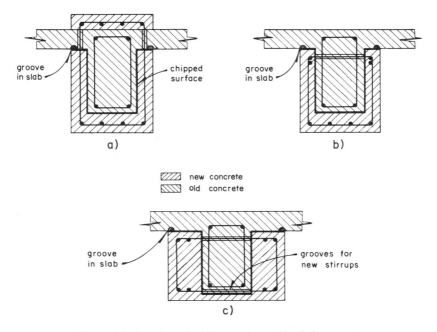

Figure 16.12. Increasing the section and reinforcement of existing beams.

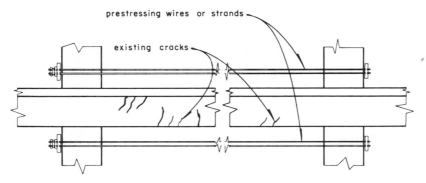

Figure 16.13. Exterior longitudinal prestressing. *After Díaz de Cossío and Martínez (1961).*

large portion of the increment in strength is associated with this change in the direction of cracking, which offsets the effects of having cut longitudinal steel in regions of tension. Figure 16.14 eloquently evinces this trait of longitudinal prestressing: a sufficient amount of it eliminates the difference in capacity between a well-reinforced beam and one whose longitudinal reinforcement has been cut in a region of tension (Díaz de Cossío and Martínez, 1961).

The loss of ductility can be compensated for by providing the beam with prestressed, exterior transverse reinforcement. We may accomplish this through the use of either straps (Lunoe and Willis, 1957) or stirrups [Díaz de Cossío and Martínez (1961); see Fig. 16.15]. The added transverse reinforcement has

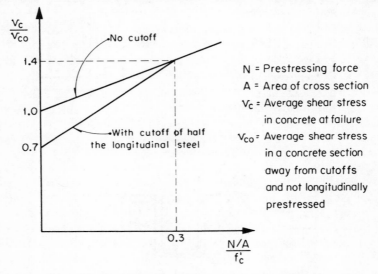

Figure 16.14. Effect of axial force on shear strength of beams with and without cutoffs of longitudinal bars. *After Díaz de Cossío and Martínez (1961).*

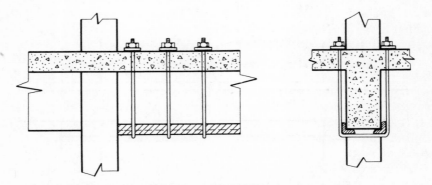

Figure 16.15. Exterior prestressed stirrups.

relatively little effect on the strength of beams that have developed transverse cracks and have not been prestressed longitudinally, although it is effective in raising the ductility of beams with inclined cracks.

The third example concerns a 10-story office building having a steel frame and brick partitions. The partitions, plaster, and other "nonstructural" components underwent extensive cracking during an earthquake of intensity 8 on the MM scale. The direct cause was the near absence of moment resistance in the frame connections. Alternate solutions of strengthening were studied. One suggested steel cross braces capable of taking story torques and shears in both directions at all stories. The other suggested strengthening all beam-column connections and increasing the section moduli of a few beams that had been underdesigned for lateral force.

In this case, rigidity was an asset because of the great length of locally prevailing ground periods and because rigidity would tend to reduce nonstructural damage in future ground motions. For a given base shear coefficient, the second solution was undoubtedly the more expensive and less rigid one. Yet it was chosen because it provided greater ductility of the frame and a greater possibility of changing office space distribution. Another factor that influenced the decision was that the second solution caused smaller axial forces in columns due to overturning moments, and there was much uncertainty about the conditions of the building's foundation. Still, the decisive considerations were based on the relatively short recurrence times associated with earthquake intensities capable of endangering the stability of this building. By designing the strengthening with this factor in mind, rather than with the possibility of damage to nonstructural members, the latter factor was not entirely well taken care of. The situation became evident five years later, when two earthquakes of intensities between 6 and 7 caused some damage to plaster and, although quite minor, to partitions as well. To compensate for this shortcoming, most of the partitions were then covered with a plastic finish and the remainder with timber.

The last example refers to a nine-story building at a site where prevailing ground periods are of the order of 2.5 sec. Earthquake damage was so extensive that a proper increase in column sizes would have entailed a serious sacrifice of functionality. By demolishing the uppermost two stories the required additions in column sizes became reasonable. This reduction in the mass of the building brought a reduction in the length of the natural periods of vibration and hence in the base shear coefficient; this made the measure doubly effective.

The study of means for strengthening a structure damaged by earthquake makes obvious many of the apparently insignificant defects in design and construction that otherwise escape unnoticed. Temptation then usually overcomes the engineer so that he ascribes all the damage that meets the eye to these details. Yet, in most cases extensive damage is really due to a gross oversight or mistake, or to several failures in design or construction. And the local, very noticeable effects are part of a syndrome.

The foregoing assertion rests on the fact that the maximum deformations of a pronouncedly nonlinear system depend little on the details of the force-deformation curve, unless the structure is exceptionally rigid. For most structures the capacity to resist earthquakes is almost synonymous with the capacity to deform.

Lastly, to undergo extensive damage may be quite the proper thing for a structure to do. Optimum design often calls for savings in initial investment at the sacrifice of the cost of failure. The engineer must guard against his tendency to overdesign the strengthening of a structure, since he may well be overly influenced by the sight of serious damage.

The solutions adopted in the strengthening of a diversity of structures are described in papers by Krishna and Chandra (1969); Arias, Arze, and Beuza (1969); Churayan and Djabua (1969); Arze (1969); and Rasskazovsky and Abdurashidov (1969).

Temporary shoring of earthquake-damaged structures constitutes a special case of strengthening. Since shoring calls for immediate action, its design must rest on the crudest analyses. For example, a structure may be so badly damaged that it looks to the experienced engineer as if it were on the verge of collapse. Probably the best estimate that he can make of the structure's capacity to withstand gravity forces is that it has a load factor insignificantly greater than 1, and that—under sustained loading and the weakening effects of subsequent seisms and of the redistribution of stresses called for by the shoring that he is designing—the existing structure may be unable to resist more than, perhaps, 70 percent of the dead load. He will, therefore, design the shoring for 30 percent of the same load, to be resisted for a period of several months. In this design he will be conscious that the consequences of collapse would be less important than under more common circumstances but, also, that the quality of execution will lie below ordinary standards.

He would also design to resist lateral and vertical forces due to wind and earthquake. He will find it difficult to decide on the earthquake intensities that he should foresee. True, the expected life span of the shoring is short, but seismicity during that period will be much higher than it is on the average, for there is the near certainty of aftershocks. (Put in other terms, when dealing with such short time intervals the assumption is untenable that earthquake occurrence is a generalized Poisson process.) Still, we can guide our judgment on this matter by establishing two bounds, however wide apart they may be. Knowing that the seismicity will be much higher than it is on the average already fixes a lower bound. At the same time, the probability is high that an earthquake of intensity slightly smaller than that of the one that caused the damage will occur in the near future, and the probability is quite small that the maximum intensity of ground motions in that interval will greatly exceed that of the last strong earthquake if the latter's recurrence period is at least of the order of several decades.

Once the intensities have been chosen, only very gross effects can be taken into consideration for the design of the shoring. This does not mean that the engineer is entitled to disregard such matters as story torques and overturning moments. It does mean that he may have to make some calculations without the benefit

16.8 Protection against Tsunamis

In Section 6.5 we spoke about the prediction of the amplitude of tsunamis. Here we shall describe some of the measures that have been successful in limiting the destruction wrought by these phenomena.

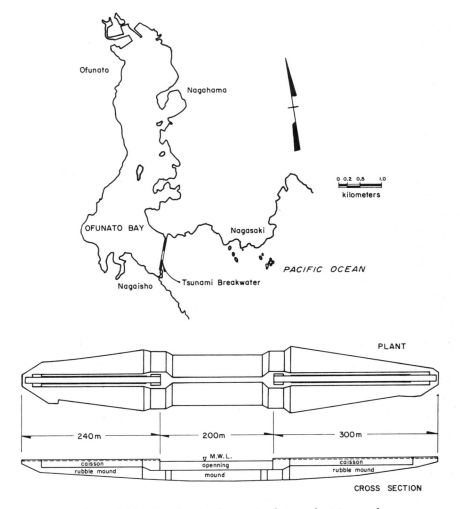

Figure 16.16. Breakwater for protection against tsunami. *After Horikawa.*

An international warning system is the most economical means for mitigating tsunami effects on lives, although it does not protect property at all. We have seen that it takes several hours for a tsunami to traverse the Pacific Ocean. Because of the time involved it has been possible for the coastal inhabitants of, for example, Hawaii and Japan to receive timely notice of a tsunami originating in the coasts of Chile, and even an idea of its magnitude, so that many lives have been saved by evacuation of towns bordering the seashore.[15]

Construction of breakwaters, illustrated in Fig. 16.16, and planting and densification of shrubbery have sometimes proved effective in nullifying or mitigating the effects of "tidal waves."

Since the amplitude of oscillations in estuaries is a function of their natural periods of vibration, measures that tend to change the latter favorably have also been effective. This has been particularly true when these measures have been accompanied by the construction of obstacles to the entrance of water into the basin, which at the same time introduce some degree of damping (Horikawa).

16.9 Urban Planning and Relocation

When laying out the master plan for a new city or for the growth of an existing one, considerations of seismicity should play an important part. The problem is one of operations research. In its simplest terms it is reduced to a comparison between the cost of building structures to resist a given set of earthquake intensities[16] and the present value of the cost of building in one site rather than in another. The possibility of soil liquefaction is one of the most critical factors to consider in these studies.[17] In reality the problem is very complex: there are questions of esthetics, of transportation under normal conditions and of transportation immediately following a strong earthquake, of hospital services under both sets of circumstances, of gas piping and the possible propagation of fire, and so on.

In coastal towns where the danger of tsunamis can be stated quantitatively, the problem is more clear-cut. But, in any case, progress in this branch of earthquake engineering can be expected to come as part of the advances in the application of operations research to city planning.

History abounds with instances of abandonment of an entire city and its rebuilding at a nearby site. Ignorance of regional and microregional seismicity and of social factors has made most of these decisions unwise. For example,

[15] For a description of the Pacific warning system see U.S. Dept. of Commerce, *U.S. Coast and Geodetic Survey*, 1965.

[16] Simplified, comparative designs of typical structures furnish the required basis.

[17] Close and McCormick (1922) give an account of the earthquake that occurred at 8:09 P.M. Shanghai time, on 16 December 1920, in Kansu Province, China. The earthquake destroyed 100,000 lives. It caused large land displacements in an area of 100 by 300 miles. In some places roads were carried $\frac{3}{4}$ mi.

partial destruction of Antigua, former capital of Central America, in 1773, led to the founding of the city of Guatemala, 44 km away. Today the ruins of Antigua are beautifully preserved; they have not suffered the effects of new, very strong motions, while Guatemala City has been frequently struck and harmed by such mishaps.

In 1854 the capital of El Salvador was almost destroyed by an earthquake. It was moved to "New San Salvador," in the town of Santa Tecla, 11 km away. The new site had a slightly lower seismicity, but the displacement of the capital was unsuccessful. Social forces converged to its former site, where it took hold anew in a matter of a few years.

Progress is being made in the area of decisions connected with relocation following earthquakes and will undoubtedly extend to urban planning, but the discipline must remain somewhat intuitive in view of the need for rapid decisions under emotional pressure.

Four synchronized vibration generators exciting an earth-filled dam. (From Keightley, 1966).

17

TESTS AND OBSERVATIONS

17.1 Introductory Note

In earthquake engineering it is of interest to record earthquake motions in terms of their relevance to the effects of these motions on structures in the area. The ground undergoes strain during an earthquake. Hence, a relatively complete description of the ground motion during an earthquake requires synchronized records at a number of points, and in three orthogonal directions at each point. The records should be obtained for sufficiently high-intensity earthquakes so that they have significant effects on structures, and the time resolutions should be adequate. The last two requirements lead to the use of recording instruments that differ from those used by seismologists. The seismological instruments are intended to record many small quakes and have a much poorer time resolution than that required in engineering as well as little ability to record exceptionally strong motions.

The cost of such sophisticated sets of instruments is high. The more versatile instruments can only be afforded at a few sites, and interconnected groups of them are even rarer. Those instruments must therefore be supplemented with large numbers of inexpensive *seismoscopes*, if information is to be gathered over an extensive area. Seismoscopes render a smaller amount of information than the more sophisticated instruments, but the relation between their records and some spectral ordinates is more direct and trustworthy. Thus, aside from covering a wide region, seismoscopes furnish a check on the records of the more expensive instruments.

Presence of a structure modifies the ground motions. In aiming to record these ground motions as they directly affect a structure, we install instruments in the structure itself. From one point of view our records then represent the disturbance; from another they constitute structural responses.

Tests of models and prototypes either tend to establish the capacity of the structure to resist certain earthquakes or attempt to determine structural characteristics that will permit computing responses to earthquakes and hence the capacity to resist these motions.

Tests having the first objective can in practice only be comprehensive when performed on models because the detailed characteristics of an earthquake are unknown before it occurs. Hence, the objective is realized in a relatively complete way only when the effects of a wide class of probable ground motions are explored. Considerations of the time and cost involved preclude comprehensive tests to be performed on prototypes. Moreover, although a satisfactory test result does not usually imply destruction, it often involves damage. Despite these limitations, some prototypes must be tested under simulated earthquake conditions, however poor the simulation may be, for one does encounter structures about which practically nothing is known. This is especially true of earthquake-damaged structures.

Matters become simple when we adopt the assumption of linear behavior. It suffices then to test the model or prototype under steady-state oscillations of arbitrary amplitude. This should be done for a sufficient number of frequencies or force configurations to allow determining the inertia, stiffness, and damping matrices. (See Appendix 1 for procedures to accomplish this calculation.) Free vibrations or other disturbances may also be used for the purpose.

The number of parameters to be established is greater for nonlinear, hysteretic structures. The corresponding tests necessarily involve excitation at various amplitudes of vibration. And definition of the pertinent parameters for structures that may deteriorate demand much more extensive testing.

In general, structural behavior depends on the excitation. Hence, there is advantage in placing instruments in structures that are expected to undergo strong seismic disturbances. The most realistic type of testing structures coincides with the most realistic manner of observing earthquake motions.

In all model testing, a question of paramount importance concerns dimensional similarity. We shall devote some space to this matter. Measuring hydrodynamic phenomena requires special techniques and involves special difficulties. These questions will be treated separately.

17.2 Seismoscopes

A seismoscope is a device that furnishes a single point in a linear spectrum of whichever earthquake excites it. It is essentially a single-degree linear system that supplies a record of the amplitude of its displacement relative to the ground. The most common means for obtaining the record uses a stylus scratching on smoked glass. This introduces a small amount of nonlinear damping. The main contribution to damping is attained by having a metallic portion of the oscillating system move within a field produced by permanent magnets. This damping is viscous. A typical seismoscope of this class is shown in Fig. 17.1.[1] Such seismo-

---

[1] Ulloa and Prince (1965) give a more complete description.

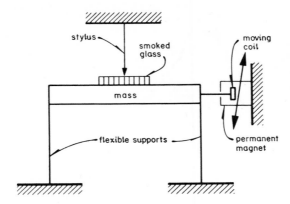

Figure 17.1. Seismoscope having a single degree of freedom.

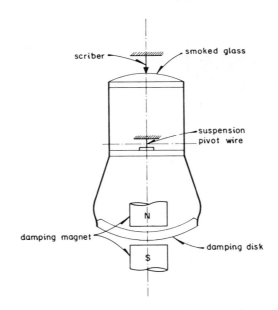

Figure 17.2. Schematic diagram of the Wilmot instrument.
After Hudson (1958).

scopes have been built and have operated successfully for natural periods of 0.2 to 2.0 sec.

A common variation, which gives the amplitudes of response in all horizontal directions, is seen in Fig. 17.2. Its natural period is 0.75 sec. By changing the dimensions or masses, other natural periods can be attained. A typical record obtained with such an instrument is shown in Fig. 17.3. (One drawback of this type lies in the practically inevitable coupling between modes of vibration in orthogonal directions.) Many other types of seismoscopes have been proposed

568 TESTS AND OBSERVATIONS Chap. 17

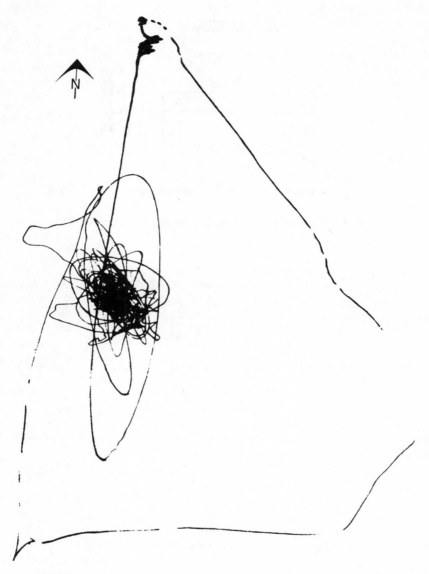

Figure 17.3. Typical record obtained with the Wilmot instrument. *Courtesy of M. Schultz.*

or used: a sphere rolling over a curved, smoked glass (Prince, 1963), cantilever beams, U-shaped tubes filled with mercury, and so on. Practical versions must be reliable and inexpensive.

In areas where information is required on local shapes of strong-motion spectra, one must install groups of seismoscopes. In each of two horizontal, orthogonal directions there should be several instruments, each with a different natural period. Where the vertical component of ground motion is of interest, vertical seismoscopes should also be installed.

Ordinarily, over a large area it is practical to place one or several accelerographs and to supplement them with a large number of seismoscopes. It is advisable that some of the seismoscopes lie alongside the accelerographs. This permits checking the spectra derived from accelerograms and gives a basis for interpreting the records of the other seismoscopes.

17.3 Accelerographs

Most strong-motion instruments in use, capable of recording ground motion as a function of time, have three, single-degree linear elements. These give graphs in the vertical and two mutually perpendicular horizontal directions. The natural periods of the elements usually lie in the range of 0.05 and 0.10 sec. The percentages of critical damping lie between 50 and 100 percent. The records very nearly represent ground accelerations for components of ground motion having periods not smaller than about one half the natural periods of the elements (see Chapter 1). The phase shifts in these ranges are practically independent of period. Most of the natural periods of interest are therefore adequately covered.

It is not a serious drawback that the natural periods of the systems somewhat exceed the value required for a flat acceleration response curve in some instances, provided that their behavior is linear with sufficient accuracy. The situation is dealt with by processing the record in such a way as to reconstruct the ground acceleration (see Problem 1.2). Approximately viscous damping is achieved either by using fins that oppose the motion of the individual elements relative to the surrounding air or by means of a magnetic field.

Records are often obtained by scratching paraffin-coated rolls of paper with a stylus or by an optical procedure. Both methods are objectionable in that the zero mark is not strictly fixed. Matters improve by tracing such a mark simultaneously with the record. A better method consists of recording on magnetic tape by using frequency modulation. This has the added advantage of ease in processing the record, since it can be fed to an analog computer or digitized for use in digital computers.

High resolution in time calls for a high speed of the paper or tape. Speeds of 1 to 2 cm/sec are common. Consequently it is impractical to obtain continuous records. Commonly used accelerographs have one or two starting devices. These operate when the ground motion has become sufficiently perceptible. The recording mechanism stops from 0.5 to 3.0 min after starting or several seconds after the starter's last contact and is ready to begin recording again.

A supplementary circuit is advisable; this will stop the recording device after a number of seconds of continuous contact of the starter. It will also prevent the instrument from continuing to record if it has tilted, in which case it would run out of paper or tape.

The starter must be adjusted so that it does not set the instrument recording under too small a disturbance (e.g., vibrations due to traffic). Ordinarily the limit is fixed so that the accelerograph altogether misses earthquakes weaker than intensity 5 or 6 in the MM scale.

Loss of the initial portions of records, due to the limited sensitivity of the starting mechanism, is an objectionable feature of present, commercially available, strong-motion instruments. The matter is critical for moderate earthquakes. For these motions it may imply loss of an important section of the record.

The combined consequences of loss of the initial portion of records and shifts of the zero acceleration reference are particularly noticeable in the computed, long-period spectral ordinates and computed ground displacements. Usually the directly computed maximum ground displacements exceed reasonable bounds, some times reaching several meters far away from the focus. Various criteria have been used to correct for these matters. No method is entirely justifiable nor satisfactory. One of these methods introduces fictitious acceleration pulses prior to the beginning of the record obtained and moves the base line wherever there are indications that this is proper (Sheth, 1959). Another criterion introduces a parabolic correction to the base line, so as to make the end velocity zero and minimize the time integral of squared velocities (Berg and Housner, 1961).

If we are concerned only with spectra in the range of relatively short periods, it does not matter very much which method we adopt to correct the record. Even the uncorrected record will be satisfactory in this range, unless a long initial portion should have been lost. But for long natural periods and for calculation of the responses of strongly nonlinear systems, the question becomes critical. It is not a purely academic one, for some structures (e.g., suspension bridges) are known to have natural periods in excess of 15 sec. In this range no trustworthy, strong-motion spectra are available.

Rather than attempting to develop better corrective criteria, efforts should obviously be directed to obtaining complete and correct records.

Time markings on accelerograms are important, since little faith can be placed on the velocity of an unwinding roll. This aspect of the problem also merits additional attention, as it has been shown that significant errors can derive from inaccurate reading of the time scale (Berg, 1963).

Peters (1968) has reported on test results to evaluate the records obtained with accelerographs in current use. Aside from the sources of error we have mentioned, he finds that the dynamic characteristics of these instruments introduce minor nonlinearities and some spurious, transient, high-frequency vibrations in the records.

Much can be said about the optimum location of accelerographs. In a city, accelerographs should be installed at every location where they record ground motions at the surface of every specific type of soil formation, if this is possible. Zones near the boundaries between various types of ground are also worth monitoring.

In order to have information applicable to the design of buried structures and of those with deep-seated foundations, buried accelerographs are desirable in some cases. Records supplied by these instruments shed light on seismic wave filtering through various soil formations. And study of soil-structure interaction is aided by the installation of accelerographs in the foundations of existing buildings and at nearby sites.

To obtain information on seismic-induced soil deformations and rotational components of motion (tilt and twist), accelerographs must be placed at several points of the ground surface and must be interconnected so that they have common time markings. Instruments are also available which respond to rotational accelerations. These accelerometers can be combined of interconnecting accelerographs.

Study of the accelerations undergone by structures at various points of the structure can be based on accelerograms. Preferably the instruments to record them should be interconnected for common time markings. The accelerations recorded are the forcing functions on the structure; they are also the structure's responses to the ground motion. This use of accelerographs will be dealt with in Section 17.5.

17.4 Other Ground-Motion Recording Instruments

Among the primitive means for estimating maximum ground acceleration, the most commonly used was the observation of reed gages and rigid blocks, such as stone columns. By noticing which blocks were unaffected by an earthquake, which were moved by it, and which tumbled over, an estimate of maximum ground acceleration could be made. The estimate was necessarily crude, as the laws governing the motion of rigid blocks are strongly nonlinear, and the behavior of these objects is difficult to interpret. The same criticism applies to the estimates based on the response of reed gages, for these instruments ordinarily consist of a series of prisms having different slenderness ratios. Modern reed gages are actually series of one-dimensional seismoscopes having different natural periods.

Traditional seismographs consist of essentially single-degree linear systems having a long natural period—of the order of several seconds. The records they produce are practially proportional to ground displacement for components of ground motion having medium to high frequency (see Chapter 1). These records are useful for obtaining response spectra in the range of long natural periods.

Use of instruments providing records practically proportional to ground velocity would have an important advantage: response spectra based on these records would be dependable over a range of natural periods wider than when using the more traditional instruments. This would apply to nonlinear spectra as well, since in the range of greatest interest, strongly nonlinear structures are more sensitive to energy—and hence to ground velocity—than to ground acceleration or displacement.

17.5 Measurement of Structural Characteristics

Ideally, for earthquake-resistant design we would measure dynamic behavior of structures up to the point of collapse. This rarely can be done, at least under controlled testing conditions. For one thing it is intrinsic to civil engineering that every system should present problems of its own, and this is especially true in dynamic behavior outside the linear range because of the large number of variables involved, especially when the soil may participate to an appreciable degree. For another, the equipment required for such testing is excessively expensive. Consequently, one usually cannot aim beyond, or at least not much beyond, the determination of natural modes and periods of vibration and percentages of damping within the range of near linear behavior.

In the following paragraphs we shall take it for granted, unless we state otherwise, that we idealize structural behavior within the range of small oscillations as linear and as though the structure had natural modes of vibration in the classical sense.

Rather coarse equipment suffices when the purpose is merely to establish the fundamental or the first few natural modes and periods of a structure in which these periods are well differentiated. Such is the case with many a building when it is not specifically desired to determine its percentages of damping. Commonly used methods of exciting vibrations under these circumstances include the following.

1. Taking advantage of wind or traffic disturbances. Typical records (Fig. 17.4) permit an accurate estimate of the first natural period of translational vibrations in two orthogonal directions and crude estimates of the first or first two harmonics in both directions. Exceptionally, in very flexible structures the record shows several well defined harmonics. Although simple inspection yields considerable information, a harmonic analysis discloses, with greater accuracy, a larger number of natural periods. Conceivably, harmonic analysis of records obtained simultaneously at various locations of a structure excited in this manner should yield the shapes of the first few natural modes of vibration.

2. Taking advantage of nearby explosions or actual earthquakes. By installing instruments on the ground near the structure we can obtain precise information about the disturbing motion. If the structure is profusely instrumented, if the motion is recorded from its very beginning, and if the instruments have accurately synchronized time markings, we can obtain very complete information on the structural characteristics. For motions within the linear

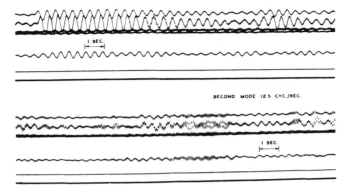

Figure 17.4. Sample records of wind-excited tower vibrations. *After Keightley, Housner, and Hudson (1961).*

range of structural behavior this information allows us to compute all the elements in the stiffness and damping matrices provided we know those in the inertia matrix, even if the system has no classical natural modes. If, however, our interest is confined to the natural frequencies, assuming that these exist in the classical sense, we need only record part of the structure's vibrations, omitting ground instrumentation, and proceed as in the foregoing paragraph. When the disturbance to be recorded consists of one or more explosions, such as those detonated for rock exploitation, the drums of the recording instruments can be set in motion shortly before the disturbance begins and there is no need for starters. The experiment can be completed in a short time. Much more patience is required when recording earthquakes.

3. Producing steady-state forced vibrations. The centrifugal force of either a single- or a double-arm eccentric vibrator may be used. In double-arm vibrators the arms rotate in opposite senses, so that the centrifugal forces compound along one direction and cancel each other at right angles thereto. By running the vibrator at a number of reasonably constant frequencies and measuring the structure's responses we obtain information with which to draw a resonance curve for excitation force proportional to the square of the excitation frequency. Interpretation of such a curve furnishes relatively accurate values of the natural periods that are well within the range of the test. But, unless the excitation frequency is kept constant within each run with extremely high accuracy, the percentages of damping derived from these curves tend to be much too high. The shapes of natural modes can be obtained by measuring the amplitudes of vibration at several locations, preferably through simultaneous recordings. In this manner it is possible to define, for example, the torsional modes (Fig. 17.5a) and vibration modes for a building that involve large flexural deformations of floor diaphragms in their own plane (Fig. 17.5b).

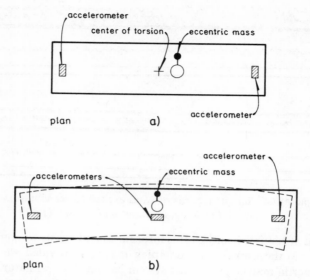

Figure 17.5. Use of several accelerometers to define natural modes involving unequal slab displacements in buildings.

4. Running a vibrator to a relatively high frequency and cutting off its motor. As the excitation frequency diminishes, it may pass through one or more of the system's natural frequencies. The response amplitude shows then one or more maximums that occur roughly at the natural frequencies. The record in Fig. 17.6 is typical of those maximums obtained in this fashion. The method is quick, and its accuracy can be improved by using certain analytical results based on the assumption that the frequency of excitation decays according to a simple rule (Keightley, Housner, and Hudson, 1961).

5. Setting up free vibrations. For this purpose, the structure must be given an initial disturbance, be it a set of displacements, velocities, or both. This is accomplished in a variety of ways. For example, in small, flexible constructions one or two persons can sway synchronously with the structure's oscillations (keeping time by watching the needle of a vibration meter or an oscilloscope); when sufficiently large amplitudes are attained, the persons stop swaying and the structure's decaying oscillations are recorded (Hudson, Keightley, and Nielsen, 1964). The record in Fig. 17.7, which shows the oscillations of a concrete tower, was obtained in this manner. In more rigid or massive structures it may be necessary to resort to the firing of rockets (Hudson, 1961b) or to pulling the structure by means of cables that are released suddenly (Del Valle and Prince, 1965). In every case a direct measurement of the fundamental period can be done on the record. When more violent means of excitation are used, one may also derive a good approximation to some higher natural periods, taking care to eliminate or ignore very high-frequency initial vibrations of local character. The shapes of the corresponding natural modes may also be

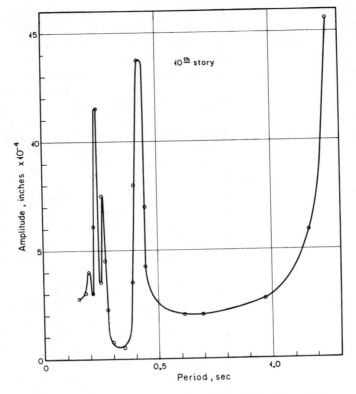

Figure 17.6. Resonance curve from forced vibration of the Alexander Building. *After Blume (1956).*

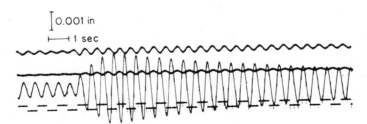

Figure 17.7. "Man-excited" vibrations of 150 ft high concrete intake tower of dam. *After Hudson, Keightley, and Nielsen (1964).*

derived from successive or simultaneous records at various points of the structure, along the lines of testing used by the aircraft industry.

Free vibrations can be used to determine damping provided vibration in the natural mode of interest can be cleaned of components in other modes, either by ignoring the first portion of the record or by proper filtering.

Damping ratios derived from forced vibration tests are systematically too high, unless the frequency of excitation is controlled with extreme accuracy. This is difficult to achieve, particularly due to instability problems arising in frequency ranges where the response amplitude is a rapidly decreasing function of frequency. The difficulty arises from oscillation of the support of the eccentric mass. To overcome it, one must normally use servomechanisms that ensure tolerances considerably better than 1: 1000 in the frequency of the disturbing force (Hudson, 1962b). Using the relations in Section 1.3 it is then possible to compute the damping ratios in several natural modes from the curve of response amplitude vs. frequency. The results obtained in this manner are usually quite accurate, except when two or more natural frequencies lie close together, in which case it is advisable to correct the results.

In order to excite higher modes it has sometimes been necessary to use several synchronized vibrators placed where they produce maximum or near maximum effects in the mode of interest (Hudson, 1962b). Such refined procedures are necessary even when it is only desired to determine natural frequencies of vibration if these lie close together. This is the case with many types of dams, for example.

Despite the limitations imposed by the inaccuracies of the older methods of producing forced vibrations, important strides were made when those methods were used carefully and were supplemented with analytical studies. The reconciliation of measured and computed periods of the Alexander Building in San Francisco stands foremost among examples of this type of work (Blume, 1956; Blume and Binder, 1960).

The practice of running a vibrator to a high frequency and letting it coast down to rest was at one time used for constructing nearly continuous curves of response as a function of the frequency of excitation under the erroneous assumption that the amplitude attained at each instant equaled the steady-state amplitude for the frequency of excitation considered. The percentages of damping derived from these curves grossly overestimated the actual damping ratios, especially when the latter were small. For this reason many reports of typical percentages of damping, particularly in buildings, recorded before 1960 are suspect. (The same criticism applies, albeit to a lesser degree, to values of damping ratios based on supposedly steady-state excitation in which frequency was not controlled with extreme accuracy.) Records obtained from run-down tests can be used in principle to estimate damping ratios if the nature of the excitation is explicitly taken into account, as pointed out previously, but there is little experience on the accuracy of such calculations.

Most of the methods described that have the purpose of exciting structural vibrations are limited to small oscillations. Tests that take the structure frankly out of the linear range are expensive because of the equipment required, the damage wrought to the structure, and the large number of tests imposed by the high number of variables.

There are exceptions to the preceding statement. First, the information furnished by records in adequately instrumented structures when these undergo a strong earthquake is ideally suited for purposes of earthquake-resistant design. It is relatively inexpensive to obtain. Still, the data are difficult to interpret when the purpose is to derive all the characteristics of structural behavior. And certainly one cannot be content with this sort of information; even in the seismically most intensively active zones, large inelastic excursions must be infrequent.

It is expedient to sacrifice some of the information that can be derived from records of structural responses to earthquakes. The advisable approach consists of proceeding by trial and error. Presumably we have a good approximation of the inertia matrix in most cases. By trying various stiffness and damping matrices, or percentages of damping in the first few natural modes, we can adjust the computed to the recorded responses. Outside the linear range, we can use various idealizations of the force-deformation relations or establish reduction factors relative to the responses that we would obtain if behavior continued to be linear indefinitely. The loss of information is not objectionable if we adjust with appropriate accuracy in the range of greatest interest from the design standpoint.

In this same vein, rather than recording accelerograms, there are often advantages in cost and in simplicity of interpretation if we record certain responses that are more directly relevant to design. In multistory buildings this is particularly well achieved by resorting to devices, such as the one sketched in Fig. 17.8 that record relative displacements between consecutive floors (Zeevaert and Newmark, 1956). The record may be obtained on a drum provided with a clock mechanism that rotates at the rate of one turn per week. The rate is so slow that an entire earthquake reduces to a single line furnishing only the maximum relative displacements. Yet this arrangement may find favor because of its simplicity and because it also gives direct information about the daily deformations undergone by the building as it responds to uneven heating from the sun.

Time records of relative displacements may be obtained by using starter

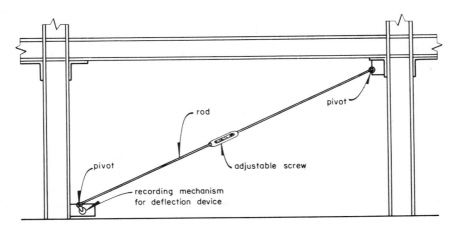

Figure 17.8. Relative deflection device. *After Zeevaert and Newmark (1956).*

mechanisms. Interconnection of these devices as well as accelerographs placed at various points of the building furnishes records with synchronized markings. Scratch gages are particularly useful in view of their very low cost and the trustworthy readings they give of maximum local strains in the structure.

There is a second set of circumstances in which the difficulties are not unsurmountable in testing prototypes considerably outside the range of linear behavior. This is the case with small structures when many of them are to be built using the same set of drawings, as with some low-cost housing projects (see Ishigaki and Hatakeyama, 1960; Ihara and Ueda, 1965).

When we are willing to use indirect information on the effects of the speed of loading, static testing of small structures and of models lends itself to the study of behavior outside the linear range. The structure or model can usually be tested even to failure at a relatively low cost either monotonically or in alternating loading (see Nakano, 1965).

17.6 Dynamic Testing of Models

If a model test is to reflect accurately the behavior of a prototype, certain dimensional requirements must be satisfied (Hudson, 1961a). Let λ denote the scale, that is, the factor by which one dimension in the model must be multiplied to obtain the corresponding dimension in the prototype. Also, let subscript l refer to linear dimensions, t to times, and f to forces.

TABLE 17.1. SCALES IN DYNAMIC MODELS

Variable	Any acceleration and density scales	Unit acceleration scale; any density scale	Unit acceleration and density scales	Unit velocity and density scales
Displacement	λ_l	λ_l	λ_l	λ_l
Velocity	$\lambda_l \lambda_t^{-1}$	$\lambda_l^{1/2}$	$\lambda_l^{1/2}$	1
Acceleration	$\lambda_l \lambda_t^{-2}$	1	1	λ_l^{-1}
Unit strain, Poisson's ratio	1	1	1	1
Stress, moduli of elasticity	$\lambda_f \lambda_l^{-2}$	$\lambda_f \lambda_l^{-2}$	λ_l	1
Weight per unit volume	$\lambda_f \lambda_l^{-3}$	$\lambda_f \lambda_l^{-3}$	1	1
Mass per unit volume	$\lambda_f \lambda_l^{-4} \lambda_t^2$	$\lambda_f \lambda_l^{-3}$	1	1
Time, (natural) period	λ_t	$\lambda_l^{1/2}$	$\lambda_l^{1/2}$	λ_l

Given λ_l, λ_t, and λ_f, the scales of all other variables of interest can be obtained. Those for the most usual variables are written in the second column of Table 17.1. (Some scales are independent of the three λ's, as is the case with unit strain and Poisson's ratio.)

Ordinarily in dynamic models, the accelerations equal those of the prototype,

Sec. 17.6   DYNAMIC TESTING OF MODELS   579

since it is not simple to change the acceleration of gravity. The third column in Table 17.1 gives the scales corresponding to this condition, which implies $\lambda_l \lambda_t^{-2} = 1$. If, besides, the model is made of materials having the same unit weights as those of the prototype or is made of the same materials, we arrive at the scales in the fourth column of the table that is derived from the conditions $\lambda_l \lambda_t^{-2} = 1$ and $\lambda_f \lambda_l^{-3} = 1$. The last column corresponds to a convenient criterion of modeling scale. Gravity effects must be simulated separately. This practice makes it easy to model relatively small seismic accelerations.

Usually the unit weights of materials in the model are the same or approximately the same as those in the prototype. Often, when it is desired to test outside the linear range, this imposes requirements difficult to fulfill concerning material strength, because the scale of strengths must be the same as the linear scale. When the latter is large, one must use very weak materials in the model. For example, correct modeling of a rockfill dam in one instance required use of a material in the impervious core, having a cohesion of 5.8 psf (2.8 gr/cm²) and no angle of internal friction (Clough and Pirtz, 1958). Such conditions demand special techniques. Modeling granular materials, with adequate simulation of grain breakage, involves use of weak materials constituting grains whose shapes are statistically similar to those of the prototype material. Mixtures of plaster and lead oxide have furnished a satisfactory material from this point of view (Nieto and Díaz, 1969).

Use of the same materials in the model as in the prototype makes it impossible to comply with the scale for moduli of elasticity. Within the linear range of material behavior this objection is not serious, provided deformations are not so large in the prototype as to introduce appreciable geometric nonlinearity. To generalize, suppose that the scales for moduli of elasticity and for unit weight are λ_e and λ_γ rather than λ_l and 1. It is desirable to have the same unit strains in the model as in the prototype, so that the displacements be scaled as λ_l. The force scale, say λ_f, must then be such that

$$1 = \frac{\lambda_f / \lambda_l^2}{\lambda_e}$$

or $\lambda_f = \lambda_e \lambda_l^2$. Hence, the scale of unit weights is λ_e / λ_l. If we wish to model deformations due to gravity loads, the scale of accelerations must be 1 as before and those of velocities and of time must be $\lambda_l^{1/2}$. When we are not concerned about the deformations, λ_t may be arbitrary, and we would use the first three scales in the second column of Table 17.1.

Some of the difficulties involved in the dynamic testing of structures influenced by the vibration of liquids were mentioned in Chapter 6. These arise when the liquid may undergo appreciable volume changes in the prototype. Special care must also be exercised when vibration will cause the liquid to flow through small orifices or porous media because the Reynolds and Froude numbers may then play important roles.

Testing of models within the linear range of behavior is justified only for very complex structures, such as a rockfill dam resting on an irregular rock formation.

Ordinarily a numerical model of the structure can be formulated and analyzed with better accuracy, at a lower cost and in a shorter time. Most of the practical applications of dynamic model testing are therefore restricted to cases in which long excursions into nonlinear behavior or even collapse are of interest.

APPENDIX 1: CALCULATION OF DYNAMIC PARAMETERS FROM NATURAL MODES OF VIBRATION

Results of dynamic tests on a system of linear behavior are considered interpreted completely if one has found the inertia, stiffness, and damping matrices of the system. This can be accomplished starting from the system's natural modes as derived from either free or forced vibrations.[1]

Suppose we have measured the shapes z_n of r of a system's N natural modes of vibration, as well as the corresponding damping coefficients, ζ_n, and undamped natural circular frequencies ω_n. According to the theory in Chapter 2 the following relation must be satisfied,

$$(\omega_n^2 \mathbf{M} - \mathbf{K})\mathbf{z}_n = 0 \qquad (A1.1)$$

where $n = 1, 2, \ldots, r$, $\mathbf{M}$ is the inertia matrix, and $\mathbf{K}$ is the stiffness matrix. Orthogonality of natural modes requires that

$$\mathbf{z}_m \mathbf{M} \mathbf{z}_n^T = 0 \qquad m, n = 1, 2, \ldots, r \qquad m \neq n \qquad (A1.2)$$

Finally, the coefficients are symmetric

$$M_{ij} = M_{ji}$$
$$C_{ij} = C_{ji} \qquad (A1.3)$$

and

$$K_{ij} = K_{ji}$$

where $\{M_{ij}\} = \mathbf{M}$, $\{K_{ij}\} = \mathbf{K}$, and $\{C_{ij}\} = \mathbf{C}$ is the damping matrix.

Matrices $\mathbf{M}$ and $\mathbf{K}$ are positive definite, and $\mathbf{C}$ is nonnegative definite. Therefore,

$$M_{ii} > 0$$
$$C_{ii} \geq 0 \qquad (A1.4)$$

and

$$K_{ii} > 0$$

[1] The method presented here is taken from the paper by Berg (1962).

Once we have solved for **M** in Eqs. A1.1 and A1.2, we can solve for **C** from
$$(2\zeta_n \omega_n \mathbf{M} - \mathbf{C})\mathbf{z}_n = 0 \tag{A1.5}$$

Equation A1.3 reduces the number of unknowns that we must solve for in Eqs. A1.1 and A1.2 and later in A1.5. Equation A1.4 serves to disclose gross errors in the solution.

Information derived from the natural modes must be supplemented by at least one more equation that establishes the scale of forces or magnitude of the unknowns. This additional datum may be one of the M_{ij}'s, the total mass of the structure, or the amplitudes of responses to a given amplitude of steady-state disturbing forces, for example.

The number of modes that must be known to find the inertia, damping, and stiffness matrices depends on the number of unknowns. This in turn depends on how the equations of motion are coupled. For the stiffness matrix there are four common situations.

1. Far coupled. Springs connect each mass to all other masses and to the base. All K's may differ from zero. There are $N(N + 1)/2$ unknown stiffnesses.

2. Close coupled. Springs connect each mass to the immediately adjacent ones and to the base. The stiffness matrix is tridiagonal with $2N - 1$ unknowns.

3. Simply coupled. Each mass is connected by springs to the two adjacent masses, and only one mass is connected to the base. The "shear building" is simply coupled. The stiffness matrix is again tridiagonal, but there are only N unknowns.

4. Uncoupled. The coordinates have been chosen so as to remove stiffness coupling. **K** is diagonal. There are N unknown stiffnesses.

Similar possibilities exist for the other two matrices.

Figure A1.1 shows the least number of modes, as a function of N, required to establish the M's and K's from Eqs. A1.1 and A1.2 for several cases of coupling. In systems with uncoupled inertia and simply coupled stiffness, two modes suffice regardless of the number of degrees of freedom.

Equations A1.1 and A1.2 and the additional magnitude equations can be put in the form
$$\mathbf{Ax} = \mathbf{b} \tag{A1.6}$$
where **A** is the matrix of natural mode data, **x** is the vector of unknown M's and K's, and **b** is a vector of zeros for all except the magnitude equations. **A** has, let us say, u rows and v columns; **x** has v terms; **b** has u terms. If $u < v$ there is no unique solution. If $u > v$, there will generally be no exact solution because errors in the data will render the system inconsistent. We can in this case write Eq. A1.6 in the form
$$\mathbf{Ax} - \mathbf{b} = \mathbf{e} \tag{A1.7}$$
where **e** is the vector of errors in Eq. A1.6. We can evaluate **x** so as to minimize

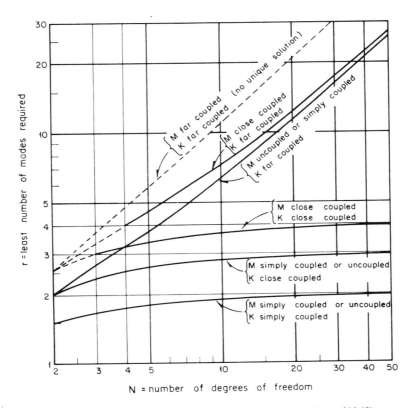

Figure A.1.1. Modes needed to find M and K. *After Berg (1962)*.

the mean-squared error or, if we feel that some equations are more reliable than others, the weighted mean-squared error. From Eq. A1.7,

$$e^T W e = (Ax - b)^T W (Ax - b)$$

where **W** is the diagonal matrix of weighting factors. To minimize the mean-squared error, we set $\partial e^T/\partial x_k = 0$, $k = 1, 2, \ldots, v$. This yields

$$A^T W A x = A^T W b$$

or

$$A'x = b' \qquad (A1.8)$$

where

$$A' = A^T W A$$

and

$$b' = A^T W b \qquad (A1.9)$$

The solution of Eq. A1.8 is unique.

Having determined the M's we can go to Eq. A1.5 and apply the same procedure to get the C's. The form of the system of equations will again be that of Eq. A1.6, but this time the x's will be the unknown damping terms, and **b** will involve the modal properties and the inertia terms which will now be known.

APPENDIX 2:
THE MODIFIED MERCALLI INTENSITY SCALE

Mercalli's (1902) improved intensity scale served as a basis for the scale advanced by Wood and Neumann (1931), known as the modified Mercalli scale and commonly abbreviated MM. The modified version is described below with some improvements by Richter (1958). The following remarks are taken almost verbatim from *Elementary Seismology*, Charles F. Richter (W. H. Freeman and Company, San Francisco, copyright © 1958).

To eliminate many verbal repetitions in the original scale, the following convention has been adopted. Each effect is named at that level of intensity at which it first appears frequently and characteristically. Each effect may be found less strongly, or in fewer instances, at the next lower grade of intensity; more strongly or more often at the next higher grade. A few effects are named at two successive levels to indicate a more gradual increase.

Masonry A, B, C, D. To avoid ambiguity of language, the quality of masonry, brick or otherwise, is specified by the following lettering (which has no connection with the conventional Class *A, B, C* construction).

Masonry A. Good workmanship, mortar, and design; reinforced, especially laterally, and bound together by using steel, concrete, etc.; designed to resist lateral forces.

Masonry B. Good workmanship and mortar; reinforced, but not designed in detail to resist lateral forces.

Masonry C. Ordinary workmanship and mortar; no extreme weaknesses like failing to tie in at corners, but neither reinforced nor designed against horizontal forces.

Masonry D. Weak materials, such as adobe; poor mortar; low standards of workmanship; weak horizontally.

Modified Mercalli Intensity Scale of 1931 (Abridged and Rewritten by C. F. Richter.)

1. Not felt. Marginal and long-period of large earthquakes.

2. Felt by persons at rest, on upper floors, or favorably placed.

3. Felt indoors. Hanging objects swing. Vibration like passing of light trucks. Duration estimated. May not be recognized as an earthquake.

4. Hanging objects swing. Vibration like passing of heavy trucks; or sensation of a jolt like a heavy ball striking the walls. Standing motor cars rock. Windows, dishes, doors rattle. Glasses clink. Crockery clashes. In the upper range of 4, wooden walls and frames crack.

5. Felt outdoors; direction estimated. Sleepers wakened. Liquids disturbed, some spilled. Small unstable objects displaced or upset. Doors swing, close, open. Shutters, pictures move. Pendulum clocks stop, start, change rate.

6. Felt by all. Many frightened and run outdoors. Persons walk unsteadily. Windows, dishes, glassware broken. Knickknacks, books, and so on, off shelves. Pictures off walls. Furniture moved or overturned. Weak plaster and masonry D cracked. Small bells ring (church, school). Trees, bushes shaken visibly, or heard to rustle.

7. Difficult to stand. Noticed by drivers of motor cars. Hanging objects quiver. Furniture broken. Damage to masonry D including cracks. Weak chimneys broken at roof line. Fall of plaster, loose bricks, stones, tiles, cornices, unbraced parapets, and architectural ornaments. Some cracks in masonry C. Waves on ponds; water turbid with mud. Small slides and caving in along sand or gravel banks. Large bells ring. Concrete irrigation ditches damaged.

8. Steering of motor cars affected. Damage to masonry C; partial collapse. Some damage to masonry B; none to masonry A. Fall of stucco and some masonry walls. Twisting, fall of chimneys, factory stacks, monuments, towers, elevated tanks. Frame houses moved on foundations if not bolted down; loose panel walls thrown out. Decayed piling broken off. Branches broken from trees. Changes in flow or temperature of springs and walls. Cracks in wet ground and on steep slopes.

9. General panic. Masonry D destroyed; masonry C heavily damaged, sometimes with complete collapse; masonry B seriously damaged. General damage to foundations. Frame structures, if not bolted, shifted off foundations. Frames racked. Conspicuous cracks in ground. In alluviated areas sand and mud ejected, earthquake fountains, sand craters.

10. Most masonry and frame structures destroyed with their foundations. Some well-built wooden structures and bridges destroyed. Serious damage to dams, dikes, embankments. Large landslides. Water thrown on banks of canals, rivers, lakes, etc. Sand and mud shifted horizontally on beaches and flat land. Rails bent slightly.

11. Rails bent greatly. Underground pipelines completely out of service.

12. Damage nearly total. Large rock masses displaced. Lines of sight and level distorted. Objects thrown into the air.

Other commonly used intensity scales include those of Rossi–Forel (Rossi, 1883; Forel 1884), Cancani (1904), Sieberg, (1923), and Medvedev (1953). The latter is known as the Soviet scale and roughly coincides with the MM intensity scale. Also roughly equivalent is the MSK scale (Medvedev and Sponheuer, 1969). The other scales are falling slowly into disuse. The same is true of the Japanese, Chilean, and other systems of intensity grading that have enjoyed some degree of popularity at national or regional levels.

A now classical piece of work on earthquake intensity and its relation with magnitude is found in a paper by Gutenberg and Richter (1942 and 1956).

APPENDIX 3: NOTATION

Symbols are defined where they are first used. The present appendix contains the principal meanings of some of the most commonly used notations. The reader is cautioned that some symbols are given more than one meaning along this work; this is done in cases where there is no question of confusion.

A = area (cm²); spectral acceleration (cm sec⁻²)
a = amplitude vector of base displacements in Chapter 2
a = amplitude of motion (cm); acceleration (cm sec⁻²); maximum ground acceleration (cm sec⁻²)
a_n = participation coefficient of the nth natural mode (dimensionless)
B = response factor (dimensionless); expected benefits in Chapter 14 (dollars)
b = vector of amplitudes of external forces
b = linear dimension (cm); benefit per unit time in Chapter 14 (dollars yr⁻¹)
C = damping matrix
C = dashpot constant (kg cm⁻¹ sec); expected initial investment in Chapter 14 (dollars)
C_{cr} = critical value of C (kg cm⁻¹ sec)
c = coefficient of variation (dimensionless); wave velocity (cm sec⁻¹); rate of interest in Chapter 14 (yr⁻¹)
D = spectral deformation or relative displacement (cm)
d = maximum ground displacement (cm)
E = modulus of elasticity (kg cm⁻²)
$E(\cdot)$ = expectation
e = base of natural logarithms (dimensionless); void ratio in Chapter 13 (dimensionless)
F = vector of generalized inertia forces
F = inertia force (kg); expected cost of failure in Chapter 14 (dollars)
$F(\cdot)$ = Fourier transform

$F_q =$ Fourier transform of ψ_q
$f =$ frequency (Hz)
$f_i = i$th natural frequency (Hz); cost of the ith mode of failure in Chapter 14 (dollars)
$G =$ modulus of rigidity (kg cm^{-2}); square root of the power spectral density (cm sec$^{-3/2}$)
$g =$ acceleration of gravity (cm sec^{-2})
$H =$ height (cm); depth of reservoir (cm); hysteretic strain energy (kg cm)
$H(\cdot) =$ Heaviside's step function
$h =$ height (cm); $\max_t |d^3x/dt^3|$ (cm sec^{-3})
$I =$ intensity in the MM scale; moment of inertia (cm^4)
$i = \sqrt{-1}$ (dimensionless); subscript usually identifying a time interval or a degree of freedom
$J =$ polar moment of inertia (cm^4); correction factor for overturning moment (dimensionless)
$J_0(\cdot) =$ Bessel's function of the first class, order zero (dimensionless)
$j =$ subscript usually identifying a time interval or a degree of freedom
$\mathbf{K} =$ stiffness matrix
$K =$ stiffness (kg cm^{-1})
$k =$ stiffness of a unit length (kg); torsional stiffness (kg cm)
$\mathbf{L} =$ vector of maximum numerical values of the design responses in Chapter 2
$L =$ length (cm); surface waves
$l =$ vector of design responses in Chapter 2
$\mathbf{M} =$ inertia matrix
$M =$ mass (kg cm^{-1} sec^2)
MM $=$ Modified Mercalli scale
$\mathcal{M} =$ moment (kg cm)
$m =$ mass per unit length or per unit volume; subscript usually identifying a natural mode of vibration
$N =$ number of degrees of freedom (dimensionless)
$n =$ subscript usually identifying a natural mode vibration
$O(\cdot) =$ order of
$P =$ force (kg); primary waves
$P(\cdot) =$ probability
$p =$ force per unit length (kg cm^{-1}); pressure or stress (kg cm^{-2})
$Q =$ viscoelastic constant (dimensionless); $\max_t |q(t)|$; spring force (kg)
$q =$ structural response
$R =$ radius (cm); $\max_t (r)$ (cm sec^{-1})
$\mathcal{R}(\cdot) =$ real part of
$r =$ radius (cm); radius in phase plane (cm sec^{-1})
$S =$ shear (kg); secondary waves; spectral intensity (ft)
$s =$ duration of ground motion (sec)
$\mathbf{T} =$ transform matrix

T = period of vibration (sec)
T_n = nth undamped natural period (sec)
T'_n = nth damped natural period (sec)
$T_{av} = 2\pi v/a$ (sec)
$T_{vd} = 2\pi d/v$ (sec)
t = time (sec)
u = change in ground velocity (cm sec^{-1})
V = spectral pseudovelocity (cm sec^{-1})
v = maximum ground velocity (cm sec^{-1})
v_p = velocity of P waves (cm sec^{-1})
v_r = velocity of Rayleigh waves in Chapter 3 (cm sec^{-1})
v_s = velocity of S waves (cm sec^{-1})
W = weight (kg); strain energy (kg cm)
X = coordinate (cm); objective function in Chapter 14 (dollars)
x = displacement (cm)
x_0 = ground displacement (cm)
$Y_0(\cdot)$ = Bessel's function of the second kind, order zero in Chapter 3 (dimensionless)
y = displacement relative to the ground (cm)
z_{in} = displacement of the ith mass in the nth natural mode (cm)
α = reflection constant in wave transmission across interphase (dimensionless); parameter associated with local seismicity (yr^{-1}); coefficient (dimensionless)
β = attenuation (cm^{-1}); correction factor due to damping (dimensionless); parameter associated with local seismicity (dimensionless); coefficient (dimensionless)
β_E = correction factor due to damping for expected response (dimensionless)
γ = weight per unit volume (kg cm^{-3})
$\Delta(\cdot)$ = increment
$\delta(\cdot)$ = Dirac's delta function
δ_{ik} = Kronecker's delta (dimensionless)
$\epsilon_{mn} = (\omega'_m - \omega'_n)/(\zeta_m \omega_m + \zeta_n \omega_n)$
$\zeta = C/C_{cr}$ = coefficient of damping (dimensionless)
$\eta = (1 - \zeta^2)^{-1/2}$ (dimensionless); damping ratio of filter or soil (dimensionless)
θ = angle (rad); harmonic time function in Chapters 2 and 3; cubical dilatation (dimensionless); $\zeta\omega/\alpha$ in Chapter 10 (dimensionless)
κ = constant in Chapter 3 (cm^{-1}); curvature in Chapter 16 (cm^{-1})
Λ = matrix of influence coefficient in Chapter 2
Λ = wavelength in Chapter 3 (cm)
λ = Lamé's constant (kg cm^{-2}); characteristic value; rate of earthquake generation (yr^{-1})
μ = Lamé's constant (kg cm^{-2}); rate of earthquake occurrence (yr^{-1}); ductility factor (dimensionless)

ν = Poisson's ratio (dimensionless); rate of earthquake occurrence (yr^{-1}); dummy variable for μ (dimensionless)
$\xi = \theta^{1/(1-\theta)}$ in Chapter 10 (dimensionless)
ρ = mass density (kg cm^{-4} sec^2)
σ = normal stress (kg cm^{-2})
$\sigma(\cdot)$ = dispersion
τ = dummy time variable (sec)
ϕ = angle of internal friction (deg); velocity potential in Chapter 6 (cm^2 sec^{-1})
$\phi(\cdot)$ = autocorrelation function
ψ = transfer function; story drift (dimensionless)
ψ_q = transfer function associated with the response $q(t)$
ω = circular frequency of vibration (rad sec^{-1})
ω_n = nth undamped natural circular frequency (rad sec^{-1})
ω'_n = nth damped natural circular frequency (rad sec^{-1})
$\nabla^2(\cdot)$ = Laplace's operator
$\ln(\cdot)$ = natural logarithm
$\log(\cdot)$ = common logarithm
$\doteq$ signifies asymptotically equal to
$\cong$ signifies approximately equal to

REFERENCES

Abramson, H. N., 1963, "Dynamic Behavior of Liquid in Moving Containers," *App. Mechs. Rev.*, *16*(7), 501–6.

Aitken, A. C., 1937, "The Evaluation of the Latent Roots and Latent Vectors of a Matrix," *Proc. Royal Soc. Edinburgh*, *62*, 269–304.

Aki, K., 1963, "Some Problems in Statistical Seismology," *Zisin*, *8* (1956), 205–208, translated by A. S. Furimoto, Hawaii, 1963.

Alford, J. L., G. W. Housner, and R. R. Martel, 1951, "Spectrum Analysis of Strong-Motion Earthquakes," *Earthq. Enrg. Res. Lab.* (rev. ed., 1964), Pasadena: California Institute of Technology.

Allen, C. R., P. St. Amand, C. F. Richter, and J. M. Nordquist, 1965, "Relationship Between Seismicity and Geologic Structure in the Southern California Region," *Bull. Seism. Soc. Am.*, *55*(4), 753–97.

Ambraseys, N. N., 1957, "Seismic Hydrodynamics and Wave Generation in Reservoirs," *Proc. Ass. Internatl. Rech. Hydrauliques*, *7*, D19.1–D19.9.

———, 1958, "The Seismic Stability of Earth Dams," doctoral thesis, University of London.

———, 1959, "A Note on the Response of an Elastic Overburden of Varying Rigidity to an Arbitrary Ground Motion," *Bull. Seism. Soc. Am.*, *49*(3) 211–20.

———, 1960a, "On the Seismic Behavior of Earth Dams," *Proc. Second World Conf. Earthq. Engrg.*, Tokyo and Kyoto, Japan, 331–58.

———, 1960b, "The Seismic Stability of Earth Dams," *Proc. Second World Conf. Earthq. Engrg.*, Tokyo and Kyoto, Japan, 1345–63.

———, 1962, "The Seismic Stability Analysis of Earth Dams," *Proc. Second Symp. Earthq. Engrg.*, Roorke, India, 11–21.

———, 1964, "The Skopje Earthquake of July 26, 1963," *Skopje Internatl. Seminar Earthq. Engrg.*, Skopje, Jugoslavia.

———, 1965, discussion of J. Takeda and H. Tachikawa, "Mechanical Properties of Sand Subjected to Dynamic Load by Shallow Footings," *Proc. Third World Conf. Earthq. Engrg.*, Auckland and Wellington, New Zealand, 1.195.

———, 1969, "Maximum Intensity of Ground Movements Caused by Faulting," *Proc. Fourth World Conf. Earthq. Engrg.*, Santiago, Chile, *1*, A-2, 154–71.

Amin, M., H. S. Ts'ao, and A. H.-S. Ang, 1969, "Significance of Nonstationarity of Earthquake Motions," *Proc. Fourth World Conf. Earthq. Engrg.*, Santiago, Chile, 1, A–1, 97–114.

Anderson, A. W., J. A. Blume H. J. Degenkolb, H. B. Hammill, E. M. Knapik, H. L. Marchand, H. C. Powers, J. E. Rinne, G. A. Sedgwick, and H. O. Sjoberg, 1952, "Lateral Forces of Earthquake and Wind," *Trans. ASCE, 117*, 716–54; 755–80.

Ang, A. H.-S., 1966, "Numerical Approach for Wave Motions in Nonlinear Solid Media," *Proc. Conf. Matrix Methods in Structural Mechs.*, Wright-Patterson Air Force Base, Dayton, Ohio, 759–78.

Ang, A. H.-S., and M. Amin, 1968, "Reliability of Structures and Structural Systems, *Proc. ASCE, 94*(EM2), 671–91.

———, 1969, "Safety Factors and Probability in Structural Design," *Proc. ASCE, 95*(ST7), 1389–1405.

Ang, A. H.-S., and J. H. Rainer, 1964, "Model for Wave Motions in Axi-Symmetric Solids," *Proc. ASCE, 90*(EM2), 195–223.

Argyris, J. M., 1965, "Continua and Discontinua," *Conf. Matrix Methods Structural Mechs.*, Wright-Patterson Air Force Base, Dayton, Ohio, 11–189.

Arias, A., and R. Husid, 1962a, "Fórmula Empírica para el Cálculo del Período Propio de Vibración de Edificios de Hormigón Armado con Muros de Rigidez," *Rev. IDIEM, 1*(1), 1–11.

———, 1962b, "Influencia del Amortiguamiento sobre la Respuesta de Estructuras Sometidas a Temblor," *Rev. IDIEM, 1*(3), 219–28.

Arias, A., R. Husid, and M. Baeza, 1963, "Períodos de Vibración de Edificios Chilenos de Hormigón Armado," *Primeras Jornadas Chilenas de Sismología e Ingeniería Antisísmica, 2*, B3.9, 1–13.

Arias, A. and L. Petit-Laurent, 1963, "Funciones de Autocorrelación y Densidades de Potencia de Acelerogramas de Movimientos Fuertes," *Primeras Jornadas Chilenas de Sismología e Ingeniería Antisísmica, 1*, B1.4, 1–25.

Arias, S., V. Arze, and J. Bauza, 1969, "Repairs on Power House and Boiler Support Structure Damaged by 1965 Earthquake, Ventanas 115 MW Steam Electric Station (Chile)," *Proc. Fourth World Conf. Earthq. Engrg.*, Santiago, Chile, 3, B–6, 31–45.

Arnold, P., P. F. Adams, and L.-W. Lu, 1966, "The Effect of Instability on the Cyclic Behavior of a Frame," *Proc. Internatl. Symp. Effects of Repeated Loading of Materials and Structures*, RILEM-Inst. Ing., 4.

Arnold, R. N., G. N. Bycroft, and G. B. Warburton, 1955, "Forced Vibrations of a Body on an Infinite Elastic Solid," *Journ. App. Mechs., 22*(3), 391–400.

Arze, E., 1969, "Seismic Failure and Repair of an Elevated Water Tank," *Proc. Fourth World Conf. Earthq. Engrg*, Santiago, Chile, 3, B–6, 57–69.

Asano, S., 1960, "Reflection and Refraction of Elastic Waves at a Corrugated Boundary Surface. Part 1, The Case of Incidence of SH Waves," *Bull. Earthq. Res. Inst.*, University of Tokyo, 38, 117–97.

———, 1961, "Reflection and Refraction of Elastic Waves at a Corrugated Boundary Surface. Part 2," *Bull. Earthq. Res. Inst.*, University of Tokyo, 39, 367–466.

Backer, S., 1966, "Fibrous Materials," in *Mechanical Behavior of Materials*, F. A. McClintock and A. S. Argon, eds., Reading, Mass.: Addison-Wesley Publishing Co., Inc., Chap. 21.

Baker, J. F., M. R. Horne, and J. Heyman, 1956, *The Steel Skeleton*, vol. 2, Cambridge: Cambridge University Press.

Barazangi, M. and J. Dorman, 1969, "World Seismicity Maps Compiled from ESSA, Coast and Geodetic Survey, Epicenter Data, 1961–1967," *Bull. Seism. Soc. of Am.*, *59*, 1, 369–80.

Barkan, D. D., 1962, *Dynamics of Bases and Foundations*, New York: McGraw-Hill Book Co., Inc.

Barlow, R. E., and F. Proschan, 1965, *Mathematical Theory of Reliability*, New York: John Wiley & Sons, Inc.

Baron, M. K., 1966, "Shear Strength of Reinforced Concrete Beams at Points of Bar Cutoff," *Proc. ACI*, *63*(1), 127–34.

Barstein, M. F., 1960, "Application of Probability Methods for Design. The Effect of Seismic Forces on Engineering Structures," *Proc. Second World Conf. Earthq. Engrg.*, Tokyo and Kyoto, Japan, 1467–82.

Battelle Memorial Institute, 1949, *Prevention of the Failure of Metals Under Repeated Stress*, New York: John Wiley & Sons, Inc.

Bažant, Z., 1965, "Stability of Saturated Sand During Earthquake," *Proc. Third World Conf. Earthq. Engrg.*, Auckland and Wellington, New Zealand, 1.6–21.

Benjamin, J. R., 1969, "Variability Analysis of Shear Wall Structures," *Proc. Fourth World Conf. Earthq. Engrg.*, Santiago, Chile, *2*, B-3, 45–52.

Benjamin, J. R., and H. A. Williams, 1958, "The Behavior of One-Story Brick Shear Walls," *Proc. ASCE*, *84*(ST4), 1723–30.

———, 1959, "Behavior of Reinforced Concrete Shear Walls," *Trans. ASCE*, *124*, 669–708.

Bennett, J. R., 1965, "Earthquake Insurance in New Zealand," *Proc. Third World Conf. Earthq. Engrg.*, Auckland and Wellington, New Zealand, 4.23–32.

Berg, G. V., 1962, "Finding System Properties from Experimentally Observed Modes of Vibration," *Proc. First Argentine Conf. Earthq. Engrg.*, San Juan, Argentina,

———, 1963, "A Study of Error in Response Spectrum Analysis," *Primeras Jornadas Chilenas de Sismología e Ingeniería Antisísmica*, B1.3, 1–11.

Berg, G. V., Y. C. Das, K. Gokhale, and A. V. Setlur, 1969, "The Koyna, India Earthquakes," *Proc. Fourth World Conf. Earthq. Engrg.*, Santiago, Chile, *3*, J–2, 44–57.

Berg, G. V., and G. W. Housner, 1961, "Integrated Velocity and Displacement of Strong Earthquake Ground Motion," *Bull. Seism. Soc. Am.*, *51*(2), 175–89.

Bertero, V. V., 1969, "Seismic Behavior of Steel Beam-to-Column Connection Subassemblages," *Proc. Fourth World Conf. Earthq. Engrg.*, Santiago, Chile, *2*, B-3, 31–44.

Bertero, V. V., and B. Bresler, 1969, "Seismic Behavior of Reinforced Concrete Framed Structures," *Proc. Fourth World Conf. Earthq. Engrg.*, Santiago, Chile, *1*, B–2, 109–24.

Bertero, V. V., G. McClure, and E. P. Popov, 1962, "Behavior of Reinforced Concrete Frames Subjected to Repeated Reversible Loads," *Structures and Materials Research*, *Series 100*, Dept. of Civ. Engrg., University of California, Berkeley, 18.

Bertero, V. V., and E. P. Popov, 1965, "Effect of Large Alternating Strains of Steel Beams," *Proc. ASCE*, *91*(ST1), 1–12.

Bielak, J., 1966, "Dynamic Response of Single-Degree-of-Freedom Bilinear System," master's thesis, Rice University, Houston, Texas.

———, 1969, "Base Moment for a Class of Linear Systems," *Proc. ASCE*, *95*, EM 5, 1053–62.

Binder, R. W., and W. T. Wheeler, 1960, "Building Code Provisions for Aseismic Design," *Proc. Second World Conf. Earthq. Engrg.*, Tokyo and Kyoto, Japan, 1843–75.

Biot, M., 1941, "A Mechanical Analyzer for the Prediction of Earthquake Stresses," *Bull. Seism. Soc. Am.*, *31*, 151–71.

———, 1943, "Analytical and Experimental Methods in Engineering Seismology," *Trans. ASCE*, *108*, 365–408.

Bisplinghoff, R. L., H. Ashley, and R. L. Halfman, 1955, *Aerolasticity*. Reading, Mass.: Addison–Wesley Publishing Co., Inc.

Blake, R. E., 1961, "Basic Vibration Theory," in *Shock and Vibration Handbook*, C. M. Harris and C. E. Crede, eds., New York: McGraw-Hill Book Co., Inc. Chap. 2.

Blume, J. A., 1956, "Period Determination and Other Earthquake Studies of a Fifteen-Story Building," *Proc. First World Conf. Earthq. Engrg.*, Berkeley, Calif., 11-1-27.

———, 1960a, "A Reserve Energy Technique for the Earthquake Design and Rating of Structures in the Inelastic Range," *Proc. Second World Conf. Earthq. Engrg.*, Tokyo and Kyoto, Japan, 1061–83.

———, 1960b, "Structural Dynamics in Earthquake Resistant Design," *Trans. ASCE*, *125*, 1088–1139.

———, 1963, "A Structural-Dynamic Analysis of Steel Plant Structures Subjected to the May 1960 Chilean Earthquakes," *Bull. Seism. Soc. Am.*, *53*(2) 439–80.

———, 1968, "Dynamic Characteristics of Multistory Buildings," *Proc. ASCE*, *94*(ST2), 377–402.

———, 1969, "Structural Dynamics of Cantilever-Type Buildings," *Proc. Fourth World Conf. Earthq. Engrg.*, Santiago, Chile, 2, A-3, 1–18.

Blume, J. A., and R. W. Binder, 1960, "Periods of a Modern Multi-Story Office Building During Construction," *Proc. Second World Conf. Earthq. Engrg.*, Tokyo and Kyoto, Japan, 1195–1205.

Blume. J. A., and D. Jhaveri, 1969, "Time-History Response of Buildings with Unusual Configurations," *Proc. Fourth World Conf. Earthq. Engrg.*, Santiago, Chile, 2, A-3, 155–70.

Blume, J. A., N. M. Newmark, and L. H. Corning, 1961, *Design of Multistory Reinforced Concrete Buildings for Earthquake Motions*, Chicago: Portland Cement Association.

Bogdanoff, J. L., J. E. Goldberg, and M. C. Bernard, 1961, "Response of a Simple Structure to a Random Earthquake-Type Disturbance," *Bull. Seism. Soc. Am.*, *51*(2), 293–310.

Bolotin, V. V., 1960, "Statistical Theory of the Aseismic Design of Structures," *Proc. Second World Conf. Earthq. Engrg.*, Tokyo and Kyoto, Japan, 1365–74.

Borges, J. F., 1956, "Statistical Estimate of Seismic Loading," preliminary publication, *Fifth Congr. Internatl. Ass. Bridge and Structural Engineering*, Lisbon, Portugal.

———, 1965, "Seismic Design Criteria for Reinforced Concrete Buildings," *Proc. Third World Conf. Earthq. Engrg.*, Auckland and Wellington, New Zealand, 4.72–89.

Borges, J. F., and A. Ravara, 1969, "Seismic Design of Traditional and Prefabricated Reinforced Concrete Buildings," *Proc. Fourth World Conf. Earthq. Engrg.*, Santiago, Chile, 3, B-5, 13–27.

Bouwkamp, J. G., 1966, "Tubular Joints under Slow-Cycle Alternating Loads," *Proc. Internatl. Symp. Effects of Repeated Loading of Materials and Structures*, RILEM-Inst. Ing., 6.

Bouwkamp, J. G., and J. F. Meehan, 1960, "Drift Limitations Imposed by Glass," *Proc. Second World Conf. Earthq. Engrg.*, Tokyo and Kyoto, Japan, 1763–78.

Brady, A. G., 1966, "Studies of Response to Earthquake Ground Motion," doctoral thesis, California Institute of Technology, Pasadena, Calif.

Brady, A. G., and R. Husid, 1966, "Distributions of Maximum Response to Random Excitation," *Proc. Internatl. Symp. Effects of Repeated Loading of Materials and Structures*, RILEM-Inst. Ing., *1*.

Bresler, B., and P. H. Gilbert, 1961, "Tie Requirements for Reinforced Concrete Columns," *Journ. ACI*, 58(5), 555–70.

Brokaw, M. P., and G. W. Foster, 1945, *Effect of Rapid Loading and Duration of Stress on the Strength Properties of Wood Tested in Compression and Flexure*, Madison, Wisc.: U.S. Department of Agriculture, Forest Products Laboratory.

Bronson, W., 1959, *The Earth Shook, the Sky Burned*, Garden City, N.Y.: Doubleday & Co., Inc.

Brune, J. N., 1968, "Seismic Moment, Seismicity, and Rate of Slip along Major Fault Zones," *Journ. Geophys. Res.*, 73(2), 777–84.

Building Code Requirements for Reinforced Concrete, 1963 (*ACI 318–63*) American Concrete Institute, Detroit.

Bullen, K. E., 1953, *An Introduction to the Theory of Seismology*, London: Cambridge University Press.

Burridge, R., and L. Knopoff, 1967, "Model of Theoretical Seismicity," *Bull. Seism. Soc. Am.*, 57(3), 341–71.

Bustamante, J. I., 1961a, "Reglamento del Distrito Federal. Estudio Comparativo entre los Métodos Estático y Dinámico de Análisis Sísmico," *Ingeniería*, 31(2), 82–95.

———, 1961b, "Torsión Dinámica en Estructuras de Edificios," *Ingeniería*, 31(4), 3–10.

———, 1964, "Dynamic Behavior of Non-Cohesive Embankment Models," doctoral thesis, University of Illinois, Urbana, Ill.

———, 1965a, "Dynamic Behavior of Noncohesive Embankment Models," *Proc. Third World Conf. on Earthq. Engrg.*, Auckland and Wellington, New Zealand, 4.596–612.

———, 1965b, "Seismic Shears and Overturning Moments in Buildings," *Proc. Third World Conf. Earthq. Engrg.*, Auckland and Wellington, New Zealand, 4.144–60.

———, 1965c, "Los Espectros Elásticos de Sismos Mexicanos y el Reglamento Propuesto para el Distrito Federal," *Primer Congreso Nacional de Ingeniería Sísmica*, Guadalajara, Mexico, Prelim. Publ., Soc. Mex. de Ing. Sísmica.

———, 1965d, *Model Study of the Dynamic Response of the San Francisco Rapid Transit Tube and Ventilation Caisson*, Instituto de Ingeniería, Mexico, unpublished report to Parsons, Brinckerhoff–Tudor–Bechtel.

Bustamante, J. I., and A. Flores, 1966a, "Hydrodynamic Pressure for Design of Dams Subjected to Earthquakes," *Joint ASCE–CICM Meeting on Structural Engineering*, Mexico.

———, 1966b, "Water Pressure on Dams Subjected to Earthquake," *Proc. ASCE*, 92(EM5), 115–27.

Bustamante, J. I., and J. González, "Simulation of Earthquakes on Firm Ground," unpublished report, Institute of Engineering, National University of Mexico, Mexico.

Bustamante, J. I., and J. Prince, 1963, Corrección de los Acelerogramas de Cuatro Macrosismos Registrados en la Ciudad de México," *Bol. Soc. Mex. Ing. Sism*, *1*(2), 61–82.

Bustamante, J. I., and L. Rapoport, 1964, "Momento de Volteo y Fuerzas Cortantes Sísmicas," *Bol. Soc. Mex. Ing. Sism.*, *2*(1), 19–31.

Bustamante, J. I., and E. Rosenblueth, 1960, "Building Code Provisions on Torsional Oscillations," *Proc. Second World Conf. Earthq. Engrg.*, Tokyo and Kyoto, Japan, 879–94.

Bustamante, J. I., E. Rosenblueth, I. Herrera, and A. Flores, 1963, "Presión Hidrodinámica en Presas y Depósitos," *Bol. Soc. Mex. Ing. Sism.*, *1*(2), 37–54.

Bycroft, G. N., 1960a, "White Noise Representation of Earthquakes," *Proc. ASCE*, *86*(EM2), 1–16.

———, 1960b, "Yield Displacements in Multistory Aseismic Design," *Bull. Seism. Soc. Am.*, *50*(3), 441–53.

Bycroft, G. N., M. J. Murphy, and K. J. Brown, 1959, "Electrical Analog for Earthquake Yield Spectra," *Proc. ASCE*, *85*(EM4), 43–64.

Cancani, A., 1904, "Sur l'Emploi d'une Double Échelle Séismique des Intensités, Empirique et Absolue," *G. Beitr. Ergänzungsband*, *2*, 281–83.

Carder, D. S., 1936, "Observed Vibrations of Buildings," *Bull. Seism. Soc. Am.*, *26*(3), 245–77.

———, 1937, "Observed Vibrations of Bridges," *Bull. Seism. Soc. Am.*, *27*(4), 267–304.

Carder, D. S., and W. K. Cloud, 1959, "Surface Motions from Large Underground Explosions," *Journ. Geophys. Res.*, *64*(10), 1471–87.

Carmona, J. S., and J. Herrera-Cano, 1969, "Periods of Buildings of Mendoza City Argentina," *Proc. Fourth World Conf. Earthq. Engrg.*, Santiago, Chile, *1*, B–1, 72–86.

Carpenter, L. D., and L.-W. Lu, 1969, "Repeated and Reversed Load Tests on Full-Scale Steel Frames," *Proc. Fourth World Conf. Earthq. Engrg.*, Santiago, Chile, *1*, B–2, 125–36.

Casagrande, A., 1936, "Characteristics of Cohesionless Soils Affecting the Stability of Slopes and Earth Fills," *Contributions to Soil Mechanics, 1925–1940*, Boston Society of Civil Engineers, 1940, pp. 257–76. Previously published in *Journ. Boston Soc. of Civil Engrs.*, 1936.

———, 1938, "The Shearing Resistance of Soils and Its Relation to the Stability of Earth Dams," *Proc. Soils and Foundation Conf.* Cambridge, Mass., Corps of Engineers, U. S. Army.

Casagrande, A., and W. L. Shannon, 1948, "Research on Stress-Deformation and Strength Characteristics of Soils and Soft Rocks Under Transient Loading," *Harvard Soil Mechanics Series*, *31*, Cambridge, Mass.: Harvard University.

Castro, G., 1969, "Liquefaction of Sands," *Harvard Soil Mechanics Series*, *81*, Cambridge, Mass.: Harvard University.

Caughey, T. K., 1960, "Classical Normal Modes in Damped Linear Dynamic Systems," *Journ. of App. Mechs.*, *27*, and *Trans. ASME*, *82*, Series E, 269–71.

Caughey, T. K., and A. H. Gray, 1963, discussion of Rosenblueth and Bustamante (1962), *Proc. ASCE*, *89*(EM2), 159–68.

Chen' Chzhen'-Chen, 1961, "On the Hydrodynamic Pressure on a Dam Caused by its Aperiodic or Impulsive Vibrations and Vertical Vibrations on the Earth Surface," *Journ. App. Math. Mechs. (P.M.M.)*, *25*(4), 1060–76.

Chi-Chang Chao, 1960, "Dynamical Response of an Elastic Half-Space to Tangential Surface Loadings," *Journ. App. Mechs.*, *27*(3), 559–67.

Chopra, A. K., 1967, "Hydrodynamic Pressures on Dams During Earthquakes," *Proc. ASCE*, *93*(EM6), 205–23.

———, 1968, "Earthquake Behavior of Reservoir-Dam Systems," *Proc. ASCE*, *94*(EM6), 1475–1500.

Chopra, A. K., M. Dibaj, R. W. Clough, J. Penzien, and H. B. Seed, 1969, "Earthquake Analysis of Earth Dams," *Proc. Fourth World Conf. Earthq. Engrg.*, Santiago, Chile, *3*, A-5, 55–72.

Chopra, A. K., and P. R. Perumalswami, 1969, "Dam-Foundation Interaction During Earthquakes," *Proc. Fourth World Conf. Earthq. Engrg.*, Santiago, Chile, *3*, A-6, 37–52.

Chopra, A. K., E. L. Wilson, and I. Farhoomand, 1969, "Earthquake Analysis of Reservoir-Dam Systems," *Proc. Fourth World Conf. Earthq. Engrg.*, Santiago, Chile, *2*, B-4, 1–10.

Churayan, A., and S. A. Djabua, 1969, "On One Method of Increasing the Seismic Stability of Brick Buildings," *Proc. Fourth World Conf. Earthq. Engrg.*, Santiago, Chile, *3*, B-6, 47–56.

Church, A. H., 1957, *Mechanical Vibrations*, New York: John Wiley & Sons, Inc.

Churchman, C. W., R. L. Ackoff, and E. L. Arnoff, 1961, *Introduction to Operations Research*, New York: John Wiley & Sons, Inc.

Cismigiu, A., E. Titaru, and M. Velkov, 1969, "Criteria for Earthquake Resistant Codes Based on Energy Concept. Draft Design Code," *Proc. Fourth World Conf. Earthq. Engrg.*, Santiago, Chile, *3*, B-5, 53–67.

Close, U., and Elsie McCormick, 1922, "Where the Mountains Walked," *National Geographic Magazine*, *41*(5), 445–64.

Clough. R. W., 1960a, "Effects of Earthquakes on Underwater Structures," *Proc. Second World Conf. Earthq. Engrg.*, Tokyo and Kyoto, Japan, 815–32.

———, 1960b, "The Finite Element Method in Plane Stress Analysis," *Proc. Second ASCE Conf. Electronic Computation*, Pittsburgh, Pa., 345–78.

———, 1961, "Dynamic Effects of Earthquakes," *Trans. ASCE*, *126*, 847–76.

———, 1962, "Earthquake Analysis by Response Spectrum Superposition," *Bull. Seism. Soc. Am.*, *52*(3), 647–60.

———, 1965, "The Finite Element Method in Structural Mechanics," in *Stress Analysis*, O. C. Zienkiewicz and G. W. Holister, eds., London: John Wiley & Sons, Ltd., Chap. 7.

Clough, R. W., and K. L. Benuska, 1966, "FHA Study of Seismic Design Criteria for High Rise Buildings," *Report HUD, TS-3*, Washington, D.C.: Federal Housing Administration.

Clough, R. W., K. L. Benuska, and E. L. Wilson, 1965, "Inelastic Earthquake Response of Tall Buildings," *Proc. Third World Conf. Earthq. Engrg.*, Auckland and Wellington, New Zealand, 2.68–89.

Clough, R. W., and A. K. Chopra, 1966, "Earthquake Stress Analysis in Earth Dams," *Proc. ASCE*, *92*(EM2), 197–211.

Clough, R. W., I. P. King, and E. L. Wilson, 1964, "Structural Analysis of Multistory Buildings," *Proc. ASCE*, *90*(ST3), 19–34.

Clough, R. W., and D. Pirtz, 1958, "Earthquake Resistance of Rock-Fill Dams," *Trans. ASCE*, *123*, 792–816.

Clyde, D. H., 1966, "Equilibrium Method for Limit Calculation of Frames," *Proc. ASCE*, *92*(EM1), 169–81.

Corley, W. G., and N. W. Hanson, 1969, "Design of Beam-Column Joints for Seismic

Resistant Reinforced Concrete Frames," *Proc. Fourth World Conf. Earthq. Engrg.*, Santiago, Chile, *2*, B–3, 69–82.

Cornell, C. A., 1967, "Bounds on the Reliability of Structural Systems," *Proc. ASCE, 93*(ST1), 171–200.

Costantino, C. J., 1967, "Finite Element Approach to Stress Wave Problems," *Proc. ASCE, 93*(EM2), 153–76.

Courant, R., and D. Hilbert, 1953, *Methods of Mathematical Physics*, Vol. 1, New York: Wiley–Interscience Publishers, Inc.

Crandall, S. H., 1956, *Engineering Analysis*, New York: McGraw-Hill Inc.

———, 1958, "Statistical Properties of Response to Random Vibration," in *Random Vibration*, S. H. Crandall, ed., Cambridge, Mass.: Massachusetts Institute of Technology Press, Chap. 4.

Crandall, S. H., W. D. Mark, and G. R. Khabbaz, 1962, "The Variance in Palmgren–Miner Damage due to Random Vibration," *Proc. Fourth U.S. Natl. Congr. App. Mechs., 1*, 119–26.

Crandall, S. H., and R. B. McCalley, Jr., 1961, "Numerical Methods of Analysis," in *Shock and Vibration Handbook*, C. M. Harris and C. E. Crede, eds., New York: McGraw-Hill Book Co., Inc., Chap. 28.

Crandall, S. H., and W. G. Strang, 1957, "An Improvement of the Holzer Table Based on a Suggestion of Rayleigh's," *Journ. App. Mechs., 24*(2), 228–30.

Cruickshank, C., 1969, "Derrumbes en Embalses," *Ingeniería, 40*(1), 33–62.

Damy, J. E., 1965, "Comentarios Sobre el Cálculo de Modos Naturales de Vibración," *Primer Congreso Nacional de Ingeniería Sísmica*, Guadalajara, Mexico, Prelim. Publ., Soc. Mex. de Ing. Sísmica.

D'Appolonia, W., 1953, "Loose Sands—Their Compaction by Vibroflotation," *ASTM Spec. Techn. Pub. 156*, 138–52.

Davis, R. E., 1960, *Model Study of Stability of Portage Mountain Dam during Earthquake*, Report to International Power and Engineering Consultants Limited.

Davison, C., 1936, *Great Earthquakes*, London: Thomas Murphy and Co.

Degenkolb, H., 1969, "Limitations and Uncertainties of Present Structural Design Methods for Lateral Force Resistance," *Proc. Fourth World Conf. Earthq. Engrg.*, Santiago, Chile, *3*, J–4, 35–49.

De Rossi, M. S., 1883, "Programma dell' Osservatorio ed Archivo Centrale Geodinamico," *Boll. Vulcan. Ital. 10*, 3–124.

Del Valle, and J. Prince, 1965, "Analytical and Experimental Studies of Vibration in Two Buildings," *Proc. Third World Conf. Earthq. Engrg.*, Auckland and Wellington, New Zealand, 2.648–62.

Den Hartog, J. P., 1956, *Mechanical Vibrations*, New York: McGraw-Hill Book Co., Inc., 37–40.

Despeyroux, J., 1960, "The Agadir Earthquake of February 29th, 1960. Behavior of Modern Buildings during the Earthquake," *Proc. Second World Conf. Earthq. Engrg.*, Tokyo and Kyoto, Japan, 521–42.

Díaz de Cossío, R., and E. Martínez, 1961, "Estudio de Marcos Sujetos a Carga Lateral," *Ingeniería, 31*(4), 9–24.

Dick, I. D., 1965, "Extreme Value Theory and Earthquakes," *Proc. Third World Conf. Earthq. Engrg.*, Auckland and Wellington, New Zealand, pp. 3.45–53.

Duke, C. M., 1969, "Techniques for Field Measurement of Shear Wave Velocity in Soils," *Proc. Fourth World Conf. Earthq. Engrg.*, Santiago, Chile, *3*, A–5, 39–54.

Duke, C. M., and D. J. Leeds, 1963, "Response of Soils, Foundations and Earth

Structures to the Chilean Earthquakes of 1960," *Bull. Seism. Soc. Am.*, 53(2), 309–57.
Duncan, W. J., 1952, "A Critical Examination of the Representation of Massive and Elastic Bodies by Systems of Rigid Masses Elastically Connected," *Quart. Journ. Mech. App. Math.*, 5(1), 97–108.
Duvall, G. E., 1962, "Concepts of Shock Wave Propagation," *Bull. Seism. Soc. Am.*, 52(4), 869–93.
Earthquake Resistant Regulations. A World List, 1966a, "Recommended Lateral Force Requirements," Seismology Committe, Structural Engineers Association of California, compiled by International Association for Earthquake Engineering, Tokyo. 319–35.
Earthquake Resistant Regulations. A World List, 1966b, "Regulations for Seismic Design, Republic of El Salvador, C.A., 21 January 1966," compiled by International Association for Earthquake Engineering, Tokyo, 39–52.
Earthquake Resistant Regulations. A World List, 1966c, "Provisions for Earthquake Resistant Design in the Federal District, Mexico," compiled by International Association for Earthquake Engineering, Tokyo, 213–20.
Earthquake Resistant Regulations. A World List, 1966d, "Norms for Design of Constructions in Acapulco, State of Guerrero, Mexico," compiled by International Association for Earthquake Engineering, Tokyo, 221–28.
Eckel, E. B., 1967, "Effects of the Earthquake of March 27, 1964 on Air and Water Transport, Communications, and Utility Systems in South-Central Alaska," *U.S. Geological Survey Professional Paper 545-B*, B21–B23.
Elorduy, J., 1967, "Sobre el Comportamiento Dinámico de Bases Rígidas Sujetas a Perturbaciones Armónicas," doctoral thesis, National University of Mexico, Mexico.
——, 1969, discussion of Fenves and Newmark (1969), *Proc. Fourth World Conf. Earthq. Engrg.*, Santiago, Chile, 4.
Elorduy, J., and E. Rosenblueth, 1968, "Torsiones Sísmicas en Edificios de Un Piso," *Segundo Congreso Nacional de Ingeniería Sísmica*, Veracruz, Mexico, 3.
Esteva, L., 1965, "Vibraciones de Marcos de Un Piso: Efectos de la Distribución de la Masa," *Primer Congreso Nacional de Ingeniería Sísmica*, Guadalajara, Mexico, Prelim. Publ., Soc. Mex. de Ing. Sísmica.
——, 1966, "Behavior Under Alternating Loads of Masonry Diaphragms Framed by Reinforced Concrete Members," *Proc. Internatl. Symp. Effects of Repeated Loading of Materials and Structures*, RILEM-Inst. Ing., 5.
——, 1968, "Bases para la Formulación de Decisiones de Diseño Sísmico," doctoral thesis, National University of Mexico, Mexico.
——, 1969, "Seismic Risk and Seismic Design Decisions," *Seminar on Seismic Design of Nuclear Power Plants*, Cambridge, Mass.: Massachusetts Institute of Technology Press.
——, 1970, "Consideraciones Prácticas en la Estimación Bayesiana de Riesgo Sísmico," *Inst. Ing.*, Mexico, 248.
Esteva, L., R. Díaz de Cossío, and J. Elorduy, 1968, "El Temblor de Caracas, Julio de 1967," *Ingeniería*, 38(3), 289–314; *Inst. Ing.*, Mexico, 168.
Esteva, L., J. Elorduy, and J. Sandoval, 1969, "Análisis de la Confiabilidad de la Presa Tepuxtepec ante la Acción de Temblores," *Inst. Ing.*, Mexico, 194.
Esteva, L., and J. A. Nieto, 1967, "El Temblor de Lima, Perú, Octubre 17, 1966," *Ingeniería*, 37(1), 45–62.
Esteva, L., O. A. Rascón, and A. Gutiérrez, 1969, "Lessons from Some Recent Earthquakes in Latin America," *Proc. Fourth World Conf. Earthq. Engrg.*, Santiago, Chile, 3, J-2, 58–73.

Esteva, L., and E. Rosenblueth, 1964, "Espectros de Temblores a Distancias Moderadas y Grandes," *Bol. Soc. Mex. Ing. Sism.*, 2(1), 1–18.

Esteva, L., R. Sánchez Trejo, and E. Rosenblueth, 1961, "Consideraciones sobre el Diseño Sísmico de Presas de Tierra y Enrocamiento," *Ingeniería*, 31(2), 68–81.

Evans, R. H., 1942, "Effects of Rate of Loading on the Mechanical Properties of Some Materials," *Journ. Inst. Civil Engrs.*, 18, 296–306.

Evison, F. F., 1963, "Earthquakes and Faults," *Bull. Seism. Soc. Am.*, 53(5), 873–91.

———, 1967, "On the Occurrence of Volume Change of the Earthquake Source," *Bull. Seism. Soc. Am.*, 57(1), 9–25.

Ewing, W. M., W. S. Jardetzky, and F. Press, 1957, *Elastic Waves in Layered Media*, New York: McGraw-Hill Book Co., Inc.

Falconer, B. H., 1964, "Niigata Earthquake, Japan 1:02 P.M., 16 June, 1964," *Internatl. Inst. of Seism. and Earthq. Engrg.*, Tokyo, Japan.

Faradji-Capón, M. J., and R. Díaz de Cossío, 1965, "Tensión Diagonal en Miembros de Concreto de Sección Circular," *Ingeniería*, 36(2), 257–80.

Fenves, S. J., 1964, *STRESS: A User's Manual: A Problem-Oriented Computer Language for Structural Engineering*, Cambridge, Mass.: Massachusetts Institute of Technology Press.

Fenves, S. J., and N. M. Newmark, 1969, "Seismic Forces and Overturning Moments in Buildings, Towers and Chimneys," *Proc. Fourth World Conf. Earthq. Engrg.*, Santiago, Chile, 3, B-5, 1–12.

Ferguson, P. M., and F. N. Matlob, 1959, "Effect of Bar Cutoff on Bond and Shear Strength of Reinforced Concrete Beams," *Proc. ACI*, 56(7), 5–25.

Ferrandon, J., 1960, "Actions Hydrodynamiques des Séismes sur les Ouvrages de Retenue," *Le Génie Civil*, 107–9.

Figueroa, J., 1963, "Características de Cuatro Macrosismos Mexicanos," *Rev. Soc. Mex. Ing. Sism.*, 1(2), 61–75.

———, 1964, "Determinación de las Constantes de la Arcilla del Valle de México por Prospección Sísmica," *Bol. Soc. Mex. Ing. Sism.*, 2(2), 57–66.

Finn, W. D. L., P. L. Bransby, and D. J. Pickering, 1970, "Effect of Strain History on Liquefaction of Sand," *Proc. ASCE*, 96 (SM6), 1917–34.

Finn, W. D. L., D. J. Pickering, and P. L. Bransby, 1971, "Sand Liquefaction in Triaxial and Simple Shear Tests," Accepted by *J. of Soil Mechs. and Found. Div., ASCE*.

Fiorato, A. E., M. A. Sozen, and W. L. Gamble, 1968, "Behavior of Five-Story Reinforced Concrete Frames with Filler Walls," Interim report to the U.S. Defense Department, University of Illinois, Urbana, Illinois.

Flores, A., 1966, "Presión Hidrodinámica en Presas Sujetas a Sismos," master's thesis, National University of Mexico, Mexico.

Flores, A., I. Herrera, and C. Lozano, 1969, "Hydrodynamic Pressures Generated by Vertical Earthquake Component," *Proc. Fourth World Conf. Earth. Engrg.*, Santiago, Chile, 2, B-4, 25–36.

Flygare, R. W., 1955, "An Investigation of Ground Accelerations Produced by Machines," mechanical engineering thesis, California Institute of Technology, Pasadena, Calif.

Forel, F. A., 1884, "Les Tremblements de Terre Étudiés par la Commission Sismologique Suisse Pendant l'Année 1881; 2me Rapport," *Arch. Sci. Phys. Nat.*, 11, 147–82.

Franklin, J. N., 1963, "Digital-Computer Simulation of a Gaussian Random Process with Given Power Spectral Density," Tech. Rept. 120, *Computing Center*, Pasadena: California Institute of Technology.

Freudenthal, A. M., 1962, "Safety, Reliability, and Structural Design," *Trans. ASCE*, *127*, 304–19.

Freudenthal, A. M., J. M. Garrelts, and M. Shinozuka, 1966, "The Analysis of Structural Safety," *Proc. ASCE*, *92*(ST1), 267–325.

Freudenthal, A. M., and F. Roll, 1957–1958, "Creep and Creep Recovery of Concrete Under High Compressive Stress," *Proc. ACI*, *54*, 1111–42.

Funahashi, I., K. Kinoshita, and H. Aoyama, 1969, "Vibration Tests and Test to Failure of a 7 Storied Building Survived a Severe Earthquake," *Proc. Fourth World Conf. Earthq. Engrg.*, Santiago, Chile, *1*, B-1, 26–43.

Furumoto, A. S., 1966, "Seismicity of Hawaii. Part 1, Frequency-Energy Distribution of Earthquakes," *Bull. Seism. Soc. Am.*, *56*(1), 1–12.

Gajardo, E., and C. Lomnitz, 1960, "Seismic Provinces of Chile," *Proc. Second World Conf. Earthq. Engrg.*, Tokyo and Kyoto, Japan, 1529–40.

Garza-Tamez, F., 1968, "Sistema para la Reducción de los Efectos Sísmicos en las Edificaciones," *Segundo Congreso Nacional de Ingeniería Sísmica*, Veracruz, Mexico 2.

Gerstle, K. H., and L. G. Tulin, 1966, "Shakedown of Continuous Concrete Beams," *Proc. Internatl. Symp. Effects of Repeated Loading of Materials and Structures*, RILEM-Inst. Ing., 5.

Glanville, W.H., 1930, "The Creep or Flow of Concrete Under Load," *Building Research Technical Paper No. 12*, London: Department of Scientific and Industrial Research.

Godden, W. G., 1965, *Numerical Analysis of Beams and Column Structures*, Englewood Cliffs, N.J.: Prentice-Hall, Inc.

Goldberg, J. E., J. L. Bogdanoff, and Z. L. Moh, 1959, "Forced Vibration and Natural Frequencies of Building Frames," *Bull. Seism. Soc. Am.*, *49*(1), 33–47.

Goldberg, J. E., J. L. Bogdanoff, and D. R. Sharpe, 1964, "The Response of Simple Nonlinear Systems to a Random Disturbance of the Earthquake Type," *Bull. Seism. Soc. Am.*, *54*(1), 263–76.

González-Flores, M., 1964, "Sistema para Eliminar los Esfuerzos Peligrosos que los Temblores Causan en las Estructuras," *Quinto Congreso Mexicano de la Industria de la Construcción*, Tijuana, B.C., Mexico.

Goodman, R. E., 1963, "The Stability of Slopes in Cohesionless Materials During Earthquake," doctoral thesis, University of California, Berkeley, Calif.

Goto, H., and H. Kameda, 1969, "Statistical Inference of the Future Ground Motion," *Proc. Fourth World Conf. Earthq. Engrg.*, Santiago, Chile, *1*, A-1, 39–54.

Goto, H., and K. Toki, 1969, "Structural Response to Nonstationary Random Excitation," *Proc. Fourth World Conf. Earthq. Engrg.*, Santiago, Chile, *1*, A-1, 130–44.

Graham, E. W., and A. M. Rodriguez, 1952, "The Characteristics of Fuel Motion which Affect Airplane Dynamics," *Journ. App. Mechs.*, *19*(3), 321–88.

Green, N. B., 1935, "Flexible 'First Story' Construction for Earthquake Resistance," *Trans. ASCE*, *100*, 645–74.

Grigorian, S. S., 1960 "On Basic Concepts in Soil Mechanics," *App. Math. and Mechs.*, *24*(6), 1604–27.

Guerrero y Torres, J., 1965, "Bandas Amortiguadoras para Muros de Partición," *Primer Congreso Nacional de Ingeniería Sísmica*, Guadalajara, Mexico.

Gumbel, E. J., 1958, *Statistics of Extremes*, New York: Columbia University Press.

Gutenberg, B., 1955, "Wave Velocities in the Earth's Crust," Spec. Paper 62, *Geol. Soc. Am.*, 19–34.

——, 1958, "Attenuation of Seismic Waves in the Earth's Mantle," *Bull. Seism. Soc. Am.*, *48*(3), 269–82.

Gutenberg, B., and C. F. Richter, 1942, "Earthquake Magnitude, Intensity, Energy, and Acceleration," *Bull. Seism. Soc. Am.*, *32*, 163–91.

———, 1954, *Seismicity of the Earth (Associated Phenomena)*, Princeton, N. J.: Princeton University Press.

———, 1956, "Earthquake Magnitude, Intensity, Energy, and Acceleration," *Bull. Seism. Soc. Am.*, *46*, 105–45.

Gzovsky, M. G., 1962, "Tectonophysics and Earthquake Forecasting," *Bull. Seism. Soc. Am.*, *52*(3), 485–505.

Hall, W. J.; H. Kihara, W. Soete, and A. A. Wells, 1967, *Brittle Fracture of Welded Plates*, Englewood Cliffs, N.J.: Prentice-Hall, Inc.

Hanson, N. W., and H. W. Conner, 1967, "Seismic Resistance of Reinforced Concrete Beam-Column Joints," *Proc. ASCE*, *93*(ST5), 533–60.

Hanson, R. D., and W. R. S. Fan, 1969, "The Effect of Minimum Cross Bracing on the Inelastic Response of Multi-Story Buildings," *Proc. Fourth World Conf. Earthq. Engrg.*, Santiago, Chile, *2*, A–4, 15–30.

Hatanaka, M., 1955, "Fundamental Considerations on the Earthquake-Resistant Properties of the Earth Dams," *Bull. Disaster Prevention Res. Inst.*, *11*, Kyoto University.

Hatano, T., 1960, "Dynamical Behavior of Concrete Under Impulsive Tensile Load," *Technical Report C-6002*, Tokyo: Central Research Institute of Electric Power Industry.

———, 1966, "An Examination on the Resonance of Hydrodynamic Pressure During Earthquakes Due to Elasticity of Water," *Trans. Japan Soc. Civil Engrs.*, *129*.

Hatano, T., and H. Tsutsumi, 1959, "Dynamical Compressive Deformation and Failure of Concrete Under Earthquake Load," *Technical Report C-5904*, Tokyo: Central Research Institute of Electric Power Industry.

Hatano, T., and H. Watanabe, 1969, "Seismic Analysis of Earth Dams," *Proc. Fourth World Conf. Earthq. Engrg.*, Santiago, Chile, *3*, A–5, 117–32.

Hausner, M., 1954, "Multidimensional Utilities," in *Decision Processes*, R. M. Thrall, C. H. Coombs, and R. L. Davis, eds., New York: John Wiley & Sons, Inc., 167–80.

Heidebrecht, A. C., and W. K. Tso, 1969, "A Research Program on the Earthquake Resistance of Shear Wall Buildings," *Proc. Fourth World Conf. Earthq. Engrg.*, Santiago, Chile, *1*, B–2, 1–13.

Heierli, W., 1962, "Inelastic Wave Propagation in Soil Columns," *Proc. ASCE*, *88*(SM6), 33–63.

L'Hermite, R., and G. Tournon, 1948, "La Vibration de Béton Frais," *Ann. Inst. Tech. Batiment Trav. Publ. 11*.

Herrera, I., 1964, "A Perturbation Method for Elastic Wave Propagation. 1, Nonparallel Boundaries," *Journ. of Geophys. Res.* *69*(18), 3845–51.

———, 1965a, "A Perturbation Method for Elastic Wave Propagation. 2, Small Inhomogeneities," *Journ. of Geophys. Res.*, *70*(4), 871–83.

———, 1965b, "Modelos Dinámicos para Materiales y Estructuras del Tipo Masing," *Bol. Soc. Mex. Ing. Sism.*, *3*(1), 1–8.

Herrera, I., and E. Rosenblueth, 1965, "Response Spectra on Stratified Soil," *Proc. Third World Conf. on Earthq. Engrg.*, Auckland and Wellington, New Zealand, 1.44–60.

Herrera, I., E. Rosenblueth, and O. A. Rascón, 1965, "Earthquake Spectrum Prediction for the Valley of Mexico," *Proc. Third World Conf. Earthq. Engrg.*, Auckland and Wellington, New Zealand, 1.61–74.

———, 1966, "Nota sobre la Velocidad Máxima del Terreno Durante un Sismo," *Rev. Soc. Mex. Ing. Sism.*, *4*(1), 26–27.

Hildebrand, F. B., 1952, *Methods of Applied Mathematics*, Englewood Cliffs, N.J.: Prentice-Hall, Inc., 254–55.

Hisada, T., K. Nakagawa, and M. Izumi, 1965, "Earthquake Response of Idealized Twenty Story Buildings Having Various Elasto-Plastic Properties," *Proc. Third World Conf. Earthq. Engrg.*, Auckland and Wellington, New Zealand, 2.168–84.

Hisada, T., and H. Sugiyama, 1966, "Effect of Repeated Loading on the Strength Properties of Wood," *Proc. Internatl. Symp. Effects of Repeated Loading of Materials and Structures*, RILEM-Inst. Ing., *3*.

Hodge, P. G., 1959, *Plastic Analysis of Structures*, New York: McGraw-Hill Inc.

Hoff, G. C., 1964, "Shock-Isolating Backpacking Materials, A Review of the State of the Art," *Proc. Symp. Soil-Structure Interaction*, Tucson: University of Arizona, 138–54.

Horikawa, K., *Tsunami Phenomena in the Light of Engineering View-Point*, Report on the Chilean Tsunami of May 24, 1960, as Observed Along the Coast of Japan, Tokyo: Committee for the Field Investigation of the Chilean Tsunami of 1960, 136–50.

Hoskins, L. M., and L. S. Jacobsen, 1934, "Water Pressure in a Tank Caused by a Simulated Earthquake," *Bull. Seism. Soc. Am.*, *24*(1), 1–32.

Housner, G. W., 1947, "Characteristics of Strong-Motion Earthquakes," *Bull. Seism. Soc. Am.*, *37*(1), 19–31.

———, 1952a, "Spectrum Intensities of Strong-Motion Earthquakes," *Proc. Symp. Earthq. and Blast Effects Structures*, C. M. Duke and M. Feigen, eds., Los Angeles: University of California, 21–36.

———, 1952b, "Bending Vibrations of a Pipe Line Containing Flowing Fluid," *Journ. of App. Mechs.*, *19*(2), 205–8.

———, 1956, "Earthquake Resistant Design Based on Dynamic Properties of Earthquakes," *Journ. ACI*, *28*(1), 85–98.

———, 1959, "Behavior of Structures During Earthquakes," *Proc. ASCE*, *85*(EM4), 109–29.

———, 1961, "Vibration of Structures Induced by Seismic Waves, Part 1," in *Shock and Vibration Handbook*, C. M. Harris and C. E. Crede, eds., New York: McGraw-Hill Book Co. Inc., Chap. 50.

———, 1962, "Fundamentos de Ingeniería Sísmica, *Ingeniería*, *32*(3), 25–55.

———, 1963, "The Dynamic Behavior of Water Tanks," *Bull. Seism. Soc. Am.* *53*(2), 381–87.

———, 1965, "Intensity of Earthquake Ground Shaking Near the Causative Fault," *Proc. Third World Conf. Earthq. Engrg.*, Auckland and Wellington, New Zealand, 3.94–115.

Housner, G. W., and A. G. Brady, 1963, "Natural Periods of Vibration of Buildings," *Proc. ASCE*, *89*(EM4), 31–65.

Housner, G. W., and D. E. Hudson, 1958, "The Port Hueneme Earthquake of March 18, 1957," *Bull, Seism. Soc. Am.*, *48*(2), 163–68.

Housner, G. W., and P. C. Jennings, 1964, "Generation of Artificial Earthquakes," *Proc. ASCE*, *90*(EM1), 113–50.

Housner, G. W., and G. D. McCann, 1949, "The Analysis of Strong-Motion Earthquake Records with the Electric Analog Computer," *Bull. Seism. Soc. Am.*, *39*(1), 47–56.

Hudson, D. E., 1958, "The Wilmot Survey Type Strong-Motion Earthquake Recorder," *Earthq. Engrg. Res. Lab.*, Pasadena: California Institute of Technology.

Hudson, D. E., 1961a, "Scale-Model Principles," in *Shock and Vibration Handbook* C. M. Harris and C. E. Crede, eds., New York: McGraw-Hill Book Co., Inc., Chap. 27.

———, 1961b, "Vibration of Structures Induced by Seismic Waves, Part II," in *Shock and Vibration Handbook*, C. M. Harris and C. E. Crede, eds., New York: McGraw-Hill Book Co., Inc., Chap. 50.

———, 1962a, "Some Problems in the Application of Spectrum Techniques to Strong-Motion Earthquake Analysis," *Bull. Seism. Soc. Am.*, 52(2) 417–30.

———, 1962b, "Synchronized Vibration Generators for Dynamic Tests of Full-Scale Structures," *Earthq. Engrg. Res. Lab.*, Pasadena: California Institute of Technology.

Hudson, D. E., W. O. Keightley, and N. N. Nielsen, 1964, "A New Method for the Measurement of the Natural Periods of Buildings," *Bull. Seism. Soc. Am.*, 54(1), 233–41.

Hult, J., 1966, *Creep in Engineering Structures*. Waltham, Mass.: Blaisdell Publishing Company.

Hunter, S. C., 1960, "Viscoelastic Waves," in *Progress in Solid Mechanics*, vol. 1, I. N. Sneddon and R. Hill, eds., Amsterdam: North-Holland Publishing Co., 1–57.

Husid, R., 1967, "Gravity Effects on the Earthquake Response of Yielding Structures," *Earthq. Engrg. Research Laboratory Report*, California Institute of Technology, Pasadena, Calif.

———, 1969, "The Effect of Gravity on the Collapse of Yielding Structures with Earthquake Excitation," *Proc. Fourth World Conf. Earth. Engrg.*, Santiago, Chile, 2, A-4, 31–43.

Hutchinson, B. G., 1965, "The Evaluation of Pavement Structural Performance," doctoral thesis, University of Waterloo, Ontario, Canada.

Idriss, I. M., 1968, "Finite Element Analysis for the Seismic Response of Earth Banks," *Proc. ASCE*, 94(SM3), 617–36.

Idriss, I. M., and H. B. Seed, 1967, "Response of Earth Banks during Earthquakes," *Proc. ASCE*, 93(SM3), 61–82.

———, 1968, "Seismic Response of Horizontal Soil Layers," *Proc. ASCE*, 94(SM4) 1003–31.

Ihara, M., and C. Ueda, 1965, "Horizontal Loading and Vibration Test on 2-Storied Concrete Structures," *Proc. Third World Conf. Earthq. Engrg.*, Auckland and Wellington, New Zealand, 2.20–36.

Iida, K., 1937, "Determination of Elastic Constants of Superficial Soil and Base-Rock at Maru-no-uti in Tokyo," *Bull. Earthq. Res. Inst.*, 37, 828–36.

———, 1963a, "Magnitude, Energy, and Generation of Tsunamis, and Catalogue of Earthquakes Associated with Tsunamis," Monograph 24, *Proc. Tsunami Meetings Associated with the Tenth Pacific Science Congress*, IUGG, 7–18.

———, 1963b, "On the Estimation of Tsunami Energy," Monograph 24, *Proc. Tsunami Meetings Associated with the Tenth Pacific Science Congress*, IUGG, 167–73.

Ishigaki, H., and N. Hatakeyama, 1960, "Experimental and Numerical Studies on Vibrations of Buildings," *Proc. Second World Conf. Earthq. Engrg.*, Tokyo and Kyoto, Japan, 1263–84.

Iwan, W. D., 1969, "The Distributed-Element Concept of Hysteretic Modeling and its Application to Transient Response Problems," *Proc. Fourth World Conf. Earthq. Engrg.*, Santiago, Chile, 2, A-4, 45–57.

Jacobsen, L. S., 1930, "Motion of a Soil Subjected to a Simple Harmonic Ground Vibration," *Bull. Seism. Soc. Am.*, 20(3), 160–95.

———, 1949, "Impulsive Hydrodynamics of Fluid Inside a Cylindrical Tank and of Fluid Surrounding a Cylindrical Pier," *Bull. Seism. Soc. Am.*, *39*(3), 189–204.

———, 1964, "Vibrational Transfer from Shear Buildings to Ground," *Proc. ASCE*, *90*(EM3), 21–38.

Jacobsen, L. S., and R. S. Ayre, 1951, "Hydrodynamic Experiments with Rigid Cylindrical Tanks Subjected to Transient Motions," *Bull. Seism. Soc. Am.*, *41*(4), 313–46.

———, 1958, *Engineering Vibrations*, New York: McGraw-Hill Book Co., Inc., 112–25.

Japan National Committee on Earthquake Engineering, 1965, "Niigata Earthquake of 1964," *Proc. Third World Conf. Earthq. Engrg.*, Auckland and Wellington, New Zealand, S78–S109.

Jauffred, F. J., 1960, "Carga Viva en Unidades de Habitación y Oficina en el Distrito Federal," *Ingeniería*, *30*(4), 60–75.

Jennings, P. C., 1963, "Response of Simple Yielding Structures to Earthquake Excitation," doctoral thesis, California, Institute of Technology, Pasadena, Calif.

———, 1964, "Periodic Response of a General Yielding Structure," *Proc. ASCE*, *90*(EM2), 131–66.

———, 1965, "Earthquake Response of a Yielding Structure," *Proc. ASCE*, *91*(EM4), 41–68.

Jennings, P. C., G. W. Housner, and N. C. Tsai, 1969. "Simulated Earthquake Motions for Design Purposes," *Proc. Fourth World Conf. Earthq. Engrg.*, Santiago, Chile, *1*, A–1, 145–60.

Jennings, R. L., and N. M. Newmark, 1960, "Elastic Response of Multi-Story Shear Beam Type Structures Subjected to Strong Ground Motions," *Proc. Second World Conf. Earthq. Engrg.*, Tokyo and Kyoto, Japan, 699–717.

Jenschke, V. A., R. W. Clough, and J. Penzien, 1965, "Characteristics of Strong Ground Motions," *Proc. Third World Conf. Earthq. Engrg.*, Auckland and Wellington, New Zealand, 3.125–92.

Johnson, A. I., 1953, "Strength, Safety and Economical Dimensions of Structures," *Bull. Div. of Bldg. Statics and Structural Engrg.*, Royal Inst. of Tech., Stockholm, Sweden, No. 12.

Johnson, J. B., and T. T. Oberg, 1929, "Fatigue Resistance of Some Aluminum Alloys," *Proc. Am. Soc. Test. Mat.*, *39*(2), 339–52.

Jordan, J., R. Black, and C. C. Bates, 1965, "Patterns of Maximum Amplitudes of P_n and P Waves over Regional and Continental Areas," *Bull. Seism. Soc. Am.*, *55*(4), 693–720.

Jorgenson, J. L., and J. E. Goldberg, 1969, "Probability of Plastic Collapse Failure," *Proc. ASCE*, *95*(ST8), 1743–61.

Joshi, R. N., 1960, "Striking Behavior of Structures in Assam Earthquakes," *Proc. Second World Conf. Earthq. Engrg.*, Tokyo and Kyoto, Japan, 2143–58.

Kanai, K., 1950, "The Effect of Solid Viscosity of Surface Layer on the Earthquake Movements," *Bull. Earthq. Res. Inst.*, University of Tokyo, *28*(1, 2) 31–35.

———, 1957, "Semi-Empirical Formula for the Seismic Characteristics of the Ground," *Bull. Earthq. Res. Inst.*, University of Tokyo, *35*(2), 309–25.

———, 1961, "An Empirical Formula for the Spectrum of Strong Earthquake Motions," *Bull. Earthq. Res. Inst.*, University of Tokyo, *39*(1), 85–96.

Kanai, K., T. Tanaka, and K. Osada, 1954, "Measurement of the Micro-Tremor. I," *Bull. Earthq. Res. Inst.* University of Tokyo, *32*(2) 199–209.

Kanai, K., T. Tanaka, and S. Yoshizawa, 1959, "Comparative Studies of Earthquake Motions on the Ground and Underground (Multiple Reflection Problem)," *Bull. Earthq. Res. Inst.*, University of Tokyo, *37*, 53–88.

Kanai, K., K. Hirano, S. Yoshizawa, and T. Asada, 1966, "Observation of Strong Earthquake Motions in Matsushiro Area. Part 1. Empirical Formulae of Strong Earthquake Motions," *Bull. Earthq. Res. Inst.*, University of Tokyo, *44*, 1269–96.

Karal, F. C., and J. B. Keller, 1964, "Elastic, Electromagnetic and Other Waves in a Random Medium," *Journ. Math. Phys.*, *5*, 537–47.

Kasiraj, I., and J. T. P. Yao, 1969, "Fatigue Damage in Seismic Structures," *Proc. ASCE*, *95*(ST8), 1673–92.

Kato, B., and H. Akiyama, 1969, "The Ultimate Strength of the Steel Structures Subjected to Earthquakes," *Proc. Fourth World Conf. Earthq. Engrg.*, Santiago, Chile, 2, A-4, 59–70.

Keightley, W. O., G. W. Housner, and D. E. Hudson, 1961, "Vibration Tests of the Encino Dam Intake Tower," *Earthq. Engrg. Res. Lab.*, Pasadena: California Institute of Technology.

Khan, F. R., and J. A. Sbarounis, 1964, "Interaction of Shear Walls and Frames," *Proc. ASCE*, *90*(ST3), 285–335.

Khanna, J., 1969, "Elastic Soil-Structure Interaction," *Proc. Fourth World Conf. Earthq. Engrg.*, Santiago, Chile, 3, A-6, 143–52.

Kishida, H., and K. Matsushita, 1969, "Soil-Structure Interaction of the Elevator Tower and of Concrete Footings," *Proc. Fourth World Conf. Earthq. Engrg.*, Santiago, Chile, 3, A-6, 101–15.

Klotter, K., 1962, "Nonlinear Vibrations," in *Handbook of Engineering Mechanics*, W. Flügge, ed., New York: McGraw-Hill Book Co., Inc., Chap. 65.

Knopoff, L., 1956, "The Seismic Pulse in Materials Possessing Solid Friction. 1. Plane Waves," *Bull. Seism. Soc. Am.*, *46*(3), 175–83.

———, 1964, "The Statistics of Earthquakes in Southern California," *Bull. Seism. Soc. Am.*, *54*(6), 1871–73.

Knopoff, L., and J. A. Hudson, 1964, "Scattering of Elastic Waves by Small Inhomogeneities," *Journ. Acoustic. Soc. Am.*, *36*, 338–43.

Kobayashi, H., 1965, "The Deflection of Tall Building due to Earthquake" *Proc. Third World Conf. Earthq. Engrg.*, Auckland and Wellington, New Zealand, 2.321–41.

Kobori, T., 1962, "Dynamical Response of Rectangular Foundations on an Elastic Space," *Proc. Japan Natl. Symp. Earthq. Engrg.*, Tokyo, Japan, 81–86.

Kobori, T., R. Minai, and Y. Inoue, 1969, "On Earthquake Response of Elasto-Plastic Structure Considering Ground Characteristics," *Proc. Fourth World Conf. Earthq. Engrg.*, Santiago, Chile, 3, A-6, 117–32.

Kodera, J., 1965, "Some Tendencies in the Failures of Bridges and their Foundations during Earthquakes," *Proc. Third World Conf. Earthq. Engrg.*, Auckland and Wellington, New Zealand, 5.86–95.

Koh, T., H. Takase, and T. Tsugawa, 1969, "Torsional Problems in Aseismic Design of High-Rise Buildings," *Proc. Fourth World Conf. Earthq. Engrg.*, Santiago, Chile, 2, A-4, 71–87.

Kompaneets, A.S., 1956, "Shock Waves in Plastic Compacting Media," *Proc. Acad. Sci. USSR*, *106*(1), 49–52.

Konishi, I., and Y. Yamada, 1965, "Earthquake Response and Earthquake-Resistant Design of Long Span Suspension Bridges," *Proc. Third World Conf. Earthq. Engrg.*, Auckland and Wellington, New Zealand, 4.312–23.

Korener, B. G., V. A. Iljichjov, and L. M. Reznikov, 1969, "Oscillations of Tower-like Structures with Account of Inertia and Elasticity of Solid Medium," *Proc. Fourth World Conf. Earthq. Engrg.*, Santiago, Chile, *3*, A-6, 167-81.

Kotsubo, S., 1959, "Dynamic Water Pressure on Dam due to Irregular Earthquakes," *Mem. Fac. Engrg.*, Kyushu University, *18*(4), 119-29.

———, 1961, "External Forces on Arch Dams during Earthquakes. A Study of the Aseismic Design of Arch Dams. 1." *Mem. Fac. Engrg.*, Kyushu University, *20*(4), 327-66.

Krishna, J., and B. Chandra, 1969, "Strengthening of Brick Buildings in Seismic Zones," *Proc. Fourth World Conf. Earthq. Engrg.*, Santiago, Chile, *3*, B-6, 11-20.

Krishnasamy, S., and A. N. Sherbourne, 1966, "Mild Steel Structures under Reversed Loading," *Proc. Internatl. Symp. Effects of Repeated Loading of Materials and Structures*, RILEM-Inst. Ing., *4*.

Lamb, H., 1945, *Hydrodynamics*, New York: Dover Publications, Inc.

Lawson, A. C., ed., 1908 and 1910, *The California Earthquake of April 18, 1906*, Report to the State Earthquake Investigation Commission, *1*(1908), *2*(1910).

Lazan, B. J., 1968, *Damping of Materials and Members in Structural Mechanics*, New York: Pergamon Press, Inc.

Lazan, B. J., and L. E. Goodman, 1961, "Material and Interface Damping," in *Shock and Vibration Handbook*, C. M. Harris and C. E. Crede, eds., New York: McGraw-Hill Book Co. Inc., Chap. 36.

Lee, K. L., and H. B. Seed, 1967, "Cyclic Stress Conditions Causing Liquefaction of Sand," *Proc. ASCE*, *93*(SM1), 47-70.

Leet, L. D., 1948, *Causes of Catastrophe: Earthquakes, Volcanoes, Tidal Waves, and Hurricanes*, New York: McGraw-Hill Book Co., Inc.

———, 1950, "Earth Waves," *Harvard Monographs in Applied Science, No. 2*, Cambridge, Mass.: Harvard University Press. Reprinted by Johnson Reprint Corporation, New York.

Lightfoot, E., 1956, "The Analysis for Wind Loading of Rigid Jointed Multi-Story Building Frames," *Civil Engrg. and Public Works Rev.*, London, *51*, 602.

Lomnitz, C., 1970, "Some Observations of Gravity Waves in the 1960 Chile Earthquake," *Bull. Seism. of Am.*, 669-70.

Lunoe, R. R., and G. A. Willis, 1957, "Applications of Steel Strap Reinforcement to Girders of Rigid Frames, Special AMC Warehouse," *Proc. ACI*, *53*(7) 669-78.

Lysmer, J., 1965, "Vertical Motion of Rigid Footings," doctoral thesis, University of Michigan, Ann Arbor, Mich.

Machlan, G. R., and W. H. Edmunds, eds., 1962, *Third Semi-Annual Polar Glass Reinforced Plastics Research and Development Conference*, Owens-Corning Fiberglass Corp.

Marguerre, K., 1960, "Matrices of Transmission in Beam Problems," in *Progress in Solid Mechanics*, I. N. Sneddon and R. Hill, eds., Amsterdam: North-Holland Publishing Co., 59-82.

Marsal, R. J., 1963a, Internal Report, Institute of Engineering, National University of Mexico, Mexico.

———, 1963b, *Informe Sobre Pruebas Triaxiales Efectuadas con Suelos Granulares y Materiales para Enrocamientos*, Internal Report, Instituto de Ingeniería, Mexico.

Marsal, R. J., E. Moreno, A. Núñez, R. Cuéllar, and R. Moreno, 1965, "Research on the Behavior of Granular Materials and Rockfill Samples," Mexico: Comisión Federal de Electricidad.

REFERENCES

Masing, G., 1926, "Eigenspannungen und Verfestigung beim Messing," *Proc. Second Internatl. Congr. App. Mechs.*, Zurich,

Maslov, N. N., 1957, "Questions of Seismic Stability of Submerged Sandy Foundations and Structures," *Proc. Fourth World Conf. Soil Mechs. and Foundation Engrg.*, *1*, 368–71.

Matsushita, K., and M. Izumi, 1965, "Some Analyses on Mechanism to Decrease Seismic Force Applied to Buildings," *Proc. Third World Conf. Earthq. Engrg.*, Auckland and Wellington, New Zealand, 4.342–59.

Matsushita, K., and M. Izumi, 1969, "Studies on Mechanisms to Decrease Earthquake Forces Applied to Buildings," *Proc. Fourth World Conf. Earthq. Engrg.*, Santiago, Chile, *2*, B–3, 117–29.

Matuo, H., and S. O–Hara, 1965, "Dynamic Pore Water Pressure Acting on Quay Walls during Earthquakes," *Proc. Third World Conf. Earthq. Engrg.*, Auckland and Wellington, New Zealand, 1.130–42.

McKaig, T. H., 1962, *Building Failures*, New York: McGraw-Hill Book Co., Inc.

Medearis, K., 1966, "Static and Dynamic Properties of Shear Structures," *Proc. Internatl. Symp. Effects of Repeated Loading of Materials and Structures*, RILEM-Inst. Ing., *6*.

Medvedev, S. V., 1953, "Novaya Seysmicheskaya Shkala, *Trudy Geophyz. Inst. Akad. Naut. USSR*, *21*, 148.

Medvedev, S. V., and W. Sponheuer, 1969, "Scale of Seismic Intensity," *Proc. Fourth World Conf. Earthq. Engrg.*, Santiago, Chile, *1*, A–2, 143–53.

Meli, R., and R. Díaz de Cossío, 1964, "Evaluación de Daños en un Miembro de Concreto Reforzado," *Rev. IMCYC*, *2*(11), 38–67.

Meli, R., and L. Esteva, 1968, "Comportamiento de Muros de Mampostería Hueca ante Carga Lateral Alternada," *Segundo Congreso Nacional de Ingeniería Sísmica*, Veracruz, Mexico,

Mercalli, G., 1902, *Boll. Soc. Sism. Ital.*, *8*, 184–91.

Miles, J. W., 1954, "On Structural Fatigue under Random Loading," *Journ. Aero. Sci.*, *21*, 753–62.

Minami, I., 1969, "On a Vibration Characteristic of Fill Dams in Earthquakes," *Proc. Fourth World Conf. Earthq. Engrg.*, Santiago, Chile, *3*, A–5, 101–15.

Minami, J. K. 1965, "Relocation and Reconstruction of the Town of Barce, Cyrenaica, Libya, Damaged by the Earthquake of 21 February, 1963," *Proc. Third World Conf. Earthq. Engrg.*, Auckland and Wellington, New Zealand, 5.96–108.

Minami, J. K., and J. Sakurai, 1969, "Some Effects of Substructure and Adjacent Soil Interaction on the Seismic Response of Buildings," *Proc. Fourth World Conf. Earthq. Engrg.*, Santiago, Chile, *3*, A-6, 71–86.

Mindlin, R. D., 1965, "Influence of Couple-Stresses on Stress Concentrations," *Exptl. Mechs.*, *3*(1), 1–7.

Miner, M. A., 1945, "Cumulative Damage in Fatigue," *Journ. App. Mechs.*, *12*(1), A159–64.

Mizuhata, K., 1969, "Low Cycle Fatigue under Multi-Axial Stress Conditions," *Proc. Fourth World Conf. Earthq. Engrg.*, Santiago, Chile, *1*, B–2, 31–46.

Monge, J. E., and L. A. Rosenberg, 1964, "Análisis Dinámico de Estructuras Fundadas en Suelo Elástico. Segunda Parte: Interacción Dinámica Entre Estructura y Suelo," *Segundo Simposio Panamericano de Estructuras*, Lima, Perú, *1*, 201–10.

Montes, R., and E. Rosenblueth, 1968, "Cortantes y Momentos Sísmicos en Chimeneas," *Segundo Congreso Nacional de Ingeniería Sísmica*, Veracruz, Mexico.

Morduchow, M., 1961, "On Classical Normal Modes of a Damped Linear System," *Journ. of App. Mechs.*, *28*, Trans. *ASME*, *88*, Series E, 458.

Müller, L., 1964, "The Rock Slide in the Vajont Valley," *Rock Mechs. Engrg., Geol.*, *2*(3–4), 148–212.

Murphy, M. J., and G. N. Bycroft, 1956, "The Response of a Nonlinear Oscillator to an Earthquake," *Bull. Seism. Soc. Am.*, *46*(1), 57–65.

Murphy, M. J., G. N. Bycroft, and L. W. Harrison, 1956, "Electrical Analog for Earthquake Shear Stresses in a Multi-Story Building," *Proc. First World Conf. Earthq. Engrg.*, Berkeley, Calif., 9-1-19.

Muto, K., 1969, "Earthquake Resistant Design of 36-Storied Kasumigaseki Building," *Proc. Fourth World Conf. on Earthq. Eng.*, Santiago, Chile, *3*, J–4, 15–33.

Naito, T., 1960, "Fifty Years of Earthquake Engineering Practice," *Proc. Second World Conf. Earthq. Engrg.*, Tokyo and Kyoto, Japan, 127–32.

Nakano, K., 1965, "Experiment on Behavior of Prestressed Concrete Four-Storied Model Structure under Lateral Force," *Proc. Third World Conf. Earthq. Engrg.*, Auckland and Wellington, New Zealand, 4.572–90.

Napedvaridze, Sh. G., 1959, "Seismostoikest Gidroteckhnicheskii Soorullenii," *Gosstroiizdat*, Moscow.

Nath, B., 1969, "Hydrodynamic Pressures on Arch Dams during Earthquakes," *Proc. Fourth World Conf. Earthq. Engrg.*, Santiago, Chile, *2*, B–4, 97–105.

National Research Council–National Academy of Engineering, 1969, *Earthquake Engineering Research*, Washington, D.C.: National Academy of Sciences.

Neumann, F., 1954, *Earthquake Intensity and Related Ground Motion*, Seattle: Washington University Press.

Newmark, N. M., 1943, "Numerical Procedure for Computing Deflections, Moments, and Buckling Loads," *Trans. ASCE*, *108*, 1161–1234.

———, 1947, "Influence Charts for Computation of Vertical Displacements in Elastic Foundations," University of Illinois, Urbana: *Bull. 367, Engrg. Exper. Sta.*

———, 1952, "A Review of Cumulative Damage in Fatigue," *Symp. Fatigue and Fracture of Metals*, W. M. Murray, ed., New York: John Wiley & Sons, Inc., 197–229.

———, 1959, "A Method of Computation for Structural Dynamics," *Proc. ASCE*, *85*(EM3), 67–94.

———, 1962, "A Method of Computation for Structural Dynamics," *Trans. ASCE*, *127*, 1406–35.

———, 1965a, "Current Trends in the Seismic Analysis and Design of High Rise Structures," *Proc. Symp. Earthq. Engrg.*, University of British Columbia, 6.1–55.

———, 1965b, "Effects of Earthquakes on Dams and Embankments," Fifth Rankine Lecture, *Géotechnique*, *15*(2), 139–60.

———, 1968, "Problems in Wave Propagation in Soil and Rock," *Symp. Wave Propagation and Dynamic Properties of Earth Materials*, University of New Mexico, Albuquerque, 7–26.

———, 1969a, "Torsion in Symmetrical Buildings," *Proc. Fourth World Conf. Earthq. Engrg.*, Santiago, Chile, *2*, A–3, 19–32.

———, 1969b, "Design Criteria for Nuclear Reactors Subjected to Earthquake Hazards," *Proc. IAEA Panel on Aseismic Design and Testing of Nuclear Facilities*, Tokyo: Japan Earthquake Engineering Promotion Society, 90–113.

———, 1970, "Current Trends in the Seismic Analysis and Design of High-Rise Structures," *Earthquake Engineering*, R. L. Wiegel, ed., Englewood Cliffs, N.J.: Prentice-Hall, Inc., 403–24.

Newmark, N. M., and W. J. Hall, 1968, "Dynamic Behavior of Reinforced and Prestressed Concrete Buildings under Horizontal Forces and the Design of Joints (Including Wind, Earthquake, Blast Effects)," *Proc. Eighth Cong. Internatl. Assoc. for Bridge and Structural Engrg.*, New York, 585–613.

———, 1969, "Seismic Design Criteria for Nuclear Reactor Facilities," *Proc. Fourth World Conf. Earthq. Engrg.*, Santiago, Chile, 2, B-4, 37–50.

Nielsen, N. N., 1969, "Dynamic Response of a 90-ft Steel Frame Tower," *Proc. Fourth World Conf. Earthq. Engrg.*, Santiago, Chile,

Nieto, J. A., and J. A. Díaz, 1969, "Material Deleznable para Modelos Dinámicos," *Ingeniería*, 39(3), 340–50.

Nieto, J. A., E. Rosenblueth, and O. A. Rascón, 1965, "Modelo Matemático para Representar la Interacción Dinámica del Suelo y Cimentación," *Primer Congreso Nacional de Ingeniería Sísmica*, Guadalajara, Mexico. Also *Ingeniería* 36(1), 117–25.

Nigam, N. C., and G. W. Housner, 1969, "Elastic and Inelastic Responses of Framed Structures during Earthquakes," *Proc. Fourth World Conf. Earthq. Engrg.*, Santiago, Chile, 2, A-4, 89–104.

Norris, C. H., and J. B. Wilbur, 1960, *Elementary Structural Analysis*, New York: McGraw-Hill Book Co., Inc.

Nuclear Reactors and Earthquakes, 1963, prepared by Lockheed Aircraft Corporation and Holmes & Narver, Inc., Washington, D.C.: U.S. Atomic Energy Commission.

Nunnally, S. W., 1966, "Development of a Liquefaction Index for Cohesionless Soils," doctoral thesis, Northwestern University, Evanston, Ill.

Ohsaki, Y., 1969, "The Effects of Local Soil Conditions upon Earthquake Damage," *Soil Dynamics Specialty Conf. Seventh Internatl. Conf. Soil Mechs. and Foundation Engrg.*, Mexico, 3–32.

Okamoto, S., K. Kato, M. Hakuno, and Y. Miyakoshi, 1961, "Observations of Earthquakes on an Arch Dam," *Trans. Japan Soc. Civil Engrs.*, 76, 1–11.

Okamoto, S., M. Hakuno, K. Kato, and M. Otawa, 1964, "Observation of Earthquakes on an Arch Dam and Its Abutment," *Trans. Japan Soc. Civil Engrs.*, 112, 20–27.

Okamoto, S., and K. Kato, 1969, "A Method of Dynamic Model Test of Arch Dams," *Proc. Fourth World Conf. Earthq. Engrg.*, Santiago, Chile, 1, B-1, 87–97.

Okamoto, S., C. Tamura, and K. Kato, 1968, "Nonlinear Behaviors of the Earth Dam during Earthquakes," *Bull. Earthq. Resis. Structure Res. Center*, University of Tokyo, 2, 1–16.

Osawa, Y., T. Tanaka, M. Murakami, and Y. Kitagawa, 1969, "Earthquake Measurements in and around a Reinforced Concrete Building," *Proc. Fourth World Conf. Earthq. Engrg.*, Santiago, Chile, 1, B-1, 1–16.

Parmelee, R. A., 1967, "Building-Foundation Interaction Effects," *Proc. ASCE*, 93(EM2), 131–52.

———, 1969, "Universities Research Reports: Northwestern University," *Report on NSF–UCEER Conf. on Earthq. Engrg. Res.*, Universities Council for Earthquake Engineering Research, Pasadena, Calif., 96–105.

Parzen, E., 1964, *Stochastic Processes*, San Francisco: Holden–Day, Inc.

Patel, V. J., and K. L. Arora, 1965, "Existence of the Critical Surface in Earth Dams during Earthquakes," *Proc. Third World Conf. Earthq. Engrg.*, Auckland and Wellington, New Zealand 4.441–48.

Patel, V. J., and S. D. Bokil, 1962, "Effect of Earthquake on Pore Pressure in Earth Dam," *Proc. Second Symp. Earthq. Engrg.*, Roorkee, India, 1–10.

Paulay, T., 1969, "The Coupling of Reinforced Concrete Shear Walls," *Proc. Fourth World Conf. Earthq. Engrg.*, Santiago, Chile, *1*, B–2, 75–90.

Peck, R. B., W. E. Hanson, and T. H. Thornburn, 1953, *Foundation Engineering*, New York: John Wiley & Sons, Inc.

Pekeris, C. L., 1955, "The Seismic Surface Pulse," *Proc. NAS*, *41*(7), 469–480.

Penzien, J., 1960, "Elasto-Plastic Response of Idealized Multi-Story Structures Subjected to a Strong Motion Earthquake," *Proc. Second World Conf. Earthq. Engrg.*, Tokyo and Kyoto, Japan, 739–60.

———, 1969, "Earthquake Response of Irregularly Shaped Buildings," *Proc. Fourth World Conf. Earthq. Engrg.*, Santiago, Chile, *2*, A–3, 75–89.

Penzien, J., and S.-C. Liu, 1969, "Nondeterministic Analysis of Nonlinear Structures Subjected to Earthquake Excitations," *Proc. Fourth World Conf. Earthq. Engrg*, Santiago, Chile, *1*, A–1, 114–29.

Penzien, J., Ch. F. Scheffey, and R. A. Parmelee, 1964, "Seismic Analysis of Bridges on Long Piles," *Proc. ASCE*, *90*(EM3), 223–54.

Peters, R. B., 1968, "Strong-Motion Accelerograph Evaluation," *Earthq. Engrg Res. Lab.*, Pasadena: California Institute of Technology.

Pfister, J. F., and E. Hognestad, 1964, "High Strength Bars as Concrete Reinforcement. Part 6, Fatigue Tests," *Journ. PCA Res. Devel. Labs.*, *6*(1), 65–76.

Poceski, A., 1969, "Response Spectra for Elastic and Elastoplastic Systems Subjected to Earthquakes of Short Duration," *Proc. Fourth World Conf. Earthq. Engrg.*, Santiago, Chile, *2*, A–4, 171–78.

Polyakov, S. V., B. E. Denisov, T. Zh. Zhunosov, V. I. Konovodchenko, and A. V. Cherkashin, 1969, "Investigations into Earthquake Resistance of Large-Panel Buildings," *Proc. Fourth World Conf. Earthq. Engrg.*, Santiago, Chile, *1*, B–1, 165–80.

Popov, E. P., 1966, "Low-Cycle Fatigue of Steel Beam-to-Column Connections," *Proc. Internatl. Symp. Effects of Repeated Loading of Materials and Structures*, RILEM-Inst. Ing., *6*.

Popov, E. P., and R. E. McCarthy, 1960, "Deflection Stability of Frames under Repeated Loads," *Proc. ASCE*, *86*(EM1), 61–78.

Popov, E. P., and R. B. Pinkney, 1969, "Reliability of Steel Beam-to-Column Connections under Cyclic Loading," *Proc. Fourth World Conf. Earthq. Engrg.*, Santiago, Chile, *2*, B–3, 15–30.

Prager, W., 1957, "On Ideal Locking Materials," *Trans. Soc. of Rheol.*, *1*, 169–75.

Prince, J., 1963, "Un Nuevo Sismoscopio," *Ingeniería*, *33*(1), 69–72.

Raiffa, H., and R. Schlaifer, 1961, *Applied Statistical Decision Theory*, Cambridge, Mass.: Harvard University Press.

Ramberg, W., and W. T. Osgood, 1943, "Description of Stress-Strain Curves by Three Parameters," *Technical Note 902. NACA*.

Raphael, J. M., 1960, "The Effect of Lateral Earthquake on a High Buttress Dam," *Proc. Second World Conf. Earthq. Engrg.*, Tokyo and Kyoto, Japan, 1791–1801.

Rascón, O. A., 1965a, "Estudios Encaminados a la Predicción de Espectros de Temblores en el Valle de México," *Ingeniería*, *35*(1), 41–58.

———, 1965b, "Efectos Sísmicos en Estructuras en Forma de Péndulo Invertido," *Rev. Soc. Mex. Ing. Sism.*, *3*(1), 8–16.

———, 1967a, "Stochastic Model to Fatigue," *Proc. ASCE*, *93*(EM3), 147–55.

———, 1967b, "Estudio Teórico y Estadístico de las Componentes de Traslación del Suelo durante un Sismo," *Ingeniería*, *37*(4), 384–88.

Rascón, O. A., and C. A. Cornell, 1969, "A Physically Based Model to Simulate Strong Earthquake Records on Firm Ground," *Proc. Fourth World Conf. Earthq. Engrg.*, Santiago, Chile, *1*, A-1, 84–96.

Rasskazovsky, V., and K. Abdurashidov, 1969, "Restoration of Stone Buildings after Earthquake," *Proc. Fourth World Conf. Earthq. Engrg.*, Santiago, Chile, *3*, B–6, 83–91.

Ravara, A. P., 1968, "Comportamento de Estructuras de Betão Armado sob a Accão dos Sismos," Tese, Latoratório Nacional de Engenharia Civil, Lisbon, Portugal.

Rayleigh, Lord, 1945, *Theory of Sound*, New York: Dover Publications, Inc., 130–31.

Rea, D., R. W. Clough, J. E. Bouwkamp, and U. Vogel, 1969, "Damping Capacity of a Model Steel Structure," *Proc. Fourth World Conf. Earthq. Engrg.*, Santiago, Chile, *1*, B–2, 63–73.

Reimbert, M., and A. Reimbert, 1962, *Les Silos Agricoles et Industriels*, Paris: Dunod.

Reissner, E., 1936, "Stationäre, Axialsymmetrische, durch eine Schüttelnde Masse Erregte Schwingungen Eines Homogenen Elastischen Halbbraumes," *Ing. Arch. 7*, 381–96.

Richard, R. M., and J. E. Goldberg, 1965, "Analysis of Nonlinear Structures: Force Method," *Proc. ASCE*, *91*(ST6), 33–48.

Richart, F. E., 1962, "Foundation Vibrations," *Trans. ASCE*, *127*(1), 863–925.

Richart, F. E., A. Brandtzaeg, and R. L. Brown, 1928, "A Study of the Failure of Concrete under Combined Compressive Stresses," University of Illinois, Urbana: *Bull. Engrg. Exper. Sta.*, *185*.

Richter, C. F., 1958, *Elementary Seismology*, San Francisco: W. H. Freeman and Company.

———, 1959, "Seismic Regionalization," *Bull. Seism. Soc. Am.*, *49*(2), 123–62.

Rinne, J. E., 1960, "Design Criteria for Shear and Overturning Moment," *Proc. Second World Conf. Earthq. Engrg.*, Tokyo and Kyoto, Japan, 1709–23.

Riznichenko, Y. V., 1962, "Seismic Magnitudes of Underground Nuclear Explosions," *Trans. O. Yu. Schmidt Inst. of Geophys.*, *15*(182), 43–69.

Robertson, I. A., 1966, "Forced Vertical Vibration of a Rigid Circular Disc on a Semi-Infinite Elastic Solid," *Proc. Camb. Phil. Soc.*, *62*, 41.

Rocha, M., 1965, *Mechanical Behavior of Rock Foundations in Concrete Dams*, Memória No. 244, Laboratório Nacional de Engenharia Civil, Lisbon, Portugal.

Rodríguez-Caballero, M., 1966, "Algunos Modelos Matemáticos para la Solución Optima del Problema de Diseño Sísmico de Estructuras," *Ingeniería*, *36* (4), 420–32.

Roesset, J. M., and R. V. Whitman, 1969, "Theoretical Background for Amplification Studies. 5, Effect of Local Soil Conditions upon Earthquake Damage," *Soils Publication 231*, R69-15, Massachusetts Institute of Technology, Cambridge, Mass.

Rojansky, V., 1948, "Gyrograms for Simple Harmonic Systems Subjected to External Forces," *Journ. of App. Phys.*, *19*(3), 297–301.

Rosenblueth, E., 1951, "A Basis for Aseismic Design," doctoral thesis, University of Illinois, Urbana, Ill.

———, 1952a, "Diseño Sísmico de las Estructuras Simples," *Ediciones ICA*, *B*, 10.

———, 1952b, "Diseño Sísmico de las Estructuras Elásticas, *Ediciones ICA*, *B*, 13.

———, 1959, "Teoría de la Carga Viva en Edificios," *Ingeniería*, *29*(4), 51–72.

———, 1960, "Aseismic Provisions for the Federal District, Mexico," *Proc. Second World Conf. Earthq. Engrg.*, Tokyo and Kyoto, Japan, 2009–26.

———, 1961, "Temblores Chilenos de Mayo 1960: Sus Efectos en Estructuras Civiles," *Ingeniería*, *31*(1), 1–31.
———, 1964a, discussion of Arias and Husid (1962b), *Rev. IDIEM*, *3*(1), 63–65.
———, 1964b, "Tratamiento Inelástico," *Diseño Sísmico de Estructuras*, Part 3 6, 8.1–8.33, Soc. Mex. Ing. Sísmica.
———, 1964c, Probabilistic Design to Resist Earthquakes," *Proc. ASCE*, *90*(EM5), 189–219.
———, 1965, "Slenderness Effects in Buildings," *Proc. ASCE*, *91*(ST1), 229–52.
———, 1966, "Consideraciones Sobre el Diseño Sísmico," *Ingeniería Civil*, *132*, 6–25.
———, 1968a, "Presión Hidrodinámica en Presas Debida a Aceleración Vertical con Refracción en el Fondo," *Segundo Congreso Nacional de Ingeniería Sísmica*, Veracruz, Mexico, *3*.
———, 1968b, "Presión Hidrodinámica en Cortinas de Gravedad," *Inst. Ing.* Publ. No. 161.
———, 1968c, "Sobre la Respuesta Sísmica de Estructuras de Comportamiento Lineal," *Segundo Congreso Nacional de Ingeniería Sísmica*, Veracruz, Mexico, *1*.
———, 1969a, "Seismicity and Earthquake Simulation," *Report on NSF–UCEER Conf. on Earthq. Engrg. Res.*, Universities Council for Engineering Research, Pasadena, Cal., 47–64.
———, 1969b, "Confiabilidad y Utilidad en Ingeniería," *Ingeniería*, *39*(3), 421–35.
Rosenblueth, E., and J. I. Bustamente, 1962, "Distribution of Structural Response to Earthquakes," *Proc. ASCE*, *88*(EM3), 75–106.
Rosenblueth, E., and R. Díaz de Cossío, 1964, "Instability Considerations in Limit Design of Concrete Frames," *Proc. Internatl. Symp. Flexural Mechanics of Reinforced Concrete*, Miami, Fla., 439–63.
Rosenblueth, E., and J. Elorduy, 1969a, "Responses of Linear Systems to Certain Transient Disturbances," *Proc. Fourth World Conf. Earthq. Engrg.*, Santiago, Chile, *1*, A–1, 185–96.
———, 1969b, "Características de los Temblores en la Arcilla de la Ciudad de Mexico," *Nabor Carrilo. El Hundimiento de la Ciudad de México y Proyecto Texcoco*, contribución de Proyecto Texcoco al Séptimo Congreso Internacional de Mecánica de Suelos e Ingeniería de Cimentaciones. Secretaría de Hacienda y Crédito Público, Fiduciaria: Nacional Financiera, S. A., Mexico, 287–328.
Rosenblueth, E., and L. Esteva, 1962, "Folleto Complementario: Diseño Sísmico de Edificios, Proyecto de Reglamento de las Construcciones en el Distrito Federal," Ediciones Ingeniería, Mexico.
Rosenblueth, E., and I. Herrera, 1964, "On a Kind of Hysteretic Damping," *Proc. ASCE*, *90*(EM4), 37–48.
Rosenblueth, E., R. J. Marsal, and F. Hiriart, 1958, "Los Efectos del Terremoto del 28 de Julio y la Consiguiente Revisión de los Criterios para el Diseño Sísmico de Estructuras," *Ingeniería*, *28*(1), 1–28.
Rosenblueth, E., and J. Prince, 1965, "El Temblor de San Salvador, 3 de Mayo 1965: Ingeniería Sísmica," *Rev. Soc. Mex. Ing. Sísm.*, *3*(2), 33–60.
Roy, H. E. H., and M. A. Sozen, 1964, "Ductility of Concrete," *Proc. Internatl. Symp. Flexural Mechanics of Reinforced Concrete*, Miami, Fla., 213–35.
Royles, R., 1966, "Mild Steel Beams under Low Cycle Fatigue Conditions," *Proc. Internatl. Symp. Effects of Repeated Loading of Materials and Structures*, RILEM-Inst. Ing., *4*.

Rubin, S., 1961, "Concepts in Shock Data Analysis," in *Shock and Vibration Handbook*, C. M. Harris and C. E. Crede, eds., New York: McGraw-Hill Book Co., Inc, Chap. 23.

Rudolph, E., and E. Tams, 1907, "Seismogramme des Nordpasifischen und Sudamerikanischen Erdheben rom 16 August 1906," *Inter. Seis. Ass.*, I.E., 98.

Ruge, A. C., 1938, "Earthquake Resistance of Elevated Water Tanks," *Trans. ASCE*, *103*, 889–949.

Rumman, W. S., 1967, "Earthquake Forces in Reinforced Concrete Chimneys," *Proc. ASCE*, *93*(ST6), 55–70.

Rüsch, H., 1960, "Researches Toward a General Flexural Theory for Structural Concrete," *Proc. ACI*, *57*(1), 1–28.

Sakurai, A., and T. Takahashi, 1969, "Dynamic Stresses of Underground Pipe Lines during Earthquakes," *Proc. Fourth World Conf. Earthq. Engrg.*, Santiago, Chile, 2, B-4, 81–96.

Salmon, E. H., 1953, *Materials and Structures*, vol. 1, London: Longman Group Limited.

Salvadori, M. G., and M. L. Baron, 1952, *Numerical Methods in Engineering*, Englewood Cliffs, N.J.: Prentice-Hall, Inc., 91–132.

Salvadori, M. G., R. Skalak, and P. Weidlinger, 1960, "Waves and Shocks in Locking Dissipative Media," *Proc. ASCE*, *86*(EM3), 77–105.

———, 1961, "Spherical Waves in a Plastic Locking Medium," *Proc. ASCE*, *87*(EM1), 1–11.

Sandi, H., 1960, "A Theoretical Investigation of the Interaction between Ground and Structure during Earthquakes," *Proc. Second World Conf. Earthq. Engrg.*, Tokyo and Kyoto, Japan, 1327–43.

Saul, W. E., J. F. Fleming, and S. L. Lee, 1965, "Dynamic Analysis of Bilinear Inelastic Multiple Story Shear Buildings," *Proc. Third World Conf. Earthq. Engrg.*, Auckland and Wellington, New Zealand, 2.533–51.

Schlaifer, R., 1959, *Probability and Statistics for Business Decisions*, New York: McGraw-Hill Book Co., Inc.

Seed, H. B., 1960, "Soil Strength during Earthquakes," *Proc. Second World Conf. Earthq. Engrg.*, Tokyo and Kyoto, Japan, 183–95.

———, 1966, "A Method for Earthquake Resistant Design of Earth Dams," *Proc. ASCE*, *92*(SM1), 13–41.

Seed, H. B., and R. E. Goodman, 1964, "Earthquake Stability of Slopes of Cohesionless Soils," *Proc. ASCE*, *90*(SM6), 43–73.

Seed, H. B., and I. M. Idriss, 1966, "An Analysis of Soil Liquefaction in the Niigata Earthquake," *Proc. ASCE 93*(SM3), 83–108.

Seed, H. B., I. M. Idriss, and F. W. Kiefer, 1968, "Characteristics of Rock Motions during Earthquakes," *Earthq. Engrg. Res. Center*, Report EERC 68-5. Berkeley: University of California.

Seed, H. B., and K. L. Lee, 1966, "Liquefaction of Saturated Sands during Cyclic Loading," *Proc. ASCE*, *92*(SM6), 105–34.

Seling, E. T., 1964, "Characteristics of Stress Wave Propagation in Soil," *Proc. Symp. Soil-Structure Interaction*, University of Arizona, Tucson, Ariz., 27–61.

Sezawa, K., 1927a, "Dispersion of Elastic Waves," *Bull. Earthq. Res. Inst.*, Univ. of Tokyo, 3, 1–18.

———, 1927b, "On the Decay of Waves in Visco-Elastic Soil Bodies," *Bull. Earthq. Res. Inst*, 3, 43–54.

Sezawa, K., and K. Kanai, 1932, "Vibrations of a Single-Storied Framed Structure," *Bull. Earthq. Res. Inst.*, Univ. of Tokyo, *10*(3), 767–802.

Shannon and Wilson Inc., 1964, "Anchorage Area Soil Studies, Alaska," Report to U.S. Army Engineer District, Anchorage, Alaska.

Sherard, J. L., 1967, "Earthquake Considerations in Earth Dam Design," *Proc. ASCE*, *93*(SM4), 377–401.

Sherard, J. L., R. J. Woodward, S. F. Gizienski, and W. A. Clevenger, 1963, *Earth and Earth-Rock Dams*, New York: John Wiley & Sons, Inc.

Sheth, R., 1959, "Effect of Inelastic Action on the Response of Simple Structures to Earthquake Motions," master's thesis, University of Illinois, Urbana, Ill.

Shibata, A., J. Onose, and T. Shiga, 1969, "Torsional Response of Buildings to Strong Earthquake Motions," *Proc. Fourth World Conf. Earthq., Engrg.*, Santiago, Chile, *2*, A-4, 123–38.

Shinozuka, M., 1964, "Probability of Structural Failure under Random Loading," *Proc. ASCE*, *90*(EM5), 147–70.

Sieberg, A., 1923, *Erdbebenkunde*, Jena, Germany: Gustav Fischer, 102–104.

Sinha, B. P., K. H. Gerstle, and L. G. Tulin, 1964a, "Stress-Strain Relations for Concrete Under Cyclic Loading," *Proc. ACI*, *61*(2), 195–211.

———, 1964b, "Response of Singly Reinforced Beams to Cyclic Loading," *Proc. ACI*, *61*(8), 1021–37.

Smith, B. S., 1962, "Lateral Stiffness of Infilled Frames," *Proc. ASCE*, *88*(ST6), 183–99.

———, 1966, "Behavior of Square Infilled Frames," *Proc. ASCE*, *92*(ST1), 381–403.

Smith, R. H., 1958, "Numerical Integration for One-Dimensional Stress Waves," doctoral thesis, University of Illinois, Urbana, Ill.

Sneddon, I. N., 1951, *Fourier Transforms*, New York: McGraw-Hill Book Inc.

Soroka, W. W., 1961, "Analog Methods of Analysis," in *Shock and Vibration Handbook*, C. M. Harris and C. E. Crede, eds., New York: McGraw-Hill Book Co. Inc., Chap. 29.

Sozen, M. A., and N. N. Nielsen, 1966, "Earthquake Resistance of Reinforced Concrete Frames," *Proc. Internatl. Symp. Effects of Repeated Loading of Materials and Structures*, RILEM-Inst. Ing., 6.

Spencer, R. A., 1968, "Stiffness and Damping of Cyclically Loaded Prestressed Concrete Members," *Segundo Congreso Nacional de Ingeniería Sísmica*, Veracruz, Mexico, *1*.

———, 1969, "The Nonlinear Response of a Multistory Prestressed Concrete Structure to Earthquake Excitation," *Proc. Fourth World Conf. Earthq. Engrg.*, Santiago, Chile, *2*, A-4, 139–54.

Steinbrugge, K. V., and R. Flores A., 1963, "The Chilean Earthquakes of May, 1960: A Structural Engineering Viewpoint," *Bull. Seism. Soc. Am.*, *53*(2) 225–307.

Sung, T. Y., 1953, "Vibrations in Semi-Infinite Solids due to Periodic Surface Loading," Symposium on Dynamic Testing of Soils, *ASTM Spec. Techn. Publ. 156*, 35–63.

Sutherland, H., and H. L. Bowman, 1958, *Structural Theory*, New York: John Wiley & Sons, Inc.

Sutherland, J. G., and L. E. Goodman, 1951, "Vibrations of Prismatic Bars Including Rotary Inertia and Shear Corrections," Rept. N6-ORI-71, T.O. 6, Project NR-064-183, pp. 1–23.

Tajimi, H., 1960, "A Statistical Method of Determining the Maximum Response of a Building Structure during an Earthquake," *Proc. Second World Conf. Earthq, Engrg.*, Tokyo and Kyoto, Japan, 781–97.

Tajimi, H., 1969, "Dynamic Analysis of a Structure Embedded in an Elastic Stratum," *Proc. Fourth World Conf. Earthq. Engrg.*, Santiago, Chile, 3, A–6, 53–64.

Takahasi, R., 1955, "A Short Note on a Graphical Solution of the Spectral Responses of the Ground," *Bull. Earthq. Res. Inst.*, University of Tokyo, 33, 259–64.

Takahasi, R., and K. Hirano, 1941, "Seismic Vibrations of Soft Ground," *Bull. Earthq. Res. Inst.*, Univ. of Tokyo, 19, 534–43.

Takahashi, T., 1969, "Vibration Studies of an Arch Dam," *Proc. Fourth World Conf. Earthq. Engrg.*, Santiago, Chile, 1, B–1, 61–71.

Tamura, R., M. Murakami, Y. Osawa, N. Miyajima, and Y. Tanaka, 1969, "A Vibration Test of a Large Model Steel Frame with Precast Concrete Panel until Failure," *Proc. Fourth World Conf. Earthq. Engrg.*, Santiago, Chile, 1, B–2, 15–30.

Tanabashi, R., and K. Kaneta, 1962, "On the Relation Between the Restoring Force Characteristics of Structures and the Pattern of Earthquake Ground Motions," *Proc. Japan Natl. Symp. Earthq. Engrg.*, Tokyo, Japan, 57–62.

Taylor, P. W., and J. M. O. Hughes, 1965, "Dynamic Properties of Foundation Subsoils as Determined from Laboratory Tests," *Proc. Third World Conf. Earthq. Engrg.*, Auckland and Wellington, New Zealand, 1.196–212.

Tezcan, S. S., 1966, "Computer Analysis of Plane and Space Structures," *Proc. ASCE*, 92(ST2), 143–73.

Thrall, R. M., 1954, "Applications of Multidimensional Utility Theory," in *Decision Processes*, R. N. Thrall, C. H. Coombs, and R. L. Davis, eds., New York: John Wiley & Sons, Inc., 181–86.

Timoshenko, S., and J. N. Goodier, 1951, *Theory of Elasticity*, New York: McGraw-Hill Book Co., Inc. 366–72.

Torroja, E., N. Esquillan, J. P. Mazure, G. Rinaldi, H. Rüsch, and F. G. Thomas, 1958, "Load Factors," *Proc. ACI*, 55, 567–72.

Torvi, A. A., A. T. Olson, and A. G. Davenport, 1966, "Dynamic Buckling in Modified I-Beams," *Proc. Internatl. Symp. Effects of Repeated Loading of Materials and Structures*, RILEM-Inst. Ing., 6.

"Trial Load Method of Analyzing Arch Dams," 1938, *Boulder Canyon Project Final Reports*, Part 5, Bull. 1, Denver, Colo.: U.S. Dept. of the Interior.

Tsuboi, C., 1958, "Earthquake Province Domain of Sympathetic Seismic Activities," *Journ. Phys. Earth*, 6(1), 35–49.

Turkstra, C. J., 1962, "A Formulation of Structural Design Decisions," doctoral thesis, University of Waterloo, Ontario, Canada.

———, 1967, "Choice of Failure Probabilities," *Proc. ASCE*, 93(ST6), 189–200.

U.S. Department of Commerce and U.S. Coast and Geodetic Survey, 1965, *Tsunami: The Story of the Seismic Sea-Wave Warning System*, Washington, D.C.: U.S. Government Printing Office.

Ulloa, A., and J. Prince, 1965, "Sismología en Presas," *Primer Congreso Nacional de Ingeniería Sísmica*, Guadalajara, Mexico,

Umemura, H., and H. Aoyama, 1969, "Evaluation of Inelastic Seismic Deflection of Reinforced Concrete Frames Based on the Tests of Members," *Proc. Fourth World Conf. Earthq. Engrg.*, Santiago, Chile, 1, B–2, 91–107.

Uniform Building Code, 1967, 1967 ed., Pasadena, Calif: International Conference of Building Officials.

Van Dorn, W. G., 1965, "Tsunamis," in *Advances in Hydroscience*, vol. 2, V. T. Chow, ed., New York and London: Academic Press, 1–48.

———, 1969, "Universities Research Reports: University of California, San Diego: Gravity Waves as a Cause of Earthquake Damage?" report on *NSF-UCEER Conf. Earthq. Engrg. Res.*, Pasadena, Calif: Universities Council for Earthquake Engineering Research, 271–73.

Veletsos, A. S., 1969, "Maximum Deformations of Certain Nonlinear Systems," *Proc. Fourth World Conf. Earthq. Engrg.*, Santiago, Chile, 2, A-4, 155–70.

Veletsos, A. S., and N. M. Newmark, 1957, "Natural Frequencies of Continuous Flexural Members," *Trans. ASCE, 122,* 249–85.

———, 1960, "Effect of Inelastic Behavior on the Response of Simple Systems to Earthquake Motions," *Proc. Second World Conf. Earthq. Engrg.*, Tokyo and Kyoto, Japan, 895–912.

———, 1964, "Design Procedures for Shock Isolation Systems of Underground Protective Structures," *Report RTD-TDR-63-3096*, vol. 3, Air Force Weapons Laboratory, Albuquerque, New Mexico.

Veletsos, A. S., N. M. Newmark, and C. V. Chelapati, 1965, "Deformation Spectra for Elastic and Elastoplastic Systems Subjected to Ground Shock and Earthquake Motions," *Proc. Third World Conf. Earthq. Engrg.*, Auckland and Wellington, New Zealand, 2.663–82.

Vlasov, V. Z., 1961, *Thin-Walled Elastic Beams*, Jerusalem, Israel: Israel Program for Scientific Translations.

Von Alven, W. H., 1964, Reliability Engineering, Englewood Cliffs, N. J.: Prentice-Hall, Inc.

Von Neumann, J., and A. Morgenstern, 1943, *Theory of Games and Economic Behavior*, Princeton, N. J.: Princeton, University Press.

Wakabayashi, M., T. Nonaka, and C. Matsui, 1969, "An Experimental Study on the Horizontal Restoring Forces in Steel Frames under Large Vertical Loads," *Proc. Fourth World Conf. Earthq. Engrg.*, Santiago, Chile, 1, B-2, 177–93.

Wen, R. K., and J. G. Janssen, 1965, "Dynamic Analysis of Elasto-Inelastic Frames," *Proc. Third World Conf. Earthq. Engrg.*, Auckland and Wellington, New Zealand, 2.713–29.

Werner, P. W., and K. J. Sundquist, 1943, "On Hydrodynamic Earthquake Effects," *Trans. Am. Geophys. Union*, 30(5), 636–57.

Westergaard, H. M., 1933a, "Water Pressures on Dams during Earthquakes," *Trans. ASCE*, 98, 418–72.

———, 1933b, "Earthquake-Shock Transmission in Tall Buildings," *Engrg. News-Record*, 111, 654–56.

White, M. P., 1942, "A Method of Calculating the Dynamic Force in a Building during an Earthquake," *Bull. Seism. Soc. Am.*, 32, 193–203.

Whitman, R. V., 1969, "Equivalent Lumped System for Structure Founded upon Stratum of Soil," *Proc. Fourth World Conf. Earthq. Engrg.*, Santiago, Chile, 3, A-6, 133–42.

Whitman, R. V., and K. A. Healy, 1962, "Shearing Resistance of Sands during Rapid Loading," *Research Report R62-113*, Cambridge, Mass.: Massachusetts Institute of Technology.

Wieckowski, J., 1958, "The Influence of Material Damping on Non-Conservative Reactions of Elastic Beams during Torsional and Longitudinal Vibrations," *Archiwum Mechaniki Stosowanej*, 10(4), 479–97.

Wiegel, R. L., 1955, "Laboratory Studies of Gravity Waves Generated by the Movement of a Submerged Body," *Trans. Am. Geophys. Union*, 36(5), 759–74.

Wiegel, R. L., 1964, *Oceanographical Engineering*, Englewood Cliffs, N. J.: Prentice-Hall, Inc.

Wiggins, J. H., 1961, *The Effect of Soft Surficial Layering on Earthquake Intensity*, doctoral thesis, Urbana: University of Illinois.

Wilbur, J. B., 1934, "A New Method for Analyzing Stresses Due to Lateral Forces in Building Frames," *Journ. Boston Soc. Civil Engrs.*, *21*(1), 45–56.

———, 1935, "Distribution of Wind Loads to the Bents of a Building," *Journ. Boston Soc. Civil Engrs.*, *22*(4), 253–59.

Wilkinson, J. H., 1960, "Householder's Method for the Solution of the Algebraic Eigenvalue Problem," *Computer Journ.*, *3*, 23–27.

Wilson, B. W., L. M. Webb, and J. A. Hendrickson, 1962, "The Nature of Tsunamis, Their Generation and Dispersion in Water of Finite Depth," Technical Report SN57-2, National Engineering Science Company, for U. S. Coast and Geodetic Survey.

Wilson, E. L., 1969, "A Method of Analysis for the Evaluation of Foundation-Structure Interaction," *Proc. Fourth World Conf. Earthq. Engrg.*, Santiago, Chile, *3*, A–6, 87–99.

Wood, H. O., and F. Neumann, 1931, "Modified Mercalli Intensity Scale of 1931," *Bull. Seism. Soc. Am.*, *21*, 277–83.

Wright, D. T. and R. Green, 1959, "Human Sensitivity to Vibration," *Report No. 7*, Dept. of Civil Engrg., Queen's University, Kingston, Ontario, Canada.

Yamada, M., 1969, "Low Cycle Fatigue Fracture Limits of Various Kinds of Structural Members Subjected to Alternately Repeated Plastic Bending under Axial Compression as an Evaluation Basis or Design Criteria for Aseismic Capacity," *Proc. Fourth World Conf. Earthq. Engrg.*, Santiago, Chile, *1*, B–2, 137–51.

Yamada, M., H. Kawamura, and S. Furui, 1966, "Low Cycle Fatigue of Reinforced Concrete Columns," *Proc. Internatl. Symp. Effects Repeated Loading of Materials and Structures*, RILEM-Inst. Ing., 6.

Yamamoto, S., and N. Suzuki, 1965, "Experimental and Theoretical Analysis of Response Against Earthquakes of Tower Structures Having Non-Uniform Sections Governed by Bending Vibrations," *Proc. Third World Conf. Earthq. Engrg.*, Auckland and Wellington, New Zealand, 2.730–47.

Yao, J. T. P., 1969, "University of New Mexico: Adaptive Systems for Seismic Structures," report on *NSF–UCEER Conf. on Earthq. Engrg. Res.*, Pasadena, Calif.: Universities Council for Earthquake Engineering Research, 142–50.

Yao, J. T. P., and W. H. Munse, 1962, "Low-Cycle Axial Fatigue Behavior of Mild Steel," *ASTM Spec. Techn. Publ. 338*, 5–24.

Yeh, H. Y., 1969, "Random Vibration of Piecewise Linear Systems Excited by Nonstationary Shot Noise," doctoral thesis, University of New Mexico, Albuquerque, New Mexico.

Yeh, H. Y., and J. T. P. Yao, 1969, "Response of Bilinear Structural Systems to Earthquake Loads," presented at the ASME Vibrations Conf., Philadelphia, Pa., *ASME Preprint 69-VIBR-20*.

Yokobori, T., 1964, *The Strength, Fracture and Fatigue of Materials*, The Netherlands: P. Noordhoff Ltd.

Zangar, C. N., 1953, "Hydrodynamic Pressures on Dams due to Horizontal Earthquakes," *Proc. Soc. Ex. Stress Anal.*, *10*(2), 93–102.

Zangar, C. N., and R. J. Haefeli, 1952, "Electric Analog Indicates Effect of Horizontal Earthquake Shock on Dams," *Civil Engrg.*, *22*(4), 278–79.

Zeevaert, L., 1963, "Mediciones y Cálculos Sísmicos durante los Temblores Registrados en la Ciudad de México en Mayo de 1962," *Bol. Soc. Mex. Ing. Sism.*, *1*(1), 1–16.

———, 1964, "Strong Ground Motions Recorded during Earthquakes of May the 11th and 19th, 1962 in Mexico City," *Bull. Seism. Soc. Am.*, *54*(1), 209–31.

Zeevaert, L., and N. M. Newmark, 1956, "Aseismic Design of Latino Americana Tower in Mexico City," *Proc. First World Conf. Earthq. Engrg.*, Berkeley, Calif, 35.1–11.

INDEX

Acceleration spectra, 11
  as basis for analysis of nonlinear systems, 325, 327, 332
  expected ordinates of, 283
  of elastic nonlinear systems, 329
  of elastoplastic systems, 340
  when natural period tends to zero, 280
  when natural frequency tends to zero, 292
Accelerations perpendicular to walls, 419
Accelerograms, 220
Accelerographs, 569
  optimum location of, 570
  starting devices for, 569
Accidental eccentricities, 490
Accidental torsions, 490, 533
Accumulated damage in fatigue, 395
Actual eccentricities, 490
Actualization, 445, 461
Aftershocks, 216, 473
Agadir 1960 earthquake, 225, 265
Air lubrication in soils, 424, 429
Aleutian Islands, 205
Alpide Belt, 251, 264, 265
Amplitude
  resonant, 8

Analog computers, 363
Analogs
  electrical, 22, 181, 182
  mechanical, 22
Analysis of common structures, 514
  choice of methods for design, 514
Anchorage in reinforced concrete, 401, 405, 411, 420
Anchorage, Alaska 1964 earthquake, 226, 228, 437
Angle of internal friction, 360, 425, 579
Angle of repose, 428
Anisotropic consolidation, 433
Anisotropy, 236, 433
Antigua, 563
Apartment buildings, 451, 461, 472
Appendages, 488
  lateral-force coefficients for, 489
  magnification factors of, 488
Attenuation, 74, 233
Arch dam, 317, 323, 461, 550
Autocorrelation function, 268, 279
Axial deformations, 515
Axial forces, 53
  in columns, 468
Axle steel, 399

Bafflers, 378
Bars, 64, 119, 169
Bar bends in reinforced concrete, 401
Bar cutoffs in reinforced concrete, 359, 401
Base line, corrections to, 300
Base rotation, 188, 241, 359
Base shear coefficients, 201, 242, 455, 468, 477, 482
Basic solutions, 276
Bauschinger effect, 398, 407, 410
Bayes' theorem, 247, 256, 259, 261
Beam-column connections, 408
  eccentric, 522
Beams, 157, 359, 386
  cantilever, 80, 472, 474
  continuous, 322
  elastoplastic, 402
  flexural, 80, 108, 115
  reinforced concrete, 399, 400, 423
  tapered, 128
Behavior of complete structures, 420
Behavior of materials under earthquake loading, 381
Behavior of structural components, 381, 399
Bent bars in reinforced concrete, 411
Bernoulli–Euler formulation of beam problem, 80, 108
Bilinear systems, 162, 330, 350
Body waves, 83, 220
  cylindrical, 87, 220, 231
  spherical, 85
Bolted connections, 410, 422
Bond in reinforced concrete, 405
Boundary conditions in hydrodynamics, 178, 190
Bounds to design responses, 323
Bounds to failure probability, 323
Braced structures, 142, 349, 353, 358, 468, 475, 541
Breakage of soil particles, 424, 426, 579
Bridges, 541
  base rotations, 241
  differential motions of supports, 541
  longitudinal strains of ground, 244

Bridges (*contd.*)
  tilting of piers, 542
  water pressures on piers, 201
Brittle failure, 162, 417, 420
Buckling, 53, 161, 449
  of steel beam flanges, 407
  of steel frames, 413
  of reinforcement in concrete, 401
Buckling modes, 54
Buildings
  apartment, 451, 461, 472
  base rotation, 241, 472
  direction of disturbance, 463
  earthquake resistant design of, 468, 477
  forced vibrations, 145
  fundamental periods, 420
  natural modes of vibration, 145, 152
  rocking, 111
  shear-beam, 30, 112, 153, 300
  strengthening, 474
  testing, 422
  vertical accelerations, 111
  X-braced frames, 142
Building codes, 444, 447, 454, 461, 473
Bulk modulus of water, 184
Buttress dam stability, 549

Calculation of dynamic parameters from natural modes of vibration, 581
Calculation of natural modes, 105, 482
California, 205, 215, 218, 421
Canals, 549
Cancanni–Sieberg intensity scale, 218
Cantilever beam, 80, 472, 474
Caracas 1967 earthquake, 399, 419
Celerity of tsunami waves, 205
Cement silos, 429
Center, 217
Characteristic equation, 33
Characteristic function, 34
Characteristic length, 394
Characteristic roots, 34, 447
Characteristic strength, 461

INDEX   625

Characteristic values, 34, 447, 461
Characteristics
 graphical construction, 72
 method of, 70, 168, 373
Chile, 205, 209
Chilean earthquakes of 1960, 226, 540
Chimney anchor bolts, 349, 350
Chimney stacks, 156, 212, 317
Churchman–Ackoff utility evaluation, 444
Circular cylindrical tank, 198
Circumpacific Belt, 225, 251, 256, 257, 259, 264, 265
Cladding, 421
Classical modes of vibration, 305
Clays, 384, 438, 439
 fissured, 438
Closely-coupled systems, 30, 43
 calculation of natural modes, 149
 idealization of soil mass, 302
 on two or more supports, 44
Coastal landslides, 203, 204
Coefficient of damping, 5
Cohesion, 579
Cohesionless soils, 360, 424
Cohesive soils, partially saturated, 429, 438
Cohesive soils, saturated, 437
Collapse, 451
Collapse of cave roofs, 215
Columns of reinforced concrete, 402, 423, 471
Comité Européen du Béton, 447
Compaction of granular materials under shaking, 299, 425, 432
Comparison of El Centro spectrum with code spectra, 480
Complete orthogonal set, 34
Complex natural mode, 35
Compression reinforcement in concrete, 401, 404, 405
Concentration of rigidity, 519
Concrete, 384, 393, 463
 confinement of, 523
 frames, 359
 plain, 398
 precast, 449

Concrete (*contd.*)
 reinforced, 399, 406, 474
Confinement of concrete, 523
Congestion of reinforcement, 406
Conjugate distribution, 253
Connections
 between steel members, 410
 between structural and nonstructural elements, 453
 bolted, 410, 422
 in reinforced concrete, 409
Conrad discontinuity, 225
Conservative nonlinear systems, 161
Conservative simple system, 17
Conservative system, 5
Consolidation, 433
Constant-Q hypothesis, 75, 286
Construction standards, 449
Continuous beams, 411
Contraction joints, 550
Convergence, 50
Corrective configurations, 144, 146
Correlations between faulting and seismicity, 249
Correlations between focal and station characteristics of earthquakes, 219, 228, 301, 462
Correlations between modes of failure, 323
Cost of failure, 444, 449, 472
Coulomb friction, 360
 at grain boundaries, 383
Couple stresses, 393
Coupled air-water waves, 203
Coupling between degrees of freedom in orthogonal directions, 317
Cover-plate cutoffs, 359
Cracks
 in arch dams, 323
 in masonry and plain concrete, 398
 in prestressed concrete, 162, 330, 422
 in reinforced concrete, 409, 419
 in shear walls, 417, 423
Creep of reinforced concrete, 406, 474
Criteria of convergence and stability, 50
Criteria, design of, 443, 460, 470
Criteria of failure, 239, 453, 462

626   INDEX

Critical damping, 5
Critical frequency, 184, 186, 190, 191
Critical values, 34
Critical void ratios, 424, 432
Critical wave frequencies, 189
Cross-braced towers, 541
Cubical dilatation, 83
Cumulative damage, 473
Curvatures of a pipe, 547
Curved members of reinforced concrete, 401
Cyclic mobility, 432, 434
Cylindrical body waves, 87, 220, 231
Cylindrical, semi-infinite reservoir, 190
Cylindrical tanks, 211
    with semispherical bottom, 199

Dam flexibility, 188
Damage of utilities, 548
Damped harmonic motion, 5
Damped natural circular frequency, 5
Damped systems, 35
Damping
    approximate treatment of in structures, 38
    coefficient of, 5
    critical, 5
    effect on expected responses, 272, 277, 292, 294, 296, 312
    effect on responses associated with given probability of failure, 302
    equivalent in linear systems, 310
    equivalent in nonlinear systems, 324
    equivalent in reservoirs, 367
    hysteretic, 301, 324
    in complete structures, 421
    in nuclear reactor facilities, 422
    in piping, 422
    in prestressed concrete, 422
    in reservoirs, 366
    in soils, 296, 428, 430, 438
    in steel structures, 422
    in submerged structures, 377
    in tanks, 378
    internal in soils, 279, 289, 299
    in wood structures, 422
    modal, 306, 421

Damping (*contd.*)
    radiation, 77, 382
    ratio, 5, 35
    structural, 35
    typical values of, 422
    uncertainty concerning, 467
Damping constant, 4
Damping influence coefficient, 28
Damping matrix, 28
Damping, nonlinear, 167
Damping ratios, 576
Damping ratios from forced vibration tests, 576
Dams, 549
    arch, 317, 323, 461, 550
    buttress, 549
    earth, 474
    ground rotation, 241
    hydrodynamic pressures against, 365, 382
    longitudinal strains of ground, 244
    masonry, 475
    rock-fill, 359, 474, 579
Dashpot constant, 96
Dead loads, 449
Decay of intensity per unit time, 301
Deep reservoirs, 369
Defense plateaus, 453, 478
Deformations of container walls, 197
Deformation spectra (see *displacement spectrum*)
    when natural frequency tends to zero, 281
    as basis for analysis of nonlinear systems, 325, 327, 332
    of earthquakes on soft ground, 293
Degree of freedom
    single, 3
Delta-squared extrapolation method, 119
Depths of foci, 251, 258
Descending branch in skeleton curve, 325
Descending branches in stress–strain and force–deformation diagrams, 382, 414
Design decision, 261
Design drift, 512

INDEX **627**

Design response, 309, 323
Design shear distributions, 371
Design spectra, 230, 462, 467, 470, 477
Design story eccentricity, 498
Deterioration, 396, 400
 of diaphragms, 414
 of masonry panels, 419
 of prestressed concrete members, 407
 of reinforced concrete frames, 411
 of reinforced concrete members, 406
Diagonal tension, 399
 (see *shear*)
Diaphragms, 414, 471
Differential horizontal displacements
   in foundations, 517
Differential motions of supports
 in buildings, 111, 241, 517
 in bridges, 244, 541
 in dams, 244
Direction of travel of earth waves,
   influence of, 195
Directional features of earthquake,
   230, 237, 471
Dilatation, 86
Dilatational disturbance, 84
Dirac delta function, 9
Direct design, 530
Dispersion, 92
Displacement
   static, 4
Displacement spectrum, 11
 (see *deformation spectra*)
Discretization, 105, 106, 360
Dissipative system, 5
Distributed-mass systems, 61
Distributed-parameter systems, 61, 105,
   106, 359
Diversion tunnels, 549
Drift, 507
Drift limitations, 507, 514
Ductility, 478
Ductility factors, 331, 511
Duhamel's integral, 15
Duration of ground motion, 229, 236,
   266
 equivalent, 229, 275, 284, 340
 influence on structural responses, 229,
   274, 310

Dynamic model testing
   dimensional requirements, 578
Dynamic soil–tunnel interaction, 547

Earth dams, 474, 550
Earth waves, types of, 220
Earthquake, causes of, 215, 236, 240
Earthquakes, characteristics of, 215, 225
Earthquakes, types of, 225
Earthquake clusters, 249
Earthquake filtering, 286, 298, 323
Earthquake-resistant design, 443, 477
Earthquake simulation, 236, 280, 299,
   302, 310, 352
Earthquake swarms, 249
Eccentric beam–column connections,
   522
Eccentricities in buildings
   accidental, 490
   actual, 490
Effects of changing a single mass or a
   stiffness element, 142
Effects of successive earthquakes, 546
Eigenvalues, 34
Eigenvectors, 34
Elastic, nonlinear materials, 448
Elastic, nonlinear systems, 161, 354
Elastoplastic materials, 349
Elastoplastic systems, 162, 300, 324,
   335, 345, 349, 353, 384, 455, 473
El Centro 1940 Earthquake, 188, 194,
   225, 241, 274, 282, 285, 327,
   333, 336, 338, 352, 358, 367,
   371, 480, 492
El Salvador, 265, 563
Electrical analogs, 22, 181, 182
Elevated tanks, 200, 211, 212
Embankments, 162
Endurance limit, 383, 394
Energy dissipation, 188
 (see *damping, hysteresis, defense
   plateaus*)
Energy feedback to ground (see
   radiation damping)
Energy methods, 105, 137, 138, 142,
   158, 212
Energy release by earthquakes, 217,
   236, 248

628 INDEX

Energy transfer into the atmosphere, 382
Epicenter, 207, 217, 221, 222, 240, 251
Epicentral distance, 217, 231, 240
Epifocus, 217
Equipment, damage to, 451
Equivalent concentrated loads, 114
Equivalent concentrations, 61, 113, 114, 131
  parabolic, 114
  polygonal, 114
Equivalent damping, 383
  in analysis of response to white noise, 310
  in elastoplastic systems, 335, 339
  in reservoirs, 191, 367, 376
  in submerged structures, 382
  viscous, in lieu of radiation damping, 296
Equivalent duration, 284, 336
Equivalent linear systems, 332, 354, 383, 475
Equivalent number of pulses, 356
Equivalent stiffness, 354, 383
Equivoluminal disturbance, 84
Erosion of downstream face, 189
Eulerian coordinates, 169
Evolution in local seismicity, 249
Excitation coefficient, 42
Expected spectral ordinates, 283, 290, 292, 301, 309, 354
Explosions, 215, 220
  as consequence of earthquakes, 451
Extrapolation procedure, 16, 129
Extreme distributions, 250
Extreme type 1 distribution, 303
Extreme type 2 distribution, 302, 322

Factors of safety, 447, 467
Failure modes, 445, 451
Failure rate
  (see *return periods*)
Fatigue, high cycle
  (see *fatigue proper*)
Fatigue, low-cycle, 395
Fatigue proper, 381, 382, 393
Fault creep, 216

Field determination of soil and rock properties, 438
Filtering of earth waves through soft mantles, 382
Finite differences
  higher order approximations, 107
Finite elements, 78, 112, 168, 296, 299, 302
Finite reservoirs, 185, 190
Fire as consequence of earthquakes, 451
Fire protection in steel frames, 453
Fire stations, 451
First natural mode, 34
Fissionable material, release of, 451
Fissures in ground, 244
Flexibility and nonlinear behavior of dams, 195
Flexibility matrix, 28
Flexible first story construction, 527
Flexible tanks, 188
Flexural beams, 79, 109, 129, 537
Flexural frames, 353
Flexural members, 399
Floating partitions, 508
Flow net, 181
Floods as consequence of earthquakes, 451
Flow structure, 432
Focal depth, 222, 240
Focal distances, 217, 229
  long, 283, 470
  moderate, 263
Focus, 216, 217, 233
Footings, 434
Force-deformation curves, 473
Fourier spectra, 11, 22, 289, 299
Fourier analysis, 197
Fourier transform, 268, 276, 286, 292, 313
Forced vibration tests, 8
Forced vibrations, 6
Foreshocks, 216
Forge hammers, 220
Foundation design, 517
Foundation rocking, 111
Fox's procedure, 107
Fracture, 394

INDEX **629**

Frames, 82, 109, 357, 396, 468
   cross-braced, 142
   discretization, 113
   flexural, 353
   nonlinear, 357
   single-story, 111
   with defense plateaus, 453
Free oscillations, 4, 32
Free vibrations, 4, 32
   of systems under static axial loads, 54
Frequency
   lowest natural circular, 34
   resonant, 8
Frequency content, 225, 301
Fresh-water tsunamis, 189, 204, 209
Friction at connections, 383
Friction at grain boundaries, 382
Fundamental mode, 34
Fundamental period, 212
   of liquids in cylindrical and
      rectangular tanks, 199
   of reservoirs, 369
   of vibration, 469
Foundation compliance
   (see *soil–structure interaction*)
Foundation yielding
   (see *soil–structure interaction*)

Gamma–1 distribution, 253, 259, 260
Gaussian processes, 268, 270, 271,
      276, 286, 351
Gaussian processes, transient, 283
Generalized configuration, 27
Generalized coordinate, 27
Generalized displacements, 27
Generalized force, 28
Geologic faults, 205, 215, 236, 249
Geologic conditions, 456
Geometrical discrepancies, 449
Goldberg, Bogdanoff, and Moh
      procedure, 145, 151, 153, 158
Gnome event, 232
Goodman diagram, 394
Grain breakage in soils, 424, 426
Grain interlocking, 428
Granular soils, 163

Graphical evaluation of responses,
      164, 169, 289, 299
Graphical methods, 17
Gravity dams, 190, 211, 475
Gravity effects, 53, 350, 358, 411,
      468, 488, 516
Gravity loads as random variables, 454
Gravity waves, 244
   in liquids, 178, 188
Ground acceleration, 216, 233, 234,
      242, 272, 274, 301, 331, 343
Ground displacement, 233, 301, 331
Ground velocity, 216, 219, 233, 234,
      242, 264, 270, 272, 274, 301, 331
Group velocity, 92, 307
Grouting in dams, 475
Guatemala City, 563
Gyrogram, 17, 164, 172

Half-cycle displacement input, 326
Half-power band width, 8
Harmonic motion
   damped, 5
   simple, 5
Halfspace, 194
Hawaii, 206
Hebgen dam, 189
Higher modes of vibration, 106, 129
   importance of, 470
Hilo, Hawaii, 206
Holzer's table, 105, 143, 149, 158
Hollow tile, 419
Honolulu, Hawaii, 205
Hooks in reinforced concrete, 401
Hospitals, 451
Housner's spectral intensity, 220, 238
Human life, value of, 450
Human reaction of earthquakes, 423
Hydraulic structures, 548
Hydrodynamics, 177, 365
Hydrodynamic forces, 188, 371
Hydrodynamic pressures, 179, 365, 382
   associated with tsunamis, 208
   distributions, 369
   in three-dimensional reservoirs, 197
   nonrectangular reservoir cross
      sections, 193, 211

Hydrodynamic spectrum, 367
Hypocenter, 217
Hysteretic damping
   effects on responses, 301
   in bolted connections, 410
   in diaphragms, 414
   in prestressed concrete, 330, 406
   in reinforced concrete, 411
   in soils, 428
   in steel structures, 421

I waves, 225
Inclined upstream face, 182
Inelastic behavior, 321, 383
Inelastic buckling, 162
Inelastic reinforced concrete frames, 411
Inelastic systems, 161, 354
Inelastic strain energy dissipation, 356
Inertia matrix, 28
Infinite reservoirs, 187
Inner core, 225
Instability, 358
Instrumental intensity scales, 237
Instruments, 565
Intake towers, 382
Intensity, 216, 230, 248, 260, 456
Intensity per unit time, 236, 269, 271, 275
Intensity scales, 218, 587
   (see also *MM intensity scale*)
Interference of waves, 92
Intergranular forces, 429
Intergranular pressures, 430
Internal damping of ground, 230, 233, 286, 291, 382
Inverted pendulums, 317, 350, 463, 533
Irrotational disturbance, 84
Isoseismals, 230, 232
Iteration procedure, 106, 117, 157
   conservative systems, 117
   damped systems, 119
   fundamental mode, 124
   for steady-state vibrations, 117
   higher natural modes, 130

Japan, 205, 265
Joints
   bolted, 422
   in reinforced-concrete frames, 420
   nailed, 422
   riveted, 422
   tubular, 411
Joints, behavior of, 408

Kanai's semiempirical formulas, 297
Kelvin body, 76
Krakatoa, 203

Lamé's constants, 83
Landslides in reservoirs, 204
Laplace operator, 112
Lateral force coefficients, 484
   (see *seismic coefficients*)
Libya 1963, 225
Linear damping, 4
Linear multidegree systems, responses of, 305
Linear system, equivalent to nonlinear, 324
Liquid core, 223
Liquid viscosity, 203
Lituya Bay, 204
Live loads, 449
Load factors, 447, 450, 454, 467, 471
Loading rate, 425
Local seismicity, 248
Locking material, 170
Load tests, 553
Logarithmic decrement, 5
Long structures, 318
Longitudinal displacements of a slender bar, 64
Longitudinal prestress, 557
Longitudinal straining of tunnel lining, 547
Longitudinal waves, 84
Loose soil, 266, 432
Loss of freeboard, 551
Love waves, 84, 92, 243

INDEX **631**

Low-cycle fatigue, 381
  of reinforced concrete members, 406
  of steel members, 408
Low-seismicity macrozone, 251, 256, 264
Low temperature, 393
Lower mode elimination, 106, 118
Lowest natural circular frequency, 34
Lumped masses, 61
Lumped parameters, 105, 299
Lumping, 106
  (see *discretization*)

Machine foundations, 100, 427
Macrozones, 251, 257, 258, 259
Macuto-Sheraton Hotel, 399
Madison Canyon, 190
Magnetoelasticity, 382
Magnification factor, 289, 290, 291, 298, 368, 488,
  for eccentricity, 496
Magnitude, 205, 207, 217, 229, 232, 236, 248, 249, 250, 260, 301
  scales, 217
Magnitude-intensity correlations, 233
Masing type, 163
  of behavior, 398
  structures, 353, 357
  systems, 163, 345
Masonry, 398, 417
  dams, 475
Mass matrix, 28
Mass, uncertainty concerning, 467
Matrices of transmission, 105
Matrix equation, 34
Matrix, tridiagonal, 149
Matrix deflection for the calculation of higher modes, 106
Maximum absolute values of responses, 44
Maximum probable intensity, 216
Maximum relative velocity in buildings, 512
Mean-squared response, 276
Mechanical analogs, 22

Medvedev, Sponheuer, and Karnik intensity scale (see *MSK-64 intensity scale*)
Membrane, 112
Metastable equilibrium of soil slopes, 428
Metals, 383, 384
Meteorites, 251
Method of characteristics, 70
Method of Rayleigh-Ritz, 137
Method of transmission matrices, 106
Mexico City, 219, 295, 298
Mexico Earthquake of 19 May, 1962, 231, 241
Microregionalization, 248, 265
Microtremors, 266
Mid-oceanic ridges, 251
Minor structures, 461
MM intensity scale, 218, 233, 264, 585
Modal analysis, 36, 239, 311
Model testing, 566
  of multistory frames with filter walls, 420
  of rockfill dams, 361
  of submerged structures, 377
  of tanks, 378
  under random loading, 302
Modeling granular materials, 428, 437, 579
Modes
  normalized, 34
Modified Mercalli (see *MM intensity scale*)
Modulus of elasticity of concrete under repeated cycles, 406
Modulus of elasticity of steel under repeated loading, 407
Moho discontinuity, 225
Mohr envelopes, 361, 428
Mohorvičić discontinuity, 225
Moment-curvature relations for reinforced concrete, 403, 406
Moment-curvature relations of steel members, 408, 411
Moment redistribution, 411
Monte Carlo analyses, 170, 228, 302, 321, 346, 487

632 INDEX

Motion of the free surface in reservoirs, 188
MSK–64 intensity scale, 218, 587
Multidegree-of-freedom systems, 33
 (see also *systems with several degrees of freedom*)
Multidegree-of-freedom systems
 axial forces, 53
 buckling, 53
 buckling mode, 54
 buildings, 30
  shear-beam, 30
 characteristic equation, 33
 characteristic function, 34
 characteristic roots, 34
 closely-coupled, 30, 43, 149
  on two or more supports, 44
 closely-coupled systems, 30, 43
 complete orthogonal set, 34
 complex natural mode, 35
 convergence, 50
 criteria of convergence and stability, 50
 critical values, 34
 damped systems, 35
 damping, 38
  approximate treatment of in structures, 38
 damping influence coefficient, 28
 damping matrix, 28
 damping ratio, 35
 damping, 35
  structural, 35
 eigenvalues, 34
 eigenvector, 34
 excitation coefficient, 42
 first natural mode, 34
 flexibility matrix, 28
 free vibrations, 32
 free vibrations of systems under static axial loads, 54
 frequency
  lowest natural circular, 34
 fundamental mode, 34
 generalized configuration, 27
 generalized coordinate, 27
 generalized displacements, 27
 generalized force, 28
 gravity loads, 53, 358, 488, 516

Multidegree-of-freedom systems (*contd.*)
 inertia matrix, 28
 lowest natural circular frequency, 34
 mass matrix, 28
 matrix equation, 34
 maximum absolute values of responses, 44
 modal analysis, 36, 239, 311
 modes, 34
  normalized, 34
 natural modes, 33, 34, 35
  complex, 35
  complete set, 34
 nonlinear behavior, 53, 322, 356, 475
 normalized modes, 34
 Nth excitation coefficient, 44
 Nth natural frequency, 33
 numerical procedure, 50
  convergence, 50
  stability, 50
 orthogonality, 34
 participation coefficients, 34, 42
 remotely-coupled, 145, 469
 responses, 44
  maximum absolute values of, 44
 simply-coupled, 30, 143, 468
 shear beam, 30
 shear-beam buildings, 30
 slenderness effects, 55
 stability, 50
 static axial loads, 54
 steady-state vibrations, 37
 stiffness influence coefficient, 28
 stiffness matrix, 28
 structural damping, 35
 transient disturbances, 41
 upper bound to maxima, 44
 weighting matrix, 34
Multiple wave reflection, 279

Nailed joints, 422
Natural frequencies, 191
 of shear beams, 108
 of cylindrical containers, 198
Natural frequency
 damped circular, 5
 undamped circular, 5

INDEX **633**

Natural modes, 33, 192, 197
  calculation of dynamic parameters from, 581
  complete set, 34
  complex, 35
  of vibration, 195
  of vibration of water in reservoirs, 190
Natural periods, 5
  equivalent, 326
  uncertainty concerning, 467
Natural periods of reservoirs, 194, 211
Nearby earthquakes, 369
New Madrid, 265
Newmark's beta method, 167
Newmark's method for steady-state vibrations, 117
Newmark's method for fundamental mode, 124
Niigata 1964 earthquake, 228, 434
Nominal dimensions, 449
Nominal loads, 448
Nominal strengths, 448
Nominal values, 447
Noncohesive soil, 363
Nonlinear behavior, 53, 188, 365, 381, 383, 386, 394, 473
  of soil, 298, 456
Nonlinear criteria of failure, 321
Nonlinear systems, 161, 473
  bilinear, 162
  elastic, 328
  graphical evaluation of responses, 164
  gyrogram, 164
  inelastic systems, 161
  locking, 170
  Masing-type systems, 163
  numerical methods, 167
  phase-plane construction, 164
  responses of, 321
  responses to transient Gaussian disturbances, 301
  strain hardening, 162
  strain softening, 162
  unloading, 170
  wave transmission, 168
    one-dimensional, 168

Nonrectangular cross sections of reservoirs, 190, 211
Nonlinearity
  due to conditions of support, 321
  due to material behavior, 321
  geometric, 321, 328
  of surface waves in tanks, 378
Nonstructural damage, 451
Nonstructural details, 453
Nonstructural elements, 525
Nonstructural materials, 453
Normalized modes, 34
Norwegian fjords, 204
Nth excitation coefficient, 44
Nth natural frequency, 33
Nuclear blasts, 218, 231
Nuclear power plants, 247, 251, 461, 470
Nuclear reactor facilities, 422
Numerical computation
  natural modes, 105
  steady-state responses, 105
Numerical evaluation of responses, 14, 50
  criteria for convergence and stability, 50
  convergence, 50
  of nonlinear systems, 167
  stability, 50
Numerical methods, 14, 16, 50, 167
  convergence, 50
  criteria for convergence and stability, 50
  extrapolation procedure for, 16
  in hydrodynamics, 192
  stability, 50
Number of load applications, 381

Objective function, 443
Octohedral shear strain criterion in fatigue, 396
Office buildings, 461
Optimization, 443
Optimum structural solution, 518
Optimum water content, 429
Order of load application, 382, 395, 396

634 INDEX

Orthogonal functions procedure, 130, 135
Orthogonality, 34, 126, 309
Out-of-phase disturbances, 197
Overtopping, 189, 190
Overturning moments, 500
   distribution over height of building, 505
   in chimney stacks, 349, 537
   in chimney stacks and towers, 537
   in gravity dams, 211, 370, 373
   in shear beams, 308, 314
   in tall buildings, 111
   in tanks, 198, 212
   in towers, 537
   reduction, in buildings, 506
Overturning of buildings, 434

P waves, 84
   reflections and refractions, 223
Palmgren–Miner criterion in fatigue, 394, 396
Participation coefficients, 34, 42, 191, 307
Particle acceleration, 86
Particle displacement, 86
Particle velocity, 86
Partitions, 453
PcS wave, 223
Penstocks, 244
Percentage of damage in fatigue, 395
Period of tsunamis, 205
Permanent deformations, 474
   of ground, 226
Permissible probability of failure, 450
Phase angle, 8
Phase changes of rocks as causes of earthquakes, 215
Phase plane, 17
Phase-plane construction, 164, 271
Phase shift
   angular, 6
Phase velocity, 74
Pipes, 546
Piping, 422
Pitching, 198
PKS waves, 223
Plastic hinges, 359, 405

Plastic strain in fatigue, 395
Plastic structures, 323, 357
Poisson's boundary condition, 178, 188, 197
Poisson distribution, 253
Poisson distribution of tsunamis, 207
Poisson process, 248, 250, 260, 269
Polymers, 383, 386
Pore pressures in soils, 425
Port Hueneme 1957 earthquake, 226
Posterior probabilities, 248
Pounding, 509
Power plants, 247
Power spectral density, 13, 268, 276, 290, 300
Precast concrete structures, 449
Preliminary design, 469
   of dams, 365
Prestige, loss of, due to failure, 443, 450
Pressures against dams, 179
   distribution, 187, 189
Prestressed concrete, 162, 406
   frames, 330
Prevailing periods, 225, 243, 301
Primary waves, 84
   (see also *P waves*)
Principal stresses, direction of, 323
Prior probabilities, 248
Probability distribution of loads, 447
Probability distribution of resistances, 447
Probability distribution of responses, 267, 272, 342, 348
Probability of failure, 272, 322, 358, 394, 445, 461
Probability of survival
   (see *probability of failure, reliability*)
Probability of yielding, 358
Processes of finite duration, 268
Provisional structures, 461
Pseudoacceleration, 11, 274, 280
Pseudovelocity, 11, 275, 280
   spectra, as basis for analysis of nonlinear systems, 325, 332
Puerto Montt, 226, 434
Pulse method, 168
PP waves, 223
PS waves, 223

Quay walls, 434

Radiation damping, 77, 203, 370, 382
Raft foundations, 434
Ramberg–Osgood force–deformation relation, 346
Random loading, 382
Ranges of parameters in behavior of materials and structural components, 381
Rates of earthquake occurrence, 249
Rates of loading, 384
  (see also *time to failure*)
Ratio elimination procedure, 130, 132
Rayleigh's method, 106, 135, 141, 158
Rayleigh quotient, 138, 141
Rayleigh waves, 84, 91, 243
Recording instruments, 565
Recording relative displacements, 577
Reduction in effective sections, 449
Reduction of earthquake stresses in buildings, 527
  flexible first story, 527
  rubber pads, 529
  suspended supports, 529
  use of rollers, 529
Reduction of overturning moments, 506, 537
Rectangular acceleration pulse, 343, 356
Recurrence period
  (see *return periods*)
Recurrence times of earthquakes, 207, 216, 263
Reed gages, 571
Reflected waves, 69, 220
Reflection of sound waves, 187, 195
Refracted waves, 69, 220
Regional seismicity, 248, 260, 265
Reinforced concrete, 399, 406, 474
  frames, 359
Reinforced masonry, 399
Reinforcement of waves, 92
Relative density of soils, 426
Relative displacements between floors, 482, 507
Reliability, 394
Relocation, 562

Remotely-coupled systems, 145, 469
Repairs of structures, 445
Reserve energy technique, 353
Reservoirs
  earthquake effects on, 365
  of nonrectangular cross sections, 373
  of rectangular cross sections, 190, 360
Reservoir slope failures, 549
Residual spectra, 13
Resins, 393
Resonance, 8, 383
Resonant amplitude, 8
Resonant frequency, 8
Response
  upper bound to, 10
Response factor, 6
Responses
  maximum absolute values of, 44
  numerical evaluation of, 14, 50, 167
Responses of circular plates
  to horizontal excitation, 99
  to rocking excitation, 99
  to torsional excitation, 100
  to vertical excitation, 98
Response spectra, 11, 22
  acceleration, 11 (also see *acceleration spectra*)
  damped, 13
  displacement, 11 (also see *deformation spectra*)
  for a velocity pulse, 13
  logarithmic, 11
  pseudoacceleration, 11, 274, 280
  pseudovelocity, 11, 275, 280
  for torsional motion, 492
  two-degree-of-freedom, 489
  undamped, 13
  velocity, 11
Retaining structures, 544
Retaining walls, 544
Restoration
  (see *repairs of structures*)
Return periods of earthquakes, 207, 216, 263
  optimal in design, 452
Return periods of tsunamis, 207
Rheological models, 386
Rigid body on soil, 94

Rigid block on an inclined plane, 363
Rigid blocks as gages, 571
Rigid, circular, cylindrical pier
   rocking and translational oscillations of, 95
Rigid circular plate
   vertical oscillations of, 94
Rigid-plastic systems, 162, 342, 354, 359, 545
Rigid plate of arbitrary shape, 94
   vibrations of, 94
Rigid structures, 277, 313, 341, 469
Roadbeds, 318
Rocks, 438
Rock slides, 210
Rockfills, 360
Rockfill dams, 162, 190, 474, 550, 579
Rocking and translational oscillations, 95
Rollers, 529
Rotary inertia, 109
Rotational disturbance, 84, 216, 241, 296
Rubber pads, 529
Run-down tests, 576

S waves, 84, 234
   reflections and refractions, 222
S wave velocity, 235, 315, 422
Salinas Valley, California 1906 earthquake, 232
San Francisco 1906 earthquake, 548
San Salvador 1965 earthquake, 219, 225, 240
Sand, 384, 426, 430, 439
Santa Tecla, 563
Scales in dynamic models, 578
Schwartz quotient, 124, 136
Scour, 543
Scratch gages, 578
Secondary waves, 84
   (also see *S waves*)
Seiches, 189
Seismic coefficients, 352, 358
Seismic exploration, 438
Seismic moment, 218
Seismic regionalization, 216

Seismic sea waves, 203
Seismicity, 247
Seismographs, 571
Seismoscopes, 565, 566
Semiinfinite beam, 75
Semiinfinite reservoir, 188, 191
Semiinfinite medium, 63, 169
Semispace, 63, 169
Sensitive soils, 437
Sensitivity, 437
Several components of ground motion, 316, 358
Sezawa body, 75, 286, 421
SH waves, 84
Shadow zone, 223
Shakedown, 411
Shallow floor systems, 519
Shallow reservoirs, 370, 376
Shear, 307, 314, 370, 373
Shear beam, 30
   nonuniform, undamped, 73
   participation coefficients, 67
   undamped uniform, 62
   uniform, 148, 313
   uniform, damped, 73
   upper limit to shear envelope, 67, 68
Shear-beam buildings, 30
Shear-beam idealizations, 67, 112
Shear beams
   natural frequencies of, 108
   natural modes of, 107, 129
   matrices of transmission applied to, 155
   transfer functions in, 306
Shear distribution, 483
Shear forces
   distribution over height of building, 505
Shear in reinforced concrete, 406
   in columns of circular cross section, 399
Shear walls, 415, 422, 515
Shear waves, 68
   (see also *S waves*)
Shearing deformations, 411, 515
Shock waves, 170
Shot noise, 269, 275, 281
Shrinkage, 399, 418

INDEX **637**

Silos, 429
Silt, 424, 429
Simple harmonic motion, 5
Simple linear system, 4, 455
Simple system, 3
Simply-coupled systems, 30, 143, 468
Simulated earthquakes, 323, 335
Single degree of freedom, 3
Single-degree structures
   (see *simple system*)
Skeleton curve, 324
Skopje 1963 earthquake, 225, 265
Slender chimneys, 540
Slender structures, 216
Slenderness effects, 55
   (also see *gravity effects*)
Sliding along failure surfaces in dams, 189
Slips at geologic faults, 542
   (also see *geologic faults*)
Slope failures, 360, 437, 474
   in reservoirs, 184, 189, 549
Sloshing, 201, 211, 377
S-N curves in fatigue, 394
Snell's law, 88
Social benefits, 443
Soft ground, 226, 239, 272, 286, 323, 462, 470
Softening structures, 352, 398
Soil–foundation interaction, 93, 219, 241, 382, 487, 533
Soil irregularities, 296
Soil layers, 67
Soil liquefaction, 244, 429, 543
Soil–structure interaction, 93, 219, 241, 382, 421, 427, 470, 487, 533
Southwell–Dunkerley method, 81, 139, 142, 154
Southwell's energy methods, 138
SP waves, 223
Space derivatives of earthquake motion, 216, 241
Spalling of concrete, 405
Spectra, 11
   acceleration, 11
   damped, 13
   displacement, 11
   Fourier, 11, 22

Spectra (*contd.*)
   logarithmic, 11
   pseudoacceleration, 11
   pseudovelocity, 11
   residual, 13
   response, 11
   undamped, 13
   velocity, 11
Spherical body waves, 85, 220, 231
Split failure in reinforced concrete, 401
Spring constant, 4
Spring constant for soil, 96
SS waves, 223
Stability, 50
Stability of buttress dams, 549
Stability of piers, 543
Stack-like structures, 537
Stacks, 241, 537
   (also see *chimney stacks*)
Staircase ramps, 523
Standard penetration tests, 426, 434, 439
Starting devices for accelerographs, 569
Static axial loads, 54
Static analysis, 315, 470
Static displacement, 4
Stationary disturbances, 311
Stationary Gaussian processes, 290, 300, 313, 348
Stationary processes, 268, 276
Stationary response of reservoirs, 187
Steady-state vibrations, 6, 37, 157
   in reservoirs, 187
Steel, 393
Steel frames, 162, 359, 421
Steel pipes, concrete-filled, 406
Steel structures, 421
Step-by-step procedures, 143, 149, 157
   for calculation of natural modes, 149
   for steady-state vibrations, 143
Stiffness degradation, 348, 353, 410, 421
Stiffness degrading
   (see *deterioration*)
Stiffness influence coefficient, 28
Stiffness matrix, 28
Stiffness, uncertainty concerning, 467
Stirrups, 401, 409
   (see *transverse reinforcement*)

Stochastic models in fatigue, 394
Stochastic processes, 268
    transient, 283
Stodola's method, 117
Story eccentricity
    design, 498
Story shears, 483
Story torques in buildings, 490
Strain energy, 325, 332
Strain gradient, 386
Strain hardening, 162, 406
Strain rate, 168, 335, 437
Strain softening, 162, 545
Stratified soil formations, 155
Strength, optimum value of, 452
Strength, uncertainty concerning, 467
Strengthening damaged structures, 552
    exterior prestressed stirrups, 558
    increasing section and reinforcement of beams, 557
    longitudinal prestress, 557
    temporary shoring, 560
Stress–concentration factors, 394
Stress gradient, 394
Stress histories, 382
Stress–reduction factor, 448
Stress–reduction factors, 454
Structural capacity, 271
Structural chracteristics, 572
    measurement of, 572
        nearby earthquakes, 572
        nearby explosions, 572
        run-down vibrator tests, 574
        setting up free vibrations with initial disturbance, 574
        steady-state forced vibrations, 573
        wind or traffic disturbances, 572
Structural damage, 451
Structural damping, 35
Structural design, objectives of, 447
Structural grade steel, 383, 384, 398
Structural synthesis, 530
Submarine landslides, 203
Submerged structures, 201, 212, 377, 382, 549
Subsidence of reservoir bottom, 189
Surfaces of sliding, 359

Surface waves, 90, 190, 244, 365, 370, 378
Surge, 208
Surge speed, 209
Suspended supports, 529
Suspension bridges, 161, 329, 543
SV waves, 84
Sweeping, 106
Symmetrical force–deformation curves, 474
Systems
    closely-coupled, 43
    damped, 35
    hardening, 328
    multidegree-of-freedom (see *multidegree-of-freedom systems*)
    remotely-coupled, 145, 469
    simple linear, 4, 455
    simply-coupled, 30, 143, 468
Systems, softening, 327
Systems with asymmetric force–deformation curves, 354
Systems with distributed mass, 61
Systems with symmetric force–deformation curves, 162, 323
Systems with several degrees of freedom
    linear, 27, 467

Taft 1952 earthquake, 352
Tall buildings, 358
Tanks, 197, 211, 241, 377, 548
Tapered beam, 128
Tapered chimneys, 539
Taut string, 64
Tectonic origin of earthquakes, 215
Temporary shoring, 560
Termination of numerical model, 77
Tests
    forced vibration, 8
Tests and observations, 565
Three-dimensional wave equation, 82
Tidal waves, 203
Tie girders, 518
Tier buildings, 67
Ties, 401, 409
    (see *transverse reinforcement*)

Tie-stirrups, 406
Tilting of piers, 542
Tilting of reservoir bottom, 189
Timber, 386, 414, 449
Time to collapse, 352, 358
Time to failure, 386, 425
Timoshenko beam, 81, 109
Torsion, accidental, 241, 472, 475, 490, 533
Torsion in buildings, 242, 472
Torsion pendulum, 22
Torsional moments, 490
Torsional oscillations of a slender bar, 64
Total hydrodynamic force, 186, 189
Towers, 241, 475, 537
Towers, stacks, and stack-like structures, 537
Traffic, vibrations due to, 266
Trans-Asiatic Belt, 251
Transfer functions, 276, 305
  at free surface of ground, 286
  for hydrodynamic pressure, 366, 373
  for pseudovelocity, 292
  in system with differential base motion, 316
  with wave refraction at bottom of reservoir, 375
Transient disturbances, 9, 41, 308
Transient Gaussian processes, 301
Transient response of reservoirs, 187
Transit tubes, 244
Translational components of ground motions, 236
Transmission matrices, 105, 147, 155, 288
Transverse stresses in tunnels and pipes, 547
Transverse waves, 69, 84
Travel times, 221
Triaxial states of stress in fatigue, 393, 396
Triaxial tests on soils, 428
Tributary mass, 212
Tsunamis, 203, 561
Tsunami energies, 205, 207
Tsunami frequencies, 207

Tsunami magnitude, 205
Tsunami runup, 206
Tunnels, 244, 546
Tunnels and pipes, 546
  curvatures, 547
  dynamic soil-tunnel interaction, 547
  longitudinal straining, 547
  transverse stresses, 547
Two-degree-of-freedom response spectra, 489

Undamped natural circular frequency, 5
Undamped uniform shear beam, 62
Uniform Building Code, 477
Uniform shear beam, 148
  shear distributions, 483
United States, earthquakes in west coast of, 275, 278
Unloading, 170
Unstable equilibrium of soil slopes, 428
Upper bound to maxima, 44, 321, 327
Urban planning, 562
Utility, 443, 450
Utilities damage, 548

Vajont, 204
Valdivia, 226
Valdivia river, 209
Valparaiso 1906 earthquake, 230
Velocity of propagation, 69
Velocity of tsunami waves, 205
Velocity of wave transmission, 428
Velocity potential, 177, 197
Velocity spectra, 11, 329, 340
Vertical accelerations, 111, 216, 296, 471
  effects on inverted pendulums, 351
  in reservoirs, 185, 366, 373, 376
Vianello's method, 105, 117
Vibration
  forced, 6
  steady-state, 6
Virtual mass of liquid, 181, 201, 212, 377
Virtual mass of soil, 97
Viscoelastic body, 74, 386